The Origin and Early Evolutionary History of Snakes

Snakes comprise nearly 4,000 extant species found on all major continents except Antarctica. Morphologically and ecologically diverse, they include burrowing, arboreal, and marine forms, feeding on prey ranging from insects to large mammals. Snakes are strikingly different from their closest lizard relatives, and their origins and early diversification have long challenged and enthused evolutionary biologists. The origin and early evolution of snakes is a broad, interdisciplinary topic for which experts in palaeontology, ecology, physiology, embryology, phylogenetics, and molecular biology have made important contributions. The last 25 years has seen a surge of interest, resulting partly from new fossil material, but also from new techniques in molecular and systematic biology. This volume summarizes and discusses the state of our knowledge, approaches, data, and ongoing debates. It provides reviews, syntheses, new data, and perspectives on a wide range of topics relevant to students and researchers in evolutionary biology, neontology, and palaeontology.

DAVID J. GOWER is a Merit Researcher in herpetology and Head of the Vertebrates Division in the Department of Life Sciences at the Natural History Museum, London. A collections-based organismal biologist with interests in a wide range of topics in natural history and evolutionary biology, he has taxon expertise in Triassic archosaurs and limbless, mostly fossorial, amphibians and reptiles.

HUSSAM ZAHER is Full Professor and Curator of Herpetology and Vertebrate Palaeontology in the Museum of Zoology of the University of São Paulo. He is an evolutionary biologist focusing on the systematics, palaeontology, and comparative anatomy of reptiles, with an emphasis on the diversity and evolutionary history of snakes and other squamates.

The Systematics Association Special Volume Series

SERIES EDITOR

Gavin Broad

Department of Life Sciences, The Natural History Museum, London, UK

The Systematics Association promotes all aspects of systematic biology by organizing conferences and workshops on key themes in systematics, running annual lecture series, publishing books and a newsletter, and awarding grants in support of systematics research. Membership of the Association is open globally to professionals and amateurs with an interest in any branch of biology, including palaeobiology. Members are entitled to attend conferences at discounted rates, to apply for grants and to receive the newsletter and mailed information; they also receive a generous discount on the purchase of all volumes produced by the Association.

The first of the Systematics Association's publications *The New Systematics* (1940) was a classic work edited by its then-president Sir Julian Huxley. Since then, more than 70 volumes have been published, often in rapidly expanding areas of science where a modern synthesis is required.

The Association encourages researchers to organize symposia that result in multi-authored volumes. In 1997 the Association organized the first of its international Biennial Conferences. This and subsequent Biennial Conferences, which are designed to provide for systematists of all kinds, included themed symposia that resulted in further publications. The Association also publishes volumes that are not specifically linked to meetings, and encourages new publications (including textbooks) in a broad range of systematics topics.

More information about the Systematics Association and its publications can be found at our website: www.systass.org.

Previous Systematics Association publications are listed after the index for this volume.

Systematics Association Special Volumes published by Cambridge University Press:

SYSTEMATICS ASSOCIATION SPECIAL VOLUME 90

The Origin and Early Evolutionary History of Snakes

DAVID J. GOWER

Natural History Museum, London

HUSSAM ZAHER

University of São Paulo

THE
Systematics
ASSOCIATION

CAMBRIDGE
UNIVERSITY PRESS

CAMBRIDGE
UNIVERSITY PRESS

University Printing House, Cambridge CB2 8BS, United Kingdom

One Liberty Plaza, 20th Floor, New York, NY 10006, USA

477 Williamstown Road, Port Melbourne, VIC 3207, Australia

314–321, 3rd Floor, Plot 3, Splendor Forum, Jasola District Centre, New Delhi – 110025, India

103 Penang Road, #05–06/07, Visioncrest Commercial, Singapore 238467

Cambridge University Press is part of the University of Cambridge.

It furthers the University's mission by disseminating knowledge in the pursuit of
education, learning, and research at the highest international levels of excellence.

www.cambridge.org
Information on this title: www.cambridge.org/9781108837347
DOI: 10.1017/9781108938891

First published 2022

Printed in the United Kingdom by TJ Books Limited, Padstow Cornwall

A catalogue record for this publication is available from the British Library.

Library of Congress Cataloging-in-Publication Data
Names: Gower, David J., 1969– editor. | Zaher, Hussam, 1965– editor.
Title: The origin and early evolutionary history of snakes / edited by David J. Gower, Natural History Museum,
    London, Hussam Zaher, University of São Paulo.
Description: Cambridge, United Kingdom ; New York, NY : Cambridge University Press, 2022. |
    Series: The Systematics Association special
volume series ; volume 90 | Includes bibliographical references and index.
Identifiers: LCCN 2022000042 (print) | LCCN 2022000043 (ebook) | ISBN 9781108837347 (hardback) |
    ISBN 9781108940535 (paperback) | ISBN 9781108938891 (epub)
Subjects: LCSH: Snakes–Evolution. | Snakes.
Classification: LCC QL666.O6 O75 2022 (print) | LCC QL666.O6 (ebook) |
    DDC 597.9613/8–dc23/eng/20220107
LC record available at https://lccn.loc.gov/2022000042
LC ebook record available at https://lccn.loc.gov/2022000043

ISBN 978-1-108-83734-7 Hardback

Additional resources for this publication at www.cambridge.org/snakes.

We dedicate our editorship of this book to the memory of Hussam's son.

In memoriam, Gaël Hingst-Zaher (16.03.2004–18.03.2020)

Contents

Colour plates can be found between pages 244 and 245.

Contributors

SELMA M. ALMEIDA-SANTOS Laboratory of Ecology and Evolution, Instituto Butantan, São Paulo, SP, Brazil

BRUNO G. AUGUSTA Museu de Zoologia da Universidade de São Paulo, SP, Brazil, *and* ISEM at Southern Methodist University, Dallas, TX, USA

HENRIQUE B. BRAZ Laboratory of Ecology and Evolution, Instituto Butantan, São Paulo, SP, Brazil

DAVID CUNDALL Biological Sciences, Lehigh University, Bethlehem, PA, USA

CHRISTOPHER A. EMERLING Biology Department, Reedley College, CA, USA

SUSAN E. EVANS Department of Cell and Developmental Biology, University College London, UK

ANTHONY R. FIORILLO ISEM at Southern Methodist University, Dallas, TX, USA, *and* Roy M. Huffington Department of Earth Sciences, Southern Methodist University, Dallas, TX, USA

GEORGIOS L. GEORGALIS Palaeontological Institute and Museum, University of Zurich, Zurich, Switzerland and Institute of Systematics and Evolution of Animals, Polish Academy of Sciences, Kraków, Poland

DAVID J. GOWER Department of Life Sciences, The Natural History Museum, London, UK

EINAT HAUZMAN Department of Experimental Psychology, Psychology Institute, University of São Paulo, SP, Brazil

JASON J. HEAD Department of Zoology and University Museum of Zoology, University of Cambridge, Cambridge, UK

ALEXANDRA F. C. HOWARD Department of Zoology and University Museum of Zoology, University of Cambridge, Cambridge, UK

FRANCES IRISH Biological Sciences, Moravian University, Bethlehem, PA, USA

MARTIN IVANOV Department of Geological Sciences, Faculty of Science, Masaryk University, Brno, Czech Republic

ASHWIN IYER Evolutionary Venomics Lab. Centre for Ecological Sciences, Indian Institute of Science, Bangalore, Karnataka, India

TIMOTHY N. W. JACKSON Australian Venom Research Unit, Department of Pharmacology & Therapeutics, University of Melbourne, Australia

LOUIS L. JACOBS ISEM at Southern Methodist University, Dallas, TX, USA, *and* Roy M. Huffington Department of Earth Sciences, Southern Methodist University, Dallas, TX, USA

GIOVANNA G. MONTINGELLI Museu de Zoologia da Universidade de São Paulo, Ipiranga, São Paulo, SP, Brazil

JOHANNES MÜLLER Museum für Naturkunde – Leibniz-Institut für Evolutions- und Biodiversitätsforschung, Berlin, Germany

LEONARDO DE OLIVEIRA Laboratório de Toxinologia Aplicada, Instituto Butantan, São Paulo, SP, Brazil

MICHAEL J. POLCYN ISEM at Southern Methodist University, Dallas, TX, USA, *and* Roy M. Huffington Department of Earth Sciences, Southern Methodist University, Dallas, TX, USA

RIVKA RABINOVICH National Natural History Collections, Institute of Earth Sciences, The Hebrew University of Jerusalem, Israel

OLIVIER RIEPPEL Negaunee Integrative Research Center, The Field Museum, Chicago, IL, USA

SARA RUANE Life Sciences Section, Negaunee Integrative Research Center, Field Museum, Chicago, IL, USA

AGUSTÍN SCANFERLA CONICET, Fundación de Historia Natural 'Félix de Azara', Ciudad Autónoma de Buenos Aires, Argentina

RYAN K. SCHOTT Department of Biology, York University, Toronto, ON, Canada, *and* Department of Vertebrate Zoology, National Museum of Natural History, Smithsonian Institution, Washington, DC, USA

BRUNO F. SIMÕES School of Biological and Marine Sciences, University of Plymouth, Drake Circus, Plymouth, UK

KRISTER T. SMITH Department of Messel Research and Mammalogy, Senckenberg Research Institute, Frankfurt am Main, Germany, *and* Faculty of Biological Sciences, Institute for Ecology, Diversity and Evolution, Goethe University, Frankfurt am Main, Germany

JEFFREY W. STREICHER Department of Life Sciences, The Natural History Museum, London, UK

KARTIK SUNAGAR Evolutionary Venomics Lab. Centre for Ecological Sciences, Indian Institute of Science, Bangalore, Karnataka, India

VIVEK SURANSE Evolutionary Venomics Lab. Centre for Ecological Sciences, Indian Institute of Science, Bangalore, Karnataka, India

PAUL TAFFOREAU European Synchrotron Radiation Facility, Grenoble, France

HONGYU YI Key Laboratory of Vertebrate Evolution and Human Origins, Institute of Vertebrate Paleontology and Paleoanthropology, Chinese Academy of Sciences, Beijing, China, *and* CAS Center for Excellence in Life and Paleoenvironment, Beijing, China

HUSSAM ZAHER Museu de Zoologia da Universidade de São Paulo, Ipiranga, São Paulo, SP, Brazil

Preface

This volume was prompted by a conference held at the Linnean Society of London in June 2019: 'A Contribution to the Origin and Early Evolution of Snakes'. The meeting covered the same subject matter and multidisciplinary approaches as this volume. Via 14 scientific talks and 3 posters, the approximately 40 attendees were exposed to review, synthesis, opinion, and new data from fields as diverse as geology, palaeobiology, anatomy, molecular genetics, phylogenetics, and reproductive, feeding, sensory, and venom biology. This volume is not a conference proceedings in the usual sense but instead a collection of chapters inspired by topics covered by and themes running through the 2019 meeting. Some of the contributing authors included here were unable to attend the 2019 meeting but were subsequently invited to contribute because the editors were looking to compile a collection of works that covered a range of particularly relevant topics. The contributions vary from substantial syntheses or reviews, to more tightly focussed case studies that illustrate particular advances and/or approaches that will have a broader relevance.

Please note that manuscripts were peer reviewed, and accepted for publication between August 2020 and August 2021. The section mark (§) is used to refer to numbered sections cited in the text. Abstracts of chapters and files of supplementary material cited in the book are freely available on the Internet at www.cambridge.org/snakes.

We are very grateful to Leanne Melbourne (Linnean Society of London) for help in organizing the 2019 meeting. Funding for the meeting was provided by the United Kingdom's Systematics Association and the Linnean Society of London. We encourage readers to support the activities of the Linnean Society (www.linnean.org) and Systematics Association (www.systass.org). Help and advice in producing the book was provided by several people, including Olivia Boult, Dominic Lewis, Jenny van der Meijden and colleagues (Cambridge University Press) and Gavin Broad (Editor-in-Chief, the Systematics Association). Mary Malin is thanked for expert copy editing. The editors are extremely grateful to the expert reviewers of each chapter, those anonymous as well as Salvador Bailon, Chris Bell, Bhart-Anjan Bhullar, Nicholas Casewell, V. Deepak, Massimo Delfino, Ron Douglas, Annelise Folie, Nick Fraser, Christopher Friesen, Timothy Jackson, Mareike Janiak, Catherine Klein, Francisca Leal, Daniel Madzia, Pedro Nunes, Giulia Pasquesi, Davide Pisani, Ana Prudente, Aleta Quinn, Martha Ramirez-Pinilla, Alan Savitzky, Anne Schulp, Kurt Schwenk, Abigail Tucker, Márton Venczel, and Giulia

Zancolli. Finally, we thank the attendees of the 2019 meeting and all 33 contributing authors to the book, several of whom also peer reviewed other chapters.

Most of this book was compiled during the COVID-19 global pandemic, which presented multiple and varied extraordinary challenges to the contributors, reviewers, and their friends and families. We are extremely grateful for their hard work, forbearance and patience under these exceptional circumstances. And we thank health practitioners, those who developed, distributed, and administered vaccinations, and key workers for their myriad indirect contributions.

We each thank our loved ones for their support, friendship, and love over many years.

1

Introduction

HUSSAM ZAHER AND DAVID J. GOWER

1.1 Introduction

The last 25 years witnessed substantial advances in the knowledge of the evolutionary history of snakes. Detailed (re)investigations of key fossils, such as the legged snakes from the Upper Cretaceous, have helped to reinvigorate debates about snake origins. Along with palaeontological evidence, new genomic and evolutionary developmental ('evo-devo') studies have provided novel insights into the interrelationships of squamates and mechanisms underlying the serpentiform body. New methods and technologies have been applied to a wealth of new and existing data, with analyses addressing evolutionary issues related to topics such as body elongation, limblessness, fossoriality, and envenomation. However, large knowledge gaps remain for some key morphological complexes and taxa.

This book was prompted by our joint organization of a conference held at the Linnean Society of London in June 2019. Without the pretence of being comprehensive in terms of coverage of topics and of range of opinions (see § 1.3), this volume's aims are to help fill some of the main knowledge gaps, to review, summarize, and synthesize recent advances and remaining debates, and to highlight fruitful avenues for future research. This is undertaken via contributions from a range of experts in squamate (especially snake) systematics, palaeontology, phylogeny, physiology, and comparative and functional anatomy and genomics.

1.2 Scope and Coverage

Other than this introduction, the volume comprises 18 chapters, written by 34 authors based in 13 countries, and covers a wide range of subjects, including temporal and spatial aspects of the fossil record, palaeobiology, phylogenetics, anatomy, sensory biology, ancestral-state estimation, the history of science, inference of function from form, genomics, venomics, feeding systems and diet, and reproductive biology. However, some important topics are not covered by this volume. For example, it includes almost no evolutionary developmental biology ('evo-devo') even though this is clearly of interest in terms of, for example, axial regionalization and loss of limbs (e.g., [1]). And there is very little coverage of recent applications (e.g., [2]) of 3D geometric morphometrics to cranial or vertebral evolution across the lizard-snake transition.

The volume is subdivided into five main parts. Part I comprises four contributions on the fossil record of snakes and their closest relatives among squamates. The first chapter of this section covers the possible origin of squamates in the Early Triassic (ca 250 million years ago (Ma)) to the earliest known unequivocal squamate fossils (Middle Jurassic: *c.* 174–163 Ma) and the further diversification of Squamata into the Cretaceous. The other three chapters in this section focus on diversity and diversification early in snake history as inferred from the Cretaceous to Miocene fossil record. Part II addresses the controversial Marine-Origin Hypothesis of snakes from a palaeontological perspective. The first chapter provides an epistemological approach to the idea of a marine origin of snakes. The three other chapters reassess morphological evidence for the Marine Hypothesis. Part III examines the rapidly growing set of molecular genetic data available for understanding aspects of the origin and early history of snakes, applied to a diversity of questions about phylogenetic relationships, inferences of diet across the lizard–snake transition, and the origin and diversification of snake venoms. Part IV assesses what some sensory systems (ears, brains, eyes) can tell us about early snake history. The ear and brain chapters focus on comparative morphology and on drawing inferences about early events in snake evolutionary history by identifying correlates of soft tissue and ecology and behaviour from more readily fossilized hard tissues. The eye chapter provides an overview of the visual system of the ancestral snake that can be inferred from anatomy, physiology, and molecular genetic data for extant snakes. Part V comprises four contributions on comparative and functional morphology of primarily soft-tissue systems of mostly extant snakes. In addition to a new perspective on feeding systems and early snake history, three of the chapters (hemipenes, sperm storage, oral glands) offer reviews of topics that have been largely overlooked in terms of what they can tell us about the ancestral snake, and snake origins and early evolutionary history.

1.3 Debate, Disagreement, and Consensus

Some of the debates over the origin of snakes, at least in the past 25 years, have been somewhat fractious. Even where the debate has been less antagonistic, there remain some sharply divided opinions. Importantly, the reader should be aware that the combined authorship (and reviewership) of this book does not represent the full span of expert opinions currently in circulation (for example, for an overview of some opposing views concerning extant and fossil anatomical evidence for snake origins, see [3], and review of that work by [4]). Disagreements about the biology of snake origins have generally been more prominent and noisy for palaeontology than neontology, perhaps because the former typically involves greater overlap between more observers of the same, imperfectly preserved specimens that are being interpreted within a less-constrained phylogenetic framework, but it is likely not difficult to find people who disagree with aspects of this volume's neontological chapters also. This volume is a good place to find a way into the palaeontological and neontological debates. Careful readers will also note some disagreements among the contributions in this volume. For example, the chapters by Head et al. (Chapter 3) and by Smith and Georgalis (Chapter 4) present different opinions on the extent to which vertebral morphology is useful for palaeobiological interpretations of snake evolution. And Yi (Chapter 13) and Scanferla (Chapter 14) disagree about the inferences on ecology that can be made from the inner ear morphology of the extinct stem snake *Dinilysia patagonica.*

The two of us also have some differing opinions and perspectives, explainable to some degree by our different trajectories. Although both were undergraduates in zoology, one of us (DJG) went on to do a PhD in archosaur palaeontology before choosing to largely move back into neontological herpetology, while the other (HZ) did a PhD in neontology and established a career in that field but became very interested in palaeontology and has worked substantially in that field. Despite our different experiences and perspectives, through co-editing this book we found ourselves sharing some views. For example, we agree that palaeontology will likely lead the way in further clarifying the closest relatives of crown snakes (and perhaps also of some of the major lineages therein), but we also believe that neontology, including molecular genetics, has a major role to play in advancing inferences of the natural history of the earliest snakes, and in clarifying some interrelationships of the major extant lineages at the base of the snake crown. We both learned much about the origin and early evolutionary history of snakes from organizing the 2019 meeting and editing this volume. We have been reminded how unusual snakes are among squamates. Although obviously partly an artefact of the idiosyncrasies of extinction and survival, in so many ways snakes are an exceptionally disparate lineage of elongate, limbless squamates.

We both owe a debt of gratitude to many colleagues over many years, as well as to previous biologists neither of us met. Two particularly important, relevant and missed colleagues deserve special mention here. The title of the 2019 meeting associated with this book, *A Contribution to the Origin and Early Evolution of Snakes,* was devised in reference to the late Garth L. Underwood's highly influential 1967 book *A Contribution to the Classification of Snakes.* This was chosen because we have long been fans of Garth's inspiring science and writing, and both benefited from his insight, generosity, and wealth of knowledge during discussions with him about snakes. One of us (HZ) also benefited greatly from many discussions about snake evolution with the late Jean-Claude Rage, who made substantial contributions to snake palaeontology.

1.4 Concluding Remarks

Although snake origins encompasses several topics that have been much-debated over many years, we are struck by how much vitality remains in this area of science. This is partly because of technological advances, especially in the generation and analysis of genetic data, and in imaging. However, it is also because of a large and diverse array of researchers recently approaching the subject, because of new fossil discoveries, and because there remains so much to be discovered about extant snakes – their anatomy, development, phylogenetic relationships, ways of life, and the genetics underpinning this diversity – information that can feed into inferences about extinct geno- and phenotypes and testing of hypotheses of events that occurred tens of millions of years ago.

This volume is replete with instances where making inferences about the most-recent common ancestor of major lineages of crown snakes is difficult because of the patchiness of comparative data for extant (especially non-caenophidian) snakes. We anticipate exciting years ahead, with rapid and substantial progress in more compellingly resolved phylogenetic relationships, discovery and documentation of fossils of stem and early crown snakes, analyses of greatly expanded genomic data, refined correlations between ossified structures and soft tissues and habits, and more complete and detailed surveys of extant snake anatomy and ecology. This joint volume will fulfil its objective if it helps to stimulate and contribute to future breakthroughs in this exciting chapter of vertebrate evolutionary history.

Acknowledgements

We thank Jeff Streicher and Jason Head for constructive critiques of an earlier draft. HZ is grateful to FAPESP for funding (grant 2018/11902-9) and Ana Bottallo Quadros for her support during organization of the 2019 conference and production of this book.

References

1. F. Leal and M. J. Cohn, Developmental, genetic and genomic insights into the evolutionary loss of limbs in snakes. *The Journal of Genetics and Development*, 56 (2018), e23077.
2. F. O. Da Silva, A.- C. Fabre, Y. Savriama, et al., The ecological origins of snakes as revealed by skull evolution. *Nature Communications*, 9 (2018), 376.
3. M. W. Caldwell, *The Origin of Snakes: Morphology and the Fossil Record* (Boca Raton, FL: CRC Press, 2020).
4. D. Cundall, Review of M. W. Caldwell, The Origin of Snakes: Morphology and the Fossil Record. *Herpetological Review*, 51 (2020), 364–368.

PART I

The Squamate and Snake Fossil Record

2

The Origin and Early Diversification of Squamates

Susan E. Evans

2.1 Introduction

Squamata, the group that encompasses snakes, lizards, and amphisbaenians, is the largest (>10,500 sp.) and most disparate group of modern reptiles. Extant squamates are distributed over all but the coldest parts of the world; range in size (snout–vent length, SVL) from millimetres to metres; and show a diversity of diets, shapes, locomotor patterns, and reproductive strategies. Snakes (Serpentes) account for roughly 35 per cent of all extant squamate species and their origin and relationships, which have long intrigued herpetologists, are the focus of this volume. This chapter aims to provide a foundation for subsequent chapters, by reviewing what is currently known of the early stages of squamate evolution and diversification.

Most researchers recognize eight extant major squamate clades (Fig. 2.1): Dibamidae; Gekkota; Scincoidea (=Scinciformata [1]), encompassing Xantusiidae, Scincidae, and Cordyliformes; Lacertoidea (=Laterata [1]), encompassing Lacertidae and Teiioidea; Amphisbaenia; Iguania; Anguimorpha; and Serpentes. Although Camp [2] considered Gekkota to be primitive squamates (part of his Ascalabota), the first comprehensive cladistic analysis [3], based on morphological characters, placed Iguania as the sister group of other squamates (Scleroglossa). Within Scleroglossa, Estes et al. [3] united Scincoidea + Lacertoidea in Scincomorpha, and Scincomorpha + Anguimorpha as Autarchoglossa. The position of three limb-reduced clades, Dibamidae, Amphisbaenia, and Serpentes, was unresolved within Scleroglossa. The topology of Estes et al. [3] remained the working hypothesis for most herpetologists until 2004 with the publication of phylogenies based on molecular data [4, 5]. The molecular trees placed Gekkota rather than Iguania as the sister group of other squamates (invalidating Scleroglossa), with Scincoidea, and then Lacertoidea (including Amphisbaenia) as successive outgroups to a Toxicofera that

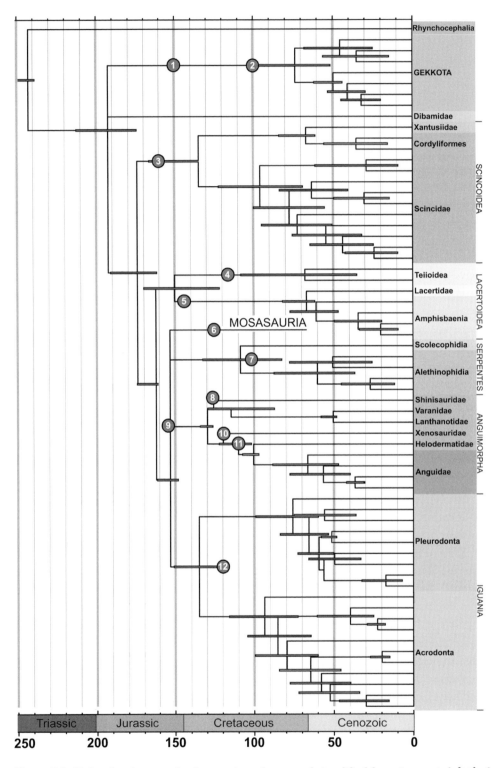

Figure 2.1 Molecular time tree for Squamata, redrawn and simplified from Jones et al. [14], showing putative early representatives of major clades discussed in the text. (1) *Eichstaettisaurus*

encompassed Iguania, Anguimorpha, and Serpentes [1]. Although morphological data sets (e.g., [6]) tend to replicate the Estes et al. [3] tree, subsequent molecular and combined evidence analyses [7–13] agree with those of Townsend et al. [4] and Vidal and Hedges [5]. The molecular tree is therefore used as the phylogenetic framework herein (Fig. 2.1).

2.2 The Early Squamate Fossil Record

Within Lepidosauria, Squamata is the sister group of Rhynchocephalia. Rhynchocephalia is represented today only by the genus *Sphenodon* (New Zealand), but the group has a relatively good fossil record through much of the Mesozoic. The occurrence of a primitive rhynchocephalian in the Middle Triassic of Germany [14] provides a latest possible age for the division of the lepidosaurian stem, although that division, and thus the origin of Squamata, probably occurred in the Early Triassic [12, 14]. However, although there are many records of Triassic rhynchocephalians, there are currently no unequivocal records of Triassic squamates. Those described last century have all been re-assigned to other reptile groups or to the lepidosaurian stem [15, 16].

Tikiguana estesi, represented by a single jaw with an acrodont dentition collected from Upper Triassic deposits in India [17], was shown to be a modern intrusion [18]. More recently, Simões et al. [19] placed three monotypic genera, *Sophineta cracoviensis* (Lower Triassic, Poland, [20]), *Megachirella wachtleri* (Upper Triassic, Italy [21]), and *Marmoretta oxoniensis* (Middle Jurassic, UK [22]) into Squamata. However, re-analyses, based on new data for *Marmoretta* [23], do not support a squamate attribution for either *Marmoretta* or *Sophineta*, and leave the position of *Megachirella* as equivocal.

Paikasisaurus indicus is based on two jaw fragments from the Lower–Middle Jurassic Kota Formation of India [24]. The designated holotype bears two teeth, neither of which is obviously pleurodont, whereas the referred specimen bears a single tooth with little resemblance to that of the holotype. *Bharatagama rebbanensis,* also from the Kota Formation, is represented by multiple jaws bearing both pleurodont and acrodont teeth [25]. It was described as a possible acrodont iguanian but could also be an aberrant rhynchocephalian [18, 26].

The earliest unequivocal squamates are from the Middle Jurassic (Bathonian) of the United Kingdom, Russia, Kyrgyzstan, and Morocco. One of the most productive Bathonian

---

Figure 2.1 (*cont.*) (Upper Jurassic, Germany); (2) *Hoburogekko* (Lower Cretaceous, Mongolia); (3) Paramacellodidae and *Saurillodon* (Upper Jurassic, Portugal); (4) *Asagaolacerta, Kuwajimalla,* Polyglyphanodontia (Lower Cretaceous, Japan); (5) *Purbicella* (Lower Cretaceous, UK); (6) *Kaganaias* (Lower Cretaceous, Japan); (7) 'Coniophis' (Albian-Cenomanian, North America) and terrestrial + aquatic snakes (Albian-Cenomanian, Algeria); (8) *Dalinghosaurus* (Lower Cretaceous, China); (9) *Dorsetisaurus* (Upper Jurassic to Lower Cretaceous, Pan-Laurasia); (10) ?*Xenostius* (Early Cretaceous, Mongolia); (11) *Primaderma* (Albian–Cenomanian, North America); (12) ?*Hoyalacerta* (Early Cretaceous, Spain). See Supplementary Figure 2.S1 for terminal taxon labels in full. (A black and white version of this figure will appear in some formats. For the colour version, please refer to the plate section.)

localities has been Kirtlington Quarry (UK, [27, 28]). In addition to the lepidosauromorph *Marmoretta*, Kirtlington yielded five named squamate genera: *Balnealacerta, Bellairsia,* cf. *Saurillodon, Oxiella,* and *Parviraptor*. Squamate bones were also reported from roughly contemporaneous deposits on the Isle of Skye, Scotland [23], and new fieldwork on Skye has yielded both isolated elements and associated material, including specimens referable to the Kirtlington genera *Balnealacerta, Bellairsia,* and *Parviraptor* [29]. These specimens both validate the original associations [27, 28], *contra* Caldwell et al. [30], and add new morphological data for phylogenetic analysis. Work on this material is in progress, but it highlights the clear morphological gap between lepidosaurian stem taxa, like *Marmoretta,* and squamates from the Middle Jurassic onward that show derived features including biradiate squamosals, loss of quadrate–pterygoid fixation, loss of quadratojugal, a synovial epipterygoid–pterygoid joint, and subdivision of the metotic fissure in the braincase.

Among other Middle Jurassic records, *Changetisaurus estesi* (Bathonian, Kyrgyzstan [31]) is a partial skeleton attributed to Squamata, but it requires re-study. Squamate remains are also recorded from the Bathonian Berezovsk Mine, western Siberia (a possible paramacellodid, two 'scincomorphs': [32]) and from similar aged deposits in Morocco ('scincomorphs', '*Parviraptor*-like' taxon [33]). An osteoderm-covered lizard was briefly described from Callovian–Oxfordian age deposits at Shishougou (Junggar Valley, China [34]), but a detailed description and analysis has yet to be published. Roughly contemporaneous deposits at Daohugou, Chinese Inner Mongolia, have also yielded associated lizard skeletons, one of general proportions and the other with elongate limbs [35, 36], but both are immature and the preservation is poor. More recently, however, a complete lizard skeleton was recovered from Daohugou equivalent deposits at Guancaishan, Liaoning Province (*Hongshanxi xiei*, [37]).

Moving into the Upper Jurassic, squamate material is known from several additional Laurasian localities, notably: Portugal (Oxfordian, Guimarota [38]), Kazakhstan (Oxfordian–Kimmeridgian, Karabastau Formation [39]), North America (Kimmeridgian, Morrison Formation [40–42]), Germany (Tithonian, Solnhofen and neighbouring localities [26, 43–48] and France (Kimmeridgian, Cerin [49]). Unfortunately, the only contemporaneous Gondwanan record is a 'paramacellodid' osteoderm from the Tendaguru Formation of Tanzania [50], but this identification is unconfirmed.

The squamate fossil record improves substantially in the Early to Middle Cretaceous (i.e., Berriasian–Cenomanian), with specimens from Europe, Africa, the Middle East, Asia, and the Americas. Concurrent with this apparent global expansion, Early Cretaceous (Barremian) specimens include some of the first known occurrences of squamate gliding [51], body elongation with limb reduction [52], viviparity [53], herbivory [54], and specialist climbing [55, 56].

2.3 The First Records of Crown-Group Squamates

Most recent molecular divergence estimates date the first radiation of crown-group squamate lineages to the Early–Middle Jurassic [12–14], with further splitting of major lineages in the Late Jurassic. As yet, however, none of the known Middle Jurassic taxa, and relatively few of those from the Late Jurassic, can be placed unequivocally into crown lineages (Fig. 2.1). There is some improvement in the Early Cretaceous, but many lizards from this period are phylogenetically problematic and behave as wild cards in analyses.

2.3.1 Gekkota

Given that gekkotans are now considered to be one of the first of the major branches to diverge from the ancestral crown squamate, they should be well represented in early squamate assemblages. Their apparent absence could be due to preservational or ecological constraints, or a failure of identification due to a lack of distinctive gekkotan characters in their early history. Several Late Jurassic and Early Cretaceous genera (notably *Ardeosaurus, Bavarisaurus, Eichstaettisaurus, Palaeolacerta, Yabeinosaurus*) were originally attributed to Gekkota (e.g., [44]), but a majority of these have been re-assigned. One exception is *Eichstaettisaurus* (Upper Jurassic, Germany [43]; Lower Cretaceous, Spain, Italy [57, 58]) which is frequently placed on the gekkotan stem (e.g., [6, 12, 59]), albeit with limited evidence. The first unequivocal stem gekkotan is *Hoburogekko suchanovi* (Aptian–Albian, Höövor, Mongolia [60, 61]) although *Norellius nyctisaurops* (Aptian–Albian, Mongolia [62, 63]) has also been assigned to Gekkota (e.g., [12], but see [9]), based primarily on the combination of notochordal vertebrae and paired parietals. In recent years, Cretaceous amber has offered an important window into gekkotan history because it is able to preserve small, delicate skeletons, often with exquisite soft tissues (e.g., *Cretaceogekko burmae* [56]). The Albian–Cenomanian amber deposits of north-western Myanmar are particularly rich in gekkotans and stem gekkotans from a rarely sampled tropical forest ecosystem [64, 65].

2.3.2 Dibamidae

The phylogenetic position of this small, specialized, and biogeographically disparate clade (*Dibamus*, South East Asia, New Guinea, Philippines; *Anelytropsis*, Mexico) has long been problematic (e.g., [3, 66]). Molecular analyses have provided greater clarity but there remains uncertainty as to whether dibamids are sister to all other squamates including gekkotans [10, 11], to gekkotans alone [13], or to all squamates except gekkotans [12]. Unfortunately, the group has no Mesozoic record. The only putative fossil dibamid is *Hoeckosaurus mongoliensis* from the Oligocene of Mongolia [67].

2.3.3 Scincoidea

Estes et al. [3] grouped scincoids and lacertoids within Scincomorpha. Consequently, many fossil lizard remains are classified simply as 'scincomorphs', making them difficult to attribute to any clade. Scincoidea is probably represented by the Jurassic-Cretaceous Paramacellodidae [68] with their body covering of rectangular osteoderms, and possibly by short-jawed Jurassic-Cretaceous taxa like *Saurillodon* (Fig.2.1). *Saurillodon* (*S. proraformis, S. henkeli*) was first described from Guimarota (Oxfordian, Portugal) based on short robust dentaries with a few large conical teeth [38, 69], resembling the jaws of limb-reduced modern scincids and amphisbaenians. Similar jaws were reported from

Anoual (Lower Cretaceous, Morocco, *Tarratosaurus anoualensis* [70]) and from Kirtlington (Middle Jurassic, UK [28]). The Kirtlington specimens were referred to *Saurillodon*, and tentatively associated with unusual elongated vertebrae. If the association is correct, then this Middle Jurassic taxon would be of considerable interest as an early long-bodied, and presumably limb-reduced, squamate. *Balnealacerta*, also from Kirtlington [28] was attributed to Paramacellodidae based on similarities in jaw and tooth structure, as were dentaries from contemporaneous localities on Skye [23]. However, no osteoderms were recovered from among the many kilos of residue from Kirtlington sorted by the author, nor are they known from Skye (e.g., in association with the new material of *Balnealacerta* [29]), and therefore attribution of these Middle Jurassic remains to Paramacellodidae remains tentative. Nonetheless, paramacellodids (including osteoderms) are recorded from Upper Jurassic deposits including those from Guimarota (Oxfordian [38, 71]), Kazakhstan (Kimmeridgian, *Sharovisaurus* [39]), and the Morrison Formation, USA (Kimmeridgian [40, 41]), as well as from the Lower Cretaceous of the United Kingdom [72], China [73], Russia [74], Spain [75], Morocco [75], and North America [76]. Most recently, a partial lizard skeleton, *Neokotus sanfranciscanus* (Valanginian, Brazil), was described as a possible Gondwanan paramacellodid [77], but no osteoderms are associated with the skeleton and the attribution is based mainly on dental morphology.

Other than paramacellodids, the well-preserved *Tepexisaurus tepexii* (Albian, Mexico [78]) may also be a scincoid [6], but other attributions have less support. Talanda [48] argued that *Ardeosaurus* (Upper Jurassic, Germany) was a scincid, but this was based on an interpretation of cranial ornamentation as representing compound osteoderms. By contrast, Simões et al. [47] recovered *Ardeosaurus* as a gekkotan, whereas other researchers (e.g., [12]) found *Ardeosaurus* to be a wild-card taxon in analyses. Finally, *Calanguban alamoi* is known from a poorly preserved skeleton from Brazil (Aptian–Albian, Crato Formation [79]). It was recovered within Cordyliformes in an analysis using the Conrad [59] matrix, but unresolved using Gauthier et al.'s [6] matrix. Neither analysis attempted to use a molecular constraint.

2.3.4 Lacertoidea

Lacertoidea encompasses lacertids, amphisbaenians, teiids and gymnophthalmids, and may be represented by *Purbicella ragei* from the Lower Cretaceous of England (Berriasian, Purbeck Limestone Group [80]). *Meyasaurus* spp. (Barremian, Spain) was proposed to lie on the teiioid stem by Evans and Barbadillo [81], but within anguimorphs by Richter [82] and Conrad [59]. *Tijubina pontei* (Aptian–Albian, Crato Formation, Brazil) was originally attributed to Teiidae [83], but subsequent researchers either failed to place it securely within the squamate crown [84, 85] or suggested that it and a second Brazilian species (*Olindalacerta brasiliensis* [86]) might be polyglyphanodonts [79]. Until recently, Polyglyphanodontia were classified as teiioids (e.g., [68, 87, 88]), a position also recovered by Pyron [12] using a combined morphological [59] and molecular matrix. However, Gauthier et al. [6] recovered polyglyphanodonts on the stem of 'Scleroglossa', whereas Pyron [12], combining the Gauthier et al. [6] matrix with a molecular dataset, found polyglyphanodonts to be the sister group of Iguania. The position of this large and important fossil group therefore remains problematic, but it is attributed to Teiioidea in

Figure 2.1. The earliest attributed polyglyphanodonts are *Kuwajimalla kagaensis* [54] and *Asagaolacerta tricuspidens* [89] from the Tetori Formation (Barremian–Aptian, Japan). An unnamed lizard specimen preserved in amber (Albian–Cenomanian, Myanmar) was also described as a crown lacertoid [64], but this attribution was based primarily on comparison with polyglyphanodonts.

Molecular divergence estimates (e.g., [12]) date the separation of stem amphisbaenians and lacertids to the Late Jurassic–Early Cretaceous, but the earliest unequivocal record of amphisbaenians is from the Palaeocene [90]. *Hodzhakulia magna* (Albian, Uzbekistan [91]; Mongolia [92]) and *Sineoamphisbaena hexatabularis* (Santonian–Campanian, China [93]) were originally classified as stem amphisbaenians, but *Hodzhakulia* is fragmentary ('scincomorph' [92]) and *Sineoamphisbaena* may be related to polyglyphanodonts [12]. The limb-reduced *Slavoia darevskii* has also been proposed as a candidate stem amphisbaenian [94]. The type material is from the Campanian of Mongolia, but older 'slavoiids' were reported from the Albian Mongolian locality of Höövor [95].

2.3.5 Anguimorpha

Parviraptorids (Middle Jurassic–Early Cretaceous, UK [27]) were originally attributed to Anguimorpha, based on jaw and tooth morphology. This attribution will be tested by new, in progress, analyses based on associated material from Skye (Middle Jurassic). The Middle Jurassic *Changetisaurus* (Bathonian, Kyrgyzstan) was also attributed to Anguimorpha [31], but this taxon needs re-examination. Currently, the earliest generally accepted anguimorph is *Dorsetisaurus purbeckensis*. As its name suggests, this taxon was first described from the Lower Cretaceous (Berriasian) Purbeck Limestone Group of England [72], but dorsetisaurs have subsequently been reported from the Upper Jurassic of Portugal (Guimarota, Oxfordian [38, 71] and the USA (Morrison Formation, Kimmeridgian [40], as well as the Early Cretaceous of Mongolia (Aptian–Albian, *Paradorsetisaurus postumus* [96] (Fig. 2.1). Additionally, a small lizard skeleton (*Indrasaurus wangi* [97] found as gut contents within an Early Cretaceous Chinese *Microraptor* has a dorsetisaur-like dentition and may be related to dorsetisaurs.

Further anguimorph taxa are recorded from the Lower Cretaceous of Mongolia (Aptian–Albian, *Xenostius futilis*, crown xenosaur [95]); Uzbekistan (Albian, xenosaur [98]); Thailand (Barremian, eggs with embryos [99]); China (Barremian, *Dalinghosaurus longidigitus*, stem shinisaur [100]); North America (Albian–Cenomanian, *Primaderma*, stem monstersaur [101, 102]); England (Wealden, indet. [103]); and Spain (Barremian–Aptian, *Arcanosaurus ibericus*, 'varanoid' [104]), although many of these records are based on isolated elements. The tiny attenuate *Barlochersaurus winhtini* from the mid-Cretaceous amber of Myanmar was also tentatively attributed to Anguimorpha [105], but a more mature specimen is needed to verify this because the skull is poorly preserved.

The relationships of mosasaurians, also frequently grouped within Anguimorpha, are discussed separately (§ 2.3.7).

2.3.6 Iguania

This major squamate clade, comprising both pleurodont and acrodont jawed taxa, is first recorded with certainty from the Late Cretaceous (e.g., Mongolia, North America [68]. Some analyses (e.g., [59]) recovered *Hoyalacerta sanzi*, a short-limbed lizard from the Lower Cretaceous of Spain (Barremian, Las Hoyas [106]) on the iguanian stem but further specimens are needed to test this attribution. Conrad [59] also recovered the enigmatic *Huehuecuetzpalli mixtecus* (Albian, Mexico [107]) as a stem iguanian, although other researchers (e.g., [6, 107]) inferred that it lies on the squamate stem. *Huehuecuetzpalli* combines primitive characters (e.g., notochordal vertebrae, trunk intercentra, retained second distal tarsal) with derived ones (e.g., retracted nares, anterior parietal foramen). A stem-iguanian position was also recovered by Pyron [12], evaluating both the Conrad [59] and the Gauthier et al. [6] matrices using time calibration and combined evidence. Such major inconsistencies in tree position are, unfortunately, rather common for Jurassic–Early Cretaceous taxa.

Xianglong zhaoi, a gliding lizard from the Lower Cretaceous (Barremian) of China was described as an acrodontan [51], but this was based on the misinterpretation of the jagged broken edge of the crushed juvenile skull as an acrodont jaw (pers. obs.). Its affinities remain unknown. The Albian-Cenomanian amber of Myanmar has also yielded possible acrodontans [64], but this attribution is based on postcranial material and needs to be confirmed from well-preserved skulls. A stem chameleon described from the same deposits [64] is actually an albanerpetontid amphibian [108]. *Jeddaherdan aleadonta*, a partial dentary from the Cenomanian of Kem Kem, in Morocco, is possibly an early acrodontan [109], but *Gueragama sudamerica* (Turonian–Campanian, Brazil [79]) is a partial dentary with teeth like those of scincoids or teiids, and is unlikely to be an acrodontan as proposed.

2.3.7 Mosasauria

This group encompasses the Mosasauroidea (Late Cretaceous mosasaurs and aigialosaurs), and several smaller, long-necked, long-bodied taxa ('dolichosaurs') from the Early–Late Cretaceous. Mosasaurians are included briefly in this review chapter for completeness (see the relevant chapters in this volume for a more detailed discussion).

Generally considered close relatives of varanids and/or lanthanotids within Anguimorpha (e.g., [110, 111; Chapter 9]), several subsequent cladistic analyses distanced mosasaurians from varanids, albeit to varying degrees (e.g., [6, 112–114]). The proposal that mosasaurians were closely related to snakes (e.g., [113, 115, 116] and subsequent papers by the same authors and their collaborators) led Lee and Caldwell [117] to resurrect Cope's [118] name Pythonomorpha for their mosasaurian + snake clade, and generated the associated, and more controversial, suggestion of a marine origin for snakes [117, 119] (see also Chapter 7).

There is general agreement as to the monophyly of Mosasauroidea, but not of Aigialosauridae ([113] versus [115, 120, 121; Chapter 8]). Both groups have their earliest representatives in the Cenomanian [122]. 'Dolichosaurs' are more problematic, mainly because their skulls are poorly known. Recent papers, often with overlapping groups of authors, disagree as to whether 'dolichosaurs' are monophyletic (e.g., [123; Chapter 9]) or paraphyletic (e.g., [122, 124]), how they are defined and diagnosed (contrast [125] with [123, 124]), and whether they are closer to snakes (forming the Ophidiomorpha of [122, 125]) or to mosasauroids [123]. Most 'dolichosaur' genera were recovered from shallow marine deposits associated with either the remnant Tethys seaway (Slovenia, Croatia, Kazakhstan, Lebanon, Palestine, United Kingdom, Germany, France, Spain) or the Western Interior Seaway (North America) [122]. However, the earliest accepted 'dolichosaurs' are from freshwater deposits: in Spain (Barremian, vertebrae originally identified as snake [126]), Japan (Barremian–Albian *Kaganaias* [52]); and Australia (Albian [127]). The enigmatic snake-like *Tetrapodophis amplectus* (Albian, Brazil [128]), also from a freshwater deposit, was recovered by Paparella et al. [123] on the stem of a Mosasauroidea + Dolichosauridae clade that is proposed to be the sister group of snakes (see also Chapter 8).

2.3.8 Pan-Serpentes

The morphological specializations of the snake skeleton and the lack of early fossil representatives have hampered attempts to resolve the relationships of snakes to extant lizards. McDowell and Bogert [110] argued that snakes were related to anguimorphs, particularly varanids and/or lanthanotids, and several cladistic analyses (e.g., [115, 116, 123, 124, 129]) supported this (snake–mosasaurian relationships notwithstanding). However, convergence between snakes and other limbless squamates (e.g., amphisbaenians, dibamids, limbless scincids, and anguids) tends to confound analyses based solely on phenotypic data (e.g., [6, 59, 116, 130]), leading to artificial groupings. Molecular and combined evidence analyses avoid this problem, and unite Serpentes, Anguimorpha, and Iguania within Toxicofera (e.g., [1]). Most analyses find Serpentes + [Iguania + Anguimorpha] (e.g., [1, 8, 10, 11, 13, 131]) but the sister-group relationship between Iguania and Anguimorpha is not strongly supported. Furthermore, this does not resolve the position of mosasaurians with respect to snakes, because their common placement within an extended Anguimorpha (e.g., [115, 116, 123, 124, 132]) is incompatible with the molecular results. Pyron [12] ran combined evidence analyses using either the matrix of Conrad [59] or that of Gauthier et al. [6] for the morphological component. Predictably, he obtained very different results. His optimal tree for the Conrad matrix placed mosasaurians as sister to varanids (i.e., within Anguimorpha) and only distantly related to snakes, whereas the equivalent tree for the Gauthier et al. [6] matrix placed mosasaurians as the sister group of snakes, as did Reeder et al. [9] (see Augusta et al. and Zaher et al., this volume, for further discussion of snake and mosasaurian relationships).

The first unequivocal snake fossils are isolated vertebrae from the Lower Cretaceous of Algeria (Albian–Cenomanian [133, 134]), reportedly from both terrestrial (*Lapparentophis* sp.) and aquatic taxa [135], and isolated vertebrae referred to '*Coniophis*' sp. from North America (terrestrial, Aptian–Cenomanian [136]). A much more extensive suite of snake fossils is recorded from the Cenomanian. This Cenomanian record is dominated (and biased) by aquatic taxa from the widespread shallow marine deposits of this period, notably the simoliophiids (or pachyophiids; see Zaher et al., this volume): *Simoliophis* spp. (France, Portugal, Egypt, Morocco, western Europe and North Africa [135, 137]), *Pachyophis woodwardi* (Bosnia-Herzegovina [138]), *Pachyrhachis problematicus* (Palestine [139–142]); *Eupodophis descouensi* (Lebanon [143]), and *Haasiophis terrasanctus* (Palestine [144]), as well as the enigmatic aquatic *Lunaophis aquaticus* (Venezuela [145]). The current Cenomanian record of terrestrial snakes is more limited, but this is likely to be a gross underestimate considering the global distribution of these taxa: *Pouitella pervetus* (France [146]), *Xiaophis myanmarensis* (Myanmar [147]), *Najash rionegrina* (Argentina [148, 149]), *Seismophis septentrionalis* (Brazil [150]), and *Norisophis begaa* (Morocco [151]).

Snakes had clearly diversified, both ecologically and geographically, by the mid-Cretaceous. This is consistent with molecular divergence estimates that place the separation of stem snakes from other toxicoferans during the Jurassic (e.g., [10, 12]). Nonetheless, there remains a significant gap in the fossil record. Filling this gap, Caldwell et al. [30] re-interpreted specimens referred to the Middle Jurassic–Early Cretaceous lizard *Parviraptor* [27] as stem snakes. Based only on jaw fragments, Caldwell et al. [30] named *Eophis underwoodi* (Middle Jurassic, UK, dentary symphysis), *Diablophis gilmorei* (Upper Jurassic, Morrison Formation, USA, maxilla), and *Portugalophis lignites* (Upper Jurassic, Portugal, maxilla). However, this revision excluded most of the non-dental elements originally attributed to *Parviraptor* [27], based on the claim that *Parviraptor* as originally described was a chimaera of several different lizard taxa. New associated parviraptorid material from the Middle Jurassic of Skye confirms the original attribution of elements, both the preserved association on the Lower Cretaceous holotype block and the inferred association from Kirtlington (Middle Jurassic). These elements include paired frontals and parietals, the latter enclosing a parietal foramen, short palatines, and vertebrae that are notochordal and amphicoelous in immature individuals, becoming procoelous with maturity [27]. Work on the new Skye material and other parviraptorids is ongoing but *Parviraptor* as originally diagnosed is not a chimaera, and phylogenetic analyses must therefore include all of the attributed skeletal elements, rather than selecting only those (maxillae, dentaries) consistent with a particular hypothesis. Preliminary analyses (work in progress) do not support the inclusion of parviraptorids within Pan-Serpentes, and the Middle Jurassic age of the first recorded parviraptorid should therefore not be used to date snake origins in molecular divergence estimates (e.g. [152]).

2.4 Discussion and Conclusion

Although the squamate stem probably extended back into the Early Triassic, morphologically diagnosable squamates are currently unknown prior to the Middle Jurassic of Eurasia and North Africa. Molecular divergence estimates predict that major squamate clades had arisen by this time but, as yet, none of the currently known Middle Jurassic squamates can be placed confidently into the crown. A few Late Jurassic taxa, notably the dorsetisaurs (Anguimorpha), the paramacellodids (Scincoidea), and perhaps *Eichstaettisaurus* (Gekkota), may lie on the stems of crown clades, but the first unequivocal snake fossils currently date from the mid Cretaceous. The near global distribution of both terrestrial (France, USA, Gondwana) and aquatic (mostly Tethyan) snakes by the Cenomanian provides strong evidence of an earlier origin, but recent reports of Jurassic snake fossils [30] are based on misconception (see above). Given the close relationship between iguanians, anguimorphs and snakes, as strongly supported by molecular data, stem iguanians and stem anguimorphs have the potential to shed light on the expected morphology of the earliest stem snakes – given that stem toxicoferans presumably resembled one another morphologically until they diverged. Taxa like the dorsetisaurs, *Huehuecuetzpalli*, *Hoyalacerta*, and perhaps *Parviraptor* may therefore provide insights into early toxicoferan morphology, and may help to refine ideas on ancestral traits (e.g., [153]). Moreover, if mosasaurians (in total or in part) are genuinely the sister group of snakes (but see Augusta et al. and Zaher et al., this volume), then stem members of that clade also need to be identified, probably from terrestrial and/or freshwater deposits.

Further progress in unravelling the early history of squamates generally, and snakes in particular, will therefore require a combination of new fossil material; accurate (and objective) re-description of existing material using CT scan technology where possible; and a concerted effort to recover early squamate material from biogeographically and ecologically diverse deposits. Unfortunately, most Jurassic and Early Cretaceous squamate taxa are phylogenetically unstable when analysed in matrices combining living and extinct taxa. The topographic position of Mesozoic squamates is sensitive to analysis variables like data input (character choice and definition, ingroup and outgroup taxa sampled, incorporation of temporal data), and the use of ordering, weighting, or molecular constraints. Moreover, apomorphic skeletal features that characterize individuals of modern clades may be absent in early or stem representatives of those clades, further complicating their phylogenetic placement, especially when incomplete. Improving tree resolution is not simply a question of adding more phenotypic characters. We need to develop a greater understanding of those characters, and of the developmental and functional relationships between them.

Acknowledgements

My thanks to David Gower and Hussam Zaher for the invitation to participate in the symposium and this volume, and to the many colleagues around the world who have worked with me on fossil lepidosaurs. My particular thanks to Professor Roger Benson, University of Oxford, who is coordinating the research on the new material from the Middle Jurassic of Skye, and to Nick Fraser and an anonymous reviewer whose comments improved the manuscript. Bruno Navarro (University of São Paulo, Brazil) drafted Fig. 2.1.

References

1. N. Vidal and S. B. Hedges, The phylogeny of squamate reptiles (lizards, snakes, and amphisbaenians) inferred from nine nuclear protein-coding genes. *Comptes Rendus Biologies*, 328 (2005), 1000–1008.

2. C. L. Camp, Classification of the lizards. *Bulletin of the American Museum of Natural History*, 48 (1923), 289–481.

3. R. Estes, K. De Queiroz, and J. Gauthier, Phylogenetic relationships within Squamata. In R. Estes, and G. Pregill, eds., *Essays Commemorating Charles L. Camp. Phylogenetic Relationships of the Lizard Families* (Stanford, CA: Stanford University Press, 1988), pp. 119–281.

4. T. M. Townsend, A. Larson, E. Louis, and J. R. Macey, Molecular phylogenetics of

Squamata: the position of snakes, amphisbaenians, and dibamids, and the root of the squamate tree. *Systematic Biology*, 53 (2004), 735–757.

5. N. Vidal and S. B. Hedges, Molecular evidence for a terrestrial origin of snakes. *Proceedings of the Royal Society of London, series B: Biological Sciences*, 271 (2004), suppl: S226–S229.

6. J. A. Gauthier, M. Kearney, J. A. Maisano, O. Rieppel, and A. D. B. Behlke, Assembling the squamate tree of life: perspectives from the phenotype and the fossil record. *Bulletin of the Peabody Museum of Natural History*, 53 (2012), 3–308.

7. J. J. Wiens, C. R. Hutter, D. G. Mulcahy, et al., Resolving the phylogeny of lizards and snakes (Squamata) with extensive sampling of genes and species. *Biology Letters*, 8 (2012), 1043–1046.

8. R. A. Pyron, F. T. Burbrink, and J. J. Wiens, A phylogeny and revised classification of Squamata, including 4161 species of lizards and snakes. *BMC Evolutionary Biology*, 13 (2013), 93.

9. T. W. Reeder, T. M. Townsend, D. G. Mulcahy, et al., Integrated analyses resolve conflicts over squamate reptile phylogeny and reveal unexpected placements for fossil taxa. *PLoS ONE*, 10 (2015), e0118199.

10. Y. Zheng and J. J. Wiens, Combining phylogenomic and supermatrix approaches, and a time-calibrated phylogeny for squamate reptiles (lizards and snakes) based on 52 genes and 4162 species. *Molecular Phylogenetics and Evolution*, 94 (2016), 537–547.

11. J. W. Streicher and J. J. Wiens, Phylogenomic analyses of more than 4000 nuclear loci resolve the origin of snakes among lizard families. *Biology Letters*, 13 (2017), 20170393.

12. R. A. Pyron, Novel approaches for phylogenetic inference from morphological data and total-evidence dating in squamate reptiles (lizards, snakes, and amphisbaenians). *Systematic Biology*, 66 (2017), 38–56.

13. F. T. Burbrink, F. G. Grazziotin, R. A. Pyron, et al., Interrogating genomic-scale data for Squamata (lizards, snakes, and amphisbaenians) shows no support for key traditional morphological relationships. *Systematic Biology*, 69 (2020), 502–520.

14. M. E. H. Jones, C. L. Anderson, C. A. Hipsley, et al., Integration of molecules and new fossils supports a Triassic origin for Lepidosauria (lizards, snakes, and tuatara). *BMC Evolutionary Biology*, 13 (2013), 208.

15. S. E. Evans, At the feet of the dinosaurs: the origin, evolution and early diversification of squamate reptiles (Lepidosauria: Diapsida). *Biological Reviews*, 78 (2003), 513–551.

16. S. E. Evans and M. E. H. Jones, The origins, early history and diversification of lepidosauromorph reptiles. In S. Bandyopadhyay, ed., *New Aspects of Mesozoic Biodiversity, Lecture Notes in Earth Sciences 132* (Berlin, Germany: Springer-Verlag, 2010), pp. 27–44.

17. P. M. Datta and S. Ray, Earliest lizard from the Late Triassic (Carnian) of India. *Journal of Vertebrate Paleontology*, 26 (2006), 795–800.

18. M. N. Hutchinson, A. Skinner, and M. S. Y. Lee, *Tikiguana* and the antiquity of squamate reptiles (lizards and snakes). *Biology Letters*, 8 (2012), 665–669.

19. T. R. Simões, M. W. Caldwell, M. Tałanda, et al., The origin of squamates revealed by a Middle Triassic 'lizard' from the Italian Alps. *Nature*, 557 (2018), 706–709.

20. S. E. Evans and M. Borsuk-Białynicka, A small lepidosauromorph reptile from the Early Triassic of Poland. *Palaeontologica Polonica*, 65 (2009), 179–202.

21. S. Renesto and R. Posenato, A new lepidosauromorph reptile from the Middle Triassic of the Dolomites (Northern Italy). *Rivista Italiana di Paleontologia i Stratigrafia*, 109 (2003), 463–474.

22. S. E. Evans, A new lizard-like reptile (Diapsida: Lepidosauromorpha) from the Middle Jurassic of Oxfordshire. *Zoological*

Journal of the Linnean Society, 103 (1991), 391–412.

23. E. F. Griffiths, D. P. Ford, R. B. J. Benson, and S. E. Evans, New information on the Jurassic lepidosauromorph *Marmoretta oxoniensis*. *Papers in Palaeontology*, 7:4 (2021), 2255–2278

24. P. Yadagiri, Lower Jurassic lower vertebrates from Kota Formation, Pranhita-Godavari valley, India. *Journal of the Palaeontological Society of India*, 31 (1986), 89–96.

25. S. E. Evans, G. V. R. Prasad, and B. Manhas, Fossil lizards from the Jurassic Kota Formation of India. *Journal of Vertebrate Paleontology*, 22 (2002), 299–312.

26. J. L. Conrad, A new lizard (Squamata) was the last meal of *Compsognathus* (Theropoda: Dinosauria) and is a holotype in a holotype. *Zoological Journal of the Linnean Society*, 183 (2018), 584–634.

27. S. E. Evans, A new anguimorph lizard from the Jurassic and Lower Cretaceous of England. *Palaeontology*, 37 (1994), 33–49.

28. S. E. Evans, Crown group lizards from the Middle Jurassic of Britain. *Palaeontographica, Abt.A* 250 (1998), 123–154.

29. E. Pancirioli, R. B. J. Benson, S. Walsh, et al., Diverse vertebrate assemblage of the Kilmaluag Formation (Bathonian, Middle Jurassic) of Skye, Scotland. *Earth and Environmental Science Transactions of the Royal Society of Edinburgh*, 111 (2020), 135–56.

30. M. W. Caldwell, R. L. Nydam, A. Palci, and S. Apesteguia, The oldest known snakes from the Middle Jurassic–Lower Cretaceous provide insights on snake evolution. *Nature Communications*, 6 (2015), 5996.

31. P. V. Fedorov and L. A. Nessov, A lizard from the boundary of the Middle and Late Jurassic of north-east Fergana, *Bulletin of St. Petersburg University, Geology and Geography*, 3 (1992), 9–14 [In Russian].

32. A. Averianov, T. Martin, P. P. Skutschas, et al., Middle Jurassic vertebrate assemblages of Berezovsk coal mine in western Siberia, (Russia). *Global Geology*, 19 (2016), 187–204.

33. H. Haddoumi, R. Allain, S. Meslouh, et al., Guelb el Ahmar (Bathonian, Anoual Syncline, eastern Morocco): first continental flora and fauna including mammals from the Middle Jurassic of Africa. *Gondwana Research*, 29 (2016), 290–319.

34. J. L. Conrad, Y. Wang, X. Xu, R. A. Pyron, and J. Clark, Skeleton of a heavily armoured and long-legged Middle Jurassic lizard (Squamata, Reptilia). *Journal of Vertebrate Paleontology Supplement, Annual Meeting Abstracts*, 73 (2013), 108.

35. S. E. Evans and Y. Wang, A juvenile lizard from the Late Jurassic/Early Cretaceous of China. *Naturwissenschaften*, 94 (2007), 431–439.

36. S. E. Evans and Y. Wang, A long-limbed lizard from the Upper Jurassic/Lower Cretaceous of Daohugou, Inner Mongolia, China. *Vertebrata Palasiatica*, 47 (2009), 21–34.

37. L. P. Dong, Y. Wang, L. Mou, G. Zhang, and S. E. Evans, A new Jurassic lizard from China. *Geodiversitas*, 41 (2019), 623–641.

38. J. Seiffert, Upper Jurassic lizards from Central Portugal. *Memoria, Serviços Geológicos de Portugal*, 22 (1973), 1–85.

39. M. K. Hecht and B. M. Hecht, A new lizard from Jurassic deposits of Middle Asia. *Paleontological Journal*, 18 (1984), 133–136.

40. D. R. Prothero and R. Estes, Late Jurassic lizards from Como Bluff Wyoming, and their palaeobiogeographic significance. *Nature*, 286 (1980), 484–486.

41. S. E. Evans and D. C. Chure, Paramacellodid lizard skulls from the Jurassic Morrison Formation at Dinosaur National Monument, Utah. *Journal of Vertebrate Paleontology*, 18 (1998), 99–114.

42. S. E. Evans and D. C. Chure, Morrison lizards: structure, relationships and biogeography. *Modern Geology*, 23 (1998), 35–48.

43. M. Cocude-Michel, Les rhynchocéphales et les Sauriens des Calcaires Lithographiques (Jurassique supérieur) d'Europe Occidentale. *Nouvelles Archives du Muséum d'Histoire naturelle de Lyon*, 7 (1963), 1–187.

44. R. Hoffstetter, Les Sauria du Jurassic supérieur et specialement les Gekkota de Bavière et de Mandchourie. *Senckenbergiana Biologie*, 45 (1964), 281–322.

45. N. J. Mateer, Osteology of the Jurassic lizard *Ardeosaurus brevipes* (Meyer). *Palaeontology*, 25 (1982), 461–9.

46. S. E. Evans, The Solnhofen (Jurassic: Tithonian) lizard genus *Bavarisaurus*: new skull material and a reinterpretation. *Neues Jahrbuch für Geologie und Paläontologie, Abhandlungen*, 192 (1994), 37–52.

47. T. R. Simões, M. W. Caldwell, R. L. Nydam, and P. Jiminez-Huidabro, Osteology, phylogeny, and functional morphology of two Jurassic lizard species and the early evolution of scansoriality in geckoes. *Zoological Journal of the Linnean Society*, 180 (2017), 216–241.

48. M. Talanda, An exceptionally preserved Jurassic skink suggests lizard diversification preceded the fragmentation of Pangaea. *Palaeontology*, 61 (2018), 659–677.

49. S. E. Evans, A re-evaluation of the late Jurassic (Kimmeridgian) reptile *Euposaurus* (Reptilia: Lepidosauria) from Cerin, France. *Geobios*, 27 (1994), 621–631.

50. W. Zils, C. Werner, A. Moritz, and C. Saanane, Tendaguru, the most famous dinosaur locality of Africa. Review, survey and future prospects. *Documenta naturae, Munich*, 97 (1995), 1–41.

51. P. P. Li, K. Q. Gao, L. H. Hou, and X. Xu, A gliding lizard from the Early Cretaceous of China. *Proceedings of the National Academy of Sciences of the U.S.A.*, 104 (2007), 5507–5509.

52. S. E. Evans, M. Manabe, M. Noro, S. Isaji, and M. Yamaguchi, A long-bodied lizard from the Lower Cretaceous of Japan. *Palaeontology*, 49(6) (2006), 1143–1165.

53. Y. Wang and S. E. Evans, A gravid lizard from the Early Cretaceous of China: insights into the history of squamate viviparity. *Naturwissenschaften*, 98 (2011), 739–743.

54. S. E. Evans and M. Manabe, A herbivorous lizard from the Early Cretaceous of Japan. *Palaeontology*, 51 (2008), 487–498.

55. S. E. Evans and L. J. Barbadillo, An unusual lizard (Reptilia, Squamata) from the Early Cretaceous of Las Hoyas, Spain. *Zoological Journal of the Linnean Society*, 124 (1998), 235–266.

56. E. N. Arnold and G. Poinar, A 100 million year old gecko with sophisticated adhesive toepads preserved in amber from Myanmar. *Zootaxa*, 1847 (2008), 62–68.

57. S. E. Evans, A. Lacasa-Ruiz, and J. Erill Rey, A lizard from the Early Cretaceous (Berriasian-Hauterivian) of Montsec. *Neues Jahrbuch für Geologie und Paläontologie, Abhandlungen*, 215 (1999), 1–15.

58. S. E. Evans, P. Raia, and C. Barbera, New lizards and rhynchocephalians from the Early Cretaceous of southern Italy. *Acta Palaeontologica Polonica*, 49 (3) (2004), 393–408.

59. J. L. Conrad, Phylogeny and systematic of Squamata (Reptilia) based on morphology. *Bulletin of the American Museum of Natural History*, 310 (2008), 1–182.

60. V. R. Alifanov, The oldest gecko (Lacertilia: Gekkonidae) from the Lower Cretaceous of Mongolia. *Paleontological Journal*, 23 (1990), 128–131.

61. J. D. Daza, A. M. Bauer, and E. D. Snively, On the fossil record of Gekkota. *Anatomical Record*, 297 (2014), 433–462.

62. J. L. Conrad and M. A. Norell, High-resolution X-ray computed tomography of an early Cretaceous gekkonomorph (Squamata) from Öösh (Övörkhangai: Mongolia). *Historical Biology*, 18 (2006), 405–431.

63. J. L. Conrad and J. D. Daza, Naming and re-diagnosing the Cretaceous gekkonomorph (Reptilia, Squamata) from Öösh (Övörkhangai, Mongolia). *Journal of*

Vertebrate Paleontology, 35 (2015), e980891.

64. J. D. Daza, E. L. Stanley, P. Wagner, A. Bauer, and D. A. Grimaldi, Mid-Cretaceous amber fossils illuminate the past diversity of tropical lizards. *Science Advances*, 2 (2016), e1501080.

65. G. Fontanarrosa, J. D. Daza, and V. Abdala, Cretaceous fossil gecko hand reveals a strikingly modern scansorial morphology: qualitative and biometric analysis of an amber-preserved lizard hand. *Cretaceous Research*, 84 (2018), 120–133.

66. A. E. Greer, The relationships of the lizard genera *Anelytropsis* and Dibamus. *Journal of Herpetology*, 19 (1985), 116–156.

67. A. Cernansky, The first potential fossil record of a dibamid reptile (Squamata; Dibamidae): a new taxon from the early Oligocene of Central Mongolia. *Zoological Journal of the Linnean Society*, 187 (2019), 782–799.

68. R. Estes and Sauria, Amphisbaenia. *Handbuch der Paläoherpetologie/ Encyclopedia of Paleontology*, Part 10A (Stuttgart: Gustav Fischer, 1983).

69. R. Kosma, The dentitions of recent and fossil scincomorphan lizards (Lacertilia, Squamata). Systematics, functional morphology, palaeoecology. Unpublished PhD Thesis, University of Hannover (2004).

70. A. Broschinski and D. Sigogneau-Russell, Remarkable lizard remains from the lower Cretaceous of Anoual (Morocco). *Annales de Paléontologie (Vert.-Invert.)*, 82 (1996), 147–175.

71. A. Broschinski, The lizards from the Guimarota mine. In T. Martin and B. Krebs, eds., *Guimarota. A Jurassic Ecosystem* (Munich: Dr. Friedrich Pfeil, 2000), pp. 59–68.

72. R. Hoffstetter, Coup d'oeil sur les Sauriens (Lacertiliens) des couches de Purbeck (Jurassique Supérieur d'Angleterre). Problemes Actuels de Paleontologie (Evolution des Vertebrates), *Colloques Internationaux du Centre National de la Recherche Scientifique*, 163 (1967), 349–371.

73. J.-L. Li, A new lizard from the late Jurassic of Subei, Gansu. *Vertebrata PalAsiatica*, 23 (1985), 13–18.

74. A. O. Averianov and P. P.Skutchas, Paramacellodid lizard (Squamata, Scincomorpha) from the Early Cretaceous of Transbaikalia. *Russian Journal of Herpetology*, 6 (1999), 115–117.

75. A. Richter, Lacertilia aus der Unteren Kreide von Uña und Galve (Spanien) und Anoual (Marokko). *Berliner geowissenschaftliche Abhandlungen*, 14 (1994), 1–147.

76. R. L. Nydam and R. L. Cifelli, Lizards from the Lower Cretaceous (Aptian-Albian) Antlers and Cloverley Formations. *Journal of Vertebrate Paleontology*, 22 (2002), 286–298.

77. J. S. Bittencourt, T. R. Simões, M. W. Caldwell, and M. C. Langer, Discovery of the oldest South American fossil lizard illustrates the cosmopolitanism of early South American squamates. *Communications Biology*, 3 (2020), 201.

78. V. H. Reynoso and G. Callison, A new scincomorph lizard from the Early Cretaceous of Puebla, Mexico. *Zoological Journal of the Linnean Society*, 130 (2000), 183–212.

79. T. R. Simões, E. Wilner, M. W. Caldwell, L. C. Weinschutz, and A. W. A. Kellner, A stem-acrodontan lizard in the Cretaceous of Brazil revises early lizard evolution in Gondwana. *Nature Communications*, 6 (2015), 9149.

80. S. E. Evans, M. E. H. Jones, and R. Matsumoto, A new lizard skull from the Purbeck Limestone Group of England. *Bulletin of the Geological Society of France*, 183 (2012), 517–524.

81. S. E. Evans and L. J. Barbadillo, Early Cretaceous lizards from las Hoyas, Spain. *Zoological Journal of the Linnean Society*, 119 (1997), 23–49.

82. A. Richter, Der problematische Lacertilier *Ilerdaesaurus* (Reptilia: Squamata) aus der Unter-Kreide von Uña und Galve. *Berliner geowissenschaftliche Abhandlungen*, 13 (1994), 135–161.

83. F. C. Bonfim-Junior and R. B. Marques, Um novo lagarto do Cretáceo do Brasil (Lepidosauria, Squamata, Lacertilia – formação Santana, Aptiano da Bacia do Araripe). *Anuario do Instituto de Geociencias*, 20 (1997), 233–240.

84. F. C. Bonfim-Junior and L. D. S. Avila, Phylogenetic position of *Tijubina pontei* Bonfim and Marques 1997 (Lepidosauria, Squamata), a basal lizard from the Santana Formation, Lower Cretaceous of Brazil. *Journal of Vertebrate Paleontology*, 22 (Supplement to 3) (2002), 37A–38A.

85. T. R. Simões, Redescription of *Tijubina pontei*, and Early Cretaceous lizard (Reptilia; Squamata) from the Crato Formation of Brazil. *Anais da Academia Brasileira de Ciencias*, 84 (2012), 1.

86. S. E. Evans and Y. Yabumoto, A lizard from the Early Cretaceous Crato Formation, Araripe Basin, Brazil. *Neues Jahrbuch für Paläontologie und Geologie, Monatshefte* 1998 (1998), 349–364.

87. V. R. Alifanov, New lizards of the Macrocephalosauridae (Sauria) from the Upper Cretaceous of Mongolia, critical remarks on the systematics of the Teiidae (sensu Estes, 1983). *Paleontological Journal*, 27 (1993), 70–90.

88. R. L. Nydam, J. G. Eaton, and J. Sankey, New taxa of transversely-toothed lizards (Squamata; Scincomorpha) and new information on the evolutionary history of 'teiids'. *Journal of Paleontology*, 81 (2007), 538–549.

89. S. E. Evans and R. Matsumoto, An assemblage of lizards from the Early Cretaceous of Japan. *Palaeontologica Electronica*, 18.2.36A (2015), 1–36.

90. R. M. Sullivan, A new middle Paleocene (Torrejonian) rhineurid amphisbaenian, *Plesiorhineura tsentasi* new genus, new species, from the San Juan Basin, New Mexico. *Journal of Paleontology*, 59 (1985), 1481–1485.

91. L. A. Nessov, Rare bony fishes, terrestrial lizards and mammals from the lagoonal zone of the litoral lowlands of the Cretaceous of the Kyzylkumy. *Yearbook of the All-Union Palaeontological Society, Leningrad*, 28 (1985), 199–219.

92. V. R. Alifanov, Lizards of the family Hodzhakuliidae (Scincomorpha) from the Lower Cretaceous of Mongolia. *Paleontological Journal*, 50 (2016), 504–513.

93. X.-C. Wu, D. B. Brinkman, and A. P. Russell, *Sineoamphisbaena hexatabularis*: an amphisbaenian (Diapsida: Squamata) from the Upper Cretaceous redbeds at Bayan Mandahu (Inner Mongolia, People's Republic of China), and comments on the phylogenetic relationships of the Amphisbaenia. *Canadian Journal of Earth Sciences*, 33 (1996), 541–577.

94. M. Talanda, Cretaceous roots of the amphisbaenian lizards. *Zoologica Scripta*, 45 (2015), 1–8.

95. V. R. Alifanov, Lizards of the families Eoxantidae, Ardeosauridae, Globauridae, and Paramacellodidae (Scincomorpha) from the Aptian-Albian of Mongolia. *Paleontological Journal*, 53 (2019), 74–88.

96. V. R. Alifanov, Lizards of the families Dorsetisauridae and Xenosauridae (Anguimorpha) from the Aptian–Albian of Mongolia. *Paleontological Journal*, 53 (2019), 183–193.

97. J. M. O'Connor, X. Zheng, L. Dong, et al., *Microraptor* with ingested lizard suggests non-specialized digestive function. *Current Biology*, 29 (2019), 2423–2429.

98. K. Q. Gao and L. A. Nessov, Early Cretaceous squamates from the Kyzylkum Desert, Uzbekistan. *Neues Jahrbüch für Geologie und Paläontologie, Abhandlungen*, 207 (1998), 289–309.

99. V. Fernandez, E. Buffetaut, V. Suteethorn, et al., Evidence of egg diversity in squamate evolution from Cretaceous anguimorph embryos. *PLoS ONE*, 10 (2015), e0128610.

100. S. E. Evans and Y. Wang, The Early Cretaceous lizard *Dalinghosaurus* from China. *Acta Palaeontologica Polonica*, 50 (2005), 725–742.

101. R. L. Cifelli and R. L. Nydam, Primitive, helodermatid-like platynotan from the Early Cretaceous of Utah. *Herpetologica*, 51 (1995), 286–291.

102. R. L. Nydam, A new taxon of helodermatid-like lizard from the Albian-Cenomanian of Utah. *Journal of Vertebrate Paleontology*, 20 (2000), 285–294.

103. S. C. Sweetman and S. E. Evans, Lepidosaurs (lizards). In D. J. Batten, ed., Palaeontological Association Field Guide to Fossils, 14. *English Wealden Fossils* (London: The Palaeontological Association, 2011), pp. 264–284.

104. A. Houssaye, J.-C. Rage, F. T. Fernandez-Baldor, et al., A new varanoid squamate from the Early Cretaceous (Barremian-Aptian) of Burgos, Spain. *Cretaceous Research*, 41 (2013), 127–135.

105. J. D. Daza, A. M. Bauer, E. L. Stanley, et al., An enigmatic miniaturized and attenuate whole lizard from the mid-Cretaceous amber of Myanmar. *Breviora*, 563 (2018), 1–18.

106. S. E. Evans and L. J. Barbadillo, A short-limbed lizard from the Lower Cretaceous of Spain. *Special Papers in Palaeontology*, 60 (1999), 73–85.

107. V. H. Reynoso, *Huehuecuetzpalli mixtecus* gen. et sp. nov.: a basal squamate (Reptilia) from the early Cretaceous of Tepexi de Rodriguez, Central Mexico. *Philosophical Transactions of the Royal Society of London, Biological Sciences*, 353 (1998), 477–500.

108. R. Matsumoto and S. E. Evans, The first record of albanerpetontid amphibians (Amphibia, Albanerpetontidae) from East Asia. *PLoS ONE*, 13 (2018), e0189767.

109. S. Apesteguia, J. D. Daza, T. R. Simões, and J.-C. Rage, The first iguanian lizard from the Mesozoic of Africa. *Royal Society Open Science*, 3 (2016), 160462.

110. S. B. McDowell and C. M. Bogert, The systematic position of *Lanthanotus* and the affinities of the anguinomorphan lizards. *Bulletin of the American Museum of Natural History*, 105 (1954), 1–142.

111. R. L. Carroll and M. De Braga, *Aigialosaurus*: mid-Cretaceous varanoid lizards. *Journal of Vertebrate Paleontology*, 12 (1992), 66–86.

112. M. W. Caldwell, R. L. Carroll, and H. Kaiser, The pectoral girdle and forelimb of *Carsosaurus marchesetti* (Aigialosauridae) with a preliminary phylogenetic analysis of mosasauroids and varanoids. *Journal of Vertebrate Paleontology*, 15 (1995), 516–531.

113. M. W. Caldwell, Squamate phylogeny and the relationships of snakes and mosasauroids. *Zoological Journal of the Linnean Society*, 125 (1999), 115–147.

114. T. R. Simões, O. Vernygora, I. Paparella, P. Jimenez-Huidobro, and M. W. Caldwell, Mosasauroid phylogeny under multiple phylogenetic methods provides new insights on the evolution of aquatic adaptations in the group. *PLoS ONE* 12 (2017), e0176773.

115. M. S. Y. Lee, The phylogeny of varanoid lizards and the affinities of snakes. *Philosophical Transactions of the Royal Society, Biological Sciences*, 352 (1997), 53–91.

116. M. S. Y. Lee, Convergent evolution and character correlation in burrowing reptiles: towards a resolution of squamate phylogeny. *Biological Journal of the Linnean Society*, 65 (1998), 369–453.

117. M. S. Y. Lee and M. W. Caldwell, *Adriosaurus* and the affinities of mosasaurs, dolichosaurs, and snakes. *Journal of Paleontology*, 74 (2000), 915–37.

118. E. D. Cope, On the reptilian orders, Pythonomorpha and Streptosauria. *Proceedings of the Boston Society of Natural History*, 12 (1869), 250–66.

119. M. S. Y. Lee, Molecular evidence and marine snake origins. *Biology Letters*, 1 (2005), 227–230.

120. M. W. Caldwell, On the aquatic squamate *Dolichosaurus longicollis* Owen, 1850 (Cenomanian, Upper Cretaceous), and the evolution of elongate necks in squamates. *Journal of Vertebrate Paleontology*, 20 (2000), 720–735.

121. A. R. Dutchak, A review of the taxonomy and systematics of aigialosaurs.

Netherlands Journal of Geosciences, 84 (2005), 221–229.

122. M. C. Mekarski, S. E. Pierce, and M. W. Caldwell, Spatiotemporal distributions of non-ophidian ophidiomorphs, with implications for their origin, radiation, and extinction. *Frontiers in Earth Science*, 7 (2019), article 24.

123. M. Paparella, A. Palci, U. Nicosia, and M. W. Caldwell, A new fossil marine lizard with soft tissues from the Late Cretaceous of southern Italy. *Royal Society Open Science*, 5 (2018), e172411.

124. S. E. Pierce and M. W. Caldwell, Redescription and phylogenetic position of the Adriatic (Upper Cretaceous; Cenomanian) dolichosaur *Pontosaurus lesinensis* (Kornhuber, 1873). *Journal of Vertebrate Paleontology*, 24 (2004), 373–386.

125. A. Palci and M. W. Caldwell, Redescription of *Acteosaurus tommasinii* Von Meyer 1860, and a discussion of evolutionary trends within the clade Ophiodiomorpha. *Journal of Vertebrate Paleontology*, 30 (2010), 94–108.

126. J.-C. Rage and A. Richter, A snake from the lower Cretaceous (Barremian) of Spain: the oldest known snake. *Neues Jahrbuch für Geologie und Paläontologie, Monatshefte*, 1995 (1995), 561–565.

127. J. D. Scanlon and S. A. Hocknull, A dolichosaurid lizard from the latest Albian (mid-Cretaceous) Winto Formation, Queensland, Australia. In M. J. Everhard, ed., *Proceedings of the Second Mosasaur Meeting, Fort Hays Studies Special Issue*, 3 (2008), 131–136.

128. D. M. Martill, H. Tischlinger, and N. R. Longrich, A four-legged snake form the Early Cretaceous of Gondwana. *Science*, 349 (2015), 416–419.

129. M. S. Y. Lee, Squamate phylogeny, taxon sampling, and data congruence. *Organisms, Diversity, and Evolution*, 5 (2005), 25–45.

130. T. R. Simões, M. W. Caldwell, and A. W. A. Kellner, A new Early Cretaceous lizard species from Brazil, and the phylogenetic position of the oldest known South American squamates. *Journal of Systematic Palaeontology*, 13 (2015), 601–614.

131. D. G. Mulcahy, B. P. Noonan, T. Moss, et al., Estimating divergence dates and evaluating dating methods using phylogenomic and mitochondrial data in squamate reptiles. *Molecular Phylogenetics and Evolution*, 65 (2012), 974–991.

132. M. S. Y. Lee, Hidden support from unpromising data sets strongly unites snakes with anguimorph 'lizards'. *Journal of Evolutionary Biology*, 22 (2009), 1308–1316.

133. R. Hoffstetter, Un serpent terrestre dans le Crétacé inférieur du Sahara. *Bulletin de la Société Géologique de France*, 1 (1959), 897–902.

134. G. Cuny, J. J. Jaeger, M. Mahboubi, and J.-C. Rage, Les plus anciens serpentes (Reptilia, Squamata) connus. Mise au point sur l'age géologiques des Serpents de la partie moyenne du Crétacé. *Comptes rendus des séances de l'Académie des Sciences Paris, Series* 2, 311 (1990), 1267–1272.

135. J.-C. Rage and F. Escuillié, The Cenomanian: stage of hind-limbed snakes. *Notebooks on Geology*, (1) (2003), (CG2003 A01 JCR-FE).

136. J. D. Gardner and R. L. Cifelli, A primitive snake from the Cretaceous of Utah. *Special Papers in Palaeontology*, 60 (1999), 87–100.

137. J.-C. Rage and D. B. Dutheil, Amphibians and squamates from the Cretaceous (Cenomanian) of Morocco. A preliminary study. *Palaeontographica Abteilung A*, 286 (2008), 1–22.

138. M. S. Y. Lee, M. W. Caldwell, and J. D. Scanlon, A second primitive marine snake: *Pachyophis woodwoodi* from the Cretaceous of Bosnia-Herzgovina. *Journal of Zoology*, 248 (1999), 509–520.

139. G. Haas, On a new snake-like reptile from the lower Cenomanian of Ein Jabrud, near Jerusalem. *Bulletin du Muséum National*

d'Histoire Naturelle de Paris, 1 (1979), 51–64.

140. M. W. Caldwell and M. S. Y. Lee, A snake with legs from the marine Cretaceous of the Middle East. *Nature*, 386 (1997), 705–709.

141. H. Zaher, The phylogenetic position of *Pachyrachis* within snakes (Squamata, Lepidosauria). *Journal of Vertebrate Paleontology*, 18 (1998), 1–3.

142. H. Zaher and O. Rieppel, The phylogenetic relationships of *Pachyrachis problematicus* and the evolution of limblessness in snakes (Lepidosauria, Squamata). *Comptes Rendus de l'Academie de Sciences, Series IIA – Earth and Planetary Science*, 329 (1999), 831–837.

143. J.-C. Rage and F. Escuillié, Un nouveau serpent bipède du Cénomanien (Crétacé). Implications phylétiques. *Comptes Rendus de l'Academie des Sciences, Series IIA - Earth and Planetary Science*, 330 (2000), 513–520.

144. E. Tchernov, O. Rieppel, H. Zaher, M. J. Polcyn, and L. L. Jacobs, A fossil snake with limbs. *Science*, 287 (2000), 2010–2012.

145. A. Albino, J. D., Carillo-Briceno, and J. M. Neenan, An enigmatic aquatic snake from the Cenomanian of northern South America. *PeerJ*, 4 (2016), DOI 10.7717/peerj .2027e2027.

146. J.-C. Rage, Un serpent primitif (Reptilia, Squamata) dans le Cénomanien (base du Crétacé supérieur). *Comptes rendus de l'Academie des Sciences, Paris*, 307 (1988), 1027–1032.

147. L. Xing, M. W. Caldwell, R. Chen, et al., A mid-Cretaceous embryonic-to-neonate snake in amber from Myanmar. *Science Advances*, 4 (2018), eaat5042.

148. S. Apesteguía and H. Zaher, A Cretaceous terrestrial limbed snake with robust hindlimbs and sacrum, *Nature*, 440 (2006), 1037–1040.

149. F. F. Garberoglio, S. Apesteguia, T. R. Simões, et al., New skulls and skeletons of the Cretaceous legged snake *Najash*, and the evolution of the modern snake body plan. *Science Advances*, 5 (2019), eaax5833.

150. A. S. Hsiou, A. M. Albino, M. A. Medeiros, and R. A. B. Santos, The oldest Brazilian snakes from the Cenomanian (early Late Cretaceous). *Acta Palaeontologica Polonica*, 59 (2013), 635–642.

151. C. G. Klein, N. R. Longrich, N. Ibrahim, S. Zouhri, and D. M. Martill, A new basal snake from the mid-Cretaceous of Morocco. *Cretaceous Research*, 72 (2017), 134–141.

152. S. M. Harrington and T. W. Reeder, Phylogenetic inference and divergence dating of snakes using molecules, morphology and fossils: new insights into convergent evolution of feeding morphology and limb reduction. *Biological Journal of the Linnean Society*, 121(2017), 379–394.

153. F. O. Da Silva, A.-C. Fabre, Y. Savriama, et al., The ecological origins of snakes as revealed by skull evolution. *Nature Communications*, 9 (2018), 376.

The First 80 Million Years of Snake Evolution

The Mesozoic Fossil Record of Snakes and Its Implications for Origin Hypotheses, Biogeography, and Mass Extinction

Jason J. Head, Alexandra F. C. Howard, and Johannes Müller

3.1 Introduction

With over 3,000 extant species, snakes are one of the great vertebrate radiations, and the origins and interrelationships of the clade have been long been intensely studied and debated (see [1] for historical overview). Hypotheses for the ecological contexts of the origin for snakes and their body form have historically been studied in the context of comparative anatomies of sensory systems and the fossil record, and include competing inferences of fossoriality based on visual and auditory apomorphies in extant taxa (e.g., [2–4]) and marine habits based on the early fossil record (e.g., [5]). These hypotheses have been revisited over the last 23 years based on new fossil discoveries and reinterpretation of anatomical data from fossil and extant taxa (e.g., [6–11]).

The interrelationships of snakes and their position within squamate phylogeny have also been intensely studied. Analyses of morphological data historically proposed a sister-taxon relationship between snakes and a clade of varanids and/or mosasauroids (e.g., [12–14]) or snake-like squamates variably including dibamids, amphisbaenians, and acontid and feyliniid skinks [15–19]. Analyses of molecular data recover a sister-taxon relationship between snakes and a clade comprising Anguimorpha+Iguania [20–23], which is similarly recovered in combined analyses [24].

Fossil evidence for snake origins and relationships has been primarily derived from comparatively well-preserved Late Cretaceous specimens (Fig. 3.1; [6–8, 25–27]). The geographic distributions, depositional environments, and morphology of these specimens

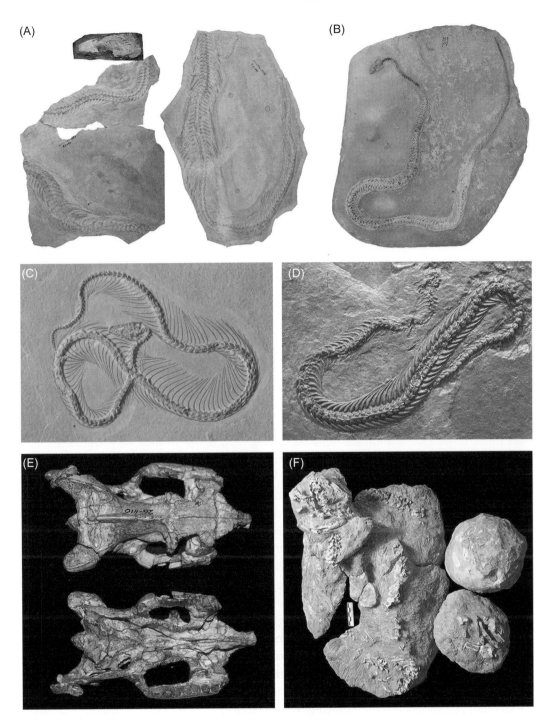

Figure 3.1 Well-preserved Cretaceous snake taxa. (A). *Pachyrhachis problematicus* (HUJ-PAL 3569, HUJ-PAL 3775); (B). *Haasiophis terrasanctus* (HUJ-PAL EJ 695); (C). *Eupodophis descouensi* (MSNM V 4014); (D). *Pachyophis woodwardi* (Cast, Oxford Museum of Natural History) (E). *Dinilysia patagonica* (MLP 26-410), skull in dorsal (top) and ventral (bottom) views; (F). *Sanajeh indicus* (GSI/GC/2901–2906). Image of *Haasiophis* courtesy of C. A. Brochu, all other images are by JJH. (A black and white version of this figure will appear in some formats. For the colour version, please refer to the plate section.).

has been used to infer early biogeographic histories for snakes, ecological contexts of snake origins, and evolutionary relationships of snakes to extant and fossil squamates (e.g., [14, 28]).

Less attention has been paid to the potential explanatory power of the rest of the Mesozoic record, which consists primarily of isolated or associated vertebral remains, due, in part, to the relative dearth of anatomical data. Here we review and discuss the Mesozoic snake record in the contexts of both molecular and morphological phylogenies, focusing on the temporal and geographic distributions of under-examined taxa. We then discuss these records with respect to competing hypotheses of the origin of the snake body form, biogeographic histories of early snakes, and survival of Cretaceous snake clades into the Cenozoic. Because the fossil record consists primarily of precloacal vertebrae, we provide an overview of vertebral osteological correlates to the snake body form and their phylogenetic significance.

3.2 Vertebral Morphology and Its Necessity for Inferring the Evolution of Snakes

The vast majority of the fossil record of snakes consists of isolated precloacal vertebrae [29, 30] and systematic identification and inferences of the evolution of the snake body form require identifying osteological correlates to the musculoskeletal and circulatory components of the elongate, limb-reduced precloacal body. Comparison of crown-snake vertebrae with those of generalized limbed squamates (Fig. 3.2) demonstrates an integrated set of changes relating to the transition from a body form in which the axial skeleton generates pulmonary ventilation and supports propulsive locomotion from the appendicular skeleton to a body form in which the axial skeleton both provides locomotion through direct contact with substrate and ventilates the lungs primarily via the dorsal ribs (e.g., [31, 32]). Although there is variability in the degree of development in these changes among extant snake clades, especially with respect to the repeated evolution of fossorial ecomorphologies [33], components of them are present across the snake crown, and the majority are present across Alethinophidia.

A primary component of the snake body form is the ventrolateral orientation of the ribs versus a more lateral position in limbed taxa. This is achieved by depressing the prezygapophyses and synapophyses relative to the centrum, reducing the angle of the prezygapophyseal articular facet, separating the synapophyseal articular facets from the lateral margin of the centrum body via a distinct apophyseal trunk that forms the ventral extent of the prezygapophyses, and angling the synapophyseal facets so that the articulation of the costovertebral joint is dorsolaterally angled (Fig. 3.2B1).

Increased stabilization of costovertebral joints is achieved through separation of a unicapitate, convex synapophyseal articular facet in limbed squamates to a bi-faceted surface consisting of a convex posterodorsal diapophysis and concave to flat anteroventral parapophysis (Fig. 3.2B1) (e.g., [33]) which stabilizes against medial rotation of the rib [32].

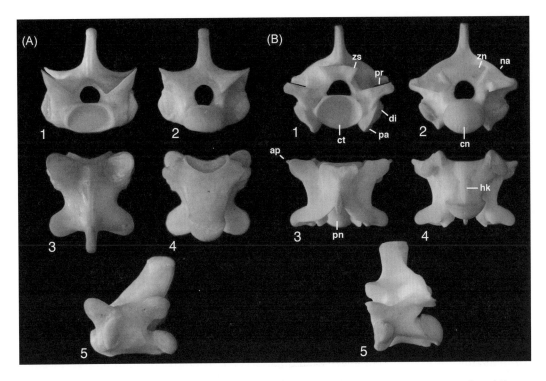

Figure 3.2 Precloacal vertebral morphology in limbed squamates and crown-group snakes. (A), *Heloderma suspectum*; (B). *Boa constrictor* in (1) anterior, (2) posterior, (3) dorsal, (4) ventral, (5) left lateral views. Abbreviations: ap, accessory process of prezygapophyses; cn, condyle; ct, cotyle; di, diapophysis; hk, haemal keel; na, neural arch; pa, parapophysis; pn, posteromedian notch of neural arch; pr, prezygapophyses; zn, zygantrum; zs, zygosphene.

Increased stabilization of intervertebral joints is achieved through lateral elongation and lowered angles of zygapophyseal articulations, greater rounding of the cotyle–condyle articulation including increased development of the ventral margin of the cotyle, and through limiting torsion between adjacent elements by the articulation of a large, wide zygosphene with dorsolaterally angled articular facets at the dorsal margin of the anterior opening of the neural canal with a corresponding deep posterior zygantrum incised into the posterior neural arch of the preceding vertebra (Fig. 3.2B1, 3.2B2, and 3.2B5) [34].

Vertebral morphology corresponds to shifts in muscle size and position in the transition to the snake body form. *Levator costae* is a small hypaxial muscle in some limbed lizards that inserts on the posterior edge of the synapophysis anteriorly and the anterodorsal edge of the following rib, posteriorly and functions to lift the ribcage [35]. In extant snakes, this muscle is proportionally much larger and inserts anteriorly on the posteroventral surface of the expanded prezygapophyseal body and/or on the lateral prezygapophyseal accessory process (Fig. 3.2B2) [36–40]. The shifted position and increased size of *levator costae* in snakes is functionally related to elevating ribs during locomotion and respiration, as well as stabilizing ribs against compressional forces during substrate contact and against movement of the skin in rectilinear locomotion [32]. Similarly, *M. interarticulares superiores* and *inferiores* are epaxial muscles that insert on the prezygapophyseal accessory process and likely serve as flexors [36–38]. *Multifidus* inserts anteriorly on the posterior edge of the neural spine and neural arch in limbed lizards, where it forms a fascicle of *transversospinalis*, which serves to dorsally flex and stabilize the vertebral column [36–41]. In snakes, the large zygantrum generally produces an arched, elevated dorsal margin of the posterior neural arch (Fig. 3.2B5) and is associated with the anterior positioning of the neural spine at the apex of a posteromedian notch that exposes the dorsal surface of the condyle in alethinophidians (Fig. 3.2B2). The incision of the posteromedian notch and the elevation of the arch provide a wider, longer surface for insertion of *multifidus*.

Osteological correlates to evolutionary changes in the architecture of the circulatory system in snakes consist of increased complexity of the ventral centrum associated with the development of the lymphatic system. In snakes, the lymphatic system of the axial skeleton consists of paired paravertebral vessels [42, 43] that extend along the ventral surface of the vertebral column and are medially anastomosed at each vertebral joint [43]. The paravertebral vessels are housed in shallow to deep impressions on the ventral centrum, walled medially by a haemal keel (Fig. 3.2B4) and laterally by the medial margins of the synapophyses and subcentral ridges that originate anteriorly at the posterior margin of the synapophyses. The haemal ridge is capped by a pronounced pleuropophyseal hypapophysis in anterior precloacal vertebrae. Posteriorly, the impressions become larger, forming subcentral paramedian lymphatic fossae [44]. The size, definition, and development of the haemal keel and subcentral ridges vary among taxa. The distribution of these characters in the fossil record of snakes thus allows for both the reconstruction of character evolution in the snake body form and hypotheses of evolutionary relationships along the snake stem and within the crown.

Although vertebral morphology is a limited dataset from which systematic hypotheses can be based, especially with respect to lower-order diagnoses within diverse clades such as Colubroidea (e.g., [45]; Colubroidea as applied in this article is equivalent to Colubroides of Zaher et al. [46]), claims that precise systematic hypotheses based on single, or small samples of, vertebrae are indefensible (e.g., [28]; Chapter 4) require explicit demonstration. Intracolumnar, individual, and ontogenetic variation in vertebral morphology certainly limit and potentially confound systematic identification; however, numerous discrete vertebral characters and character combinations diagnose snakes at all taxonomic levels [29, 33, 45, 47–49], suggesting that the utility of isolated vertebrae requires assessment on a case-by-case basis instead of as a general statement.

3.3 The Mesozoic Fossil Record of Snakes

3.3.1 Jurassic and Early Cretaceous Records

A sister-taxon relationship between Serpentes and Anguimorpha or Anguimorpha+Iguania requires divergence of the snake total clade at no younger than late Jurassic based on the fossil record of taxa referred to *Dorsetisaurus* as an anguimorph (e.g., [50]); however, very few pre-Late Cretaceous fossils have been referred to snakes. A single vertebra from the Barremian of Spain initially identified as a snake [51] was subsequently referred to a non-snake squamate [52].

Parviraptorids, a clade of phylogenetically enigmatic squamates [18, 53, 54] represented by disarticulated but associated cranial and postcranial elements from the late Jurassic and Early Cretaceous of North America and Europe [50, 53] have been reinterpreted as stem snakes on the basis of cranial and postcranial morphologies [54]. Cranial characters comparable with snakes include elongation of the maxilla and reduction of the maxillary facial process, recurved teeth, and tooth implantation. Conversely, neurocranial, basicranial [50], and appendicular skeletal anatomies [54] including large limbs and a robust pelvic girdle, are consistent with a more generalized, limbed squamate body form. Multiple snake-like vertebral characters were reported for parviraptorids [54], but these are not apparent in any of the referred taxa [50, 53].

A partial squamate braincase from the Valanginian Kirkwood Formation of South Africa [55] has been reinterpreted as a stem snake ('snake lizard' p.28 in [28]) possibly referable to Madtsoiidae [28]. Characters uniting the specimen with Serpentes have not been demonstrated, and the braincase retains a large, free-ending triangular prootic alar process as in the parviraptorid *Diablophis gilmorei* [50] and multiple squamate clades, as well as an anteriorly open *sella turcica*, elevated *dorsum sellae*, and neurovascular plumbing of the parabasisphenoid shared with most non-snake squamates [55–57]. These characters are consistent with a membranous anterior braincase and preclude presence of an enclosed bony braincase formed by ventral down-growths of the parietal to fully (or nearly) contact the prootic and parabasisphenoid, which is an apomorphy of extant snakes and Cretaceous taxa for which the skull is known (e.g., [3, 7, 11, 26, 58–61]). Morphology of the Kirkwood Formation specimen is potentially consistent with a phylogenetic position similar to that proposed for parviraptorids as basal-most divergences along the snake stem, but not with allocation to Madtsoiidae, because such a placement requires either homoplastic evolution of the enclosed bony braincase shared by Cenozoic madtsoiids and crown snakes (e.g., [10, 62]), or secondary loss and reversion to the unossified, membraneous anterolateral braincase in most squamates [57].

Tetrapodophis amplectus was described as a stem snake on the basis of a nearly complete, articulated skeleton of an elongate squamate with small fore- and hindlimbs preserved in laminated limestone referred to the Aptian Crato Formation of Brazil [63]. The specimen is extremely problematic regarding provenance and systematic interpretations. There is no history of its discovery or collection and referral to the Crato Formation is based solely on lithologic characters (fine-grained lamination and fish coprolites) common to shallow marine carbonate platforms. Thus, in the absence of additional evidence, the assigned spatiotemporal history can only be considered a tentative hypothesis. Described osteological synapomorphies uniting *Tetrapodophis* with snakes are either not visible on the specimen as figured (e.g., splenial intramandibular joint, fig. S2 of [63]), or were not tested in phylogenetic analysis of the specimen (p.8 of supplementary material of [63]). Preliminary re-examination of the specimen contradicted most initial anatomical interpretations and concluded that *Tetrapodophis* is not a stem snake [28]. Unfortunately, the specimen is privately owned and unavailable for further examination. Thus, hypotheses surrounding *Tetrapodophis* cannot be tested and the taxon is not considered further.

3.3.2 Late Cretaceous Records: Cenomanian to Turonian

Definitive fossil snakes have been recovered from the earliest Late Cretaceous, primarily Cenomanian, of Africa, the Middle East, South America, North America, Europe, and South Asia (Fig. 3.3A; Supplementary Table 3.S1). The oldest records are referred here to *Lapparentophis* based on isolated vertebrae from El Kohol, Algeria, dated to latest Albian [64]. Early Late Cretaceous records constrained by precise radiometric dates include a record of '*Coniophis*' based on two precloacal vertebrae from the Cedar Mountain Formation of Utah [65], dated to no younger than 98.39 ± 0.07 Ma [66], and *Xiaophis myanmarensis*, based on a partial precloacal skeleton, from the Angbamo amber mines dated to no younger than 98.8 ± 0.6 Ma [67], though the collection history, and thus verifiable provenance, of the specimen from commercial amber mines is undescribed. Cenomanian records from South America include the stem snakes *Najash rionegrina* based on multiple partial skeletons from the Candeleros Formation of Argentina [8], *Seismophis septentrionalis* based on two precloacal vertebrae from the Alcântara Formation of Brazil [68], and *Lunaophis aquaticus* based on multiple precloacal vertebrae from the La Luna Formation of Venezuela [69]. European Cenomanian records consist of *Pouitella pervetus*, known from a single precloacal vertebra, from the early to middle Cenomanian of Brézé (Maine-et-Loire), France [70], and *Simoliophis rochebrunei*, known from numerous vertebrae from early to middle Cenomanian shallow marine sediments of France, Portugal, and Spain ([71], summarized in [72, 73]).

The richest Cenomanian snake records come from Africa and the Middle East. African records consist of: *Lapparentophis defrennei*, based on a single precloacal vertebra, from In Akhamil, Algeria [74]; *Lapparentophis ragei*, based on two precloacal vertebrae and *Norisophis begaa*, based on three precloacal vertebrae, from the Ifezouane Formation in the Kem Kem beds of Morocco [75, 76]; *Simoliophis* cf. *S. libycus*, Madtsoiidae 'gen. et sp. nov.', ?Nigerophiidae indet., and Serpentes indet., based on multiple precloacal vertebrae from the Kem Kem beds [77]; *Simoliophis*, based on multiple precloacal and caudal vertebrae and recognized as a distinct species [76], from the Bahariya Formation of Egypt [78]; and *Simoliophis libycus* from the Mizdah Formation of Libya [79]. Shallow marine carbonate *Lagerstätten* of the Middle East and southern Europe preserve partial and complete skeletons of simoliophiid snakes consisting of *Pachyrhachis problematicus* based on two partial skeletons (Fig. 3.1A; [6, 80]) and *Haasiophis terrasanctus*, based on a complete skeleton (Fig. 3.1B; [7]) from the early to middle Cenomanian Bet-Meir Formation at 'Ein Yabrud, Palestinian Territories, *Eupodophis descouensi*, based on two complete and two partial skeletons from the early Cenomanian of Al Nammoura, Lebanon (Fig. 3.1C; [25, 60]), as well as *Pachyophis woodwardi*, based on two partial skeletons (Fig. 3.1D; 5, 81], and *Mesophis nopcsai*, based on a single partial skeleton [82] from the late Cenomanian of East Bosnia-Herzegovina.

Turonian through Santonian records are much less dense. The North American record consists of vertebrae referred to *Coniophis* [83]. European records are limited to the

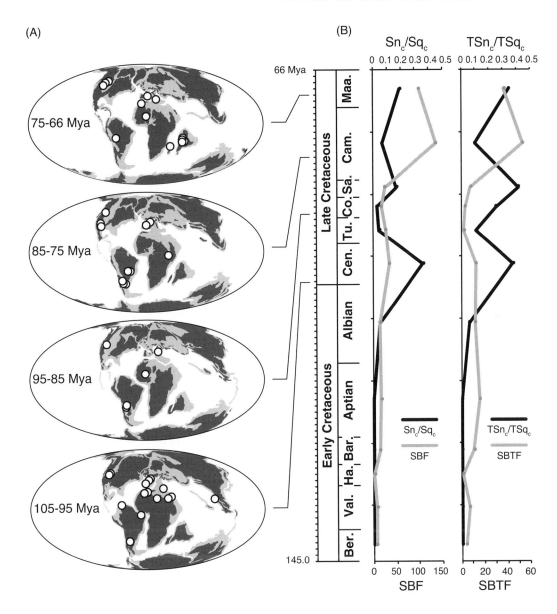

Figure 3.3 Spatial and temporal distribution of the Cretaceous fossil record of snakes. (A). Locations of snake localities binned at 10 Mya intervals. Palaeogeographic reconstructions based on www .deeptimemaps.com. Individual location points may represent multiple records. (B). Density of the published snake fossil record (Sn_c/Sq_c, TSn_c/TSq_c) compared with numbers of squamate-bearing formations through the Cretaceous. Record and formation values are for data binned at age resolution and plotted at age Mya midpoints. Plots are for the entire squamate record (left) and only the terrestrial record (right). Abbreviations: SBF, squamate-bearing formations; Sn_c, snake collections; Sq_c, squamate collections; SBTF, squamate-bearing terrestrial formations; TSn_c, terrestrial snake collections; TSq_c, terrestrial squamate collections.

youngest known record of Simoliophiidae, based on a single partial skeleton comparable to *Simoliophis,* and to *Pachyophis* from the Turonian-aged Dubovac quarry from Bosnia-Herzegovina [84]. African records are restricted to the In Beceten Formation of Niger [85] consisting of *Madtsoia* sp. ([86] after [87]). The maximum age of the In Beceten Formation is constrained by underlying Turonian limestones, whereas the minimum age is constrained by an upper Maastrichtian marine unit [88, 89]. Although precise locality information is lacking for the In Beceten collections, they are estimated to be Coniacian to Santonian in age [89]. If this is accurate, it represents the oldest unambiguous occurrence of Madtsoiidae. Reports of Cenomanian madtsoiids [77] require confirmation, because they have not been documented in detail and occur in the same beds as non-madtsoiid stem taxa that possess similar vertebral character combinations as madtsoiids [75]. The South American record consists of occurrences of *Dinilysia,* including *D. patagonica,* represented by multiple partial skulls, skeletons, and vertebrae, from the Santonian Bajo de la Carpa Formation (Fig. 3.1E; [90–92]), and *Dinilysia* sp., represented by vertebrae from the Campanian Anacleto Formation, of Argentina [93].

3.3.3 Late Cretaceous Records: Campanian to Maastrichtian

Latest Cretaceous records consist of distributions across all continents except for Antarctica, Australia, and Asia (Fig. 3.3A, Supplementary Table 3.S2), and document diversification of Madtsoiidae and the first records of unambiguous crown snake clades. Madtsoiids are the most species-rich, and geographically widespread Cretaceous snake clade, including over a dozen named taxa with Campanian occurrences in Africa and Europe, Campanian–Maastrichtian occurrences in South America, and Maastrichtian occurrences in Europe, India, and Madagascar (Supplementary Table 3.S2). Late-surviving stem snakes include *Sanajeh indicus,* based on a partial skeleton from the Maastrichtian Lameta Formation of India (Fig. 3.1F; [26]), and *Coniophis precedens,* based on precloacal vertebrae with referred cranial elements [94, 95] from the Maastrichtian Lance and Hell Creek formations of North America. Vertebral morphology consistent with fossorial habits has been used to identify *Coniophis,* in addition to the *C. precedens* record, throughout the Late Cretaceous of North America, as well as the Campanian of Africa (e.g., [65, 83, 96, 97]). These records likely reflect homoplastic ecomorphology in stem and crown taxa, and not a transcontinental, long-lived monophyletic genus [98].

Nigerophiidae and Palaeophiidae have first occurrences in the Campanian and Maastrichtian of Africa, Madagascar, and India [87, 96, 99, 100]. Specimens of ? Nigerophiidae were reported from the Cenomanian Kem Kem beds of Morocco [77]; however their taxonomic identity is poorly constrained. As noted [77], some referred materials are anatomically similar to elements identified as anterior precloacal vertebrae of *Simoliophis* by [78]. A Cenomanian nigerophiid record would represent an at least 15 million-year range extension for the clade (Fig. 3.4).

Among extant snake clades, scolecophidians, comprising Typhlopoidea, Leptotyphlopidae, and Anomalepididae as either a clade (e.g., [19]) or a paraphyletic assemblage with respect to Alethinophidia [101], have not been previously recognized prior to the Palaeocene [49, 102, 103], despite a history that must have extended into the Cretaceous based on first occurrences

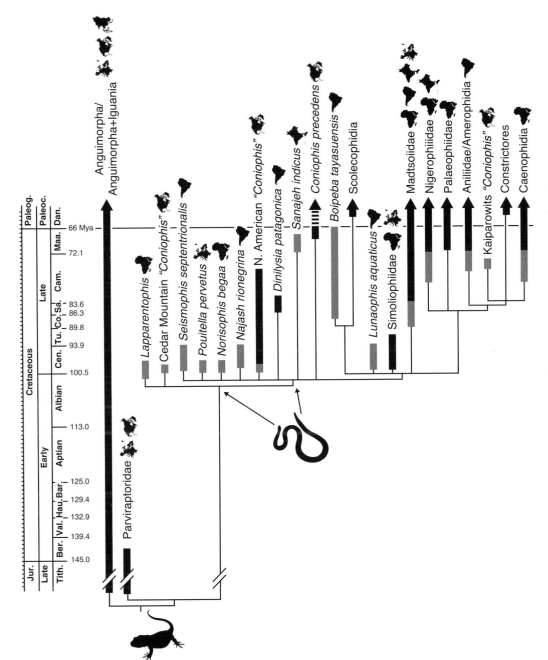

Figure 3.4 Generalized, time-calibrated phylogeny of Cretaceous snakes. Thick black bars indicate known temporal ranges, thick grey lines indicate uncertainly in temporal ranges. Dashed line for *Coniophis precedens* indicates preliminary referral of Palaeocene *Coniophis* sp. specimens to the *C. precedens* lineage. Continent and landmass silhouettes indicate Cretaceous geographic distributions. Snake silhouette indicates possible and definitive phylogenetic origins of the snake body form. Consensus topology for *Najash*, *Dinilysia*, Madtsoiidae, Simoliophiidae, Scolecophidia, and alethinophidian taxa based on [11, 104, 139]. Positions of all other taxa based on vertebral characters are discussed in text.

of alethinophidians. *Boipeba tayasuensis*, based on a single precloacal vertebra from the Adamantina Formation, estimated to be between Coniacian and Maastrichtian in age, has been proposed to represent the first Cretaceous scolecophidian record, specifically as a stem typhlopoid [104].

Unambiguous crown-clade records include *Australophis anilioides*, represented by ten precloacal vertebrae, from the late Campanian–early Maastrichtian of Argentina [105]. *Australophis* represents the oldest known record of Aniliidae and Amerophidia based on discrete vertebral apomorphies [98]. Although 'Boidae' has been used to identify Cretaceous snake vertebrae as a form taxon (see [49]), the oldest record of Constrictores (Booidea+Pythonoidea, sensu [106]) is *Titanoboa cerrejonensis*, represented by multiple partial skeletons, from the middle Palaeocene Cerrejón Formation of Colombia [107]. An incomplete precloacal vertebra from the upper Campanian Kaiparowits Formation of Utah previously referred to *Coniophis* sp. [83] may represent a Cretaceous record of Constrictores or Macrostomata of [19], however. The specimen possesses prezygapophyses with small but distinct accessory processes and prezygapophyseal angles consistent with small-bodied booids. Additionally, the specimen possesses strongly asymmetrically sized subcentral foramina as in the pythonoid *Xenopeltis* among alethinophidians [33], and has a low anterior neural spine with a concave dorsal margin similar to *Xenopeltis*. Regardless of whether the specimen can be included in Constrictores or more inclusively Macrostomata, preserved morphology provides evidence for crown snakes in North America by at least 76 Mya [108].

The late Maastrichtian *Cerberophis rex* from the Hell Creek Formation of Montana has been identified as alethinophidian [109] based on a precloacal vertebra displaying extensive osteoarthrosis [110] of the only preserved synapophysis that includes hypertrophic bone on the ventral prezygapophyses and interzygapophyseal ridge. In combination with incomplete preservation of the neural arch and a plesiomorphic ventral surface of the centrum relative to crown alethinophidians, pathologies of the specimen restrict a systematic assignment with respect to either crown or stem Serpentes. A second reported alethinophidian, the 'Lance snake' was recognized based on an incomplete precloacal vertebra from the Lance Formation of Montana [109] and distinguished from the stem snake *Coniophis precedens* on the basis of differentiated ('double-headed': p.39 of supplementary material of [109]) synapophyses and a 'large differentiated posterior condyle' (p.39 of supplementary material of [109]). However, these characters are present in the type specimen of *C. precedens* (e.g., [28, 94, 95: fig. 1]), and so it is here referred to *C. precedens* (Supplementary Table 3.S2).

The oldest reported occurrence of Caenophidia, including Colubroidea, is from the Wadi Milk Formation of Sudan [96]. The Wadi Milk Formation at Wadi Abu Hashim includes the richest and taxonomically most diverse Cretaceous fossil snake assemblages, and its age estimation and faunal composition require additional consideration. The Formation was originally dated as Albian–Santonian, with vertebrate-bearing layers dated as Cenomanian, based on water-well palynological samples, macrofloral remains, and shark biostratigraphy [111–114]. Subsequent U-Pb dating of detrital zircons from the Wadi Milk Formation at Jebel el Gamman gives a maximum age of 79.2 (+/- 2.4) Mya [115], consistent with radiometric, palynological, and vertebrate age estimates for the laterally equivalent Shendi Formation to the East [115–117] and with apparent interfingering of the Wadi Milk Formation with the Campanian–Maastrichtian Kababish Formation to the West [118]. The biostratigraphic utility of Wadi Milk Formation vertebrates is limited due to poor preservation and taxonomic ambiguity (e.g., [119]); however, presence and absence of indicator and index taxa additionally support at least a Campanian age. The Wadi Milk Formation lungfishes *Protopterus nigeriensis* and *Lavocatodus humei* ([96] after [120]) are known from the Coniacian to Campanian of Niger and Egypt and the Campanian of Egypt, respectively ([120, 121]; see also [122] for issues surrounding validity of Cenomanian *L. humei*). *Simoliophis*, which has been used as an index taxon for the African Cenomanian [123] based on its near-ubiquity in earliest Late Cretaceous snake faunas (e.g., [124]), is absent despite a rich and diverse fossil snake record [96]. Based on these radiometric, stratigraphic, and biostratigraphic data, the Wadi Milk Formation, and its constituent snake record, is best estimated as Campanian, no older than ~79.2 Mya.

Snake taxa identified from the Wadi Milk Formation consist of two lapparentophiid-grade stem snake morphologies, a madtsoiid, a possible palaeophiid, *Coniophis dabiebus*, the nigerophiid *Nubianophis afaahus*, caenophidians consisting of indeterminate colubroids and the russellophiid *Krebsophis*, and two indeterminate taxa [96]. Initial identifications provided support for a Cenomanian age of the Formation which in turn has been used as evidence for Cenomanian ages of other early Late Cretaceous snake-bearing formations (e.g., [76, 124]).

Reassessment of this record does not support some previous identifications, however. Recognition of lapparentophiids was based on a single precloacal vertebra and a single caudal vertebra [96]. Referral of the precloacal specimen was based on the absence of prezygapophyseal accessory processes, highly angled prezygapophyses, and the absence of parazygantral foramina. None of these characters are diagnostic for *Lapparentophis* individually or in combination; the absences of prezygapophyseal accessory processes and parazygantral foramina are plesiomorphic for snakes, and the prezygapophyseal angles of the slightly distorted specimen are 17° and 26°, within the ranges seen in crown taxa (e.g., [98]), but much lower than in *Lapparentophis* species (~35° in *L. defrennei*, ~40° in *L. ragei*: fig. 2 of [76]). The specimen additionally lacks unique characters of *Lapparentophis*, including a comparatively massive, nearly circular cotylar-condylar articulation and upswept posterior neural arch (e.g., [74, 76]), and possesses derived characters absent in *Lapparentophis*, including a wide zygosphene with low angled articular facets, a deep, well-defined posteromedian notch of the neural arch, and a well-defined, wide haemal keel. Based on these characters, this specimen is more likely a derived stem snake. The caudal vertebra similarly lacks diagnostic features of *Lapparentophis* and was referred solely on plesiomorphies including a narrow zygosphene, lack of prezygapophyseal accessory processes, and high-angled zygapophyses. As with the precloacal vertebra, there is no justification for assigning the specimen to Lapparentophiidae, and its identity, beyond that of stem snake, remains unknown.

Madtsoiids from the Wadi Milk Formation are fragmentary but notable in including large specimens with centra up to 3 cm in length [96], comparable to centrum lengths of large-bodied Campanian to Maastrichtian taxa with body lengths estimated to be 5–8 m (e.g., [87, 125, 126]). There is no other definitive Cenomanian record of Madtsoiidae, and the early Late Cretaceous snake record is limited to taxa < 2 m total body length (e.g., [104, 127]).

Among caenophidians, The Wadi Milk Formation preserves the oldest and only Mesozoic record of Russellophiidae. Russellophiids were originally recognized as 'pré-Colubridé' (p. 47 of [128]) caenophidians on the basis of vertebral proportions and characters including a low zygosphenial centre [128], with revised apomorphies including an elongate centrum, small and circular cotylar–condylar articulation, and well-defined paramedian lymphatic grooves forming the haemal keel used to assign Russellophiidae to the caenophidian total clade [45]. *Krebsophis thobanus* from Wadi Milk differs from species of *Russellophis* in possessing a neural spine restricted to a small post at the apex of the posterior margin of the neural arch, but uniquely shares ventrolaterally angled prezygapophyses, a compressed prezygapophyseal buttress, an anterior tubercle present ventral to prezygapophyseal articular facet, dome-like arching of the posterior neural arch, and a highly elongate, narrow centrum with *Russellophis* [96, 129, 130].

Six incomplete precloacal vertebrae from the Wadi Milk Formation were referred to Colubroidea incertae sedis based on gracile morphology, broad neural canal, zygosphene that is wider than the cotyle, and potentially a long, thin, tall neural spine [96]. The colubroid identity of those specimens has been considered tentative [131, 132], in part due to the inordinately early Cenomanian stage age previously established for the Wadi Milk Formation [131]. The revised age for Wadi Milk is more consistent with other first occurrences of crown snake clades and reduces an age discrepancy for the first occurrence of Colubroidea by ~ 20 million years.

3.4 Discussion

3.4.1 Systematics of Cretaceous Snakes and the Ecological Context of Snake Origins

A phylogenetic topology for inferring ecological and biogeographic origins of snakes based on the fossil record is hampered by missing data for most named taxa, and the main focus of Cretaceous snake phylogenetics has been the litigation of primary homologies in support of alternate phylogenetic hypotheses of relatively well-preserved taxa (e.g., [28] and references therein, [133]). Despite these limitations, several generalized interrelationships are recognized that form a coarse topology. For example *Najash* and *Dinilysia* are recognized as being stemward of simoliophiids, which are stemward of crown Macrostomata, despite no resolution for the interrelationships of these taxa with respect to scolecophidians (e.g., [11, 26, 27]). A consensus framework based on recent phylogenetic analyses (Fig. 3.4) allows for preliminary hypotheses of relationships among taxa recognized by vertebral morphology.

Lapparentophis and a precloacal vertebra referred to *Coniophis* from the Cedar Mountain Formation of Utah are hypothesized to be sister taxa to all other Late Cretaceous snake taxa and the snake crown. Both retain plesiomorphic vertebral morphologies shared within limbed squamates, including prezygapophyses that extend dorsal to the neural canal, zygosphenial roof at an angle approaching or exceeding 30°, absence of pronounced prezygapophyseal base, and a small neural arch lacking a posteromedian notch (Fig. 3.1A; [33, 65]). *Lapparentophis* additionally retains a narrow zygosphene with high-angled articular facets, and both taxa possess apomorphically large cotylar–condylar articular facets, possibly indicating monophyly.

The majority of Cenomanian snakes form a polytomy of stem taxa possessing lower-angled prezygapophyses with greater development of the prezygapophyseal base. *Seismophis* possesses a morphology similar to *Najash* [68], and both share retention of only a shallow, poorly defined posteromedian notch with *Pouitella* and multiple specimens of North American '*Coniophis*' throughout its Late Cretaceous record [65, 83]. *Najash* has been recognized variably as the sister taxon to all other known snake taxa (e.g., [8]); crownward of *Dinilysia* along the snake stem [11], the sister taxon to *Dinilysia* on the snake stem [134], potentially as the sister to Madtsoiidae [126], and as sister to all other alethinophidians [135].

Dinilysia has similarly been considered a stem alethinophidian [26, 135], though most analyses recover it as a stem snake (e.g., [27, 28 and references therein]). *Dinilysia* shares greater lateral extension of the prezygapophyses and a more acute posteromedian notch with *Norisophis* and more derived snakes to the exclusion of the aforementioned taxa.

Maastrichtian stem snakes include late-surviving *Sanajeh indicus* and *Coniophis precedens*. *Sanajeh* was placed in Madtsoiidae within crown Alethinophidia based on characters including parazygantral foramina, absence of prezygapophyseal accessory processes, and secondarily derived loss of a divided trigeminal foramen [26]. That hypothesis was based on multiple anatomical misidentifications by one of us (JJH), including misidentifying a large, externally and posterodorsally displaced stapedial footplate as a supratemporal [11], because nobody is perfect. Subsequent analysis places the taxon on the snake stem, as sister to a large-bodied madtsoiid clade [126].

Coniophis precedens was erected on a single anterior precloacal vertebra [94]. An expanded hypodigm combining dissociated cranial and vertebral elements from multiple screenwash localities was used to place *C. precedens* as the most stemward snake based on unambiguous snake vertebral characters and tooth implantation morphology of maxillary and dentary elements [95]. The presence of interdental ridges forming shallow thecae that constitute tooth bases are consistent, though not exclusive to snakes [136]; however, cranial elements otherwise retain multiple squamate plesiomorphies (e.g., large nasal capsule fossae on the medial surface of the maxillary facial process) relative to parviraptorids [54], which conversely possess more plesiomorphic vertebral morphologies than *C. precedens* [50]. As a result, either (1) parviraptorids are not stem snakes but possess homoplastic cranial characters with snakes; (2) *Coniophis precedens* is not a stem snake but possesses homoplastic vertebral characters with snakes; or (3) the hypodigm of *Coniophis* is a chimaera composed of snake vertebrae and non-snake squamate cranial elements. Until associated skeletons of *C. precedens* are discovered, the taxon should be conservatively diagnosed by vertebral morphology, which is consistent with inclusion on the snake stem based on the absence of prezygapophyseal accessory processes and a shallow posteromedian notch. One specimen included in the hypodigm of *C. precedens*, YPM-PU 16845

[95], is not consistent with being a posterior precloacal element. Instead, its lower, more elongate shape and the size and lateral position of small paramedian lymphatic channels are more reminiscent of scolecophidians.

The composition of the snake crown with respect to several Cretaceous clades that are currently considered crownward of *Dinilysia* and *Najash* is unresolved. *Lunaophis* possesses an elongate, narrow centrum with prominent subcentral ridges and strongly differentiated diapophyseal and parapophyseal articular facets of the synapophyses [69], suggesting phylogenetic positions either adjacent to, or within, the snake crown. *Boipeba* has been proposed to be nested with Scolecophidia as a stem typhlopoid [104], however the presence of a well-developed neural spine and a short, wide centrum are plesiomorphic with respect to crown scolecophidians, regardless of monophyly or paraphyly of constituent taxa. As a result, the taxon may be a stem scolecophidian or a proximal sister taxon to the snake crown.

Simoliophiids, united primarily by the presence and morphology of pachyostosis in mid-trunk precloacal vertebrae and pachyosteosclerosis of ribs in some constituent species [137], have the most disparate phylogenetic hypotheses among Cretaceous taxa. Systematic placement ranges from a clade or grade of sister taxa to all other snakes (e.g., [6, 10, 138]), to a clade nested well within Alethinophidia as macrostomatans (e.g., [7, 19]). Combined molecular-morphological analysis places Simoliophiidae as sister to crown Alethinophidia [139], and recent Bayesian analysis of morphology places simoliophiids as sister to crown Serpentes [11].

Madtsoiidae has been inferred as nested within Alethinophidia [26, 134] or Macrostomata [140], or as a stem clade [126] either stem- or crownward of simoliophiids [10, 138]. Madtsoiidae monophyly is supported by the presence of parazygantral and paracotylar foramina, though the latter are not continuously distributed within referred taxa and occur also in the majority of alethinophidian clades [135]. Parazygantral foramina are additionally not uniquely apomorphic for madtsoiids, occurring variably in macrostomatans [138], as well as *Seismophis*, *Norisophis*, *Dinilysia*, and *Najash*, all which possess plesiomorphic vertebral and cranial morphologies with respect to madtsoiids and crown snakes (e.g., [27]). Although vertebral morphology is extremely similar among giant taxa (e.g., [125, 126, 141] and cranial elements indicate monophyly of Cenozoic Australian taxa [10, 142, 143], hypotheses of monophyly of all taxa referred to Madtsoiidae and/or hypotheses that unite madtsoiids with stem taxa such as *Dinilysia* and *Najash* should be subjected to further scrutiny.

Though considered crown snakes [144] the interrelationships of palaeophiids and nigerophiids are poorly constrained. Palaeophiids possess the complete suite of vertebral apomorphies of Alethinophidia (though see [145]; contra [146]), whereas nigerophiids possess all but prezygapophyseal accessory processes.

Despite poor resolution for the majority of Cretaceous taxa, the generalized consensus phylogeny presented here provides a test for competing hypotheses for the ecological contexts of snake origins. The inference of *Pachyrhachis* as sister to all other snakes combined with a hypothesis of monophyly of snakes with mosasauroids provided support for marine origins of snakes [6, 28], which has been contested by alternative phylogenies (e.g., [7] and Chapter 9; see also [28] for discussion). Subsequent placement of simoliophiids either within the snake crown or high along the snake stem nests them deeply within an otherwise terrestrial clade, falsifying a marine origin hypothesis regardless of the relationships of snakes to mosasauroids. Inner ear and stapedial morphologies of the more stemward taxa *Najash* and *Dinilysia,* including relatively massive vestibular spaces and footplates, are consistent with those seen in extant fossorial snakes and other squamates ([127, 147]; Chapter 13; but see also Chapter 14). Multiple stem taxa, including *Norisophis* and species referred to '*Coniophis*', possess vertebral morphologies consistent with fossoriality in extant squamates, especially reduction of the neural spine as a component of shallowing the transverse section of the body. Thus both cranial and postcranial morphology of stem snakes indicates fossoriality as a locomotory habit across the snake stem. However, the poor resolution of interrelationships limits the ability to robustly reconstruct specific ecologies toward the point of origin of the snake body form.

3.4.2 Origin of the Snake Body Form

The complete and partial skeletons of simoliophiids, *Najash*, and *Xiaophis* constrain the evolution of the elongate, limbless, or limb-reduced snake body form to no younger than early Cenomanian, but the timing, tempo, and mode of body-form origin in snakes is poorly constrained between those records and the divergence of the total clade from short bodied, limbed sister taxa (Anguimorpha, Anguimorpha+Iguania, or Parviraptoridae) by the late Jurassic. The phylogenetic hypothesis provided here informs on the sequence of character acquisition in the evolution of the snake body form, both in vertebral morphology and its soft tissue correlates. The plesiomorphies retained by *Lapparentophis* as the possible sister to all other snakes may also indicate a shorter, limbed squamate-like body form. The high-angled, tall prezygapophyses suggest a lizard-like transverse aspect and the small zygosphene–zygantrum is additionally more similar to the condition in limbed squamates than other snakes [33]. Among more crownward stem snakes, vertebral morphologies of *Seismophis* and species of '*Coniophis*' with lower-angled zygapophyses and larger, lower-angled zygosphene-zygantra, are similar to *Najash*, and indicate that the snake body form was present among all taxa within the polytomy (Fig. 3.4). The presence of a well-defined haemal keel and subcentral ridges defining channels for the paravertebral lymphatic system in *Seismophis* and *Norisophis* provide additional support for those taxa possessing a derived body form.

Several of the osteological correlates to muscle reorganization associated with specialized locomotory mode in extant snakes are absent in stem taxa. The absence of accessory processes or prezygapophyseal bases indicates insertions of *levator costae* and the *M. interarticulares* complex similar to the condition in limbed squamates, and the absence of a posteromedian notch suggests a smaller *multifidus.* With the exception of *Boipeba* and many scolecophidians, all Late Cretaceous snake taxa possess at minimum a slight differentiation of the synapophyses into parapophyseal and diapophyseal facets. These character combinations suggest some capacity for stem snakes to brace the costovertebral joint against lateral compression based on joint morphology, but with less capacity for stabilizing, elevating, or flexing ribs via *levator costae* or *M. interarticulares.* Limited capacity to stabilize the costovertebral joint likely precluded rectilinear locomotion and would potentially have restricted concertina locomotion [32]. Based on stem-snake vertebral morphology, the full repertoire of locomotory regimes in extant snakes may be restricted to the crown, but stem taxa were still able to occupy fossorial, terrestrial, and aquatic habits while retaining plesiomorphic musculoskeletal morphologies.

As a preliminary test of whether the first occurrence of the snake body form in the fossil record approximates the timing of origin or is an artifact of insufficient sampling of Early Cretaceous records, we compared snake fossil density, standardized as the ratio of published snake collections to the total published squamate collection record, with the number of squamate-bearing formations, binned at the age level for the Cretaceous (Fig. 3.3B, Supplementary Table 3.S3). We examined both the entire published fossil record and only terrestrial records separately, to account for preservation differences between marine and terrestrial depositional environments. We excluded parviraptorids because there is no anatomical evidence to support a snake body form in that clade.

Sampling potential for the squamate fossil record, proxied as the number of squamate-bearing formations, increases gradually from the Berriasian to Santonian for the total record, with elevated sea levels and the origin and initial diversifications of mosasauroids driving an increase in squamate-bearing marine formations from the Cenomanian (Fig. 3.3B). For the terrestrial record, the number of formations is generally consistent from the Barremian to the Cenomanian. A pronounced increase in formations from the Campanian in both marine and terrestrial records is commensurate with increasing availability of the sedimentary rock record (e.g., [148]). Some of the variation in formation number is due to differences in duration of age ranges, notably the short intervals of the Hauterivian, Coniacian, and Santonian, whereas the longer ages of the Aptian, Albian, Cenomanian, Campanian, and Maastrichtian allow for direct comparisons of the snake fossil record.

Comparing the density of snake collections through time with the number of squamate-bearing formations does not reveal correlation through the Early and early Late Cretaceous. As noted, fossil evidence of the snake body form achieves a transcontinental distribution by the early Cenomanian, but the number of squamate-bearing formations does not increase commensurately between the Barremian and Cenomanian for either the entire or only terrestrial records. Linear regression of the total number of squamate collections onto the total number of collections per age demonstrates a significant, positive relationship for both the total and only terrestrial records ($R^2 = 0.99$, $p <0.00$; $R^2 = 0.96$, $p < 0.00$), indicating that overall fossil record density is dependent upon sampling the rock record. However, the proportion of snakes in the squamate record is not dependent upon sampling ($R^2 = 0.11$, $p = 0.29$; $R^2 = 0.03$, $p = 0.58$). As a result, the absence of evidence for the snake body form from the Early Cretaceous cannot be explained *prima facia* as sampling artifact and is instead more consistent with a latest Early Cretaceous or early Late Cretaceous origin and subsequent rapid transcontinental distribution within a period of ~ 1–2 million years. This timing postdates much of Gondwanan continental fragmentation and greatly postdates fragmentation of Pangea. As a result, the geologically simultaneous occurrences of snake fossils in both terrestrial and marine depositional environments in Africa, South America, North America, and insular Asia, combined with the absence of any evidence for the snake body form in the Early Cretaceous or Jurassic, precludes vicariance as a biogeographic mode and instead suggests rapid dispersal soon after the evolution of the snake body form near the beginning of the Late Cretaceous.

3.4.3 Cretaceous Biogeography and Snake Origins

Hypotheses for the biogeographic origins of snakes focus primarily on the Gondwanan landmass based on distributions of extant taxa as well as the comparatively rich Cretaceous records from South America and Africa (e.g., [11, 63, 67, 86]). Range evolution models based on extant taxa and a subset of the Cretaceous record have estimated a high likelihood for a Laurasian origin for the total clade and a Gondwanan origin for the snake crown [149]. That history is not necessarily inconsistent with reinterpretation of the fossil records discussed here; however, it justifiably does not account for the majority of Cretaceous taxa that have not been placed in a phylogenetic context.

Although there is no Mesozoic record from mainland Asia, the North American Cretaceous record samples the breadth of snake phylogeny, from among the most stem-ward records to within the crown, if the interpretation of the 'Coniophis' record proposed here is accurate. That, combined with the near simultaneous occurrence of snake fossils across Gondwanan and Laurasian landmasses, raises the possibility of a biogeographic history distinct from that currently considered based on better-preserved, more phylogenetically resolved Gondwanan records.

3.4.4 Survival across the K-Pg Extinction Event

Despite well-established patterns of mass extinction, faunal replacement, and recovery in terrestrial and marine ecosystems following the Chicxulub impactor event at the Cretaceous–Palaeogene boundary, the effects of the K-Pg event on the evolutionary histories of snakes remain poorly understood. Only a single study has documented patterns of known first and last occurrence in snakes, as components of a larger analysis of squamates, across the K-Pg boundary in a stratigraphically refined composite section based on multiple sedimentary basin records within the Western Interior of North America [109]. That study reported survivorship of Coniophis precedens across the K-Pg event, based on unverified elements from the Torrejonian of Montana [95, 150] (Fig. 3.4).

None of the diverse Cretaceous snake clades go extinct at or near the K-Pg event (Fig. 3.4). Madtsoiidae, the most species-rich Mesozoic clade, crosses the boundary, maintains the majority of its continental-scale geographic distribution through the Palaeogene (e.g., [96,125, 151–153]), and demonstrates survivorship across small and giant body sizes [125]. Palaeophiidae and Nigerophiidae, both aquatic and marine clades, have known first occurrences by the Maastrichtian and Campanian respectively, and both have records that extend into the Palaeogene, with palaeophiids diversifying into a near-global marine radiation by the Eocene (e.g., [30, 154]).

Among unambiguous crown taxa, the occurrence of Australophis in the Campanian of Argentina [105] representing Aniliidae and potentially Amerophidia, and Krebsophis, and potentially Colubroidea, representing Caenophidia in the Campanian of Sudan [96], constrain those clades as well as divergences of clades without recognized Mesozoic records to the Late Cretaceous. Thus the oldest recognized record of Constrictores is currently Titanoboa from the middle Palaeocene of Colombia [107], but the total clade is constrained to no younger than Campanian based on a sister-taxon relationship to Caenophidia. The 'Coniophis' specimen from the Kaiparowits Formation may provide evidence for a Late Cretaceous record of Constrictores (Fig. 3.4). There is no unambiguous fossil record of Uropeltoidea, the clade of Old World pipe snakes and shield-tail snakes [98], but they are similarly constrained by either a sister-taxon relationship to Aniliidae [19] or by being nested within Constrictores as the sister taxon to the clade of pythonids, loxocemids, and xenopeltids [155].

The fossil record of scolecophidians is extremely limited, but monophyly of Alethinophidia relative to scolecophidians as either grade or clade, constrains them to Campanian as well. Stratigraphic resolution for *Boipeba* is limited to Upper Cretaceous, but well within the minimum divergence timing for scolecophidians required by the first occurrences of Alethinophidia. If *Boipeba* is nested within a monophyletic Scolecophidia [104], then all three primary subclades, Typhlopoidea, Leptotyphlopidae, and Anomalepididae, include species that survived the K-Pg extinction.

Ecological reorganization has been proposed as a mechanism underpinning the diversification and adaptive radiations of species-rich extant clades such as birds, mammals, fishes, and anurans following the catastrophic environmental changes caused by the K-Pg event [156–160]. There is some evidence for faunal change and diversification of extant snake clades following the Cretaceous, including widespread distribution of crown-group Constrictores by the middle Eocene (e.g., [107, 161, 162]), and dispersal of caenophidians ranging from South America to India by the early Eocene [129, 161]. The major radiations within hyper-diverse Colubroidea are younger, generally constrained to late Palaeogene to early Neogene based on fossil-calibrated molecular phylogenetic analyses and the fossil record (e.g., [48, 163, 164]). Some analyses have constrained the origins and divergences of the caenophidian and colubroid total clades to the early Palaeogene (e.g., [132, 134]), which would permit hypotheses of adaptive radiation following the K-Pg event comparable to hypotheses for mammals. However, those studies did not include the Wadi Milk Formation caenophidian records as a temporal calibration, and did not test the potential for the described colubroid specimens to constrain the origin of that clade prior to the K-Pg boundary. As a result, either the Wadi Milk Formation record requires re-evaluation, or molecular dating analyses [132] require more accurate fossil calibrations [45, 48].

Overall, there is little evidence that the K-Pg extinction event had a pronounced effect on the diversification of snakes from the Mesozoic into the Cenozoic. The absence of any extinction signal in phylogenetic structure or divergence-timing estimation, combined with the perseverance of taxa across terrestrial, marine, aquatic, and fossorial habits, and body sizes approaching 10 m (e.g., [87, 126, 153, 165]), suggests that snakes may have been unusual among vertebrates in crawling untroubled across one of the most pronounced mass extinction events in Earth history.

3.5 Conclusions

Although extremely incomplete, the Cretaceous fossil record of snakes demonstrates the origin and evolution of the snake body form in terrestrial habitats at or near the beginning of the Late Cretaceous, followed by rapid transcontinental geographic dispersal, crown-clade origins, and high survivorship of latest Cretaceous taxa into the Palaeogene. This record, in conjunction with genomic, phenotypic, and behavioural and ecological datasets from extant taxa (e.g., [149]) will continue to provide new hypotheses on the tempo, mode, and ecological contexts for the early history and diversification of one of the most successful vertebrate clades.

Further discovery from the fossil record will increasingly provide the majority of new data, however, as extant taxon sampling approaches saturation between and among clades that constitute the primary divisions of the snake crown. That being said, advancing our understanding of snake evolution using the fossil record will require advances in both analytical technique (e.g., [166, 167]) and research conduct. With respect to the latter, advancement will require avoiding fossils with dubious to unknown provenance due to commercial, often unethical, collection histories, and refraining from highly selective referral of disassociated skeletal elements to create stem taxa for high-impact publication. Similarly, shifting away from a discourse tenor in which disagreements are labeled as adversarial and opposing hypotheses are labeled as myths (see discussion in [28] and references therein), will be necessary for upcoming generations of research to successfully examine the origin and evolution of snakes from their fossil record with fresh perspective and new ideas.

Acknowledgements

For access to specimens, we thank the late E. Tchernov (Hebrew University, Jerusalem), D. J. Mohabey (Geological Survey of India), M. Reguero (Museo de La Plata), and C. L. Dal Sasso and G. Teruzzi (Museo di Storia Naturale di Milano). We thank C. A. Brochu (University of Iowa) for images of *Haasiophis*. We thank the editors for the invitation to contribute to this volume, and especially D. J. Gower for his editorial patience. We thank D. J. Field (University of Cambridge), D. C. Evans (Royal Ontario Museum), R. Benson (Oxford University), and S. E. Evans (University College, London) for discussion on various aspects of this MS and A. Folie (Royal Belgian Institute of Natural Sciences) and C. J. Bell (University of Texas at Austin) for constructive reviews. This research was funded by a Natural Environment Research Council award (NE/S000739/1) to JJH.

References

1. J. J. Head, K. de Queiroz, and H. W. Greene, Pan-Serpentes. In K. de Queiroz, P. D. Cantino, and J. A. Gauthier, eds., *Phylonyms: A Companion to the PhyloCode* (Berkeley: CRC Press, 2020), pp. 1130–1134.

2. G. L. Walls, Ophthalmological implications for the early history of the snakes. *Copeia*, 1940 (1940), 1–8.

3. A. D'a. Bellairs and G. Underwood, The origin of snakes. *Biological Reviews*, 26 (1951), 193–237.

4. O. Rieppel, A review of the origin of snakes. *Evolutionary Biology*, 22 (1988), 37–130.

5. F. Nopcsa, *Eidolosaurus* und *Pachyophis*. Zwei neue Neocom-Reptilien. Palaeontographica, 65 (1923), 99–154.

6. M. W. Caldwell and M. S. Y. Lee, A snake with legs from the marine Cretaceous of the Middle East. *Nature*, 386 (1997), 705–709.

7. E. Tchernov, O. Rieppel, H. Zaher, M. J. Polcyn, and L. L. Jacobs, A fossil snake with limbs. *Science*, 287 (2000), 2010–2012.

8. S. Apesteguía and H. Zaher, A Cretaceous terrestrial snake with robust hindlimbs and a sacrum. *Nature*, 440 (2006), 1037–1040.

9. C. L. Caprette, M. S. Lee, R. Shine, A. Mokany, and J. F. Downhower, The origin of snakes (Serpentes) as seen through eye anatomy. *Biological Journal of the Linnean Society*, 81 (2004), 469–482.

10. J. D. Scanlon, Skull of the large non-macrostomatan snake *Yurlunggur* from the Australian Oligo-Miocene. *Nature*, 439 (2006), 839–842.

11. F. F. Garberoglio, S. Apesteguía, T. R. Simões, et al., New skulls and skeletons of the Cretaceous legged snake *Najash*, and the evolution of the modern snake body plan. *Science Advances*, 5 (2019), p.eaax5833.

12. C. L. Camp, Classification of the lizards. *Bulletin of the American Museum of Natural History*, 48 (1923), 289–481.

13. M. S. Y. Lee, The phylogeny of varanoid lizards and the affinities of snakes. *Philosophical Transactions of the Royal Society of London, B*, 352 (1997), 53–91.

14. M. S. Y. Lee, Convergent evolution and character correlation in burrowing reptiles: Towards a resolution of squamate relationships. *Biological Journal of the Linnean Society*, 65 (1998), 369–453.

15. J. -C. Rage, La phylogénie des Lépidosauriens (Reptilia): une approche cladistique. *Comptes Rendus, Académie des Sciences, Paris*, 294 (1982), 563–566.

16. J. Hallermann, The ethmoidal region of *Dibamus taylori* (Squamata: Dibamidae), with a phylogenetic hypothesis on dibamid relationships within Squamata. *Zoological Journal of the Linnean Society*, 122 (1998), 385–426.

17. S. E. Evans and L. J. Barbadillo, An unusual lizard from the Early Cretaceous of Las Hoyas, Spain. *Zoological Journal of the Linnean Society*, 124 (1998), 235–265.

18. J. L. Conrad, Phylogeny and systematics of *Squamata* (*Reptilia*) based on morphology. *Bulletin of the American Museum of Natural History*, 310 (2008), 1–182.

19. J. A. Gauthier, M. Kearney, J. A. Maisano, O. Rieppel, and A. D. B. Behlke, Assembling the squamate tree of life: perspectives from the phenotype and the fossil record. *Bulletin of the Peabody Museum of Natural History*, 53 (2012), 3–308.

20. J. J. Wiens, C. A. Kuczynski, T. Townsend, et al., Combining phylogenomics and fossils in higher-level squamate reptile phylogeny: molecular data change the placement of fossils. *Systematic Biology*, 59 (2010), 674–688.

21. J. J. Wiens, C. R. Hutter, D. G. Mulcahy, et al., Resolving the phylogeny of lizards and snakes (Squamata) with extensive sampling of genes and species. *Biology Letters*, 8 (2012), 1043–1046.

22. R. A. Pyron, F. T. Burbrink, and J. J. Wiens, A phylogeny and revised classification of Squamata, including 4161 species of lizards and snakes. *BMC Evolutionary Biology*, 13 (2013), 93.

23. J. W. Streicher and J. J. Wiens, Phylogenomic analyses of more than 4000 nuclear loci resolve the origin of snakes among lizard families. *Biology Letters*, 13 (2017), 20170393.

24. T. R. Simões, M. W. Caldwell, M. Tałanda, et al., The origin of squamates revealed by a Middle Triassic lizard from the Italian Alps. *Nature*, 557 (2018), 706–709.

25. J.-C. Rage and F. Escuillié, Un nouveau serpent bipède du Cénomanien (Crétacé). Implications phylétiques. *Comptes Rendus de l'Académie des Sciences - Série IIA - Earth and Planetary Science*, 330 (2000), 513–520.

26. J. A. Wilson, D. Mohabey, S. Peters, and J. J. Head, Predation upon hatchling sauropod dinosaurs by a new basal snake from the Late Cretaceous of India. *PLOS Biology*. 8 (2010), 1–5 doi:10.1371/journal. pbio.1000322.g005.

27. H. Zaher and C. A. Scanferla, The skull of the Upper Cretaceous snake *Dinilysia patagonica* Smith-Woodward, 1901, and its phylogenetic position revisited. *Zoological Journal of the Linnean Society*, 164 (2012), 194–238.

28. M. W. Caldwell, *The Origin of Snakes. Morphology and the Fossil Record* (Boca Raton, FL: CRC Press, 2020).

29. J. -C. Rage, *Encyclopedia of Paleoherpetology, part 11, Serpentes* (Stuttgart: Gustav Fischer Verlag, 1984).

30. J. A. Holman, *Fossil Snakes of North America. Origin, Evolution, Distribution, Paleoecology* (Indianapolis: Indiana University Press, 2000).

31. C. Gans, Locomotion of limbless vertebrates: Pattern and evolution. *Herpetologica*, 42 (1986), 33–46.

32. J. G. Campano, Reaction forces and rib function during locomotion in snakes. *Integrative and Comparative Biology*, 60 (2020), 215–231.

33. R. Hoffstetter and J. -P. Gasc, Vertebrae and ribs of modern reptiles. In C. Gans and T. S. Parsons, eds., *Biology of the Reptilia*, Vol. 1, *Morphology* (London: Academic Press, 1969), pp. 201–310.

34. B. R. Moon, Testing and inference of function from structure: Snake vertebrae do the twist. *Journal of Morphology*, 241 (1999), 217–225.

35. R. L. Cieri, The axial anatomy of monitor lizards (Varanidae). *Journal of Anatomy*, 233 (2018), 636–643.

36. W. Mosauer, The myology of the trunk region of snakes and its significance for ophidian taxonomy and phylogeny. *Publications of the University of California at Los Angeles in Biological Sciences*, 1 (1935), 81–120.

37. W. Auffenberg, The vertebral musculature of *Chersydrus* (Serpentes). *Quarterly Journal of the Florida Academy of Sciences*, 29 (1966), 155–162.

38. J. -P. Gasc, L'interprétation fonctionnelle de l'appareil musculo-squelettique de l'axe vertébral chez les Serpents (Reptilia).

Mémoires du Museum National d'Histoire Naturelle Série A, 83 (1974), 1–182.

39. J. -P. Gasc, Axial musculature. In C. Gans and T. S. Parsons, eds., *Biology of the Reptilia*, Vol. 11, *Morphology F* (London: Academic Press, 1981), pp. 355–435.

40. D. A. Penning, Quantitative axial myology in two constricting snakes: *Lampropeltis holbrooki* and *Pantherophis obsoletus*. *Journal of Anatomy*, 232 (2018), 1016–1024.

41. D. Ritter, Axial muscle function during lizard locomotion. *Journal of Experimental Biology*, 199 (1996), 2499–2510.

42. S. W. Chapman and R. E. Conklin, The lymphatic system of the snake. *Journal of Morphology*, 58 (1935), 385–417.

43. G. Ottaviani and A. Tazzi, The lymphatic system. In C. Gans and T. S. Parsons, eds., *Biology of the Reptilia*, Vol. 6, *Morphology H* (London: Academic Press, 1977), pp. 315–462.

44. T. C. Laduke, The fossil snakes of Pit 91, Racho La Brea, California. *Contributions in Science, Natural History Museum of Los Angeles County*, 424 (1991), 1–28.

45. J. J. Head, K. Mahlow, and J. Müller, Fossil calibration dates for molecular phylogenetic analysis of snakes 2: Caenophidia, Colubroidea, Elapoidea, Colubridae. *Palaeontologia Electronica*, 19.2.2FC (2016), 1–21.

46. H. Zaher, F. G. Grazziotin, J. E. Cadle, et al., Molecular phylogeny of advanced snakes (Serpentes, Caenophidia) with an emphasis on South American xenodontines: a revised classification and descriptions of new taxa. *Papéis Avulsos de Zoologia*, 49 (2009), 115–153.

47. J. B. Slowinski, A phylogenetic analysis of *Bungarus* (Elapidae) based on morphological characters. *Journal of Herpetology*, 28 (1994), 440–446.

48. J. J. Head, Snakes of the Siwalik Group (Miocene of Pakistan): systematics and relationship to environmental change. *Palaeontologia Electronica*, 8 (2005), 16A.

49. J. J. Head, Fossil calibration dates for molecular phylogenetic analysis of snakes

1: Serpentes, Alethinophidia, Boidae, Pythonidae. *Palaeontologia Electronica*, 18 (2015), 1–17.

50. S. E. Evans, *Parviraptor* (Squamata: Anguimorpha) and other lizards from the Morrison Formation at Fruita, Colorado. *Museum of Northern Arizona Bulletin*, 60 (1996), 243–248.

51. J. -C. Rage and A. Richter, A snake from the Lower Cretaceous (Barremian) of Spain: The oldest known snake. Neues *Jarbuch für Geologie und Paläontologie, Monatshefte, Stuttgart*, II.9 (1994), 561–565.

52. J. -C. Rage and F. Escuillié, The Cenomanian: stage of hindlimbed snakes. *Carnets de Géologie, Maintenon, Article* 2003/01 (2003), 1–11.

53. S. E. Evans, A new anguimorph lizard from the Jurassic and lower Cretaceous of England. *Palaeontology*, 37 (1994), 33–49.

54. M. W. Caldwell, R. L. Nydam, A. Palci, and S. Apesteguía, The oldest known snakes from the Middle Jurassic-Lower Cretaceous provide insights on snake evolution. *Nature Communications*, 6 (2015), 1–11.

55. C. F. Ross, H.-D. Sues, and W. J. De Klerk, Lepidosaurian remains from the lower Cretaceous Kirkwood Formation of South Africa. *Journal of Vertebrate Paleontology*, 19 (1999), 21–27.

56. O. Rieppel and H. Zaher, The braincases of mosasaurs and *Varanus*, and the relationships of snakes. *Zoological Journal of the Linnean Society*, 129 (2000), 489–514.

57. S. E. Evans, The skull of lizards and tuatara. In C. Gans, A. S. Gaunt and K. Adler, eds., *Biology of the Reptilia*, Vol. 20, *Morphology H* (Ithaca, NY: Society for the Study of Amphibians and Reptiles, 2008), pp. 1–347.

58. R. Estes, T. H. Frazzetta, and E. E. Williams, Studies on the fossil snake *Dinilysia patagonica* Woodward: Part I. Cranial morphology. *Bulletin of the Museum of Comparative Zoology*, 140 (1970), 25–74.

59. G. Haas, *Pachyrhachis problematicus* Haas, snakelike reptile from the Lower Cenomanian: ventral view of the skull.

Bulletin du Muséum national d'Histoire naturelle, Serie 4, 2 (1980), 87–104.

60. O. Rieppel and J. J. Head, New specimens of the fossil snake genus *Eupodophis* Rage and Escuillié, from the mid-Cretaceous of Lebanon. *Memorie della Società Italiana di Scienze Naturali e Museo Civico di Storia Naturale di Milano*, 23 (2004), 1–26.

61. D. Cundall and F. Irish, The snake skull. In C. Gans, A. S. Gaunt, and K. Adler, eds., *Biology of the Reptilia*, Vol. 20, *Morphology H* (Ithaca, NY: Society for the Study of Amphibians and Reptiles, 2008), pp. 349–692.

62. J. D. Scanlon, Cranial morphology of the Plio−Pleistocene giant madtsoiid snake *Wonambi naracoortensis*. *Acta Palaeontologica Polonica*, 50 (2005), 139–180.

63. D. M. Martill, H. Tischlinger, and N. R. Longrich, A four-legged snake from the Early Cretaceous of Gondwana. *Science*, 349 (2015), 416–419.

64. G. Cuny, J. -J. Jaeger, M. Mahboubi, and J. -C. Rage, Les plus anciens Serpents (Reptilia, Squamata) connus. Mise au point sur l'âge géologique des Serpents de la partie moyenne du Crétacé. *Comptes Rendus des Séances de l'Académie des Sciences*, t.311 (1990), 1267–1272.

65. J. D. Gardner and R. L. Cifelli, A primitive snake from the Cretaceous of Utah. In D.M. Unwin, ed., *Cretaceous Fossil Vertebrates* (London: The Paleontological Association, 1999), pp. 87–100.

66. R. L. Cifelli, J. I. Kirkland, A. Weil, A. L. Deino, and B. J. Kowallis, High-precision 40Ar/39Ar geochronology and the advent of North America's Late Cretaceous terrestrial fauna. *Proceedings of the National Academy of Sciences, USA*, 94 (1997), 11163–11167.

67. L. Xing, M. W. Caldwell, R. Chen, et al., A mid-Cretaceous embryonic-to-neonate snake in amber from Myanmar. *Science Advances*, 4 (2018), eaat5042.

68. A. Hsiou, A. M. Albino, M. A. Medeiros, and R. A. B. Santos, The oldest Brazilian snakes

from the Cenomanian (early Late Cretaceous). *Acta Palaeontologica Polonica*, 59 (2014), 635–642.

69. A. M. Albino, J. D. Carrillo-Briceño, and J. M. Neenan, An enigmatic aquatic snake from the Cenomanian of Northern South America. *PeerJ*, 4 (2016), e2027.

70. J. -C. Rage, Un serpent primitif (Reptilia, Squamata) dans le Cénomanien (base du Crétacé supérieur). *Comptes Rendus de l'Académie des Sciences de Paris, Série* II, 307 (1988), 1027–1032.

71. H. E. Sauvage, Sur l'existence d'un Reptile du type Ophidien dans les couches a Ostrea columba des Charentes. *Comptes Rendus Hebdomadaires des Séances de l'Académie des Sciences*, 7 (1880), 1–2325.

72. R. Vullo, J. -C. Rage, and D. Neraudeau, Anuran and squamate remains from the Cenomanian (Late Cretaceous) of Charentes, western France. *Journal of Vertebrate Paleontology*, 31 (2011), 279–291.

73. J. -C. Rage, R. Vullo, and D. Néraudeau, The mid-Cretaceous snake *Simoliophis rochebrunei* Sauvage, 1880 (Squamata: Ophidia) from its type area (Charentes, southwestern France): Redescription, distribution, and palaeoecology. *Cretaceous Research*, 58 (2016), 234–253.

74. R. Hoffstetter, Un serpent terrestre dans le Crétacé inférieur du Sahara. *Bulletin de La Société Géologique de France*, 7 (1960), 1–58.

75. C. G. Klein, N. R. Longrich, N. Ibrahim, S. Zouhri, and D. M. Martill, A new basal snake from the mid-Cretaceous of Morocco. *Cretaceous Research*, 72 (2017), 134–141.

76. R. Vullo, A new species of *Lapparentophis* from the mid-Cretaceous Kem Kem beds, Morocco, with remarks on the distribution of lapparentophiid snakes. *Comptes Rendus Palevol*, 18 (2019), 765–770.

77. J. -C. Rage and D. B. Dutheil, Amphibians and squamates from the Cretaceous (Cenomanian) of Morocco - A preliminary study, with description of a new genus of pipid frog. *Palaeontographica Abteilung A*, 285 (2008), 1–22.

78. F. Nopcsa, Ergbenisse der Forschungsreisen Prof. E. Stromers in den Wüsten Ägyptens, II. Wirbeltier-Reste de rBaharîje-Stufe (unterstes Cenoman). 5. Die Symoliophis-Reste. *Abhandlungen der Bayerischen Akademie der Wissenschaften, Mathematisch-naturwissenschaftliche Abteilung*, 30 (1925), 1–27.

79. L. A. Nessov, V. I. Zhegallo, and A. O. Averianov, A new locality of Late Cretaceous snakes, mammals and other vertebrates in Africa (western Libya). *Annales de Paléontologie*, 84 (1998), 265–274.

80. G. Haas, On a new snakelike reptile from the Lower Cenomanian of Ein Jabrud, near Jerusalem. *Bulletin Du Muséum National d'Histoire Naturelle*, 1 (1979), 51–64.

81. A. Houssaye, Rediscovery and description of the second specimen of the hind-limbed snake *Pachyophis woodwardi* Nopcsa, 1923 (Squamata, Ophidia) from the Cenomanian of Bosnia Herzegovina. *Journal of Vertebrate Paleontology*, 30 (2010), 276–279.

82. J. Bolkay, *Mesophis nopcsai* n.g. n.sp. ein neues, schlangenähnliches Reptil aus de runteren Kreide (Neocom) von Bilek-Selista (Ost-Hercegovina). *Glasnik zemaljskog Muzeja u Bosni i Hercegovini*, 37 (1925), 125–135.

83. R. L. Nydam, Lizards and Snakes from the Cenomanian through Campanian of Southern Utah: filling the Gap in the fossil record of Squamata from the Late Cretaceous of the Western Interior of North America. In A. L. Titus and M. A. Loewen, eds., *At the Top of the Grand Staircase: The Late Cretaceous of Southern Utah* (Bloomington & Indianapolis: Indiana University Press, 2013), pp. 370–423.

84. D. Đurić, D. Radosavljević, D. Petrović, M. Radonjić, and P. Vojnović, A new evidence for pachyostotic snake from Turonian of Bosnia-Herzegovina. *Annales*

Géologiques de La Péninsule Balkanique, 78 (2017), 17–21.

85. F. de Broin, E. Buffetaut, J. -C. Koeniguer, et al., La fauna de Vertébrés continentaux du gisement d'In Beceten (Sénonien du Niger). *Comptes Rendus Hebdomadaires Des Seances des Séances de l'Académie des Sciences de Paris, Série D*, 279 (1974), 469–472.

86. J. -C. Rage, Les continents péri-atlantiques au Crétacé Supérieur: Migrations des faunes continentales et problèmes paléogéographiques. *Cretaceous Research*, 2 (1981), 65–84.

87. T. C. Laduke, D. W. Krause, J. D. Scanlon, and N. J. Kley, A Late Cretaceous (Maastrichtian) snake assemblage from the Maevarano Formation, Mahajanga Basin, Madagascar. *Journal of Vertebrate Paleontology*, 30 (2010), 109–138.

88. R. T. J. Moody and P. J. C. Sutcliffe, The Cretaceous deposits of the Iullemmeden Basin of Niger, central West Africa. *Cretaceous Research*, 12 (1991), 137–157.

89. L. M. V. Meunier and H. C. E. Larsson, *Trematochampsa taqueti* as a nomen dubium and the crocodyliform diversity of the Upper Cretaceous in Beceten Formation of Niger. *Zoological Journal of the Linnean Society*, 182 (2018), 659–680.

90. A. Smith-Woodward, On some extinct reptiles from Patagonia, of the Genera *Miolania*, *Dinilysia*, and *Genyodectes*. *Proceedings of the Zoological Society of London*, 1 (1901), 169–184.

91. M. W. Caldwell and A. M. Albino, Exceptionally Preserved Skeletons of the Cretaceous Snake *Dinilysia patagonica* Woodward, 1901. *Journal of Vertebrate Paleontology*, 22 (2003), 861–866.

92. M. W. Caldwell and J. Calvo, Details of a new skull and articulated cervical column of *Dinilysia patagonica* Woodward, 1901. *Journal of Vertebrate Paleontology*, 28 (2008), 349–362.

93. C. A. Scanferla and J. I. Canale, The youngest record of the Cretaceous snake genus *Dinilysia* (Squamata, Serpentes).

South American Journal of Herpetology, 2 (2007), 76–81.

94. O. C. Marsh, Notice of New Reptiles from the Laramie Formation. *American Journal of Science*, 43 (1892), 449–453.

95. N. R. Longrich, B. A. S. Bhullar, and J. A. Gauthier, A transitional snake from the Late Cretaceous period of North America. *Nature*, 488 (2012), 205–208.

96. J. -C. Rage and C. Werner, Mid-Cretaceous (Cenomanian) snakes from Wadi Abu Hashim, Sudan: The earliest snake assemblage. *Palaeontologia Africana*, 35 (1999), 85–110.S.

97. L. Wick and T. A. Shiller, New taxa among a remarkably diverse assemblage of fossil squamates from the Aguja Formation (lower Campanian) of West Texas. *Cretaceous Research*, 114 (2020), 104516.

98. J. J. Head, A South American snake lineage from the Eocene Greenhouse of North America and a reappraisal of the fossil record of 'anilioid' snakes. *Geobios*, (2020), doi.org/10.1016/j. geobios.2020.09.005.

99. J. -C. Rage and G. Wouters, Découverte du plus ancien palaeopheidé (Reptilia, Serpentes) dans le Maestrichtien du Maroc. *Geobios*, 12 (1979), 293–6.

100. A. C. Pritchard, J. A. McCartney, D. W. Krause, and N. J. Kley, New snakes from the Upper Cretaceous (Maastrichtian) Maevarano Formation, Mahajanga Basin, Madagascar. *Journal of Vertebrate Paleontology*, 34 (2014), 1080–1093.

101. A. Miralles, J. Marin, D. Markus, et al.. Molecular evidence for the paraphyly of Scolecophidia and its evolutionary implications. *Journal of Evolutionary Biology*, 31 (2018), 1782–1793.

102. J. A. Schiebout, C. A. Rigsby, S. D. Rapp, J. A. Hartnell, and B. R. Standhardt, Stratigraphy of the Cretaceous-Tertiary and Paleocene-Eocene transition rocks of Big Bend National Park, Texas. *The Journal of Geology*, 95 (1987), 359–375.

103. M. Augé and J.-C. Rage, Herpetofaunas from the upper Paleocene and lower

Eocene of Morocco. *Annales de Paléontologie*, 92 (2006), 235–253.

104. T. S. Fachini, S. Onary, A. Palci, et al., Cretaceous blind snake from Brazil fills major gap in snake evolution. *iScience*, (2020), 101834.

105. R. O. Gómez, A. M. Baez, and G. W. Rougier, An anilioid snake from the Upper Cretaceous of northern Patagonia. *Cretaceous Research*, 29 (2008), 481–488.

106. G. L. Georgalis and K. T. Smith, Constrictores Oppel, 1811 – the available name for the taxonomic group uniting boas and pythons. *Vertebrate Zoology*, 70 (2020), 291–304.

107. J. J. Head, J. I. Bloch, A. K. Hastings, et al., Giant boid snake from the Palaeocene neotropics reveals hotter past equatorial temperatures. *Nature*, 457 (2009), 715–717.

108. E .M. Roberts, A. L. Deino, and M. A. Chan, [40]Ar/[39]Ar age of the Kaiparowits Formation, southern Utah, and correlation of contemporaneous Campanian strata and vertebrate faunas along the margin of the Western Interior Basin. *Cretaceous Research*, 26 (2005), 307–318.

109. N. R. Longrich, B. A. S. Bhullar, and J. A. Gauthier, Mass extinction of lizards and snakes at the Cretaceous–Paleogene boundary. *Proceedings of the National Academy of Sciences*, 109 (2012), 21396–21401.

110. R. Isaza, M. Garner, and E. Jacobson, Proliferative osteoarthritis and osteoarthrosis in 15 snakes. *Journal of Zoo and Wildlife Medicine*, 31 (2000), 20–27.

111. E. Schrank, Palynology of the elastic Cretaceous sediments between Dongola and Wadi Muqaddam, northern Sudan. *Berliner geowissenschaftliche Abhandlungen, (A)*, 120 (1990), 149–168.

112. E. Schrank, Nonmarine Cretaceous correlations in Egypt and northern Sudan: palynological and palaeobotanical evidence. *Cretaceous Research*, 13 (1992), 351–368.

113. C. Werner, Die kontinentale Wirbeltierfauna aus der unteren Oberkreide des Sudan (Wadi Milk Formation). *Berliner Geowissenschaftliche Abhandlungen, (E)*, 13 (1994), 221–249.

114. R. Vullo and D. Néraudeau, When the 'primitive' shark *Tribodus* (Hybodontiformes) meets the 'modern' ray *Pseudohypolophus* (Rajiformes): the unique co-occurrence of these two durophagous Cretaceous selachians in Charentes (SW France). *Acta Geologica Polonica*, 58 (2008), 249–255.

115. P. C. Owusu Agyemang, E. M. Roberts, R. Bussert, D. Evans, and J. Müller, U–Pb detrital zircon constraints on the depositional age and provenance of the dinosaur-bearing Upper Cretaceous Wadi Milk Formation of Sudan. *Cretaceous Research*, 97 (2019), 52–72.

116. A. A. M. Eisawi, Palynological evidence of a Campanian–Maastrichtian age of the Shendi Formation (Shendi Basin, central Sudan). *American Journal of Earth Sciences*, 2 (2015), 206–210.

117. K. A. O. Salih, D. C. Evans, R. Bussert, N. Klein, M. Nafi, and J. Müller, First record of *Hyposaurus* (Dyrosauridae, Crocodyliformes) from the Upper Cretaceous Shendi Formation of Sudan. *Journal of Vertebrate Paleontology*, 36 (2016), p.e1115408.

118. N. Klein, R. Bussert, D. Evans, et al., Turtle remains from the Wadi milk formation (upper cretaceous) of northern Sudan. *Palaeobiodiversity and Palaeoenvironments*, 96 (2016), 281–303.

119. O. W. M. Rauhut, A dinosaur fauna from the Late Cretaceous (Cenomanian) of northern Sudan. *Palaeontologia Africana*, 35 (1999), 61–84.

120. K. M. Claeson, H. M. Sallam, P. M. O'Connor, and J. J. Sertich, A revision of the Upper Cretaceous lepidosirenid lungfishes from the Quseir Formation, Western Desert, central Egypt. *Journal of Vertebrate Paleontology*, 34 (2014), 760–766.

121. M. Martin, *Protopterus nigeriensis* nov. sp., l'un des plus anciens protoptères—Dipnoi

(In Beceten, Sénonien du Niger). *Comptes Rendus de l'Académie des Sciences, Series IIA, Earth and Planetary Science*, 325 (1997), 635–638.

122. O. Otero, Current knowledge and new assumptions on the evolutionary history of the African lungfish, *Protopterus*, based on a review of its fossil record. *Fish and Fisheries*, 12 (2011), 235–255.

123. J. -C. Rage and H. Cappetta, Vertebrates from the Cenomanian, and the geological age of the Draa Ubari fauna (Libya). *Annales de Paléontologie*, 88 (2002), 79–84.

124. N. Ibrahim, P. C. Sereno, D. J. Varricchio, et al., Geology and paleontology of the Upper Cretaceous Kem Kem Group of eastern Morocco. *ZooKeys*, 928 (2020), 1–216.

125. D. M. Mohabey, J. J. Head, and J. A. Wilson, A new species of the snake *Madtsoia* from the Upper Cretaceous of India and its paleobiogeographic implications. *Journal of Vertebrate Paleontology*, 31 (2011), 588–595.

126. R. O. Gómez, F. F. Garberoglio, and G. W. Rougier, A new Late Cretaceous snake from Patagonia: Phylogeny and trends in body size evolution of madtsoiid snakes. *Comptes Rendus Palevol*, 18 (2019), 771–781.

127. H. Yi and M. A. Norell, The burrowing origin of modern snakes. *Science Advances*, 1 (2015), p.e1500743.

128. J. -C. Rage, Un caenophidien primitif (Reptilia, Serpentes) dans l'Éocène inférieur. *Compte Rendu Sommaire des Séances de la Société Géologique de France*, XVII °2 (1975), 46–48.

129. J. -C. Rage, A. Folie, R. S. Rana, et al., A diverse snake fauna from the early Eocene of Vastan Lignite Mine, Gujarat, India. *Acta Palaeontologica Polonica*, 53 (2008), 391–403.

130. J. -C. Rage and M. Augé, Squamate reptiles from the middle Eocene of Lissieu (France). *A landmark in the middle Eocene of Europe*. Geobios, 43 (2010), 253–268.

131. J. J. Head, P. A. Holroyd, J. H. Hutchison, and R. L. Ciochon, First report of snakes (Serpentes) from the late middle Eocene Pondaung Formation, Myanmar. *Journal of Vertebrate Paleontology*, 25 (2005), 246–250.

132. H. Zaher, R. W. Murphy, et al., Large-scale molecular phylogeny, morphology, divergence-time estimation, and the fossil record of advanced caenophidian snakes (Squamata: Serpentes). *PloS One*, 14 (2019), p.e0216148.

133. H. Zaher and O. Rieppel, On the phylogenetic relationships of the Cretaceous snakes with legs, with special reference to *Pachyrhachis problematicus* (Squamata, Serpentes). *Journal of Vertebrate Paleontology*, 22 (2002), 104–109.

134. H. Zaher and K. T. Smith, Pythons in the Eocene of Europe reveal a much older divergence of the group in sympatry with boas. *Biology Letters*, 16 (2020), p.20200735.

135. Ş. Vasile, Z. Csiki-Sava, and M. Venczel, A new madtsoiid snake from the Upper Cretaceous of the Haţeg Basin, western Romania. *Journal of Vertebrate Paleontology*, 33 (2013), 1100–1119.

136. H. Zaher and O. Rieppel, Tooth implantation and replacement in squamates, with special reference to mosasaur lizards and snakes. *American Museum Novitates*, 3271 (1999), 1–19.

137. A. Houssaye, 'Pachyostosis' in aquatic amniotes: a review. *Integrative Zoology*, 4 (2009), 325–340

138. M. S. Y. Lee and J. D. Scanlon, Snake phylogeny based on osteology, soft anatomy and ecology. *Biological Reviews*, 77 (2002), 333–401.

139. S. M. Harrington and T. W. Reeder, Phylogenetic inference and divergence dating of snakes using molecules, morphology and fossils: New insights into convergent evolution of feeding morphology and limb reduction. *Biological Journal of the Linnean Society*, 121 (2017), 379–394.

140. O. Rieppel, A. G. Kluge, and H. Zaher, Testing the phylogenetic relationships of the Pleistocene snake *Wonambi naracoortensis* Smith. *Journal of Vertebrate Paleontology*, 22 (2002), 812–829.

141. J. P. Rio and P. D. Mannion, The osteology of the giant snake *Gigantophis garstini* from the upper Eocene of North Africa and its bearing on the phylogenetic relationships and biogeography of Madtsoiidae. *Journal of Vertebrate Paleontology*, 37 (2017), p.e1347179.

142. J. D. Scanlon, *Nanowana* gen. nov., small madtsoiid snakes from the Miocene of Riversleigh: Sympatric species with divergently specialised dentition. *Memoirs of the Queensland Museum*, 41 (1997), 393–412.

143. J. D. Scanlon and M. S. Y. Lee, The Pleistocene serpent *Wonambi* and the early evolution of snakes. *Nature*, 403 (2000), 416–420.

144. P. B. Snetkov, Vertebrae of the sea snake *Palaeophis nessovi* Averianov (Acrochordoidea, Palaeophiidae) from the Eocene of Western Kazakhstan and phylogenetic analysis of the superfamily Acrochordoidea. *Paleontological Journal*, 45 (2011), 305–313.

145. A. Houssaye, J. -C. Rage, N. Bardet, et al., New highlights about the enigmatic marine snake *Palaeophis maghrebianus* (Palaeophiidae; Palaeophiinae) from the Ypresian (Lower Eocene) phosphates of Morocco. *Palaeontology*, 56 (2013), 647–661.

146. S. Bajpai and J. J. Head, An early Eocene palaeopheid snake from Vastan Lignite Mine, Gujarat, India. Gondwana. *Geological Magazine*, 22 (2008), 85–90.

147. A. Palci, M. N. Hutchinson, M. W. Caldwell, and M. S. Lee, The morphology of the inner ear of squamate reptiles and its bearing on the origin of snakes. *Royal Society Open Science*, 4 (2017), 170685.

148. P. M. Barrett, A. J. McGowan, and V. Page, Dinosaur diversity and the rock record. *Proceedings of the Royal Society, B*, 276 (2009), 2667–2674.

149. A. Y. Hsiang, D. J. Field, T. H. Webster, et al., The origin of snakes: Revealing the ecology, behavior, and evolutionary history of early snakes using genomics, phenomics, and the fossil record. *BMC Evolutionary Biology*, 15 (2015), 87.

150. R. Estes, Middle Paleocene lower vertebrates from the Tongue River Formation, southeastern Montana. *Journal of Paleontology*, 50 (1976), 500–520.

151. R. Hoffstetter, Nouvelles récoltes de serpents fossiles dans l'Éocène Supérieur du désert Libyque. *Bulletin du Muséum national d'Histoire naturelle*, 33 (1961), 326–331.

152. J. -C. Rage, Fossil snakes from the Palaeocene of São José de Itaboraí, Brazil. Part I. Madtsoiidae, Aniliidae. *Palaeovertebrata*, 27 (1998), 109–144.

153. J. -C. Rage, G. Métais, A. Bartolini, et al., First report of the giant snake *Gigantophis* (Madtsoiidae) from the Paleocene of Pakistan: paleobiogeographic implications. *Geobios*, 47 (2014), 147–153.

154. J. -C. Rage, S. M. T. Bajpai, J. G. Thewissen, and B. N. Tiwari, Early Eocene snakes from Kutch, Western India, with a review of the Palaeophiidae. *Geodiversitas*, 25 (2003), 695–716.

155. R. G. Reynolds, M. L. Niemiller, and L. J. Revell, Toward a Tree-of-Life for the boas and pythons: Multilocus species-level phylogeny with unprecedented taxon sampling. *Molecular Phylogenetics and Evolution*, 71 (2014), 201–213.

156. W. A. Clemens, Evolution of the mammalian fauna across the Cretaceous–Tertiary boundary in northeastern Montana and other areas of the western interior. *Geological Society of America, Special Paper*, 361 (2002), 217–245.

157. M. Friedman, Explosive morphological diversification of spiny-finned teleost fishes in the aftermath of the end-Cretaceous extinction. *Proceedings of the Royal Society, B*, 277 (2010), 1675–1683.

158. Y. J. Feng, D. C. Blackburn, D. Liang, et al., Phylogenomics reveals rapid, simultaneous

diversification of three major clades of Gondwanan frogs at the Cretaceous-Paleogene boundary. *Proceedings of the National Academy of Sciences, USA,* 114 (2017), E5864–E5870.

159. M. E. Alfaro, B. C. Faircloth, R. C. Harrington, et al., Explosive diversification of marine fishes at the Cretaceous–Palaeogene boundary. *Nature Ecology and Evolution,* 2 (2018), 688–696.

160. D. J. Field, A. Bercovici, J. S. Berv, et al., Early evolution of modern birds structured by global forest collapse at the end-Cretaceous mass extinction. *Current Biology,* 28 (2018), 1825–1831.

161. J. -C. Rage, Fossil snakes from the Palaeocene of São José de Itaboraí, Brazil. Part III. Ungaliophiinae, Booids incertae sedis, and Caenophidia. Summary, update, and discussion of the snake fauna from the locality. *Palaeovertebrata,* 36 (2008), 37–73.

162. A. Scanferla and K. T. Smith, Exquisitely preserved fossil snakes of Messel: Insight into the evolution, biogeography, habitat preferences and sensory ecology of early boas. *Diversity,* 12 (2020), 100.

163. R. A. Pyron and F. T. Burbrink, Extinction, ecological opportunity, and the origins of global snake diversity. *Evolution,* 66 (2012), 163–178.

164. J. A. McCartney, S. N. Bouchard, J. A. Reinhardt, et al., The oldest lamprophiid (Serpentes, Caenophidia) fossil from the late Oligocene Rukwa Rift Basin, Tanzania and the origins of African snake diversity. *Geobios,* (2020), doi.org/10.1016/j.geobios.2020.07.005.

165. G. G. Simpson, A new fossil snake from the *Notostylops* beds of Patagonia. *Bulletin of the American Museum of Natural History,* 67 (1933), 1–22.

166. F. O. Da Silva, A. C. Fabre, Y. Savriama, et al., The ecological origins of snakes as revealed by skull evolution. *Nature Communications,* 9 (2018), https://doi.org/10.1038/s41467-017-02788-3.

167. A. Watanabe, A. C. Fabre, R. N. Felice, et al., Ecomorphological diversification in squamates from conserved pattern of cranial integration. *Proceedings of the National Academy of Sciences, USA,* 116 (2019), 14688–14697.

The Diversity and Distribution of Palaeogene Snakes

A Review with Comments on Vertebral Sufficiency

KRISTER T. SMITH AND GEORGIOS L. GEORGALIS

4.1 Introduction

The first clear fossil indications of modern clades of snakes are in the Palaeogene. Many major groups, including 'scolecophidians' (blindsnakes), probably originated earlier, and some extinct groups sometimes allied with Alethinophidia, such as Madtsoiidae and Pachyophiidae, together with Palaeophiidae, Nigerophiidae and *incertae sedis* alethinophidians ([1–4]; Chapter 9), are well known already in the Cretaceous. Molecular studies indicate that the radiation of alethinophidians was well underway in the early Cenozoic [5–7].

It is well known that the vast majority of snake fossils are isolated trunk vertebrae (e.g., [8, 9]). This can scarcely surprise, because snakes have so very many of them. A recurring refrain of later twentieth century commentaries on the snake fossil record – echoing Cuvier's opinion at the dawn of comparative anatomy [10] – is that isolated trunk vertebrae are insufficient for detailed systematic study [11–13]. This too can scarcely surprise, because it is difficult to imagine that a vertebra, however exactingly it may be examined, can have enough diagnostic features to distinguish the >3,800 extant species, much less their extinct relatives – especially when ontogenetic, intracolumnar, and intraspecific variation are considered. Thus it is, that at the close of the twentieth century scarcely a handful of fossil snakes were known from more than a few broken bits, and most work on early fossil snakes was conducted without any really detailed understanding of their phylogenetic affinities. And so from the Palaeogene (Palaeocene, Eocene, and Oligocene epochs), palaeontology could demonstrate only a high morphological and alpha taxonomic diversity.

Several conceptual and material advances have given new insight into the phylogenetic, rather than only phenotypic, diversity of Palaeogene snakes. First, there has been a shift away from dichotomous keys (e.g., boas vs pythons, natricines vs. colubrines) and towards the use of apomorphies in identifying major lineages [13–15]. In any bifurcating key, one of the character states is usually primitive, the other derived, and the former only tells one what a species is *not*. In addition, a dichotomous key tends to elide crucial phylogenetic nuance about the level at which characters are diagnostic, even if apomorphic.

Second, there has been increased emphasis on reconstructing the entire vertebral column, and if possible the skull [15–19]. This delivers a better understanding of intracolumnar variation, to assist in alpha taxonomy, and of anatomy. A more complete skeletal reconstruction enhances the discovery of phylogenetically useful information. Even – or especially – caudal vertebrae have proven useful (e.g., [20, 21]).

Third, quantification is growing in prominence. Serious effort at quantification actually began with Auffenberg's [22] monograph on the fossil snakes of Florida, but this pioneering aspect of his work was not widely adopted. Quantification has three major advantages: it lends precision to otherwise vague terms (like 'moderately vaulted') and hence contributes to reproducibility; it allows for population-level variation to be measured and analysed; and it allows for detailed analysis of intracolumnar and ontogenetic change [22–24].

Finally, many new discoveries in the Palaeogene have illuminated the early history of snakes. These discoveries include associated material (partial or complete skeletons) and material from geographically under-represented areas, particularly Gondwanan continents [19, 25–30]. Gondwanan fossils are especially important if much of the early history of snakes transpired there.

In this chapter, we bring together in one place all described Palaeogene fossil snakes and place currently valid taxa in a defensible scheme. We examine the spatiotemporal distribution of snakes in the Palaeogene to document patterns and highlight where more work is needed. Finally, we address the nature of the snake fossil record by examining the sufficiency of vertebrae to study phylogenetic affinity and species diversity.

4.2 Taxonomic Framework of Palaeogene Snakes

Extant snakes have traditionally been divided into two groups, Scolecophidia (the fossorial blindsnakes) and Alethinophidia (everything else) (e.g., [31]). Due especially to the use of molecular data, some parts of the snake tree are in a state of flux [5, 32, 33]. For a framework, we follow Burbrink et al. [34] on the phylogeny of extant snake groups. As in many other recent studies, they found Scolecophidia to be paraphyletic, and 'Anilioidea' to be polyphyletic, with *Anilius* more closely related to Tropidophiidae (forming the clade Amerophidia) than to Cylindrophiidae and Uropeltidae.

For larger groups of extinct snakes (Madtsoiidae, Palaeophiidae, Nigerophiidae, Russellophiidae) we must rely on morphology. According to some phylogenetic analyses Madtsoiidae is a stem snake clade [35–38], but here we follow the alternative interpretation [39–41] that madtsoiids are crown snakes on the stem of Alethinophidia.

Palaeophiidae is usually considered to contain *Archaeophis*, *Palaeophis* and *Pterosphenus* (e.g., [8]). The latter two genera are known exclusively from vertebrae and are united by more or less laterally compressed centra, reduced prezygapophyses, pterapophyses developed to some extent, and vertical rather than slanted cotylar and condylar edges [18]. Their affinities remain unclear (see [42] for a review]. They might be related to Acrochordidae [43–45] or to Cretaceous marine snakes [31, 46]. Further study of the skull of *'Archaeophis' turkmenicus* (see [47]) may bring clarity. We conservatively regard Palaeophiidae here as a monophyletic, *incertae sedis* lineage of the total clade of Alethinophidia.

Nigerophiidae was erected for a single genus, *Nigerophis*, but subsequently additional taxa have been described, including some from the Cretaceous ([48]; see below). Rage [8, 49, 50] regarded Nigerophiidae as close to both Palaeophiidae (see also [51]) and especially Acrochordidae. Head et al. [3] noted, however, that synapomorphies of Nigerophiidae and Acrochordidae have not been presented. We consider Nigerophiidae as an *incertae sedis* lineage of the total clade of Alethinophidia.

Russellophiidae was erected for a single genus, *Russellophis*, which remains the only valid russellophiid genus in the Palaeogene. However, Russellophiidae has also been reported from the Late Cretaceous [2]. Rage [52] considered it an early caenophidian family (see also [8]). Head et al. [3] gave the following apomorphies as supporting its position in the total clade of Caenophidia: elongate centrum; small, circular cotyle and condyle; and well-developed paralymphatic grooves defining a haemal keel. We follow the latter authors, who distinguished between crown and stem.

For the purposes of this review we adopt the higher-level taxonomic hierarchy given in Table 4.1. The lowest-level entities (usually species) are treated individually and alphabetically in Supplementary Appendix 4.S1, where details on the attribution of individual species to higher taxa can also be found.

4.3 The Distribution of Palaeogene Snakes

4.3.1 Distribution of Localities in Space and Time

In this section we summarize the distribution of localities that have yielded Palaeogene fossil snakes. For detailed lists with references, see Supplementary Appendix 4.S2.

Africa (Fig. 4.1) was for a long time poorly known. Various localities had been discovered on British, French, and German expeditions in the twentieth century. These represent marginal marine environments and the Trans-Saharan Seaway, which once divided West Africa from the remainder of the continent (see [56]). Palaeocene and Eocene are dominant. However, new Eocene and Oligocene localities are yielding surprising snake assemblages.

The Neotethyan part of Asia (Fig. 4.2) has yielded principally marine localities with aquatic snakes recovered on Russian expeditions. However, rare terrestrial localities in Mongolia and Southeast Asia have yielded tantalizing specimens. Most surprising in recent years have been Eocene assemblages with mixtures of aquatic and terrestrial snakes from Indo-Pakistan. Insofar as this subcontinent was docking with Asia, it is likely that they preserve a mixture of Gondwanan and Laurasian faunal components. At the same time, the absence of terrestrial snakes from eastern Asia at a time when rich mammalian faunas are

Table 4.1. Here we give what we feel is a defensible classification of Palaeogene fossil snakes.

Valid taxa are all listed, whereas unnamed distinct or potentially distinct or indeterminate forms are subsumed into composite entries for simplicity. Named species that are currently considered junior synonyms or nomina dubia are not included. The lowest level entries in the classification are treated alphabetically in Supplementary Appendix 4.S1; thus, the classification can also be seen as a key to Supplementary Appendix 4.S1. An inverted triangle (▽) before a taxon name indicates a total clade, i.e., the crown clade plus its stem [53]. For instance, ▽Serpentes would include both crown snakes (including extinct ones) as well as stem snakes like *Najash rionegrina* [54]. It is important to be clear on this point because the distinction is frequently not observed in the literature [55] and the assignment of a species to some higher taxon is potentially misleading. However, in most cases total-clade names have not formally been erected, hence our informal approach. Where there is positive evidence that a genus name is inappropriate (e.g., because of non-monophyly), we put quotes around it. The lack of quotation marks should not be construed to mean that there is positive evidence for the monophyly of most genera.

Serpentes Linnaeus, 1758

'Scolecophidia' Duméril and Bibron, 1844

'Scolecophidia' *incertae sedis*

▽Alethinophidia Nopsca, 1923

▽Alethinophidia *incertae sedis*

†*Afrotortrix draaensis* Rage et al., 2021

†*Amaru scagliai* Albino, 2018

'anilioids' *incertae sedis*

†*Coniophis* sp.

†*Eoanilius europae* Rage, 1974

†*Eoanilius oligocenicus* Szyndlar, 1994

†*Falseryx neervelpensis* Szyndlar et al., 2008

†*Goinophis minusculus* Holman, 1976

†*Hoffstetterella brasiliensis* Rage, 1998

†*Kataria anisodonta* Scanferla et al., 2013

†*Platyspondylia germanica* Szyndlar and Rage, 2003

†*Platyspondylia lepta* Rage, 1974

†*Platyspondylia sudrei* Rage, 1988

†*Rottophis atavus* (von Meyer, 1855)

†*Szyndlaria aureomontensis* Rage and Augé, 2010

†*Tuscahomaophis leggetti* Holman and Case, 1992

†*Vectophis wardi* Rage and Ford, 1980

Table 4.1. (*cont.*)

†Madtsoiidae Hoffstetter, 1961
†*Alamitophis tingamarra* Scanlon, 2005
†*Gigantophis garstini* Andrews, 1901
†*Madtsoia bai* Simpson, 1933
†*Madtsoia camposi* Rage, 1998
†Madtsoiidae indet.
†*Patagoniophis australiensis* Scanlon, 2005
†*Platyspondylophis tadkeshwarensis* Smith et al., 2016
Amerophidia Vidal et al., 2007
Aniliidae Fitzinger, 1826
†*Borealilysia carinata* (Hecht, 1959)
†*Borealilysia gunnelli* Head, 2021
▽Tropidophiidae Brongersma, 1951
Tropidophiidae indet.
▽Constrictores Oppel, 1811 sensu Georgalis and Smith (2020)
Booidea sensu Pyron et al. (2014)
†*Bavarioboa* Szyndlar and Schleich, 1993
†*Bavarioboa bachensis* Szyndlar and Rage, 2003
†*Bavarioboa crocheti* Szyndlar and Rage, 2003
†*Bavarioboa herrlingensis* Szyndlar and Rage, 2003
†*Bavarioboa minuta* Szyndlar and Rage, 2003
†*Bavarioboa vaylatsae* Szyndlar and Rage, 2003
†*Phosphoroboa filholii* (Rochebrune, 1880)
▽Boidae Gray, 1825 sensu Pyron et al. (2014)
†*Corallus priscus* Rage, 2001
†*Eoconstrictor fischeri* (Schaal, 2004)
†*Eoconstrictor spinifer* (Barnes, 1927)
†*Titanoboa cerrejonensis* Head et al., 2009
Charinaidae Gray, 1849
†*Calamagras angulatus* Cope, 1873

Table 4.1. (*cont.*)

†*Calamagras murivorus* Cope, 1873
†*Calamagras weigeli* Holman, 1972
†*Messelophis variatus* Baszio, 2004
†*Rageryx schmidi* Smith and Scanferla, 2021
†*Rieppelophis ermannorum* (Baszio and Schaal, 2004)
'erycines' *incertae sedis*
†*Crythiosaurus mongoliensis* Gilmore, 1943
†*Bransateryx vireti* Hoffstetter and Rage, 1972
†*Cadurceryx filholi* Hoffstetter and Rage, 1972
†*Cadurceryx pearchi* Holman et al., 2006
†*'Calamagras' gallicus* Rage, 1977
Pythonoidea sensu Wallach et al. (2014)
Loxocemidae Cope, 1861
†*Ogmophis compactus* Lambe, 1908
Messelopythonidae Smith and Scanferla, 2021
†*Messelopython freyi* Zaher and Smith, 2020
†*Palaeopython cadurcensis* (Filhol, 1877)
†*Palaeopython ceciliensis* Barnes, 1927
†*Palaeopython helveticus* Georgalis and Scheyer, 2019
†*Palaeopython schaali* Smith and Scanferla, 2021
Pythonidae Fitzinger, 1826
†*Morelia riversleighensis* (Smith and Plane, 1985)
Pythonoidea *incertae sedis*
▽Constrictores *incertae sedis*
†*Anilioides nebraskensis* Holman, 1976
†*Boavus* sp.
†*Boavus affinis* Brattstrom, 1955
†*Boavus brevis* Marsh, 1871
†*Boavus idelmani* Gilmore, 1938
†*Boavus occidentalis* Marsh, 1871

Table 4.1. (*cont.*)

†*Cadurcoboa insolita* Rage, 1978
†*Calamagras* sp.
†'*Calamagras*' *platyspondyla* Holman, 1976
†'*Calamagras*' *primus* Hecht, 1959
†'*Calamagras*' *turkestanicus* Danilov and Averianov, 1999
†*Cheilophis huerfanoensis* Gilmore, 1938
†*Chubutophis grandis* Albino, 1993
†*Conantophis alachuaensis* Holman and Harrison, 2000
†*Coprophis dakotaensis* Parris and Holman, 1978
†*Dawsonophis wyomingensis* Holman, 1979
†*Dunnophis* sp.
†*Dunnophis cadurcensis* Rage, 1974
†*Dunnophis matronensis* Rage, 1973
†*Dunnophis microechinus* Hecht, 1959
†*Geringophis* sp.
†*Geringophis depressus* Holman, 1976
†*Geringophis robustus* Holman and Harrison, 2001
†*Geringophis vetus* Holman, 1982a
†*Hechtophis austrinus* Rage 2001
†*Helagras orellanensis* Holman, 1983
†*Helagras prisciformis* Cope, 1883a
†*Hordleophis balconae* Holman, 1996
†*Huberophis georgiensis* Holman, 1977b
†*Itaboraiophis depressus* Rage, 2008
†*Ogmophis* sp.
†*Ogmophis oregonensis* Cope, 1883b
†*Ogmophis voorhiesi* Holman, 1977b
†"*Palaeopython*" *neglectus* Rochebrune, 1884
†*Paleryx rhombifer* Owen, 1850
†*Paraepicrates brevispondylus* Hecht, 1959
†*Paraplatyspondylia batesi* Holman and Harrison, 1998

Table 4.1. (*cont.*)

†*Paraungaliophis pricei* Rage, 2008

†*Paulacoutophis perplexus* Rage, 2008

†*Plesiotortrix edwardsi* Rochebrune, 1884

†*Pterygoboa* sp.

†*Rukwanyoka holmani* McCartney et al., 2014

†*Sanjuanophis froehlichorum* Sullivan and Lucas, 1988

†*Tallahattaophis dunni* Holman and Case, 1988

†*Totlandophis americanus* Holman and Harrison, 2001

†*Totlandophis thomasae* Holman and Harrison, 1998

†*Waincophis australis* Albino, 1987

†*Waincophis cameratus* Rage, 2001

†*Waincophis pressulus* Rage, 2001

†Palaeophiidae Lydekker, 1888b

†*Archaeophis proavus* Massalongo, 1859

†*'Archaeophis' turkmenicus* Tatarinov, 1963

†*Palaeophis* sp.

†*Palaeophis africanus* Andrews, 1924

†*Palaeophis casei* Holman, 1982b

†*Palaeophis colossaeus* Rage, 1983b

†*Palaeophis ferganicus* Averianov, 1997

†*Palaeophis grandis* (Marsh, 1869)

†*Palaeophis littoralis* Cope, 1868a

†*Palaeophis maghrebianus* Arambourg, 1952

†*Palaeophis nessovi* Averianov, 1997

†*Palaeophis oweni* Zigno, 1881

†*Palaeophis tamdy* (Averianov, 1997)

†*Palaeophis toliapicus* Owen, 1841

†*Palaeophis typhaeus* Owen, 1850

†*Palaeophis vastaniensis* Bajpai and Head, 2007

†*Palaeophis virginianus* Lynn, 1934

†*Palaeophis zhylan* (Nessov, 1984)

†*Pterosphenus* sp.

Table 4.1. (*cont.*)

†*Pterosphenus biswasi* Rage et al., 2003
†*Pterosphenus kutchensis* Rage et al., 2003
†*Pterosphenus schucherti* Lucas, 1899
†*Pterosphenus schweinfurthi* (Andrews, 1901)
†*Pterosphenus sheppardi* Hoffstetter, 1958
▽Caenophidia Hoffstetter, 1939
†*Anomalophis bolcensis* (Massalongo, 1859)
▽Colubroides Zaher et al., 2009
†'*Coluber*' *cadurci* Rage, 1974
▽Colubriformes *incertae sedis*
▽Colubroidea *incertae sedis*
▽Elapidae *incertae sedis*
†*Floridaophis auffenbergi* Holman, 1999
Lamprophiidae *incertae sedis*
†*Natrix mlynarskii* Rage, 1988b
†*Nebraskophis oligocenicus* Holman, 1999
†*Procerophis sahnii* Rage et al., 2008
†*Texasophis bohemiacus* Szyndlar, 1987
†*Texasophis galbreathi* Holman, 1984
†*Texasophis hecki* Böhme, 2008
†Thaumastophiidae Zaher et al., 2021
†*Thaumastophis missiaeni* Rage et al., 2008
†*Renenutet enmerwer* McCartney and Seiffert, 2016
†*Headonophis harrisoni* Holman, 1993
†Nigerophiidae Rage, 1975b
†*Amananulam sanogoi* McCartney et al., 2018
†'*Nessovophis*' *zhylga* Averianov, 1997
†*Nigerophis mirus* Rage, 1975b
†*Woutersophis novus* Rage, 1980
†Russellophiidae Rage, 1978
†*Russellophis crassus* Rage et al., 2008
†*Russellophis tenuis* Rage, 1975a

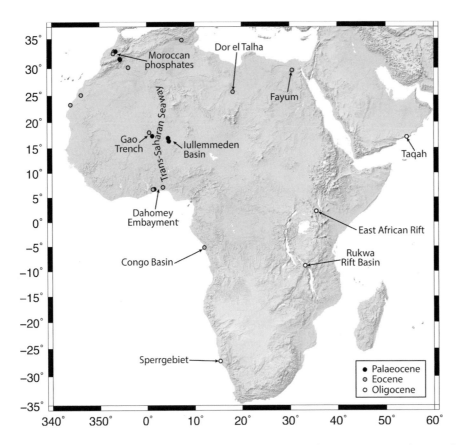

Figure 4.1 Palaeogene snake localities in Africa. See Supplementary Appendix 4.S3 for colour version.

known (e.g., [57]) remains one of the most conspicuous points about the Palaeocene and Eocene there. Although the occurrence of carnivorous anguimorph lizards (e.g., [58]) possibly explains the absence of snakes, these animals are well known in North America and Europe [59, 60] and co-preserved in many different localities [61].

Australia (Fig. 4.3) remains poorly known because the Palaeogene is so poorly represented, with only two confidently assigned localities. It was previously thought that many Riversleigh localities were Oligocene in age [62]. Possible early occurrences of '*Laticauda*' from the RSO site [63] and of *Wonambi barriei* from the CS site [64] are early Miocene according to recent radiometric dating [65]. Specimens referable to *Yurlunggur* from 'Carl Creek' and to *Yurlungurr* [sic, for *Yurlunggur*] cf. *camfieldensis* (a vertebra) from the Kangaroo Well fauna of the Ulta Limestone, whose assignment to the late Oligocene was based on biochronology [66], may also be closer to middle and early Miocene, respectively [65]. Undescribed snake records from the D-Site, Hiatus, and White Hunter localities [67] could not yet be radiometrically dated but are correlated biochronologically with the late Oligocene [65]. All three species from the early Eocene of Tingamarra are referred to genera first described from South America; these may have evolved (by dispersal or vicariance) as a

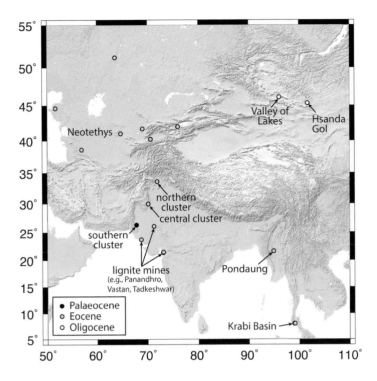

Figure 4.2 Palaeogene snake localities in Asia. See Supplementary Appendix 4.S3 for colour version.

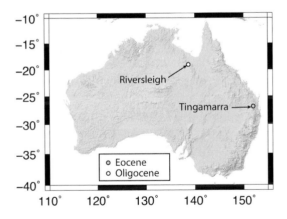

Figure 4.3 Palaeogene snake localities in Australia. See Supplementary Appendix 4.S3 for colour version.

result of an original connection between the two continents, probably via Antarctica [19]. On the other hand, the Palaeogene fauna of western Antarctica (i.e., the James Ross Basin) is similar to that of South America but unlike that of East Antarctica, suggesting the presence of barriers (geographic or climatic) that limited exchanges [68].

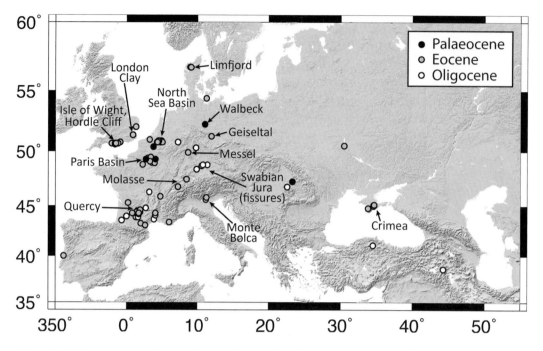

Figure 4.4 Palaeogene snake localities in Europe. See Supplementary Appendix 4.S3 for colour version.

Europe (Fig. 4.4) has been most intensely studied of all areas due to the long history of palaeontology there. With the exception of the fissure fills of Walbeck, the Palaeocene is dominated by marginal marine environments that preserve aquatic and terrestrial taxa. Marginal marine environments are also documented in the Eocene, together with fully marine environments around the Black Sea, but terrestrial environments are increasingly represented, especially with the fissure fills of Quercy [69]. Additionally, the *Konservat-Lagerstätten* of Messel (e.g., [70]), Geiseltal (e.g., [24, 71]), and Monte Bolca (e.g., [72]) are Eocene. Fissure fills are also abundant at Oligocene localities.

North America (Fig. 4.5) has also been intensely studied, beginning in the nineteenth century with Leidy, Cope, and Marsh. However, the North American record has not been critically revised, unlike the European one (e.g., [24, 73]). Some fully terrestrial Palaeocene localities are known from the Rocky Mountain interior, whereas others are known from the Gulf Coast and East Coast. The latter preserve abundant aquatic snakes. A similar distribution of localities is known from the Eocene. With the exception of the John Day Basin, Oligocene localities are restricted to the western edge of the Great Plains and marginal marine environments or fissure fills in peninsular Florida.

Finally, South America (Fig. 4.6) was for a long time unusual in that the Palaeocene record there seemed better known than that of the rest of the Palaeogene. However, newer geochronological studies indicate that most localities compiled here, including Itaboraí, pertain to the lower Eocene [74, 75] Prominent remaining Palaeocene localities are Cerrejón and Tiupampa. The Eocene is principally represented by terrestrial localities in

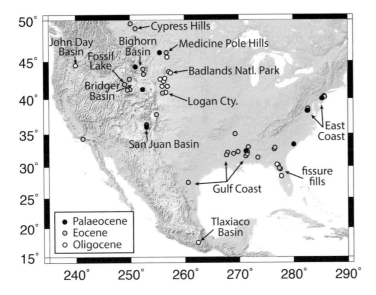

Figure 4.5 Palaeogene snake localities in North America. A single locality on Ellesmere Island, Canada, at 79 °N latitude is not shown. See Supplementary Appendix 4.S3 for colour version.

the San Jorge Basin. However, a marine site from the coast of Ecuador is also known. Thus far, snakes are known from only a single Oligocene locality in the San Jorge Basin. Insofar as South America was isolated throughout much of the Cenozoic [76], perhaps endemic snakes arose there, as among mammals [77].

4.3.2 Distribution of Taxa in Space and Time

Given the apparent age of their divergences (e.g., [32]), their present diversity, and their distribution in the subtropics to tropics on all continents except Antarctica, it is certain that the minor Palaeogene records of scolecophidians capture only very little of their history. Presently tropical squamates were widespread in middle latitudes in the early Palaeogene (e.g., [78]), so it seems unlikely that scolecophidians were not more widely distributed. Their small size doubtlessly contributes to their under-representation [8].

The present diversity of 'anilioids', fossorial non-caenophidian alethinophidians, is centred on the Indian subcontinent and Southeast Asia in the form of Uropeltidae, *Anomochilus* and *Cylindrophis*, whereas *Anilius* itself is found in the Neotropics. Curiously, no fossil representatives have been reported from the Palaeogene there. An adequate interpretation of 'anilioid' biogeography must await a final resolution of their relationships to other snakes as well as fossil discoveries. Thus far, Palaeogene fossils are known from North Africa, North and South America and possibly Europe (depending on the status of *Eoanilius*) and central and southwestern Asia.

Tropidophiidae represents a special problem. Among extant taxa, this name applies strictly to the South American and Caribbean genera *Tropidophis* and *Trachyboa*, but sometimes fossils are assigned to 'Tropidophiidae *sensu lato*' (i.e., including the booid

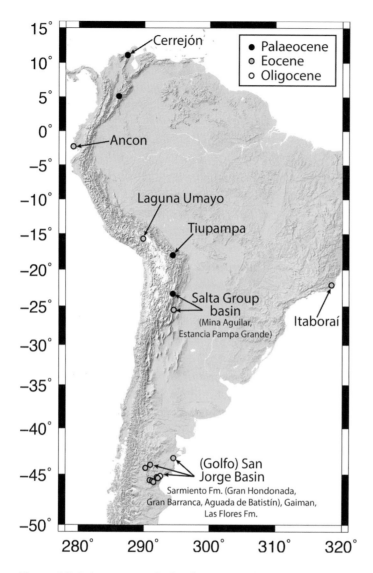

Figure 4.6 Palaeogene snake localities in South America. See Supplementary Appendix 4.S3 for colour version.

clade Ungaliophiinae). Insofar as 'Tropidophiidae *sensu lato*' is polyphyletic, discussion of its biogeography is not meaningful. The best case for the fossil occurrence of Tropidophiidae *sensu stricto* is in the Fayum [30]. From Europe and Anatolia (*Platyspondylia, Szyndlaria, Falseryx*) there are other plausible tropidophiids; the morphological evidence for these assignments is reviewed in detail in Supplementary Appendix 4. S1. This is an important problem, for if confirmed, it would represent a large range extension and call into question the assignment of numerous Constrictores *incertae sedis*, because there are no published vertebral synapomorphies of the latter group [79].

For Constrictores (sensu [79]), booids have the most detailed record. Several booid clades – Ungaliophiinae, Charinainae, and Neotropical boas – are represented by well-preserved specimens from the Eocene of Europe, and Neotropical boas are known also from the South American Palaeocene. This record has been interpreted as representing dispersal to Europe of several lineages from the New World in association with warm climates in the earliest Eocene, although the route of dispersal is uncertain in the absence of phylogenetically well-placed fossils from the Palaeogene of North America and Africa [41]. Other major lineages (*Candoia, Calabaria,* and Sanziniidae as well as Bolyeriidae and *Xenophidion*) presently lack a fossil record, so palaeontology contributes little to discussions of their origin [80].

'Erycines' have played a very prominent role in the fossil history of booids in Europe and North America (e.g., [20, 50]), culminating in Holman's [9] attribution of most smaller Palaeogene booids to this group. As reviewed by Smith [15], there is no published apomorphic evidence for erycines in North America prior to the Miocene. However, the discovery of a stem representative of Charinainae in Messel [81] is consistent with an early divergence of the group, and continued study of the North American Palaeogene with a focus on apomorphic traits will probably produce older relatives of North American 'erycines'. It is more difficult to interpret the significance of other Palaeogene 'erycines' from Europe. Although apomorphic evidence from caudal vertebrae strongly suggest an alliance of these with 'erycines', it is unclear to which of the two clades (Erycidae or Charinainae) these species pertain due to 'erycine' polyphyly (e.g., [82]).

The two extant non-pythonid pythonoid genera, *Xenopeltis* and *Loxocemus*, are Laurasian, and the stem of the latter is known from late Eocene fossils in North America. Pythonidae occurs today from Africa through southern Asia to Australia. However, there have been tantalising hints that fossil pythonids may have been present in the Eocene of Europe (Supplementary Appendix 4.S1). Scanlon [83] recognized these taxa could play a key role in interpreting pythonoid biogeography. A specimen from Messel, which had been illustrated and mentioned as a possible pythonid [84], was later described [85] as a stem pythonid, *Messelopython freyi*. Analyses in the latter study confirmed a Laurasian origin of Pythonoidea and Pythonidae, conditioned on the occurrence of *M. freyi* in Europe.

For many years J.-C. Rage's reports of unusual taxa in the early Palaeogene were the only hints of the origins of Caenophidia (e.g., [49]). Particularly fascinating have therefore been reports of new taxa, not all named, from Gondwana. These include: a diverse snake assemblage from the lower Eocene of India that contained taxa (*Thaumastophis, Procerophis*) suggestive of Caenophidia [26]; the potential stem caenophidian *Kataria* from the Palaeocene of Tiupampa [28]; and an assemblage from the Oligocene of Tanzania dominated by Colubriformes, unlike any known contemporaneous assemblage [29, 86]. These discoveries, coupled with the absence of caenophidians (except perhaps aquatic snakes such as Nigerophiidae) in the early Palaeogene of Laurasia, are consistent with their origin in Gondwana. Future discoveries there have great potential to reconcile the early Palaeogene divergence times (as inferred from molecular studies) with the fossil record.

4.4 The Ecology of Palaeogene Snakes

Isolated snake vertebrae are fairly common as fossils, whereas articulated material is rare [50]. The axial musculoskeletal system in snakes is more intimately associated with locomotion than in perhaps any other tetrapod group [87]. Thus, for most of the history of snake palaeontology, deductions based on vertebral morphology most commonly provided hints at the way of life of fossil taxa (e.g., [88]). Mosauer's [89] study of myology remains a landmark. Osteological studies include those of Johnson [90] and Schaal et al. [91]. Yet it has not yet been possible to fully disentangle phylogenetic from adaptive features in snake vertebrae, a subject with enormous potential importance to the interpretation of fossil snakes. Rare cranial elements, especially the quadrate bone, have occasionally provided insight as well (e.g., [16]). More recently, quantitative studies of body proportions [41], made possible by whole snake fossils, and of the inner ear ([92–94]; Chapter 13), made possible by μCT scans, have provided further insight.

Palaeogene snakes occupied a variety of terrestrial habitats. Fossoriality has been widely assumed for fossil 'anilioids' and supposed 'erycines' having reduced neural spines and depressed neural arches [9, 16]. This assumption may be correct in many cases, but study of the Messel snake assemblage, which preserves complete skeletons, invites caution. *Rageryx* has vertebral features considered typical for 'fossorial' snakes, yet was inferred to be ground-dwelling based on body proportions [41] and shows no evidence of fossorial adaptations in the skull, including in the inner ear [81]. Body proportions of *Messelophis variatus*, whose vertebrae resemble *Dunnophis* in their elongation, extremely reduced neural spines and depressed neural arches, approach those of arboreal snakes [41]. In Messel, *Eoconstrictor* was also determined to be ground dwelling, most likely a generalist [41]. The proportion of trunk to caudal vertebrae has been examined in a statistical fashion [15], but the method has not yet been applied outside the original locality, the Medicine Pole Hills.

Palaeophiidae, Nigerophiidae, and Russellophiidae are widely considered to be aquatic on the basis of derived features of the ribs and vertebrae. In *Archaeophis proavus*, according to Janensch [94], the following features indicate an aquatic mode of life (by comparison with extant aquatic snakes, especially Acrochordidae): lateral compression of the trunk, low position of the synapophyses, length and shape of the ribs (distinctly curved only proximally), and loss of gastrosteges. Janensch [94] even suggested that *A. proavus* might have had a ventral skin-fold in place of gastrosteges, like *Acrochordus*. Janensch [94] found the same rib form in *Anomalophis bolcensis*. Owen [88] and Marsh [95] similarly took *Palaeophis* to be aquatic. Houssaye et al. [96, 97] additionally documented osteosclerosis, an adaptation related to buoyancy in aquatic tetrapods, in *Palaeophis maghrebianus*, *Russellophis tenuis*, and *Nigerophis mirus*.

The extent to which these taxa were inhabitants of brackish water, near-shore environments or the open ocean remains unclear, but geological context brings insight. The general distribution of Palaeophiidae closely tracks past coastlines. Janensch [94] considered the limestone of Monte Bolca, preserving *Anomalophis bolcensis* and *Archaeophis proavus*, to be marine. *Pterosphenus* spp. have been described from mangrove or estuarine deposits [98] and from localities that were 10^2–10^3 km from shore [99]. *Palaeophis* is chiefly documented from marginal marine habitats but some records are probably further from the palaeoshoreline of the Neotethys (see [100]). It has been noted [101] that russellophiids are found solely in fluviatile or deltaic sediments, except for two localities in France (Saint-Maximin and Bretou), largely consistent with aquatic habitats.

Faunal associations were emphasized by Parmley and DeVore [102] for *Palaeophis* and *Pterosphenus*. In the Hardie Mine various vertebrate taxa (gars, trionychid turtles, crocodylians) are very commonly found, but also a high diversity of sharks, including forms that do not commonly enter brackish waters. Parmley and DeVore [102] reconstructed the areas as 'open marine, nearshore shallow-water habitat (similar to shallow inlet bays found today along the Gulf of Mexico coastline)'. Holman et al. [103] summarized evidence that palaeophiids co-occurred with terrestrial animals in the United States and concluded that species of *Palaeophis*, especially the smaller ones, were estuarine in habits but that *Pterosphenus* may have inhabited the open ocean. Rage et al. [18], reviewing the occurrences of *Pterosphenus*, suggested that it was a coastal dweller (marine, brackish, and freshwater near coast) and perhaps favoured mangroves.

A single palaeophiid species (*Palaeophis africanus*) obtained a very broad (transatlantic) distribution [102]. On the whole, the very broad distribution of Palaeophiidae was taken as an indication of true marine, not freshwater or mere brackish water, preferences [104].

The preponderance of aquatic snakes in the early Palaeogene is one of the most salient aspects of the Cenozoic snake fossil record. At least four major lineages (Palaeophiidae, Nigerophiidae, Russellophiidae, *Anomalophis*) are known; five, if Archaeophiinae is determined to be distinct from Palaeophiinae. The principal problem with interpretation is that their interrelationships are poorly understood. Three main hypotheses stand out. (1) If all are related to Acrochordidae, then this lineage possibly underwent an adaptive radiation in the early Palaeogene. If, on the other hand, Nigerophiidae, Russellophiidae, and *Anomalophis* are distributed around the base of Caenophidia, then either (2) snakes invaded marginal marine environments multiple times in the latest Cretaceous and the Palaeogene, or (3) the ancestor of Caenophidia was aquatic. Further discoveries clarifying the phylogenetic relations of these taxa will be crucial to evaluating these hypotheses.

In rare cases where complete specimens are known, data on trophic ecology can be gleaned from gut contents. Such specimens are generally confined to *Konservat-Lagerstätten*, most notably Messel [70]. There, the boid relative *Eoconstrictor* is known to have consumed crocodylians [105] and basilisk lizards [106]. Although these animals are ectothermic, *Eoconstrictor* is inferred to have possessed infrared-imaging pit organs in the upper jaw [41].

The evolution of body size has not been studied in detail in Palaeogene snakes. However, the early Palaeogene is notable for the very large size of some taxa. *Titanoboa cerrejonensis* is the largest known snake, living or extinct [23], and some palaeophiids and madtsoiids, together with *Chubutophis grandis* [25] and *Palaeopython cadurcensis* [24], also obtained extremely large sizes [96, 107–111]. It has been noted [112] that Madtsoiidae of the Cenozoic in South America are all larger than those of the Cretaceous. On the other end of the spectrum, some taxa scarcely exceeded 50 cm in total length [81, 113], and others, like *Eoanilius* and some scolecophidians, may have been smaller still. It has been suggested

[73] that the small size of *Eoanilius* and *Platyspondylia* possibly allowed them to survive the 'Grande Coupure' extinction event at the Eocene–Oligocene boundary.

4.5 The Problem of Vertebral Sufficiency

The widespread apprehension that fossil snake vertebrae are in many cases not sufficient for detailed systematic study (see Head et al., Chapter 3, for an alternative view) finds some support in the catalogue here. In particular, most constrictors and extinct snake taxa known solely from vertebrae (like Nigerophiidae) have poorly constrained relationships; the case for Archaeophiinae might change only when the skulls of *Archaeophis proavus* and *'Archaeophis' turkmenicus* are fully analysed. Yet the question of vertebral sufficiency involves two distinct issues. (1) Do snake vertebrae accurately record phylogenetic affinity? (2) Do snake vertebrae accurately record species diversity? We address each of these questions in a preliminary fashion below, recognizing that a more robust – but also far more difficult – approach would involve tests with extant species.

4.5.1 Do Snake Vertebrae Accurately Record Phylogenetic Affinity?

We operationalize this question by addressing the following: (1) Are taxa based on cranial or both cranial and vertebral material more precisely placed phylogenetically than taxa based solely on vertebral material? (2) Are vertebrae indirectly associated with phylogenetic precision?

Because of its high palaeodiversity, we focus here solely on Constrictores. If a species could be assigned only to Constrictores, its placement was considered 'imprecise'. If it could be assigned to a finer level, its placement was considered 'precise'. Note that we considered only direct information. For instance, *Bavarioboa* is assigned (precisely) to Booidea because of the ectopterygoid-pterygoid articulation of *B. crocheti*, but those elements are not known for other species. Thus, only one of the species is precisely assigned based on direct data. We counted the number of vertebral specimens and number of cranial elements in the hypodigm (or best-known population).

The results of this classification are shown in Figure 4.7A. Of the taxa considered, 72 per cent (=61/85) are known only from vertebrae. A few could be assigned precisely on the basis of possessing 'erycine' features in caudal vertebrae. Where cranial elements are known, the mean number of elements is 6.9 and the median 2.5. The stark difference arises because a few taxa are known from more or less complete skulls. Of the 85 species considered, vertebral counts could be estimated for 82. Most precisely identified taxa tend to be found in the upper right of the graph (many vertebrae and skull elements). Thus, we conducted nested stepwise multiple logistic regression in R v3.5 [114], the full model being: *Precision* ~ $N_{cranial} + \ln(N_{vert})$. We compared models using the Akaike Information Criterion (AIC).

The logistic regression of vertebral number (AIC = 71.69) and of cranial element number (AIC = 66.08) on precision both yielded individually significant, positive coefficients ($p = 0.00084$ and $p = 0.00109$, respectively), and both models were much better fitting than the null model in which neither variable plays a role (AIC = 82.94). However, the cranial-only model is clearly superior to the vertebral model by AIC. Furthermore, the full model with both

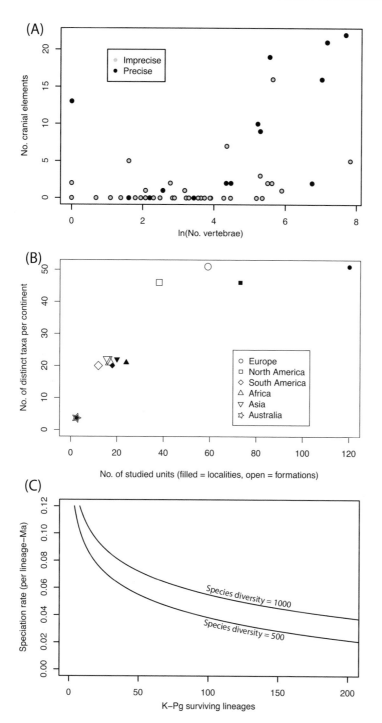

Figure 4.7 (A) Relationship between vertebral count, number of cranial elements, and precision of phylogenetic placement in Palaeogene Constrictores. Each point ($N = 85$) represents a species or unnamed local population. (B) Relationship between the number of formations and of localities and Palaeogene snake palaeodiversity in different geographic regions. In many cases, the formation of a locality is unknown, and these were listed as 'Unknown formation 1' etc. (see text). (C) Combinations of survivorship and speciation rate that would produce a cumulative total of 500 or 1000 snake species by the end of the Palaeogene.

explanatory variables (AIC = 66.06) was no better than the cranial-only model, and in the full model only the coefficient of cranial element number was statistically significant ($p = 0.015$).

We interpret these results as follows. Cranial elements contribute significantly to phylogenetic placement in Constrictores, whereas vertebrae do not. However, a large number of vertebrae is a strong predictor of the preservation of cranial elements. In fact, the only case in which substantial cranial material is present without vertebrae is the isolated skull comprising the holotype of *Crythiosaurus mongoliensis*. Assuming these results for Constrictores are representative of snakes as a whole, we can affirmatively answer the first question posed at the beginning of the section. In answer to the second question, we find that a large vertebral sample size is associated with a greater probability that cranial elements are preserved and so is indirectly associated with precision.

4.5.2 Do Snake Vertebrae Accurately Record Species Diversity?

We operationalize this question by addressing the following: (1) How does palaeodiversity relate to sampling intensity in different geographic regions? (2) Is extrapolated palaeodiversity of Palaeogene snakes biologically plausible?

Regarding the first question, Raup [115] first clearly showed that observed palaeodiversity covaries with sampling intensity of sedimentary rock in which fossils are found. The number of formations is often taken as an estimate of sampling intensity (e.g., [116]). Insofar as a geological formation is a lithologic unit with a particular character deposited at a particular place and time, it typically represents a particular broad-scale environment. Although a formation may include many facies, which are more closely tied to habitat, formations are easier to count than facies. Hence, it is thought, the larger the number of formations sampled, the more complete the habitat sampling, and – one expects – the larger the number of possible species. Secular variation in rock preservation can thus lead to artifactual fluctuations in palaeodiversity, so-called 'megabiases' [117].

We evaluated two covariates of palaeodiversity: number of localities and number of formations. The former values are quite firm in our data set, the latter squishy. The formation is frequently unknown in Europe, and especially for fissure-fills, which are quite common in our data set. Where formation name is unclear, we counted 'unknown formation 1' etc. In some cases, as in the Paris and North Sea Basins, this may overestimate the number of distinct formations. As expected, the number of localities is always greater than the number of formations, because each formation is represented by at least one locality.

Despite the foregoing caveats, the relationship between palaeodiversity and both number of formations and number of localities is qualitatively similar: a convex-up curve (Fig. 4.7B). Europe, followed by North America, are most intensively studied, Australia the least (linked to its meagre Palaeogene outcrop). Remarkably, Africa, Asia, and South America plot very close together. Viewing geographic areas (continents) as samples of an equivalent global fauna (cf. [118]), the number of taxa rises with increased sampling, but more slowly. There are many convex upward or monotonically increasing functions, some asymptotic, some not, so it is not very meaningful to try to fit curves to these data. Nevertheless, they suggest that palaeodiversity flattens at more intense sampling, as expected.

The second subquestion in this section concerns unknown true Palaeogene species diversity, which we approached as follows. We first attempted to extrapolate regional and global palaeodiversity for the Palaeogene based on our observed samples of that diversity. We then compared the order of magnitude of this extrapolated palaeodiversity with lineage-through-time plots and with theoretical values computed with plausible speciation rates.

We used two biodiversity-theoretic methods to estimate on the basis of the reference sample – a taxon-by-locality occurrence table – for each geographic area the *minimum* palaeodiversity of that area; we also did this for the whole world. The first is the Chao2 method, which takes account of the taxa present only in one or two localities [119]. The second is the more generalized incidence-based coverage estimator (ICE), which takes account of *all* rare taxa, where 'rare' is some arbitrary, user-defined cut-off, κ [119]. We followed the recommended default $\kappa = 10$ but note that estimated minimum diversity, S_{ICE}, was effectively stable at $\kappa > 2$. All calculations were performed in EstimateS v9.0 [120].

Results are shown in Table 4.2. Unlike for raw data (Fig. 4.7B), minimum palaeodiversity is highest in South America, followed by Europe. Africa, Asia, and North America were fairly similar, and Australia clearly last. In some cases (Australia and South America, possibly Europe), S_{ICE} has not reached a plateau, implying that it could rise with the discovery of new localities. World minimum palaeodiversity S_{Chao2} was 573.5 species (95 per cent C.I. 390.8–903.5), and world S_{ICE} was 558.8 species. Note that these estimators are not distributive: for instance, S_{Chao2} (A+B) $\neq$ S_{Chao2} (A) + S_{Chao2} (B).

Table 4.2. Minimum palaeodiversity of Palaeogene snakes in six geographic areas and the world as a whole.

N_{loc} is the number of localities, S_{obs} the observed species diversity, S_{Chao2} the Chao2 estimate of minimum species diversity (with 95 per cent confidence interval), and S_{ICE} the ICE estimate of minimum species diversity. For Chao2, the 'classic' rather than 'bias-corrected' formula was used following recommendation in program documentation due to the coefficient of variation. Some valid taxa were excluded from the incidence-based table because they come from imprecisely known localities: 'Palaeopython' neglectus and Plesiotortrix edwardsi. All precisely identified species were used, whereas all taxa with qualifiers like 'cf.', 'aff.' and so forth were excluded as formally unidentified. Thus, where a locality did not have precisely identified taxa, it was not included. Overall, 162 localities were thus included.

Geographic region	$N_{loc.}$	$S_{obs.}$	S_{Chao2}	95% C.I. (Chao2)	S_{ICE}
Africa	12	21	69.2	33.7–203.9	69.2
Asia	10	22	82.2	38.3–244.4	91.7
Australia	2	4	7.0	4.5–23.1	7.0
Europe	79	51	191.2	92.2–528.4	105.0
North America	52	46	98.6	65.8–185.6	109.3
South America	7	20	210.0	100.6–467.6	182.9
World	162	164	573.5	390.8–903.5	558.8

These estimates are one order of magnitude lower than extant snake species diversity, which is 3,879 species according to the Reptile Database ([121], accessed 7 May 2021). However, true diversity in the Palaeogene must have been different. First of all, it was much earlier in the radiation of snakes, and especially Colubriformes (which comprise most of modern species diversity, >3,150 species), so that standing diversity would have been much lower. Moreover, the 'scolecophidian' lineages, with about one-sixth of modern species diversity, can be discounted, because they are thought largely to be indistinguishable on the basis of vertebrae (e.g., [8]). It is the non-caenophidian alethinophidians ('Henophidia'), with $c.$ 10^2 species today, that dominate all Palaeogene assemblages in the New World and Europe.

What is a plausible value for the cumulative species diversity of snakes in the Palaeogene? Figure 4.7C describes how cumulative diversity is influenced by two variables: instantaneous speciation rate (lineage-Ma^{-1}) and the number of lineages surviving the K–Pg extinction. (We are not concerned with extinction or net diversification here.) There is a swath bounded by cumulative diversity of 500 and 1,000 species corresponding, in part, to plausible values of these two variables. Assuming standing diversity of early-branching alethinophidians at the K–Pg boundary was 100 species, there would be 17 surviving lineages given an 83 per cent extinction rate, as for squamates in North America [4]. Accordingly, speciation rates between about 0.08 and 0.11 would yield 500–1,000 species. Comparable speciation rates are found in the angiosperm clade Fagales, the crown of which originated in the Early Cretaceous and evolved into 1,370 species today but lacks a hyperdiverse subclade like Colubriformes [122]. In sum, it is plausible that true cumulative species diversity is of the same order of magnitude as extrapolated (minimum) palaeodiversity.

If many species cannot be distinguished based on vertebrae, and snake vertebrae are by far the most common fossils, then one could conclude that snake palaeodiversity will always be lower than true species diversity for the same time slice. However, the rough calculations above (extrapolations of palaeodiversity and computations of cumulative lineage diversity) suggest that the two values may not diverge as much as commonly conceived. Indeed, they raise the prospect that – with sufficient sample size and attention to intracolumnar variation – palaeontological morphospecies may not diverge so very greatly from biological species [17]. Integrative taxonomic studies incorporating DNA data (e.g., [123]) will provide more careful assessments of individual clades in the future. Additionally, the possibility must be entertained that other snake clades besides Colubriformes – for example, Constrictores – underwent a more rapid diversification in the Palaeogene than their present diversity would suggest.

4.6 Conclusions

The idea that a single mid-trunk vertebra can generally be used to identify the species of its possessor as well as establish its phylogenetic relationships is patently untenable. Taxa based solely on vertebrae are generally insufficient for precise phylogenetic placement, but given a large sample size and careful attention to variation, they may more closely approximate biological species than commonly believed. We see grounds for optimism:

1. Even if snake palaeodiversity was generally lower than true biological diversity, this would only limit comparisons with extant species diversity. It would not suggest that comparison *among* fossil time horizons does not yield meaningful conclusions, nor would it suggest limits on the study of disparity through time.

2. We believe that the palaeodiversity of snake taxa based solely on vertebrae will more closely approach true species diversity when sample size is large, so that all aspects of variation can be considered.

3. Many species based principally on vertebrae could be placed phylogenetically when cranial elements could be associated with them. Cranial elements are much rarer than vertebrae in a snake skeleton, but high vertebral sample size predicts the occurrence of cranial elements. It is incumbent on snake specialists to go themselves through screen-washed concentrate in order to recognize these elements.

There is the well-known story of the gambler who evening after evening enters a saloon to play cards and loses every time. Taking pity on him, a regular pulls him aside to ask: 'Don't you know the game is crooked?' 'I know', he responds, 'but it's the only game in town'. To be cognisant of the limitations of snake vertebrae allows us to adjust our expectations and make the most of the snake fossil record.

Acknowledgements

We thank the editors, Dave Gower and Hussam Zaher, for the opportunity to contribute to this volume and the Linnean Society and Systematics Association for underwriting the symposium on which it is based. Hussam Zaher and Jason Head kindly permitted us to include in-press work in this review. We thank Agustín Scanferla for insight into South American geology. The manuscript benefited from the close reading and constructive criticism of two anonymous reviewers.

References

1. J.-C. Rage and G. Wouters, Découverte du plus ancien Palaeopheidé (Reptilia, Serpentes) dans le Maestrichtien du Maroc. *Geobios*, 12 (1979), 293–296.

2. J.-C. Rage and C. Werner, Mid-Cretaceous (Cenomanian) snakes from Wadi abu Hashim, Sudan: The earliest snake assemblage. *Palaeontologia Africana*, 35 (1999), 85–110.

3. J. J. Head, K. Mahlow, and J. Müller, Fossil calibration dates for molecular phylogenetic analysis of snakes 2: Caenophidia, Colubroidea, Elapoidea, Colubridae. *Palaeontologia Electronica*, 19.2.2FC (2016), 1–21.

4. N. R. Longrich, B. -A. S. Bhullar, and J. A. Gauthier, Mass extinction of lizards and snakes at the Cretaceous-Paleogene boundary. *Proceedings of the National Academy of Sciences, USA*, 109 (2012), 21396–21401.

5. A. Y. Hsiang, D. J. Field, T. H. Webster, et al., The origin of snakes: revealing the ecology, behavior, and evolutionary history of early snakes using genomics, phenomics, and the fossil record.

BMC Evolutionary Biology, 15 (2015), 87.

6. S. M. Harrington and T. W. Reeder, Phylogenetic inference and divergence dating of snakes using molecules, morphology and fossils: new insights into convergent evolution of feeding morphology and limb reduction. *Biological Journal of the Linnean Society*, 121 (2017), 379–394.

7. H. Zaher, R. W. Murphy, et al., Large-scale molecular phylogeny, morphology, divergence-time estimation, and the fossil record of advanced caenophidian snakes (Squamata: Serpentes). *PloS ONE*, 14 (2019), e0216148.

8. J.-C. Rage, *Serpentes (Handbuch der Paläoherpetologie, v. 11)* (Stuttgart: Gustav Fischer Verlag, 1984).

9. J. A. Holman, *Fossil Snakes of North America: Origin, Evolution, Distribution, Paleoecology* (Bloomington, Indiana: Indiana University Press, 2000).

10. G. Cuvier, *Recherches sur les Ossemens Fossiles, oú l'on Rétablit les Caractères de Plusierus Animaux dont les Révolutions du Globe Ont Détruit les Espèces*. Vol. 4 (Paris: E. d'Ocagne: 1823).

11. S. B. McDowell, A catalogue of the snakes of New Guinea and the Solomons, with special reference to those in the Bernice P. Bishop Museum. Part III. Boinae and Acrochordoidea (Reptilia, Serpentes). *Journal of Herpetology*, 13 (1979), 1–92.

12. A. G. Kluge, *Calabaria* and the phylogeny of erycine snakes. *Zoological Journal of the Linnean Society*, 107 (1993), 293–351.

13. J. J. Head, Phylogenetic significance of vertebral morphology in snakes: Implications for interpreting the fossil record. *Journal of Vertebrate Paleontology*, 22 (2002), 63A.

14. C. J. Bell, J. J. Head, and J. I. Mead, Synopsis of the herpetofauna from Porcupine Cave. In A. D. Barnosky, ed., *Biodiversity Response to Climate Change in the Middle Pleistocene: The Porcupine Cave Fauna from Colorado* (Berkeley, California:

University of California Press, 2004), pp. 117–126.

15. K. T. Smith, New constraints on the evolution of the snake clades Ungaliophiinae, Loxocemidae and Colubridae (Serpentes), with comments on the fossil history of erycine boids in North America. *Zoologischer Anzeiger*, 252 (2013), 157–182.

16. J. -C. Rage, Les serpents des phosphorites du Quercy. *Palaeovertebrata*, 6 (1974), 274–303.

17. Z. Szyndlar, Fossil snakes from Poland. *Acta Zoologica Cracoviensia*, 28 (1984), 1–156.

18. J.-C. Rage, S. Bajpai, J. G. M. Thewissen, and B. N. Tiwari, Early Eocene snakes from Kutch, western India, with a review of the Palaeophiidae. *Geodiversitas*, 25 (2003), 695–716.

19. J. D. Scanlon, Australia's oldest known snakes: *Patagoniophis, Alamitophis*, and cf. *Madtsoia* (Squamata: Madtsoiidae) from the Eocene of Queensland. *Memoirs of the Queensland Museum*, 51 (2005), 215–235.

20. R. Hoffstetter and J. -C. Rage, Les *Erycinæ* fossiles de France *(Serpentes, Boidæ)*: compréhension et histoire de la sous-famille. *Annales de Paléontologie (Vertébrés)*, 58 (1972), 82–124.

21. Z. Szyndlar and W. Böhme, Redescription of *Tropidonotus atavus* von Meyer, 1855 from the upper Oligocene of Rott (Germany) and its allocation to *Rottophis* gen. nov (Serpentes, Boidae). *Palaeontographica A*, 240 (1996), 145–161.

22. W. Auffenberg, The fossil snakes of Florida. *Tulane Studies in Zoology*, 10 (1963), 127–213.

23. J. J. Head, J. I. Bloch, A. K. Hastings, et al., Giant boid snake from the Palaeocene neotropics reveals hotter past equatorial temperatures. *Nature*, 457 (2009), 715–717.

24. G. L. Georgalis, M. Rabi, and K. T. Smith, Taxonomic revision of the snakes of the genera *Palaeopython* and *Paleryx* (Serpentes, Constrictores) from the Paleogene of Europe. *Swiss Journal of Palaeontology*, 140 (2021), 18.

25. A. M. Albino, Snakes from the Paleocene and Eocene of Patagonia (Argentina): paleoecology and coevolution with mammals. *Historical Biology*, 7 (1993), 51–69.

26. J.-C. Rage, A. Folie, R. S. Rana, et al., A diverse snake fauna from the early Eocene of Vastan Lignite Mine, Gujarat, India. *Acta Palaeontologica Polonica*, 53 (2008), 391–403.

27. J.-C. Rage, M. Pickford, and B. Senut, Amphibians and squamates from the middle Eocene of Namibia, with comments on pre-Miocene anurans from Africa. *Annales de Paléontologie*, 99 (2013), 217–242.

28. A. Scanferla, H. Zaher, F. E. Novas, C. de Muizon, and R. Céspedes, A new snake skull from the Paleocene of Bolivia sheds light on the evolution of macrostomatans. *PLoS ONE*, 8 (2013), e57583.

29. J. A. McCartney, N. J. Stevens, and P. M. O'Connor, The earliest colubroid-dominated snake fauna from Africa: perspectives from the late Oligocene Nsungwe Formation of southwestern Tanzania. *PLoS ONE*, 9 (2014), e90415.

30. J. A. McCartney and E. R. Seiffert, A late Eocene snake fauna from the Fayum Depression, Egypt. *Journal of Vertebrate Paleontology*, 36 (2016), e1029580.

31. R. Hoffstetter, Squamates de type moderne. In J. Piveteau, ed., *Traité de Paléontologie*, Vol 5 (Paris: Masson, 1955), pp. 605–662.

32. N. Vidal, J.-C. Rage, A. Couloux, and S. B. Hedges, Snakes (Serpentes). In S. B. Hedges and K. Kumar, eds., *The Timetree of Life* (Oxford: Oxford University Press, 2009), pp. 390–397.

33. F. T. Burbrink and B. I. Crother, Evolution and taxonomy of snakes. In R. D. Aldridge and D. M. Sever, eds., *Reproductive Biology and Phylogeny of Snakes* (Boca Raton, Florida: CRC Press, 2011), pp. 19–53.

34. F. T. Burbrink, F. G. Grazziotin, R. A. Pyron, et al., Interrogating genomic-scale data for Squamata (lizards, snakes, and amphisbaenians) shows no support for key traditional morphological relationships. *Systematic Biology*, 69 (2020), 502–520.

35. M. S. Y. Lee and J. D. Scanlon, Snake phylogeny based on osteology, soft anatomy and ecology. *Biological Reviews*, 77 (2002), 333–401.

36. J. D. Scanlon, Skull of the large non-macrostomatan snake *Yurlunggur* from the Australian Oligo-Miocene. *Nature*, 439 (2006), 839–842.

37. J. L. Conrad, Phylogeny and systematics of Squamata (Reptilia) based on morphology. *Bulletin of the American Museum of Natural History*, 310 (2008), 1–182.

38. M. W. Caldwell, R. L. Nydam, A. Palci, and S. Apesteguía, The oldest known snakes from the Middle Jurassic-Lower Cretaceous provide insights on snake evolution. *Nature Communications*, 6 (2015), 5996.

39. J. A. Wilson, D. M. Mohabey, S. E. Peters, and J. J. Head, Predation upon hatchling dinosaurs by a new snake from the Late Cretaceous of India. *PLoS Biology*, 8 (2010), e100322.

40. H. Zaher and C. A. Scanferla, The skull of the Upper Cretaceous snake *Dinilysia patagonica* Smith-Woodward, 1901, and its phylogenetic position revisited. *Zoological Journal of the Linnean Society*, 164 (2012), 194–238.

41. A. Scanferla and K. T. Smith, Exquisitely preserved fossil snakes of Messel: insight into the evolution, biogeography, habitat preferences and sensory ecology of early boas. *Diversity*, 12 (2020), 100.

42. G. L. Georgalis, L. Del Favero, and M. Delfino, Italy's largest snake - redescription of *Palaeophis oweni* from the Eocene of Monte Duello, near Verona. *Acta Palaeontologica Polonica*, 65 (2020), 523–533.

43. L. A. Nessov, Paleogene sea snakes as indicators of water mass peculiarities on the east of Tethys Ocean [in Russian]. *Vestnik St. Petersberg University, Series* 7, 2 (1995), 3–9.

44. E. A. Zvonok and P. B. Snetkov, New findings of snakes of the genus *Palaeophis*

Owen, 1841 (Acrochordoidea: Palaeophiidae) from the middle Eocene of Crimea. *Proceedings of the Zoological Institute of the Russian Academy of Sciences*, 316 (2012), 392–400.

45. V. Wallach, K. L. Williams, and J. Boundy, *Snakes of the World: A Catalogue of Living and Extinct Species* (Boca Raton, London and New York: CRC Press, 2014).

46. F. B. Nopcsa, Die Familien der Reptilien. *Fortschritte der Geologie und Paläontologie*, 2 (1923), 1–210.

47. L. P. Tatarinov, The cranial structure of the lower Eocene sea snake *'Archaeophis' turkmenicus* from Turkmenia. *Paleontological Journal*, 22 (1988), 73–79.

48. T. C. LaDuke, D. W. Krause, J. D. Scanlon, and N. J. Kley, A Late Cretaceous (Maastrichtian) snake assemblage from the Maevarano Formation, Mahajanga basin, Madagascar. *Journal of Vertebrate Paleontology*, 30 (2010), 109–138.

49. J.-C. Rage, L'origine des Colubroïdes et des Acrochordoïdes (Reptilia, Serpentes). *Comptes Rendus de l'Académie des Sciences, D*, 286 (1978), 595–597.

50. J.-C. Rage, Fossil history. In R. A. Seigel, J. T. Collins, and S. S. Novak, eds., *Snakes: Ecology and Evolutionary Biology* (New York: Macmillan, 1987), pp. 51–76.

51. J. A. Holman and G. R. Case, A puzzling new snake (Reptilia: Serpentes) from the late Paleocene of Mississippi. *Annals of the Carnegie Museum*, 61 (1992), 197–205.

52. J.-C. Rage, Un caenophidien primitif (Reptilia, Serpentes) dans l'Éocène inferieur. *Comptes Rendu Sommaire des Séances de la Société Géologique de France*, 2 (1975), 46–47.

53. P. D. Cantino and K. de Queiroz, International Code of Phylogenetic Nomenclature, v. 4c (2010). Downloaded from: https://www.ohio.edu/phylocode/PhyloCode4c.pdf.

54. S. Apesteguía and H. Zaher, A Cretaceous terrestrial snake with robust hindlimbs and a sacrum. *Nature*, 440 (2006), 1037–1040.

55. J. J. Head, Fossil calibration dates for molecular phylogenetic analysis of snakes 1: Serpentes, Alethinophidia, Boidae, Pythonidae. *Palaeontologia Electronica*, 18.1.6FC (2015), 1–17.

56. M. A. O'Leary, M. L. Bouaré, K. M. Claeson, et al., Stratigraphy and paleobiology of the Upper Cretaceous–lower Paleogene sediments from the Trans-Saharan Seaway in Mali. *Bulletin of the American Museum of Natural History*, 436 (2019), 1–177.

57. Y. Q. Wang, J. Meng, C. K. Beard, et al., Early Paleogene stratigraphic sequences, mammalian evolution and its response to environmental changes in Erlian Basin, Inner Mongolia, China. *Science China Earth Sciences*, 53 (2010), 1918–1926.

58. L.-P. Dong, S. E. Evans, and Y. Wang, Taxonomic revision of lizards from the Paleocene deposits of the Qianshan Basin, Anhui, China. *Vertebrata PalAsiatica*, 54 (2016), 243–268.

59. R. Estes, *Sauria Terrestria, Amphisbaenia (Handbuch der Paläoherpetologie, v. 10A)* (Stuttgart: Gustav Fischer Verlag, 1983).

60. G. L. Georgalis, *Necrosaurus* or *Palaeovaranus*? Appropriate nomenclature and taxonomic content of an enigmatic fossil lizard clade (Squamata). *Annales de Paléontologie*, 103 (2017), 293–303.

61. G. L. Georgalis and T. M. Scheyer, A new species of *Palaeopython* (Serpentes) and other extinct squamates from the Eocene of Dielsdorf (Zurich, Switzerland). *Swiss Journal of Geosciences*, 112 (2019), 383–417.

62. M. Archer, H. Godthelp, S. Hand, and D. Megirian, Fossil mammals of Riversleigh, northwestern Queensland: preliminary overview of biostratigraphy, correlation and environmental change. *Australian Zoologist*, 25 (1989), 29–65.

63. J. D. Scanlon, M. S. Y. Lee, and M. Archer, Mid-Tertiary elapid snakes (Squamata, Colubroidea) from Riversleigh, northern Australia: early steps in a continent-wide adaptive radiation. *Geobios*, 36 (2003), 573–601.

64. J. D. Scanlon and M. S. Y. Lee, The Pleistocene serpent *Wonambi* and the early evolution of snakes. *Nature*, 403 (2000), 416–420.

65. J. Woodhead, S. J. Hand, M. Archer, et al., Developing a radiometrically-dated chronologic sequence for Neogene biotic change in Australia, from the Riversleigh World Heritage Area of Queensland. *Gondwana Research*, 29 (2016), 153–167.

66. D. Megirian, P. Murray, L. Schwartz, and C. von der Borch, Late Oligocene Kangaroo Well Local Fauna from the Ulta Limestone (new name), and climate of the Miocene oscillation across central Australia. *Australian Journal of Earth Sciences*, 51 (2004), 701–741.

67. M. Archer, D. A. Arena, M. Bassarova, et al., Current status of species-level representation in faunas from selected fossil localities in the Riversleigh World Heritage Area, northwestern Queensland. *Alcheringa Supplement*, 1 (2006), 1–17.

68. M. Reguero, F. Goin, C. Acosta Hospitaleche, T. Dutra, and S. Marenssi, *Late Cretaceous/Paleogene West Antarctica Terrestrial Biota and Its Intercontinental Affinities* (Dordrecht: Springer, 2013).

69. S. Legendre, B. Sigé, G. Astruc, et al., Les phosphorites du Quercy: 30 ans de recherche. Bilan et perspectives. *Geobios*, 30, supplement 1(1997), 331–345.

70. K. T. Smith, S. F. K. Schaal, and J. Habersetzer, eds., *Messel: An Ancient Greenhouse Ecosystem* (Stuttgart: Schweizerbart, 2018).

71. G. Krumbiegel, H. Haubold, and L. Rüffle, *Das eozäne Geiseltal : ein mitteleuropäisches Braunkohlenvorkommen und seine Pflanzen- und Tierwelt* (Wittenberg: Ziemsen, 1983).

72. M. Friedman and G. Carnevale, The Bolca Lagerstätten: shallow marine life in the Eocene. *Journal of the Geological Society, London*, 175 (2018), 569–579.

73. Z. Szyndlar and J.-C. Rage, *Non-erycine Booidea from the Oligocene and Miocene of Europe* (Krakow: Polish Academy of Sciences, 2003).

74. M. O. Woodburne, F. J. Goin, M. Bond, et al., Paleogene land mammal faunas of South America; a response to global climatic changes and indigenous floral diversity. *Journal of Mammalian Evolution*, 21 (2014), 1–73.

75. M. O. Woodburne, F. J. Goin, M. S. Raigemborn, et al., Revised timing of the South American early Paleogene land mammal ages. *Journal of South American Earth Sciences*, 54 (2014), 109–119.

76. R. Estes and A. Báez, Herpetofaunas of North and South America during the Late Cretaceous and Cenozoic: Evidence for interchange? In F. G. Stehli and S. D. Webb, eds., The Great American Interchange. Topics in Geobiology, *Vol. 4* (New York: Plenum Press, 1985), pp. 139–197.

77. G. G. Simpson, *Splendid Isolation: The Curious History of South American Mammals* (New Haven, Conn.: Yale University Press, 1980).

78. K. T. Smith, A new lizard assemblage from the earliest Eocene (zone Wa0) of the Bighorn Basin, Wyoming, USA: Biogeography during the warmest interval of the Cenozoic. *Journal of Systematic Palaeontology*, 7 (2009), 299–358.

79. G. L. Georgalis and K. T. Smith, Constrictores Oppel, 1811 - the available name for the taxonomic group uniting boas and pythons. *Vertebrate Zoology*, 70 (2020), 291–304.

80. B. P. Noonan and P. T. Chippindale, Dispersal and vicariance: The complex evolutionary history of boid snakes. *Molecular Phylogenetics and Evolution*, 40 (2006), 347–358.

81. K. T. Smith and A. Scanferla, A nearly complete skeleton of the oldest definitive erycine boid (Messel, Germany). *Geodiversitas*, 43 (2021), 1–24.

82. T. P. Wilcox, D. J. Zwickl, T. A. Heath, and D. M. Hillis, Phylogenetic relationships of the dwarf boas and a comparison of Bayesian and bootstrap measures of

phylogenetic support. *Molecular Phylogenetics and Evolution*, 25 (2002), 361–371.

83. J. D. Scanlon, *Montypythonoides*: the Miocene snake *Morelia riversleighensis* (Smith and Plane, 1985) and the geographical origin of pythons. *Memoirs of the Association of Australasian Palaeontologists*, 25 (2001), 1–35.

84. Z. Szyndlar and W. Böhme, Die fossilen Schlangen Deutschlands: Geschichte der Faunen und ihrer Erforschung. *Mertensiella*, 3 (1993), 381–431.

85. H. Zaher and K. T. Smith, Pythons in the Eocene of Europe reveal a much older divergence of the group in sympatry with boas. *Biology Letters*, 16 (2020), 20200735.

86. J. A. McCartney, S. N. Bouchard, J. A. Reinhardt, et al., The oldest lamprophiid (Serpentes, Caenophidia) fossil from the late Oligocene Rukwa Rift Basin, Tanzania and the origins of African snake diversity. *Geobios*, 66–67 (2021), 67–75.

87. J.-P. Gasc, Snake vertebrae – a mechanism or merely a taxonomist's toy? In A. d. A. Bellairs and C. B. Cox, eds., *Morphology and Biology of Reptiles. Linnean Society Symposium Series* Number 3 (London: Academic Press, 1976), pp. 177–190.

88. R. Owen, *Monograph on the Fossil Reptilia of the London Clay*. Part II. Crocodilia, Ophidia (London: The Palaeontographical Society, 1850).

89. W. Mosauer, The myology of the trunk region of snakes and its significance for ophidian taxonomy and phylogeny. *Publications of the University of California at Los Angeles in Biological Sciences*, 1 (1935), 81–120.

90. R. G. Johnson, The adaptive and phylogenetic significance of vertebral form in snakes. *Evolution*, 9 (1955), 367–388.

91. S. Schaal, S. Baszio, and J. Habersetzer, Differenzierung von Schlangenarten anhand qualitativer und quantitativer Merkmale sowie konventioneller Streckenmaße und Indizes. *Courier Forschungsinstitut Senckenberg*, 255 (2005), 133–169.

92. H. Yi and M. A. Norell, The burrowing origin of modern snakes. *Science Advances*, 1 (2015), e1500743.

93. A. Palci, M. N. Hutchinson, M. W. Caldwell, and M. S. Y. Lee, The morphology of the inner ear of squamate reptiles and its bearing on the origin of snakes. *Royal Society Open Science*, 4 (2017), 170685.

94. W. Janensch, Über *Archaeophis proavus* Mass., eine Schlange aus dem Eocän des Monte Bolca. *Beiträge zur Paläontologie und Geologie Österreich-Ungarns und des Orients*, 19 (1906), 1–33.

95. O. C. Marsh, *Introduction and Succession of Vertebrate Life in America* (New Haven, Connecticut: Unknown, 1877).

96. A. Houssaye, J.-C. Rage, N. Bardet, et al., New highlights about the enigmatic marine snake *Palaeophis maghrebianus* (Palaeophiidae; Palaeophiinae) from the Ypresian (Lower Eocene) phosphates of Morocco. *Palaeontology*, 56 (2013), 647–61.

97. A. Houssaye, A. Herrel, R. Boistel, and J.-C. Rage, Adaptation of the vertebral inner structure to an aquatic life in snakes: Pachyophiid peculiarities in comparison to extant and extinct forms. *Comptes Rendus Palevol*, 18 (2019), 783–799.

98. J. W. Westgate, Paleoecology and biostratigraphy of marginal marine Gulf Coast Eocene Vertebrate localities. In G. F. Gunnell, ed., *Eocene Biodiversity: Unusual Occurrences and Rarely Sampled Habitats* (New York: Kluwer Academic, 2001), pp. 263–297.

99. J. H. Hutchison, *Pterosphenus* cf. *P. schucherti* Lucas (Squamata, Palaeophidae) from the late Eocene of peninsular Florida. *Journal of Vertebrate Paleontology*, 5 (1985), 20–23.

100. A. O. Averianov, Paleogene sea snakes from the eastern part of Tethys. *Russian Journal of Herpetology*, 4 (1997), 128–142.

101. S. Duffaud and J.-C. Rage, Les remplissages karstiques polyphasés (Éocène, Oligocène,

Pliocène) de Saint-Maximin (Phosphorites du Gard) et leur apport à la connaissance des faunes européennes, notamment pour l'Éocène moyen (MP 13). 2.– Systématique: amphibiens et reptiles. In J. -P. Aguilar, S. Legendre, and J. Michaux, eds., *Actes du Congrès BiochroM'97* (Mémoire Travaux EPHE 21) (Montpellier: Institut Montpellier, 1997), pp. 729–35.

102. D. Parmley and M. DeVore, Palaeopheid snakes from the late Eocene Hardie Mine local fauna of central Georgia. *Southeastern Naturalist*, 4 (2005), 703–722.

103. J. A. Holman, D. T. Dockery, III, and G. R. Case, Paleogene snakes of Mississippi. *Mississippi Geology*, 11 (1990 [1991]), 1–12.

104. W. Janensch, *Pterosphenus Schweinfurthi* Andrews und die Entwicklung der Palaeophiden. *Archiv für Biontologie*, 1 (1906), 311–350.

105. H. W. Greene, Dietary correlates of the origin and radiation of snakes. *American Zoologist*, 23 (1983), 431–441.

106. K. T. Smith and A. Scanferla, Fossil snake preserving three trophic levels and evidence for an ontogenetic dietary shift. *Palaeobiodiversity and Palaeoenvironments*, 96 (2016), 589–599.

107. O. C. Marsh, Description of a new and gigantic fossil Serpent (*Dinophis grandis*) from the Tertiary of New Jersey. *American Journal of Science, Series* 2, 48 (1869), 397–400.

108. C. W. Andrews, *A Descriptive Catalogue of the Tertiary Vertebrata of the Fayûm, Egypt* (London: Longmans, 1906).

109. J.-C. Rage, *Palaeophis colossaeus* nov. sp. (le plus grand Serpent connu?) de l'Eocene du Mali et le probleme du genre chez les Palaeopheinae. *Comptes Rendus de l'Académie des Sciences (Série II)*, 296 (1983), 1741–1744.

110. J. P. Rio and P. D. Mannion, The osteology of the giant snake *Gigantophis garstini* from the upper Eocene of North Africa and its bearing on the phylogenetic relationships and biogeography of Madtsoiidae. *Journal of Vertebrate Paleontology*, 37 (2017), e1347179.

111. J. A. McCartney, E. M. Roberts, L. Tapanila, and M. A. O'Leary, Large palaeophiid and nigerophiid snakes from Paleogene Trans-Saharan Seaway deposits of Mali. *Acta Palaeontologica Polonica*, 63 (2018), 207–220.

112. J.-C. Rage, Fossil snakes from the Palaeocene of São José de Itaboraí, Brazil. Part III. Ungaliophiinae, booids *incertae sedis*, and Caenophidia. Summary, update, and discussion of the snake fauna from the locality. *Palaeovertebrata*, 36 (2008), 37–73.

113. S. Schaal and S. Baszio, *Messelophis ermannorum* n. sp., eine neue Zwergboa (Serpentes: Boidae: Tropidopheinae) aus dem Mittel-Eozän von Messel. *Courier Forschungsinstitut Senckenberg*, 252 (2004), 67–77.

114. R Core Team. *R: A Language and Environment for Statistical Computing* (Vienna: R Foundation for Statistical Computing, 2016).

115. D. M. Raup, Taxonomic diversity during the Phanerozoic. *Science*, 177 (1972), 1065–1071.

116. R. B. J. Benson, R. J. Butler, J. Lindgren, and A. S. Smith, Mesozoic marine tetrapod diversity: mass extinctions and temporal heterogeneity in geological megabiases affecting vertebrates. *Proceedings of the Royal Society of London B*, 277 (2009), 829–834.

117. M. Kowalewski and K. W. Flessa, Improving with age: The fossil record of lingulide brachiopods and the nature of taphonomic megabiases. *Geology*, 24 (1996), 977–980.

118. M. L. Rosenzweig, *Species Diversity in Space and Time* (Cambridge, UK: Cambridge University Press, 1995).

119. N. J. Gotelli and A. Chao, Measuring and estimating species richness, species diversity, and biotic similarity from sampling data. In S. A. Levin, ed., *Encyclopedia of Biodiversity* (2nd ed), Vol 5 (Waltham, Mass.: Academic Press, 2013), pp. 195–211.

120. R. K. Colwell. *Estimate S v9.0 (PC)*. 9.0 ed. (Storrs, Connecticut: University of Connecticut; 2013).

121. P. Uetz, P. Freed, and J. Hosek, The Reptile Database, www.reptile-database.org. (2021).

122. Y. W. Xing, R. E. Onstein, R. J. Carter, T. Stadler, and H. P. Linder, Fossils and a large molecular phylogeny show that the evolution of species richness, generic diversity, and turnover rates are disconnected. *Evolution*, 68 (2014), 2821–2832.

123. F. Pokrant, C. Kindler, M. Ivanov, et al. Integrative taxonomy provides evidence for the species status of the Ibero-Maghrebian grass snake *Natrix astreptophora*. *Biological Journal of the Linnean Society*, 118 (2016), 873–888.

5

Miocene Snakes of Eurasia

A Review of the Evolution of Snake Communities

MARTIN IVANOV

5.1 Introduction

The main features of today's Eurasian snake fauna were assembled during the Miocene (~23–5 Ma), and so this epoch is critical for understanding the origins and early evolution of this fauna. The origin of Miocene snakes in Eurasia can be traced back to the Palaeogene. The last epoch characterized in Eurasia by mid-latitude tropical conditions and rapid diversification of snake clades is the early and middle Eocene. However, worldwide cooling since the Early Eocene Climatic Optimum (EECO; 53.3–49.1 Ma) with a short-term warming ~40.7–39.6 Ma (MECO; Middle Eocene Climatic Optimum) [1, 2] and a subsequent steep drop in temperatures around the Eocene–Oligocene transition (33.9 Ma) resulted in a remarkable extinction period, the 'Grande Coupure' (~33.4 Ma) (e.g., [3]). Several squamate lineages temporarily left Europe or became extinct long before the complete physical separation of Europe and North America but retreat of some populations into refugia in other parts of Europe cannot be excluded [4]. The snake families Palaeophiidae, Anomalophiidae, Russellophiidae, and Nigerophiidae did not survive into the Oligocene (for Palaeogene snakes see Chapter 4). The evolution of Cenozoic European snake faunas was last summarized by Rage [3]; with more detailed discussions of the evolution of Neogene snakes restricted to particular taxa [5–7] or particular areas [8–10].

Given the importance of the Miocene for understanding the origins and evolution of Eurasian snakes, the main aim of this chapter is a comprehensive review of Eurasian Miocene snakes in the context of contemporary taxonomy [11–13], chronostratigraphy, and palaeoclimatological and palaeogeographical evolution. This review is based on data collected from almost 150 Eurasian localities covering almost 18 Ma (Fig. 5.1, Supplementary Table 5.S1).

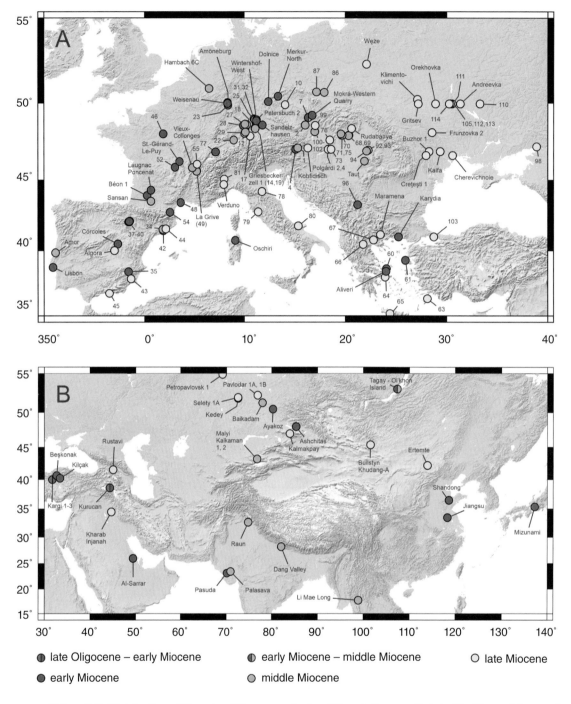

Figure 5.1 Distribution of Eurasian Miocene snake localities, A, Europe; B, Asia. Positions of the most important localities are highlighted. For locality name indicated by a number, coordinates and updated ages see Supplementary Table 5.S1. (A black and white version of this figure will appear in some formats. For the colour version, please refer to the plate section).

5.2 Eurasian Snakes at the Late Oligocene to Early Miocene Transition

The Oligocene climate was warmer than present, with prevailing subtropical conditions in mid-latitudes in most of Eurasia. Small representatives of Booidea and Amerophidia (Tropidophiidae, Aniliidae; sensu [13]) became dominant. 'Ancient' survivors beyond the 'Grande Coupure' occurred sympatrically with new invaders. These mostly Asiatic immigrants penetrated into Europe as early as MP 21–MP 22 as a result of the closure of the Turgai Sea connection in the East. Early Oligocene snakes from Asia are still poorly known, so we can only speculate about dispersal routes of these first invaders into Europe. Dispersal via a wide continental connection north of the Parathethys and subsequent penetration across the marshy terrestrial barrier between the North Sea and the Parathethys, the Mazury–Mazowsze continental bridge (34 Ma), seems plausible [14] although a southern dispersal route from Anatolia across numerous islands in southeastern Europe (today's Balkan Peninsula) cannot be ruled out. The Turkish early/late to late Oligocene herpetofauna from Kavakdare (MP 25) and Kocayarma (MP 26–MP 27) indicates a clear affinity with European amphibian and reptile communities [15, 16].

During the Miocene (23.03–5.3 Ma), the epicontinental Parathethys sea had flooded most of southern Central and Eastern Europe. In the Middle East, a continental connection between Eurasia and Africa originated at the Oligocene–Miocene transition as a result of a collision between the Eurasian and Afro-Arabian Plates, enabling exchanges of terrestrial fauna approximately 20–18 Ma [17]. The early Miocene was characterized by the return of a warm and humid climate. In continental environments, mean annual temperature (MAT) estimated from the palaeobotanical record quicky increased in Central Europe after the relatively cold early Aquitanian from ~16 °C to 18 °C and temperatures >16 °C persisted during the Burdigalian. The coldest monthly mean temperature (CMMT) was ~6 °C in Central Europe [18]. Eastern Asia (Eastern China) was cooler with MAT 13.5 °C and CMMT 5–6 °C during the early Miocene [19].

In Asiatic Eurasia, latest Oligocene (latest Chattian, MP 30; 23.2–23.03 Ma) to earliest Miocene (Aquitanian, MN 1–MN 2; 23.03–20.44 Ma) snakes occur rarely in several localities in Anatolia (Figs. 5.1 and 5.2) which was then an island separated from the rest of Asia [20]. The occurrence of Anilioidea (*Eoanilius* cf. *oligocenicus*, for current taxonomy see Chapter 4), Tropidophiidae (*Falseryx*), Booidea (*Bavarioboa, Albaneryx*) and 'Colubridae' (*Texasophis*; for taxonomy of fossil 'Colubridae' see Supplementary Table 5.S2) (Fig. 5.3) as well as other amphibians and reptiles (e.g., *Latonia*, palaeobatrachids, *Eopelobates, Salamandra,* and *Pseudopus*) in the late Oligocene to early Miocene of Anatolia prove communication between Asiatic and European herpetofaunas [16, 21, 22].

In southern Asia, the oldest fossil of Acrochordidae, the sister group of all other extant caenophidians [12, 13], is *Acrochordus* cf. *dehmi* from the earliest Miocene of Kharinadi Formation, Gujarat, Western India (Fig. 5.3; [23]). The first distinct *A. dehmi* from the Kamlial Formation of the Siwalik Group on the Potwar Plateau, Pakistan is of similar or younger age. Therefore, the origin and early diversification of this genus, known from numerous localities

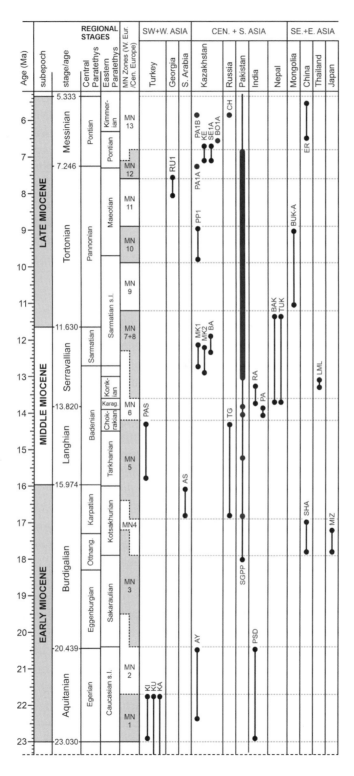

Figure 5.2 Miocene time scale and biostratigraphy (for sources see Supplementary Figure 5.S1) with stratigraphical position of Asian snake localities. **Turkey:** KI, Kilçak 0, 3A, 3B; KU, Kurucan; KA, Kargi 1–3; PAS, Paşalar. **Georgia:** RU1, Rustavi 1; **Saudi Arabia:** AS, Al-Sarrar. **Kazakhstan:** AY, Ayakoz; MK1 and MK2, Malyi Kalkaman 1 and 2; BA, Baikadam; PP1, Petropavlovsk 1; PA1A and PA1B, Pavlodar 1A and 1B; KE, Kedey; SE1A, Selety 1A; BO1A, Borki 1A. **Russia:** TG, Tagay (Togay), Ol'khon Island, Baykal Lake; CH, Cherlak. **Pakistan:** SGPP, Siwalik Group–Potwar Plateau; **India:** PSD, Pasuda; PA, Palasava; RA, Raun. **Nepal:** BAK, Babai Khola (Dang Valley), TUK, Tui Khola (Dang Valley). **Mongolia:** BUK-A, Builstyn Khudang-A. **China:** SHA, Shantung; ER, Ertemte. **Thailand:** LML, Li Mae Long. **Japan:** MIZ, Mizunami.

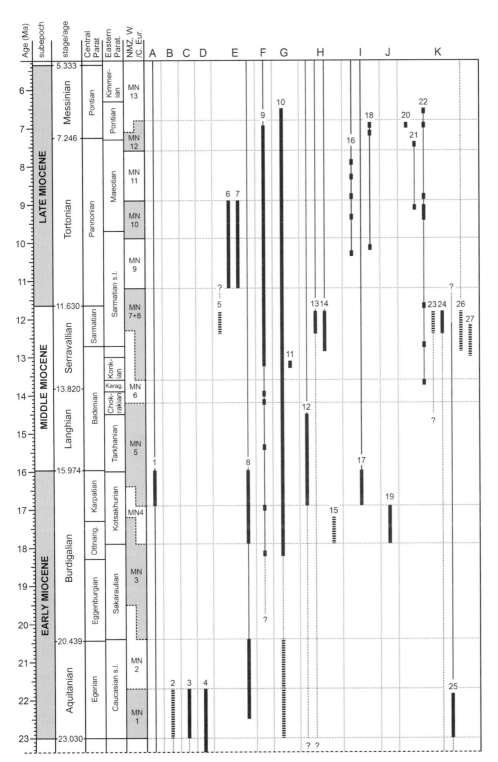

Figure 5.3 Miocene stratigraphic ranges of snake taxa (mostly extinct genera and species) reported from Asia. A, Scolecophidia; B, Anilioidea; C, Tropidophiidae; D, Boidae; E, Erycidae; F, Pythonidae; G, Acrochordidae; H, Viperidae; I, Elapidae; J, Natricidae; K, 'Colubridae'; (1) Scolecophidia indet.; (2) *Eoanilius oligocenicus*; (3) *Falseryx* sp.; (4) *Bavarioboa* sp.; (5) *Albaneryx volynicus*; (6) *Eryx* sp. 1; (7) *Eryx* sp. 2; (8) *Eryx* sp. + 'Eryx-Gongylophis' group; (9) *Python* sp.; (10) *Acrochordus dehmi*; (11) *Acrochordus*, different from *A. dehmi*; (12) Viperidae ('Oriental' vipers); (13) *Vipera* sp. ('European' vipers, 'V. aspis' complex); (14) *Gloydius* sp.; (15) *Trimeresurus* sp.; (16) *Bungarus* sp.; (17) *Naja* sp.; (18) Elapidae? indet. similar to *Ophiophagus*; (19) '*Mionatrix' diatomus*; (20) *Gansophis potwarensis*; (21) *Chotaophis padhriensis*; (22) *Sivaophis downsi*; (23) *Hierophis hungaricus*; (24) *Texasophis bohemiacus*; (25) *Texasophis* sp.; (26) *Elaphe dione*; (27) '*Elaphe*' sp. Dashed pattern = uncertain occurrence. Fossil snake occurrences according to [21–23, 27, 28, 60, 64, 83, 84, 87, 104].

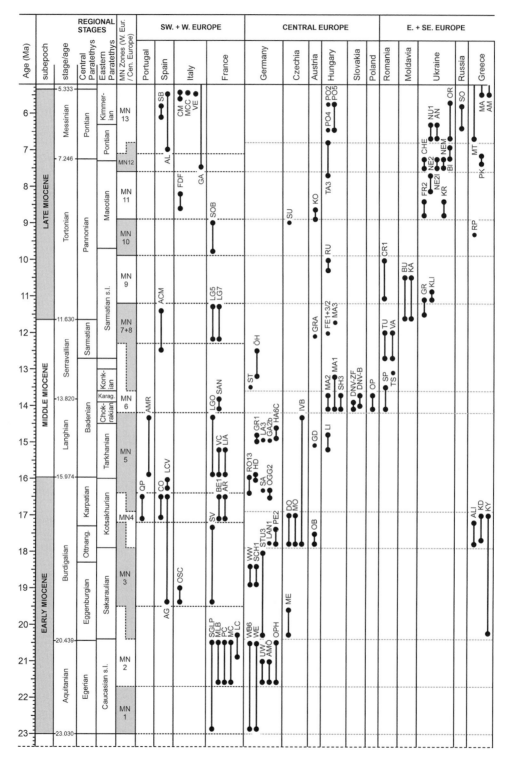

Figure 5.4 Miocene time scale and biostratigraphy (for sources see Supplementary Figure 5.S1; black dashed lines = Western Europe; grey dashed lines = Central Europe) with stratigraphical positions of European snake localities. **Portugal:** QP, Quinta das Pedreiras (Lisbon); AMR, Amor.

in Siwalik Group of Pakistan, Nepal, and northern India as well as from early Miocene of Thailand, can be placed before the Palaeogene–Neogene boundary [23–30].

European late Oligocene snakes are well documented only from France and Germany. Despite the low number of localities around the Oligocene–Miocene transition, the late Oligocene to early Miocene (MP 29–MN 2) decrease of snake diversity, named the 'Dark Period' [6], is conspicuous in comparison with the rich communities of MP 28. Anilioids were Oligocene survivors as documented by the likely permanent presence of *Eoanilius* up to the early Langhian, MN 5 [31]. The last *E. oligocenicus* has been reported from the early Miocene (early MN 3) of Oschiri, Italy (Fig. 5.4) [32, 33]. The tropidophiid genus *Falseryx* first appeared in the early Oligocene of France (MP 22) but left Europe for about 10 Ma as documented by its uncertain appearance in the late Aquitanian (MN 2a) of Amöneburg, Germany. *Falseryx*, which invaded Europe probably from southwestern Asia, can be unambiguously distinguished from the erycid genus *Bransateryx* only by the morphology of caudal vertebrae, so Čerňanský et al. [34] assumed that trunk vertebrae from various collections previously attributed to erycines might actually belong to tropidophiids.

The early Aquitanian decrease of temperatures in Central Europe with MAT ~16 °C and CMMT ~5.5 °C continued towards the Western Asia with MAT ~14–16 °C and CMMT ~2–5 °C [18, 35]. The distribution of Amerophidia and Booidea was affected by this change and most Oligocene Boidae and Tropidophiidae did not survive the Oligocene–Miocene transition at mid-latitudes. The Erycidae genus *Bransateryx* is represented rarely by *B. vireti* in MN 1 of Paulhiac, France [36] and two coeval German localities, Weisenau, MN 2 - old collection ([37]; though, caudal vertebrae are missing) and Weißenburg 6, MN 1–MN 2 (*B.* cf. *vireti*; Ivanov, unpublished data). *Bransateryx* cf. *vireti* from Germany indicates that

---

Figure 5.4 (*cont.*) **Spain:** AG, Agramón; CO, Córcoles; LCV, Les Cases de la Valenciana; ACM, Abocador de Can Mata; AL, Algora; SB, Salobreña. **Italy:** OSC, Oschiri; FDF, Fosso della Fittaia; CM, Cava Monticino; MCC, Moncucco Torinese; VE, Verduno; GA, Gargano 'terre rosse'. **France:** SGLP, St.-Gérand-le-Puy; MLB, Montaigu-le-Blin; PC, Poncenat; MC, Marcoin; LC, Laugnac; SV, Serre de Verges; BE1, Béon 1; AR, Artenay; LGO, La Grive – old levels; VC, Vieux-Collonges; LIA, L'Isle d'Abeau; SAN, Sansan; LG5, LG7, La Grive 5, 7; SOB, Soblay. **Germany:** WE, Weisenau; WB6, Weißenburg 6; AMÖ, Amöneburg; UW, Ulm-Westtangente; OPH, Oppenheim/Nierstein; WW, Wintershof-West; STU3, Stubersheim 3; SCH1, Schnaitheim 1; PE2, Petersbuch 2; LAN1, Langenau 1; RO13, Rothenstein 13; OGG 2, Oggenhausen 2; HD, Häder; GR1, Griesbeckerzell 1a+1b; SA, Sandelzhausen; LA3, Laimering 3; GA2b, Gallenbach 2b; HA6C, Hambach 6C; ST, Steinheim a. Albuch; ÖH, Öhningen; **Czechia:** ME, Merkur-North; DO, Dolnice; MO, Mokrá-Western Quarry; IVB, Ivančice near Brno; SU, Suchomasty. **Austria:** OB, Oberdorf; GD, Grund; GRA, Gratkorn; KO, Kohfidisch; **Hungary:** LI, Litke 1, 2; MA1, MA2, MA3, Mátraszőlős 1, 2, 3; SH3, Sámsonháza 3; FE, Felsőtárkány 1+3/2; RU, Rudabánya; TA3, Tardosbánya 3; PO2,4,5, Polgárdi 2, 4, 5; **Slovakia:** DNV-ZF, Devínska Nová Ves – Zapfe's fissures; DNV-B, Devínska Nová Ves – 'Bonanza'; **Poland:** OP, Opole. **Romania:** SU, Subpiatră 2; TS, Tăşad; TU, Tauţ; VA, Vârciorog; CR1, Creţeşti 1. **Moldavia:** BU, Buzhor; KA, Kalfa. **Ukraine:** GR, Gritsev; KLI, Klimentovichi; FR2, Frunzovka 2; KR, Krivoy Rog; NE2, Novoelizabetovka 2 (lower layer); NE2l, Novoelizabetovka 2 (bone-bearing lens); CHE, Cherevichnoie; NEM, Novaya Emetovka 2; BI, Bielka; NU1, Novoukrainka 1; AN, Andreevka; OR, Orekhovka. **Russia:** SO, Solnechnodolsk. **Greece:** ALI, Aliveri; KD, Karydia; KY, Kymi; RP, Ravin de la Pluie; PK, Pikermi; MT, Maritsa; MA, Maramena 1–3; AM, Ano Mentochi.

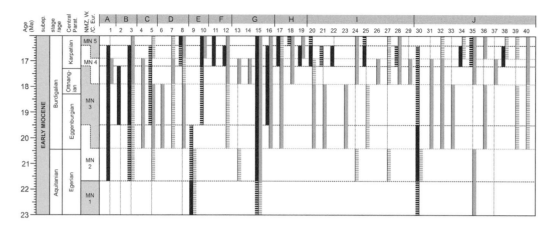

Figure 5.5 Early Miocene stratigraphic ranges of snake taxa (mostly extinct genera and species) in Western (black) and Central + Eastern (grey) Europe. A, Scolecophidia; B, Anilioidea; C, Tropidophiidae; D, Boidae; E, Erycidae; F, Pythonidae; G, Viperidae; H, Elapidae; I, Natricidae; J, 'Colubridae'; (1) Scolecophidia indet.; (2) *Eoanilius oligocenicus*; (3) *Eoanilius* sp.; (4) *Falseryx petersbuchi*; (5) *Falseryx* sp.; (6) *Bavarioboa*, morphotype 1; (7) *Bavarioboa hermi*; (8) *Bavarioboa* sp.; (9) *Bransateryx vireti*; (10) *Eryx* sp.; (11) *Python europaeus*; (12) *Python* sp.; (13) *Vipera antiqua*; (14) *Vipera platyspondyla*; (15) *Vipera* sp. ('European' vipers, '*V. aspis*' complex); (16) Viperidae ('Oriental' vipers); (17) '*Micrurus*' *gallicus*; (18) *Naja romani*; (19) *Naja* sp.; (20) *Natrix sansaniensis*; (21) *Natrix merkurensis*; (22) '*Natrix*' *natricoides*; (23) *Natrix* morphotype 1; (24) *Natrix* sp.; (25) *Neonatrix europaea*; (26) *Neonatrix nova*; (27) *Neonatrix* sp.; (28) *Palaeonatrix lehmani*; (29) *Palaeonatrix* sp.; (30) '*Coluber*' *cadurci*; (31) '*Coluber*' *dolnicensis*; (32) '*Coluber*' *caspioides*; (33) '*Coluber*' *suevicus*; (34) '*Coluber*' *pouchetii*; (35) '*Coluber*' sp.; (36) *Hierophis hungaricus*; (37) *Texasophis bohemiacus*; (38) *Texasophis meini*; (39) *Elaphe* sp.; (40) *Telescopus* sp. Dashed pattern = uncertain occurrence. Data according to: taxa 1–12: [6, 34, 44, 66]; taxa 13–16 [5]; taxon 17 [39, 44, 51]; taxa 18–19 [9, 39]; taxa 20–40 [7, 34].

populations of *B. vireti* were simultaneously evolving both in Central and Western Europe since the late Oligocene, MP 27 [6]. The disappearance of *Bransateryx* from Central Europe and its survival up to MN 2 in France (Saint-Gérand-le-Puy and Poncenat) [36, 38] might be explained by lower temperatures and increased humidity in Central Europe and lack of arid or semi-arid environments. The extant genus *Eryx* replaced *Bransateryx* in southern Europe probably as early as MN 3 (Fig. 5.5).

The first Viperidae, of the '*aspis*' complex, appeared in the early Miocene of Weisenau (MN 2; *Vipera* cf. *V. antiqua*) and St.-Gérand-le-Puy, France (MN 1 or MN 2a) [5, 39]. The exact biostratigraphic age of Weisenau is still debated [3], so isolated fangs from MN 2 of Oppenheim/Nierstein, Germany [40] corresponding in size to fangs of the '*V. aspis*' complex, represent the first stratigraphically unquestionable record of Viperidae. Rage [3] accepted that elapoid fangs reminiscent of *Naja* spp. and *Bungarus* from Oppenheim/Nierstein might represent immature viperids but the presence of still-undescribed vertebrae closely similar to those of natricids and non-*Naja* elapoids, reported by Kuch et al. [40], indicate that elapoids were indeed present in MN 2 of Central Europe.

Colubroidea (sensu [12, 13]) are largely unknown in the earliest Miocene (Fig. 5.5) with the exception of '*Coluber*' *cadurci* from Aquitanian of Weisenau and France [39, 41]. Two morphotypes of *Coluber* s.l. as well as cf. *Natrix* sp. and possibly the geologically oldest *Neonatrix* have been reported from Amöneburg, MN 2a [34].

5.3 Early and Middle Burdigalian (Early Miocene): A Rapid Change in Composition of European Snake Communities

The early Burdigalian (~late MN 2 and most of MN 3; 20.4–18.2 Ma) warming, interrupted in the Paratethys area by the Early Ottnangian Cooling (EOC) event (18.1–17.8 Ma; [42]), was followed in the late middle Burdigalian (late Ottnangian; 17.5–17.3 Ma) by rapid warming in western Central Europe (NAFB, North Alpine Foreland Basin) with paratropical temperatures ~22–24 °C and mean annual precipitation (MAP) ~830–1350 mm. Even during the coldest month of the year the temperature probably did not fall below 16.7 °C [43].

Eurasian snakes underwent a significant change during MN 3. Early Burdigalian warming is evidenced by the tropidophiid *Falseryx petersbuchi* at Bissingen 1, and Wintershof-West ([44]; pers. obs.). The Miocene dispersal of *Falseryx* in Central Europe most likely slightly preceded that of boid genus *Bavarioboa* (Fig. 5.5) [34]. *Bavarioboa* which disappeared from the European fossil record in the late Oligocene for almost 5 million years, occurred again in Central Europe in the early Burdigalian but Miocene representatives were different from those that inhabited France and Germany in the Oligocene [6]. I consider early Miocene *Bavarioboa* from Central Europe as an Asian newcomer from the east or southeast. The first Miocene appearance of *Bavarioboa* in Europe is in the early Burdigalian (MN 3a) of Merkur-North (or Ahníkov), Czechia [45]. The presence of this genus in slightly younger Schnaitheim 1 and Wintershof-West localities, Germany (both MN 3b; Fig. 5.4) indicates that *Bavarioboa*, in Schnaitheim 1 probably *B. hermi* (pers. obs.), spread rapidly in Central Europe together with other non-erycid Booidea, before the EOC. A large pythonid (centrum length >10 mm, width 13–14 mm) from the early Miocene (MN 3/4) of Kymi, Greece can be attributed to *Python* (*P. euboicus*; [6,46]). Although the biostratigraphical age is not certain, it might represent the first Miocene dispersal of *Python* into Europe either from Asia or Africa because the '*Gomphotherium* land bridge' existed in the Burdigalian [20, 47].

'Oriental vipers' characterized by large dimensions of relatively short and wide vertebrae and massive hypapophyses, have their first European occurrence in Wintershof-West [44] and Schnaitheim 1 (Ivanov, unpublished). Among Elapoidea, only small representatives with centrum length rarely reaching 3 mm have been reported in MN 3. These resemble recent coral snakes such as *Calliophis* or have been attributed to the genus '*Micrurus*' [44, 45] (for explanation of quotation marks see Supplementary Table 5.S2). Recent molecular studies recover a well-supported *Calliophis* clade (excluding *C. bivirgatus* [12]) as sister group to all other extant Elapidae [12, 48], and are consistent with an Asian origin of coral snakes. Although several morphologically different types of coral snakes likely occurred in NAFB, large elapids, usually attributed to the genus *Naja*, did not appear at mid-latitudes in

Western or Central Europe before the beginning of the Miocene Climatic Optimum (MCO), because *Naja* is more thermophilic than most 'Oriental vipers' [39, 49–51].

Colubroid snakes underwent a first massive dispersal to Central Europe in MN 3 as reported from Merkur-North (MN 3a) and Wintershof-West (MN 3b) [44,45]. 'Colubridae' and Natricidae are highly dominant in Merkur-North, with several taxa making their first appearance: '*Coluber*' *dolnicensis*, '*C*'. *suevicus*, '*C*'. *caspioides*, *Natrix merkurensis* and *N. sansaniensis* (Fig. 5.5; for explanation of quotation marks in generic names see Supplementary Table 5.S2). However, Wintershof-West provides a colubroid assemblage of different composition, with common Boidae and Viperidae ('Oriental vipers') also occurring.

The Burdigalian was a crucial period for the evolution of European snake faunas with MN 2 and early MN 3 snake communities markedly different from those of late MN 3 and MN 4. Although non-erycid Booidea have been rarely documented from the early Burdigalian sites, large representatives of Boidae and Pythonidae, together with other highly thermophilic snakes, became common soon after the EOC. This change, linked with a strong increase in snake diversity, was quite rapid and lasted probably no more than several hundred thousand years.

5.4 Miocene Climatic Optimum (MCO): The 'Golden Age' of Eurasian Snake Faunas

The MCO (~17–14.5 Ma; [52,53]) represents, after the late Ottnangian warming event, the warmest Miocene period. Palaeobotanical records of xyloflora in combination with occurrences of highly thermophilic reptiles (*Python*) indicate slight lowering of MAT to 18.6–20.8 °C and CMMT 8.1–13.3 °C from the late Karpatian to early Badenian [43, 54]. Forests changed to subtropical semi-deciduous, later (early Badenian) to oak–laurel and finally (late Badenian) to dry deciduous, consistent with palaeobotanic data from other regions of Central Europe [43, 55]. This coincides with a marked seasonality, up to six months per year with rainfall < 60 mm (14.7–14.5 Ma) [52, 56]. In Central Paratethys (Korneuburg Basin, Austria) the presence of the alligatoroid *Diplocynodon* and possibly cordylid lizards indicate a warm climate with MAT ~17 °C and CMMT ~3–8 °C, which persisted in this area up to the late Karpatian [57-59].

Late early and early middle Miocene snakes are currently almost unknown from Asia (Fig. 5.2). The wide distribution of *Python* in southern and southwestern Asia (Fig. 5.3) is documented by its continual occurrence in the Siwalik Group from the middle Burdigalian to the latest Miocene (~18 Ma–6.78 Ma; [28]), as well as the presence of a 'medium sized *Python*' from Al-Sarrar, Saudi Arabia [60], and its westernmost Asian record in the earliest middle Miocene (late MN 5) of Paşalar, Turkey [61]. A possibly close relationship between Anatolian *Python* and a pythonid from Kymi, MN 3/4 (*P. euboicus*) cannot be excluded given that faunal exchanges have been documented between Anatolia and the Greek mainland via the landmass connection that existed during the early and early middle

Miocene [61]. The Al-Sarrar locality also yielded erycids of the 'Eryx-Gongylophis' group but the oldest Asiatic *Eryx*, described on the basis of a fragmentary posterior caudal vertebra, is from the Aquitanian (MN 1 to MN 2) of Ayakoz, Kazakhstan [62] and precedes its first appearance in southern Europe (Fig. 5.3).

Caenophidian clades are represented by indeterminate viperids, 'colubrids' and elapids of the *Naja*-group from Al-Sarrar [60]. Indeterminate small-sized Viperidae and larger specimens resembling extant *Montivipera xanthina* have been reported from the late early Miocene of Zaisan Basin, Kazakhstan [63]. A colubroid, *Mionatrix diatomus* from the late early Miocene (MN 4) of Shandong, China [64] is inadequately defined and needs revision (Supplementary Table 5.S2). The allocation of a fragmentary vertebra from the late early Miocene (17.9–17.2 Ma) of Mizunami, Japan to cf. *Trimeresurus* sp. [65] cannot be considered the earliest unambiguous Crotalinae [30].

The European snake fauna of the late early and early middle Miocene (MN 4 and MN 5) was highly diversified (Figs. 5.5 and 5.6). The continuous occurrence of Scolecophidia in France since the latest Oligocene (MP 30) up to the early Miocene (MN 4) indicates the western provenance of early Miocene scolecophidians from Central Europe.

The late Burdigalian (MN 4 and early MN 5) was closely linked with a rapid dispersal of large Boidae and Pythonidae in Central and Western Europe as a result of onset of the MCO. *Bavarioboa hermi* became widespread in Europe from MN 4 [6, 66]. Despite the fact that the late Burdigalian *B. ultima* from Rothenstein 13, Germany does not represent a fully adult specimen [6], it appears that up to the early Badenian, when the MCO culminated, the average size of Booidea was increasing [67]. The largest Eurasian constrictor was *Python*, whose first indisputable European appearance is from MN 4 in France and Czechia [51, 66]. Pythonids are among the most thermophilic European Miocene squamates and their occurrence is closely linked to the warmest MCO phases. The youngest European python record, *Python* sp. from the Langhian (early Badenian; MN 5) of Griesbeckerzell 1a and 1b, Germany [54], is similar to *P. europaeus* documented from several French localities: Béon 1, MN 4 [51], La Grive P & B and Vieux-Collonges, both MN 5 [6, 49]. I assume that large Boidae and Pythonidae did not survive in Europe after the Langhian–Serravallian transition as a result of deterioration of climatic conditions. Although the extinct sand boa *Albaneryx deltpereti* has been reported from the late Langhian of France [68], erycids of the extant genus *Eryx* appeared in Spain as early as the middle Burdigalian, MN 4 or ? MN 3 [69, 70]. The return of Erycidae into Central Europe (Fig. 5.6), documented by caudal vertebrae of *Eryx* from Gallenbach 2b and Laimering 3, Germany (late MN 5), resulted most likely from increased impact of seasonality with arid periods 14.7–14.5 Ma [52]. Tropidophiidae became extinct in Eurasia shortly after the MCO, as documented by the last uncertain occurrence of *Falseryx* in Steinheim (13.6–13.4 Ma; [6]).

'Oriental vipers' and large elapids became common in Europe. 'Oriental vipers' and 'European vipers' of the 'V. aspis' complex (Fig. 5.5) inhabited Europe sympatrically during the late Burdigalian [31,51]. Large Viperinae 'Oriental vipers' or *Daboia* (see Supplementary Table 5.S2) have been identified from Griesbeckerzell 1b [54]. Their vertebrae are substantially larger than those of the other known 'Oriental vipers' and resemble vertebrae from Vieux-Collonges, France [49, 71].

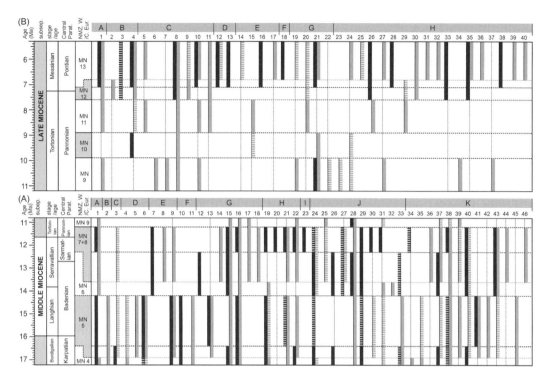

Figure 5.6 Latest early to late Miocene stratigraphic ranges of snake taxa (mostly extinct genera and species) in Western (black) and Central + Eastern (grey) Europe. **(A) Latest early and middle Miocene (MN 5–MN 7+8) snakes:** A, Scolecophidia; B, Anilioidea; C, Tropidophiidae; D, Boidae; E, Erycidae; F, Pythonidae; G, Viperidae; H, Elapidae; I, Dipsadidae; J, Natricidae; K, 'Colubridae'; (1) Scolecophidia indet.; (2) *Eoanilius* sp.; (3) *Falseryx* sp.; (4) *Bavarioboa hermi*; (5) *Bavarioboa ultima*; (6) *Bavarioboa* sp.; (7) *Albaneryx depereti*; (8) *Albaneryx volynicus*; (9) *Eryx* sp.; (10) *Python europaeus*; (11) *Python* sp.; (12) *Vipera aegertica*; (13) '*Vipera*' sp. or *Daboia* sp.; (14) *Macrovipera ukrainica* + *Macrovipera* sp.; (15) Viperidae ('Oriental' vipers); (16) *Vipera* sp. ('European' vipers, '*V. aspis*' complex); (17) Viperidae '*V. berus*' complex; (18) Crotalinae gen. et sp. indet. 1 and 2; (19) '*Micrurus*' *gallicus*; (20) '*Micrurus*' sp.; (21) *Naja romani*; (22) *Naja* sp.; (23) '*Paleoheterodon*' *arcuatus*; (24) *Natrix sansaniensis*; (25) *Natrix rudabanyaensis*; (26) '*Natrix*' *natricoides*; (27) *Natrix longivertebrata*; (28) *Natrix* sp.; (29) *Neonatrix europaea*; (30) *Neonatrix crassa*; (31) *Neonatrix* sp.; (32) *Palaeonatrix silesiaca*; (33) *Palaeonatrix lehmani*; (34) '*Coluber*' *dolnicensis*; (35) '*Coluber*' *caspioides*; (36) '*Coluber*' *suevicus*; (37) '*Coluber*' *pouchetii*; (38) '*Coluber*' sp.; (39) *Hierophis hungaricus*; (40) *Texasophis meini*; (41) *Texasophis* sp.; (42) *Zamenis kohfidischi*; (43) '*Elaphe*' sp.; (44) *Zamenis* sp.; (45) *Telescopus* sp.; (46) *Coronella miocaenica*. **(B) Late Miocene (MN 9–MN 13) snakes:** A, Scolecophidia; B, Erycidae; C, Viperidae, D, Psammophiidae; E, Elapidae; F, Dipsadidae; G, Natricidae; H, 'Colubridae'; (1) Scolecophidia indet.; (2) *Albaneryx* sp.; (3) *Eryx jaculus*; (4) *Eryx* sp.; (5) *Macrovipera gedulyi*; (6) *Macrovipera sarmatica*; (7) *Macrovipera* sp.; (8) Viperidae ('Oriental' vipers); (9) *Vipera meotica*; (10) *Vipera* sp. ('European' vipers, '*V. aspis*' complex); (11) Viperidae '*V. berus*' complex; (12) *Malpolon* sp.; (13) *Psammophis* sp.; (14) '*Micrurus*' sp.; (15) *Naja romani*; (16) *Naja iberica*; (17) *Naja* sp.; (18) '*Paleoheterodon*' *arcuatus*; (19) *Natrix rudabanyaensis*; (20) *Natrix longivertebrata*; (21) *Natrix* sp.; (22) *Neonatrix* sp.; (23) '*Coluber*' *pouchetii*; (24) '*Coluber*' sp.; (25) *Dolichophis* sp.; (26) *Platyceps planicarinatus*; (27) *Hierophis hungaricus*; (28) *Zamenis algorensis*; (29) *Zamenis kohfidischi*; (30) '*Elaphe*' *kormosi*; (31) *Zamenis praelongissimus*; (32) *Zamenis szyndlari*;

The elapid genus *Naja*, first documented in Europe (Fig. 5.5) by *N. romani* from the early Miocene (MN 4a) of Petersbuch 2, Germany [39, 72] appeared simultaneously in several late Burdigalian and Langhian (MN 4 and MN 5) sites in Germany, Austria, France, and Spain [31, 50, 51, 70, 73, 74]. The origin of Elapidae is still debated because the geologically oldest elapids are from Africa though molecular data place elapid origins in Asia 35–30 Ma [75]. A posterior trunk vertebra of Elapid morphotype A from the late Oligocene Nsungwe Formation (~24 Ma), Tanzania is similar to extant *Naja haje* and *N. nigricollis* [75] but the absence of cranial elements and caudal vertebrae prevent unambiguous distinction from the Asiatic complex of *Naja*. Despite a close affinity of *N. romani* with the Asiatic complex of *Naja* [72], the first *N. romani* might have dispersed into Europe either from Asia or Africa. '*Micrurus*' cf. *gallicus* from Sámsonháza, Hungary (MN 6; [10]) is likely the last occurrence of this species in Central Europe, although small elapids might have survived in Central Europe up to the earliest late Miocene [50].

A close early Miocene communication of snake faunas between Central and Western Europe is also documented by the distribution of colubroids. Several natricids that first appeared in Central Europe in the early Burdigalian (MN 3) have their relatives in the middle Burdigalian (MN 4) of France including *Natrix sansaniensis*, *N.* aff. *merkurensis* and *Neonatrix europaea*. On the other hand, 'colubrines' reported from the latest early Miocene of France such as '*Coluber*' *pouchetii* and *Texasophis meini*, probably appeared in the latest early and earliest middle Miocene localities of Central Europe [31, 51, 76]. Several of the above mentioned extinct species inhabited Europe during the late Burdigalian and Langhian such as *N. sansaniensis* and '*C.*' *pouchetii* from Sansan (MN 6), France and '*C.*' cf. *caspioides* from Oberdorf, Austria (MN 4) and Griesbeckerzell 1a (MN 5), Germany [54, 68, 77]. Only '*C.*' *pouchetii* survived for about 7.5 Ma, as documented by its last occurrence in the early late Miocene (MN 9) of Rudabánya, Hungary [50]. The early Badenian (MN 5) deposits from Grund, Austria provide the first occurrence of *Zamenis kohfidischi* ([73]; for a new allocation of extinct European *Elaphe* species into genus *Zamenis* see Supplementary Table 5.S2). This first possible occurrence of fossil *Zamenis* coincides with its likely divergence time estimated from molecular data ~8.5–15.0 Ma [78].

Texasophis and *Neonatrix* were considered to be of North American origin [79]. The oldest evidence of *Neonatrix* is *N. elongata* from the early Miocene of South Dakota and Texas [80, 81]. However, if ?*Neonatrix* sp. appeared in Central Europe in MN 2 [34] and *N. europaea* first appeared in MN 3 [44], we cannot ascertain whether the European representatives of *Neonatrix*, including *N. nova* from Dolnice (MN 4), Czechia [76], indeed came from North America or originated in Asia. According to Szyndlar [50], European representatives of *Neonatrix* might have represented a different lineage of tiny snakes and

Figure 5.6 (*cont.*) (33) '*Elaphe*' sp.; (34) *Telescopus bolkayi*; (35) *Telescopus* sp.; (36) *Coronella miocaenica*; (37) *Coronella* sp.; (38) *Hispanophis coronolloideus*; (39) *Periergophis micros*; (40) *Paraxenophis spanios*. Dashed pattern = uncertain occurrence. Data according to: taxa A1–11 and B1–4 [6, 8, 10, 54, 67–69, 82, 125]; A12–18 and B5–11 [5, 10, 91, 110]; B12–13 [7, 126]; A19–20 and B14 [10, 49, 67, 93]; A21–22 and B15–17 [9, 10, 31, 50, 105–107, 122]; A23 and B18 [79, 126]; A24–46 and B19–40 [7, 54, 93].

he recommended the use of the generic name in quotation marks (Supplementary Table 5. S2). A peculiar 'colubrid' snake from Greek localities Karydia and possibly also Aliveri (both MN 4a) differs from all known taxa by the presence of 'paracentral ridges' in precaudal vertebrae. Although the skeletal anatomy of extant African and Asian snakes is poorly documented, the absence of 'paracentral ridges' in all European extinct and extant snakes indicates that this snake could represent a short-lived radiation with African (via '*Gomphotherium* Landbridge') or Asian origin [82].

5.5 Late Middle Miocene Snakes from Eurasia: A Step towards Modern Snake Faunas in Europe

The late Langhian to early Serravallian (~14–13.5 Ma) was marked by a sharp drop in MAT by 7 °C (15.4–14.8 °C) and CMMT ~11 °C [52]. The climate became much drier in the late Serravallian (13.5–11.5 Ma) with MAP ~150–200 mm in the driest period. This cooling led to the disappearance of large Booidea and Pythonoidea from Europe and caused extinction of other highly thermophilic reptiles such as alligatoroids, chameleons, and giant tortoises in Central Europe [52]. Snake diversity decreased regardless of the low number of Serravallian localities.

After the MCO, Eurasian Booidea were represented mainly by erycids with the exception of southern and southeastern Asia where tropical conditions favoured pythonids [26, 28]. Erycids (Figs. 5.3 and 5.6) were represented by the extinct genus *Albaneryx* and extant *Eryx*. *Albaneryx depereti* inhabited France [36, 67] during the late Langhian and Serravallian (~14.2–12.2 Ma), whereas slightly younger *A. volynicus* was probably distributed from Central and Eastern Europe as far as Kazakhstan [10, 83, 84]. The last known *Albaneryx* is from the late Tortonian (Maeotian, MN 12) of Ukraine [85].

Serravallian snake communities from Asia are well documented only from Li Mae Long, Thailand, 13.3–13.12 Ma [26, 86] and Baikadam and Malyi Kalkaman 1, Kazakhstan, 13–12 Ma (Fig. 5.3; [84]). The Thailand snake community includes Pythonidae (*Python* sp.), Acrochordidae (*Acrochordus*, different from *A. dehmi*), Xenodermidae, Viperidae, Natricidae, and 'Colubridae', indicating humid conditions with a dominance of aquatic forms and presence of arboreal species [26]. Kazakh localities provide an interesting mix of snakes closely related to those that inhabited Europe during the latest early and earliest middle Miocene, such as *Texasophis bohemiacus*, '*Coluber*' (= *Hierophis*) cf. *hungaricus* and vipers of the '*V. aspis*' complex, together with snakes that probably never appeared in Central and Western Europe such as *Elaphe* aff. *dione* and the genus *Gloydius,* which represents the geologically oldest unambiguous Crotalinae (Fig. 5.6; [84]). Thus, possible homogeneity of snake communities over a large part of Eurasia (except southern Asia) throughout the middle Miocene [87] has not been demonstrated. Serravallian snakes from the Siwalik Group of Pakistan are represented by indeterminate erycids, the 'colubrid' *Sivaophis downsi*, with unusually tall neural spines, that survived up to the latest Miocene (~6.35 Ma), and a Colubroid morphotype A [28].

In Europe, the oldest known representatives of the genus *Macrovipera* have been reported from the Serravallian (MN 7+8) of Tauţ and Vârciorog, Romania [10] with the first diversification of this genus (*M. ukrainica, M. sarmatica*) around the middle–late Miocene transition [88, 89]. Palaeontological data thus support *Macrovipera* and *Montivipera* splitting about 15–13 Ma, as inferred from molecular data [90]. The only European Crotalinae (Fig. 5.6) comes from the late middle to early late Sarmatian (s.l.) of Gritsev [91], with the two completely preserved maxillae differing from a small maxilla of slightly older *Gloydius* from Kazakhstan. In any case, crotalines were largely distributed in Asia and Eastern Europe during the late middle and early late Miocene and a number of isolated vertebrae, formerly attributed to 'viperines', might actually belong to Crotalinae.

Among elapoids, '*Micrurus*' cf. *gallicus* from the French sites La Grive L7 and La Grive M, both MN 7 + 8 [92] represents the last known occurrence of '*Micrurus*' in Western Europe (Fig. 5.6). However, in Central/southeastern Europe the fossil record of small Elapoidea resembling *Micrurus* or *Calliophis* extends up to the early late Miocene, MN 9 [50], with an uncertain last occurrence of '*Micrurus*' from the Miocene–Pliocene transition in Maramena, Greece [93]. Large elapids represented by the genus *Naja* were distributed accross Europe as indicated by occurrences of *Naja romani* in France and Ukraine [71, 94, 95].

By far the most diverse European snake fauna, from La-Grive-Saint-Alban fissure fillings in France, yielded several colubroid taxa for which a North American origin was proposed, including *Texasophis meini*, '*Paleoheterodon*' *arcuatus* (generic allocation is questionable: Supplementary Table 5.S2), *Neonatrix europaea* and *N. crassa* [79]. A preliminary study of Colubroidea from La Grive L7 and M fissure fillings [92] reported rather diversified communities including '*C.*' cf. *dolnicensis*, several morphotypes attributed to '*Coluber*' sp. (= *Coluber* s.l.), '*Elaphe*' sp., *T. meini*, *Natrix* cf. *sansaniensis*, and several morphotypes of *Natrix* sp. (including *N.* aff *longivertebrata* sensu [96]) and *Neonatrix* sp.

In Central and Eastern/southeastern Europe, middle to earliest late Miocene Colubroidea are reported mainly from Hungary, Romania, and Ukraine [e.g., 7-9, 97]. 'Colubrines' are represented by several taxa (Fig. 5.6) including '*C.*' *pouchetii*, '*C.*' *suevicus*, '*Coluber*' sp., *Hierophis hungaricus*, '*Elaphe*' sp., *Texasophis* cf. *meini*, *Zamenis* sp. and the first appearance of *Coronella miocaenica* [10, 37, 98]. Natricids are represented by the last, uncertain occurrence of *N. sansaniensis*, but *N. rudabanyaensis* probably first appeared in the fossil record here [10, 98]. *Natrix* cf. *longivertebrata* reported from the earliest late Miocene of Gritsev might represent the first appearance of the extinct *N. longivertebrata* but the postcranial skeleton remains undescribed [99]. *Palaeonatrix* is represented by the only known occurrence of *P. silesiaca* from Opole, Poland, the genus *Neonatrix* was documented from Devínska Nová Ves, Slovakia [7, 9, 100].

5.6 Late Miocene Climatic Oscillations and Their Relation to the Composition of Eurasian Snake Communities

There was a strong climatic change in Europe about 10.7–9.7 Ma. In Central and Western Europe, rather high MAT ~14–17 °C [34] were accompanied by increased MAP to 1500 mm.

This 'wash-house' climate persisted for about 1 Ma [101]. Favorable climatic conditions are reflected in the composition of the early Vallesian (MN 9) herpetofauna. However, a sudden change occurred approximately 9.7–9.5 Ma when the MAP dropped to values similar to or even lower than those of recent times. This change led to an ecosystem collapse, the 'Vallesian Crisis', that resulted in the retreat of temperate evergreen forests and expanding steppe biomes [102]. From about 8 Ma, several further decreases in precipitation intensity occurred in Europe, the most substantial about 7 Ma [101]. During the latest Miocene (Messinian), an extensive aridification resulted in the formation of a continental connection between Africa and the Iberian Peninsula, allowing extensive faunal exchange. This connection existed until the beginning of the Pliocene [102].

The only diverse snake fauna known from the late Miocene of Asia is from the Siwalik Group of Pakistan, where the thermophilic assemblage of a highly humid environment included *Python*, *Acrochordus dehmi*, *Gansophis potwarensis*, *Bungarus*, *Chotaophis padhriensis*, *Sivaophis downsi* and a Colubroid morphotype B. *Bungarus* from the Siwalik Group occurs close to its extant westernmost distribution in southeastern Iran [103], indicating a palaeobiogeographic barrier between South Asia and Europe ~10 Ma [28]. Late Miocene (Vallesian) localities from Mongolia (Fig. 5.2) provide two morphologically different representatives of *Eryx* and two morphotypes of large-sized Colubrinae [103]. However, the posterior trunk vertebra of 'Natricinae indet.' from Builstyn-Khudang A (figs. 2a–e in [104]), with probably low neural spine and short prezygapophyseal processes could be attributed to elapoids rather than natricids.

In Central Europe, the early late Miocene (MN 9) site of Rudabánya, Hungary (Fig. 5.4) preserves a rich snake community from rather warm and humid conditions closely preceding the onset of the 'Vallesian Crisis'. A diverse colubroidean fauna includes extinct 'C.' *pouchetii*, *Hierophis hungaricus*, *Telescopus bolkayi*, *Natrix rudabanyaensis* and cf. *Neonatrix* sp., large elapids of the extant genus *Naja*, small indeterminate elapoids similar to *Micrurus*, and 'Oriental vipers' [50]. Ophidian localities coeval with the 'Vallesian Crisis' (Fig. 5.4; Supplementary Table 5.S1) are rare, including Ravin de la Pluie, Greece [105] and Suchomasty, Czechia (early MN 10). The Greek locality provided only several fragmentary vertebrae of indeterminate 'colubrines' and *Naja* cf. *romani,* whereas a small booid (probably an erycid) and indeterminate 'Colubrinae' occur in Suchomasty (pers. obs.).

A short-term improvement in climatic conditions occurred about 9–8.5 Ma [101] resulting in a quick recovery of the vertebrate communities as documented rarely from the early Turolian (MN 11) Kohfidisch site, Austria [106-108]. Scolecophidians have their last known occurrence in Central Europe but persisted in the Apennine Peninsula to the latest Miocene [109]. 'Oriental vipers' are represented by *Macrovipera gedulyi*, originally described from Kohfidisch as *Vipera burgenlandica* Bachmayer & Szyndlar, 1985 but recently synonymized with *Macrovipera gedulyi* (Bolkay, 1913) known from the latest Miocene of Hungary (Supplementary Table 5.S2; [110]). Elapidae are represented by *Naja romani* (Fig. 5.6) with the largest known specimens having a mean centrum length of 9.8 mm [108]. It is interesting that comparably large *Naja* specimens did not occur in the stable, high temperature environment known from the late early Miocene. Among Colubroidea, extinct *Zamenis kohfidischi* and *Platyceps planicarinatus* occur in Kohfidisch (for *Platyceps*, currently inhabiting the Balkan Peninsula [103], see Supplementary Table 5.S2). Natricids are represented by *Natrix longivertebrata* [106–108].

Ophidian localities from the Tortonian–Messinian transition (MN 12) are distributed in Eastern/southeastern Europe with the exception of Tardosbánya 3, Hungary, where extinct ratsnakes, '*Elaphe*' *kormosi* and *Zamenis praelongissimus* (Supplementary Table 5.S2), have been documented as well as small viperines similar to the extant *Vipera ursinii* ([5, 111]; M. Venczel, unpublished data). Several coeval localities from Ukraine yielded 'Oriental vipers' and small vipers of the '*V. aspis*' complex including *V. meotica* from Cherevichnoie and possibly Orekhovka, MN 13 [85, 97].

The latest Miocene (MN 13) is characterized by increased snake diversity (Fig. 5.6). 'Oriental vipers' occurred for the last time in Central Europe, as documented by *Macrovipera gedulyi* from Polgárdi 2, 4 and 5 [112, 113]. However, in southwestern Europe, 'Oriental vipers' have been documented in the late Miocene to early Pliocene of the Balearic Islands [114, 115] and persisted in the Iberian Peninsula until the early Pleistocene [116]. Extinct *Macrovipera kuchurganica* from Kuchurgan, Ukraine documents the persistence of 'Oriental vipers' in Eastern Europe up to the early Pliocene [88]. 'Oriental vipers' occur around the southeastern border of Europe today with *Macrovipera* inhabiting Greek islands of the Cyclades Archipelago in the Aegean Sea and Cyprus (*Macrovipera schweizeri, M. lebetina*) and *Montivipera* inhabiting northeastern Greece and several Greek islands as well as European part of Turkey [103]. Although 'European vipers' of the '*V. aspis*' complex appeared in Europe first and likely have been sympatric with 'Oriental vipers' since MN 3, fossil records indicate that 'Oriental vipers' were probably more common in Europe throughout most of the Miocene. Small vipers of the '*V. berus*' complex first appeared in Central Europe at the beginning of the late middle Miocene and became the dominant viperine group in Central Europe during the Pliocene and ultimately in the early Pleistocene [9, 97].

The Hungarian localities Polgárdi 2, 4, and 5 yielded the last known occurrences of *Hierophis hungaricus*, *Zamenis szyndlari* (see Supplementary Table 5.S2) and *Coronella miocaenica* as well as '*E.*' *kormosi* and *Z. praelongissimus*, that became extinct in the early Pliocene [111-113]. The parabasisphenoid of *Natrix* cf. *longivertebrata* from Polgárdi 4 is intermediate between *N. longivertebrata* and *N. natrix* [112] and precaudal vertebrae lack the strongly elongated centrum typical of *N. longivertebrata* from Rębielice Królewskie 1A (type locality, MN 16; [117]). Therefore, I assume that Polgárdi sites might contain the oldest occurrences of the extant grass snake *N. natrix*. This is in agreement with an inferred late Miocene divergence of *N. natrix* from *N. astreptohora* 10.6-9.6 Ma [118]. The latest Miocene to earliest Pliocene (MN 13-MN 14) snake assemblage from Maramena, Greece documents affinity with colubroids from Central Europe, including the last, uncertain occurrences of *H. hungaricus* and *N. rudabanyaensis*. This southward range constriction of Central European snakes has been documented for lizard groups such as agamids, cordylids, varanids, some anguids (*Pseudopus*) and amphisbaenians [119-121]. However, snake species including *Periergophis micros* and *Paraxenophis spanios* completely differ from known European taxa. The affinities of these snakes with other caenophidian lineages, such as dipsadids, lamprophiids, or psammophiids cannot be excluded, so an African or Asiatic origin currently seems most likely [93].

The dispersal of African lineages into southern Europe most probably resulted from an extensive aridification of the Mediterranean area during the Messinian Salinity Crisis around 5.96–5.33 Ma. An African origin is presupposed for several snake lineages documented from the latest Miocene and early Pliocene of Iberian Peninsula, including *Eryx primitivus* from Gorafe and Moreda (MN 15), *Psammophis* from Salobreña, MN 13, *Malpolon mlynarskii* from Layna, MN 15 and *Naja iberica* from Algora, MN 13 [69, 122-124]. However, it is not known whether these endemic lineages penetrated Europe directly via a terrestrial connection that appeared due to the closure of the Straight of Gibraltar or first invaded southeastern Europe and then migrated to the west. Elapoid snakes were distributed only in southern regions of Europe. The last known *Micrurus*-like snake occurred in Maramena (MN 13–14) whereas the genus *Naja* reported from Solnechnodolsk, Russia [125] persisted in southern regions of Europe up to the late Pliocene [9, 72].

5.7 Summary

The Miocene was the most-recent crucial epoch in the origins of the major elements of the modern Eurasian snake fauna. Although knowledge of Asian Miocene snakes is still limited, there was a clear communication between European and Asiatic snake communities since the Turgai Strait closure in the early Oligocene. Palaeogeographic evolution of the Paratethys area played an important role in the origin of at least two main dispersal routes for Asiatic (or African) newcomers into Europe. The northern route via the Mazury-Mazowsze continental bridge probably played a more important role in the early Oligocene. The southern dispersal route from Anatolia across numerous islands situated in southeastern Europe (today's Balkan Peninsula) also existed but a massive snake invasion from Asia (or Africa) was possible as late as the Burdigalian. The Burdigalian was a key period for the evolution of the European snake fauna, with MN 2 and early MN 3 snake communities markedly different from those of late MN 3 and MN 4. Tropidophiidae (*Falseryx*), Boidae (*Bavarioboa*), and 'Oriental vipers', known since the early Burdigalian, became common after the early Ottnangian Cooling (EOC). This increase in snake diversity correlates with the late Ottnangian warming (17.5–17.3 Ma). This change was quite rapid and lasted probably no more than a few hundred thousand years. The Miocene Climatic Optimum was associated with a substantial dispersal of non-erycid Booidea and 'Oriental vipers' in Europe. The first appearance of large elapids of the genus *Naja* during the middle Burdigalian (Ottnangian) coincides in Europe with the first indisputable appearance of Pythonidae. The highly thermophilic genus *Python* inhabited Europe for about 3 Ma and disappeared at the end of the Langhian due to rapid cooling (14–13.5 Ma). However, *Naja* survived up to the late Pannonian in Central Europe and in southern Europe even up to the late Pliocene. The climatically less-demanding 'Oriental vipers' (*Macrovipera*) disappeared from Central Europe in the latest Miocene but persisted in Western Europe until the early Pleistocene. There was a communication of the snake assemblages between Asia and Europe at mid-latitudes throughout the early and middle

Miocene. However, the east-west composition of this fauna was not homogeneous, at least since the middle Miocene, as evidenced by the snake fauna of Kazakhstan (middle Miocene; ~13–12 Ma) which includes taxa that probably never significantly penetrated into Europe, including the geologically oldest Crotalinae. Miocene snakes from South and Southeast Asia resemble those that currently inhabit this region. Increasing aridity at the end of the Miocene and the existence of a continental connection between Eurasia and Africa led to the invasion of snake lineages from Africa and/or Southwest Asia. Although several lineages dispersed in southern Europe (Psammophiidae), others represent probably restricted occurrences such as 'colubrid' snakes (*Periergophis*, *Paraxenophis*) from the Miocene–Pliocene transition of southeastern Europe. In Central and Eastern Europe, the last notable diversification occurred of 'colubrid' snakes (*Zamenis*) closely related to extant relatives.

Acknowledgements

I am deeply grateful to Salvador Bailon (Centre national de la recherche scientifique, Institut écologie et environnement, INEE), Márton Venczel (Tarii Crisurilor Museum, Oradea) and Massimo Delfino (Dipartimento di Scienze della Terra, Università di Torino) for their constructive reviews that improved my manuscript. I thank Krister T. Smith (Department of Messel Research and Mammalogy, Senckenberg Research Institute) who helped to export datasets into maps. I also would like to express my gratitude to both editors of the volume, Hussam Zaher and David Gower, for the invitation to contribute to the meeting and this volume, and to the latter for helping to improve the English translation of this chapter. This study was funded through the Specific research project at the Faculty of Science at Masaryk University, Brno (MUNI/A/1394/2021).

References

1. S. M. Bohaty and J. C. Zachos, Significant Southern Ocean warming event in the late middle Eocene. *Geology*, 31 (2003), 1017–1020.

2. S. M. Bohaty, J. C. Zachos, F. Florindo, and M. L. Delaney, Coupled greenhouse warming and deep-sea acidification in the middle Eocene. *Paleoceanography*, 24 (2009), PA2207.

3. J.-C. Rage, Mesozoic and Cenozoic squamates of Europe. *Palaeobiodiversity and Palaeoenvironments*, 93 (2013), 517–543.

4. T. J. Cleary, R. B. J. Benson, S. E. Evans, and P. M. Barrett, Lepidosaurian diversity in the Mesozoic–Palaeogene: the potential roles of sampling biases and environmental drivers. *Royal Society Open Science*, 5 (2018), 171830.

5. Z. Szyndlar and J.-C. Rage, Fossil record of the true vipers. In G. W. Schuett, M. Höggren, M. E. Douglas and H. W. Greene, eds., *Biology of the Vipers* (Eagle Mountain: Eagle Mountain Publishing, 2002), pp. 419–444.

6. Z. Szyndlar and J.-C. Rage, *Non-erycine Booidea from the Oligocene and Miocene of Europe* (Kraków: Institute of Systematics and Evolution of Animals, Polish Academy of Sciences, 2003).

7. Z. Szyndlar, Early Oligocene to Pliocene Colubridae of Europe: a review. *Bulletin de*

la Société Géologique de France, 183 (2012), 661–681.

8. Z. Szyndlar, A review of Neogene and Quaternary snakes of Central and Eastern Europe. Part I: Scolecophidia, Boidae, Colubridae. *Estudios geológicos*, 47 (1991), 103–126.

9. Z. Szyndlar, A rewiew of Neogene and Quaternary snakes of Central and Eastern Europe. Part II: Natricinae, Elapidae, Viperidae. *Estudios geológicos*, 47 (1991), 237–266.

10. M. Venczel, Middle-late Miocene snakes from the Pannonian Basin. *Acta Palaeontologica Romaniae*, 7 (2011), 343–349.

11. R. A. Pyron, R. G. Reynolds, and F. T. Burbrink, A taxonomic revision of boas (Serpentes: Boidae). *Zootaxa*, 3846 (2014), 249–260.

12. H. Zaher, R. W. Murphy, J. C. Arredondo, et al., Large-scale molecular phylogeny, morphology, divergence-time estimation, and the fossil record of advanced caenophidian snakes (Squamata: Serpentes). *PloS ONE*, 14 (2019), e0216148.

13. F. T. Burbrink, F. G. Grazziotin, R. A. Pyron, et al., Interrogating genomic-scale data for Squamata (lizards, snakes, and amphisbaenians) shows no support for key traditional morphological relationships. *Systematic Biology*, 69 (2020), 502–520.

14. M. Ivanov, Changes in the composition of the European snake fauna during the Early Miocene and at the Early/Middle Miocene transition. *Paläontologische Zeitschrift*, 74/4 (2001), 563–573.

15. A. Čerňanský, D. Vasilyan, G. L. Georgalis, et al., First record of fossil anguines (Squamata; Anguidae) from the Oligocene and Miocene of Turkey. *Swiss Journal of Geosciences*, 110 (2017), 741–751.

16. D. Vasilyan, Z. Roček, A. Ayvazyan, and L. Claessens, Fish, amphibian and reptilian faunas from latest Oligocene to middle Miocene localities from Central Turkey. *Palaeobiodiversity and Palaeoenvironments*, 99 (2019), 723–757.

17. R. L. Bernor, M. Brunet, L. Ginsburg, et al., A consideration of some major topics concerning Old World Miocene Mammalian chronology, migrations and paleogeography. *Geobios*, 20 (1987), 431–439.

18. V. Mosbrugger, T. Utescher, and D. L. Dilcher, Cenozoic continental climatic evolution of Central Europe. *PNAS*, 102 (2005), 14964–14969.

19. Q-Q Zhang, T. Smith, J. Yang, and C.-S. Li, Evidence of a cooler continental climate in East China during the warm early Cenozoic. *PLoS ONE*, 11 (2016), e0155507.

20. F. Rögl, Circum-Mediterranean Miocene palaeogeography. In G. E. Rössner and K. Heissig, eds., *The Miocene Land Mammals of Europe* (München: Verlag Dr. Friedrich Pfeil, 1999), pp. 39–48.

21. Z. Szyndlar and I. Hoşgör, *Bavarioboa* sp. (Serpentes, Boidae) from the Oligocene/ Miocene of eastern Turkey with comments on connections between European and Asiatic snake faunas. *Acta Palaeontologica Polonica*, 57 (2012), 667–671.

22. E. Syromyatnikova, G. L. Georgalis, S. Mayda, T. Kaya, and G. Saraç, A new early Miocene herpetofauna from Kılçak, Turkey. *Russian Journal of Herpetology*, 26 (2019), 205–224.

23. J. J. Head, D. M. Mohabey, and J. A. Wilson, *Acrochordus* Hornstedt (Serpentes, Caenophidia) from the Miocene of Gujarat, western India: temporal constraints on dispersal of a derived snake. *Journal of Vertebrate Paleontology*, 27 (2007), 720–723.

24. R. Hoffstetter, Les serpents du Néogène du Pakistan (couches des Siwaliks). *Bulletin de la Société Géologique de France*, 7 (1964), 467–474.

25. R. M. West, J. H. Hutchison, and J. Munthe, Miocene vertebrates from the Siwalik Group, western Nepal. *Journal of Vertebrate Paleontology*, 11 (1991), 108–129.

26. J-C. Rage and L. Ginsburg, Amphibians and squamates from the Early Miocene of Li Mae Long, Thailand: the richest and most diverse herpetofauna from the Cainozoic of Asia. In Z. Roček and S. Hart, eds.,

Herpetology '97 (Prague: Ministry of Environment of the Czech Republic, 1997), pp. 167–168.

27. J.-C. Rage, S. S. Gupta, and G. V. R. Prasad, Amphibians and squamates from the Neogene Siwalik beds of Jammu and Kashmir, India. *Paläontologische Zeitschrift*, 75 (2001), 197–205.

28. J. Head, Snakes of the Siwalik Group (Miocene of Pakistan): Systematics and relationships to environmental change. *Palaeontologia Electronica*, 8.1.18A (2005), 1–33.

29. K. L. Sanders, Mumpuni, A. Hamidy, J. J. Head, and D. J. Gower, Phylogeny and divergence times of filesnakes (*Acrochordus*): inferences from morphology, fossils and three molecular loci. *Molecular Phylogenetics and Evolution*, 56 (2010), 857–867.

30. J. J. Head, K. Mahlow, and J. Müller, Fossil calibration dates for molecular phylogenetic analysis of snakes 2: Caenophidia, Colubroidea, Elapoidea, Colubridae. *Palaeontologia Electronica*, 19 (2016), 1–21.

31. Z. Szyndlar, Snake fauna (Reptilia: Serpentes) from the Early/Middle Miocene of Sandelzhausen and Rothenstein 13 (Germany). *Paläontologische Zeitschrift*, 83 (2009), 55–66.

32. B. Mennecart, D. Zoboli, L. Costeur, and G. L. Pillola, On the systematic position of the oldest insular ruminant *Sardomeryx oschiriensis* (Mammalia, Ruminantia) and the early evolution of the Giraffomorpha. *Journal of Systematic Palaeontology* (published online June 2018): https://doi .org/10.1080/14772019.2018.1472145.

33. M. Venczel and B. Sanchíz, Lower Miocene Amphibians and Reptiles from Oschiri (Sardinia, Italy). *Hantkeniana*, 5 (2006), 72–75.

34. A. Čerňanský, J.-C. Rage, and J. Klembara, The Early Miocene squamates of Amöneburg (Germany): The first stages of modern squamates in Europe. *Journal of Systematic Palaeontology*, 13/2 (2015), 97–128.

35. M. S. Kayseri-Özer, Spatial distribution of climatic conditions from the Middle Eocene to Late Miocene based on palynoflora in Central, Eastern and Western Anatolia. *Geodinamica Acta*, 26/ 1–2 (2013), 122–157.

36. R. Hoffstetter and J.-C. Rage, Les Erycinae fossiles de France (Serpentes, Boidae). Compréhension et histoire de la sous-famille. *Annales de Paléontologie (Vertébrés)*, 58/1 (1972), 81–124.

37. Z. Szyndlar and W. Böhme, Die fossile Schlangen Deutschlands: Geschichte der Faunen und ihrer Erforschung. *Mertensiella*, 3 (1993), 381–431.

38. J. Müller, Untermiozäne Kieferfragmente von Schlangen (Reptilia: Serpentes: Erycinae) aus der französischen Lokalität Poncenat. *Neues Jahrbuch für Geologie und Paläontologie, Mh.*, 1998/2 (1998), 119–128.

39. Z. Szyndlar and H. H. Schleich, Description of Miocene Snakes from Petersbuch 2 with comments on the Lower and Middle Miocene Ophidian faunas of Southern Germany. *Stuttgarter Beiträge zur Naturkunde, B* 192 (1993), 1–47.

40. U. Kuch, J. Müller, C. Mödden, and D. Mebs, Snake fangs from the Lower Miocene of Germany: evolutionary stability of perfect weapons. *Naturwissenschaften*, 93 (2006), 84–87.

41. J.-C. Rage, The oldest known colubrid snakes. The state of the art. In Z. Szyndlar, ed., A Festschrift in honour of Professor Marian Młynarski on the occasion of his retirement, *Acta zoologica cracoviensia*, 31/ 13 (1988), 457–474.

42. P. Grunert, A. Tzanova, M. Harzhauser, and W. E. Piller, Mid-Burdigalian Paratethyan alkenone record reveals link between orbital forcing, Antarctic ice-sheet dynamics and European climate at the verge to Miocene Climate Optimum. *Global and Planetary Change*, 123(Part A) (2014), 36–43.

43. M. Böhme, A. Bruch, and A. Selmeier, The reconstruction of Early and Middle Miocene climate and vegetation in

Southern Germany as determined from the fossil wood flora. *Palaeogeography, Palaeoclimatology, Palaeoecology*, 253 (2007), 91–114.

44. V. Paclík, M. Ivanov, and A. H. Luján, Early Miocene snakes from the locality of Wintershof-West (Germany). In M. Marcola, O. Mateus and M. Moreno-Azanza, eds., *Abstract book of the XVI Annual Meeting of the European Association of Vertebrate Palaeontology, Caparica, Portugal June 26th–July 1st* (Lisbon, 2018), p. 144.

45. M. Ivanov, The oldest known Miocene snake fauna from Central Europe: Merkur-North locality, Czech Republic. *Acta Palaeontologica Polonica*, 47/3 (2002), 513–534.

46. F. Römer, Über *Python Euboïcus*, eine fossile Riesenschlange aus tertiärem Kalkschiefer von Kumi auf der Insel Euboea [On *Python Euboïcus*, a fossil giant snake from the Tertiary shale of Kumi in the island of Euboea]. *Zeitschrift der Deutschen Geologische Gesellschaft*, 22 (1870), 582–590.

47. G. L. Georgalis, M. K. Abdel Gawad, S. M. Hassan, et al., Oldest co-occurrence of *Varanus* and *Python* from Africa–first record of squamates from the early Miocene of Moghra Formation, Western Desert, Egypt. *PeerJ*, 8 (2020), e9092.

48. A. Figueroa, A. D. McKelvy, L. L. Grismer et al., A species-level phylogeny of extant snakes with description of a new colubrid subfamily and genus. *PLoS ONE*, 11 (2016), e0161070.

49. M. Ivanov, Snakes of the lower/middle Miocene transition at Vieux-Collonges (Rhône; France), with comments on the colonization of western Europe by colubroids. *Geodiversitas*, 22/4 (2000), 559–588.

50. Z. Szyndlar, Snake fauna from the Late Miocene of Rudabánya. *Palaeontographia Italica*, 90/2003 (2005), 31–52.

51. J.-C. Rage and S. Bailon, Amphibians and squamate reptiles from the late early Miocene (MN 4) of Béon 1 (Montréal-du-Gers, southwestern France). *Geodiversitas*, 27/3 (2005), 413–441.

52. M. Böhme, The Miocene Climatic Optimum: evidence from the ectothermic vertebrates of Central Europe. *Palaeogeography, Palaeoclimatology, Palaeoecology*, 195 (2003), 389–401.

53. J. C. Zachos, G. R. Dickens, and R. E. Zeebe, An early Cenozoic perspective on greenhouse warming and carbon-cycle dynamics. *Nature*, 451 (2008), 279–283.

54. M. Ivanov and M. Böhme, Snakes from Griesbeckerzell (Langhian, Early Badenian), North Alpine Foreland Basin (Germany), with comments on the evolution of snake fauna in Central Europe during the Miocene Climatic Optimum. *Geodiversitas*, 33/3 (2011), 411–449.

55. A. A. Bruch, T. Utescher, C. Alcalde Olivares, et al., Middle and Late Miocene spatial temperature patterns and gradients in Europe – preliminary results based on palaeobotanical climate reconstructions. *Courier Forschungsinstitut Senckenberg*, 249 (2004), 15–27.

56. M. Böhme, A. Ilg, A. Ossig, and H. Küchenhoff, New method to estimate paleoprecipitation using fossil amphibians and reptiles and the middle and late Miocene precipitation gradients in Europe. *Geology*, 34/6 (2006), 425–428.

57. M. Böhme, Lower Vertebrates (Teleostei, Amphibia, Sauria) from the Karpatian of the Korneuburg Basin – palaeoecological, environmental and palaeoclimatical implications. *Beiträge zur Paläontologie*, 27 (2002), 339–353.

58. P. M. Tempfer, Amphibians and Reptiles of the Karpatian Central Paratethys. In R. Brzobohatý, I. Cicha, M. Kováč and F. Rögl, eds., *The Karpatian. A Lower Miocene Stage of the Central Paratethys* (Brno: Masaryk University, 2003), pp. 285–290.

59. N. Doláková and M. Slamková, Palynological Characteristics of Karpatian

Sediments. In R. Brzobohatý, I. Cicha, M. Kováč and F. Rögl, eds., *The Karpatian. A Lower Miocene Stage of the Central Paratethys* (Brno: Masaryk University, 2003), pp. 325–337.

60. H. Thomas, S. Sen, M. Khan, B. Battail, and G. Ligabue, The Lower Miocene fauna of Al-Sarrar (Eastern province, Saudi Arabia), *Atlal, Journal of Saudi Arabian Archaeology*, 5 (1982), 109–136.

61. G. L. Georgalis, S. Mayda, B. Alpagut, et al., The westernmost Asian record of pythonids (Serpentes): the presence of *Python* in a Miocene hominoid locality of Anatolia. *Journal of Vertebrate Paleontology*, (2020), https://doi.org/10.1080/02724634.2020 .1781144.

62. D. V. Malakhov, The early Miocene herpetofauna of Ayakoz (Eastern Kazakhstan). *Biota*, 6/1–2 (2005), 29–35.

63. V. M. Chkhikvadze, Preliminary results of studies on tertiary amphibians and squamate reptiles of the Zaisan Basin. In I. Darevsky, ed., Questions of herpetology, *The Sixth All-Union Herpetological Conference* (Tashkent: Nauka, 1985) pp. 234–235. (in Russian)

64. A. Sun, Notes on fossil snakes from Shanwang, Shangtung. *Vertebrata Palasiatica*, 4 (1961), 310–312.

65. J. A. Holman and M. Tanimoto, Cf. *Trimeresurus* Lacépède (Reptilia: Squamata: Viperidae: Crotalinae) from the Early Miocene of Japan. *Acta zoologica cracoviensia*, 47/1 (2004), 1–7.

66. M. Ivanov, A. Čerňanský, I. Bonilla Salomón, and A. H. Luján, Early Miocene squamate assemblage from the Mokrá-Western Quarry (Czech Republic) and its palaeobiogeographical and palaeoenvironmental implications. *Geodiversitas*, 42/20 (2020), 343–376.

67. J.-C. Rage and Z. Szyndlar, Latest Oligocene-Early Miocene in Europe: Dark Period for booid snakes. *Comptes Rendus Palevol*, 4 (2005), 428–435.

68. M. Augé and J.-C. Rage, Les Squamates (Reptilia) du Miocéne moyen de Sansan

(Gers, France). *Mémoires du Muséum national d'Histoire naturelle*, 183 (2000), 263–313.

69. Z. Szyndlar and H. H. Schleich, Two species of the genus *Eryx* (Serpentes; Boidae; Erycinae) from the Spanish Neogene with comments on the past distribution of the genus in Europe. *Amphibia – Reptilia*, 15 (1994), 233–248.

70. Z. Szyndlar and F. Alférez, Iberian snake fauna of the Early/Middle Miocene transition. *Revista Española de Herpetología*, 19 (2005), 57–70.

71. Z. Szyndlar and J.-C. Rage, Oldest fossil vipers (Serpentes: Viperidae) from the Old World. *Kaupia*, 8 (1999), 9–20.

72. Z. Szyndlar and J.-C. Rage, West Palearctic cobras of the genus *Naja* (Serpentes: Elapidae): interrelationships among extinct and extant species. *Amphibia – Reptilia*, 11 (1990), 385–400.

73. P. M. Miklas-Tempfer, The Miocene Herpetofaunas of Grund (Caudata; Chelonii, Sauria, Serpentes) and Mühlbach am Manhartsberg (Chelonii, Sauria, Amphisbaenia, Serpentes), Lower Austria. *Annalen des Naturhistorischen Museums in Wien, A* 104 (2003), 195–235.

74. G. Daxner-Höck, P. M. Miklas-Tempfer, U. B. Göhlich, et al., Marine and terrestrial vertebrates from the Middle Miocene of Grund (Lower Austria). *Geologica Carpathica*, 55/2 (2004), 191–197.

75. J. A. McCartney, N. J. Stevens, and P. M. O'Connor, The Earliest Colubroid-Dominated Snake Fauna from Africa: Perspectives from the Late Oligocene Nsungwe Formation of Southwestern Tanzania. *PLoS ONE*, 9(3) (2014), e90415.

76. Z. Szyndlar, Snakes from the Lower Miocene locality of Dolnice (Czechoslovakia). *Journal of Vertebrate Paleontology*, 7 (1987), 55–71.

77. Z. Szyndlar, Vertebrates from the Early Miocene lignite deposits of the opencast mine Oberdorf (Western Styrian Basin, Austria). *Annalen des Naturhistorischen Museums in Wien, A* 99 (1998), 31–38.

78. D. Salvi, J. Mendes, S. Carranza, and D. J. Harris, Evolution, biogeography and systematics of the western Palaearctic *Zamenis* ratsnakes. *Zoologica Scripta*, 47 (2018), 441–461.

79. J.-C. Rage and J. A. Holman, Des Serpents (Reptilia, Squamata) de type Nord-Américain dans le Miocéne francais. *Evolution paralléle ou dispersion?* Géobios, 17 (1984), 89–104.

80. J. A. Holman, Reptiles of the Egelhoff local fauna (Upper Miocene) of Nebraska. *Contributions from the Museum of Paleontology, the University of Michigan*, 24 (1973), 125–134.

81. J. A. Holman, *Fossil Snakes of North America. Origin, Evolution, Distribution, Paleoecology* (Bloomington: Indiana University Press, 2000).

82. G. L. Georgalis, A. Villa, M. Ivanov et al. Early Miocene herpetofaunas from the Greek localities of Aliveri and Karydia – bridging a gap in the knowledge of amphibians and reptiles from the early Neogene of southeastern Europe. *Historical Biology*, 31 (2019), 1045–1064.

83. G. A. Zerova. The first find of a fossil Sand Boa of the genus *Albaneryx* (Serpentes, Boidae) in the USSR. *Vestnik Zoologii*, 5 (1989), 30–35.

84. M. Ivanov, D. Vasilyan, M. Böhme, and V. S. Zazhigin, Miocene snakes from northeastern Kazakhstan: new data on the evolution of snake assemblages in Siberia. *Historical Biology*, 31/10 (2019), 1284–1303.

85. Z. Szyndlar and G. A. Zerova, Miocene snake fauna from Cherevichnoie (Ukraine, USSR), with description of a new species of *Vipera. Neues Jahrbuch für Geologie und Paläontologie, A*, 184 (1992), 87–99.

86. Y. Chaimanee, C. Yamee, B. Marandat, and J.-J. Jaeger, First middle Miocene rodents from the Mae Moh Basin (Thailand): Biochronological and paleoenvironmental implications. *Bulletin of Carnegie Museum of Natural History*, 39 (2007), 157–63.

87. J.-C. Rage and I. G. Danilov, A new Miocene fauna of snakes from eastern Siberia, Russia.

Was the snake fauna largely homogenous in Eurasia during the Miocene? *Comptes Rendus Palevol*, 7 (2008), 383–390.

88. G. A. Zerova, A. N. Lungu, and V. M. Chkhikvadze, Large fossil vipers from northern Black Seaside and Transcaucasus. *Trudy zoologicheskogo Instituta Akademii Nauk S.S.S.R.*, 158/1986 (1987), 89–99. (in Russian)

89. G. A. Zerova, *Vipera* (*Daboia*) *ukrainica* - a new viper (Serpentes; Viperidae) from the Middle Sarmatian (Upper Miocene) of the Ukraine. *Neues Jahrbuch für Geologie und Paläontologie, A*, 184 (1992), 235–249.

90. J. Šmíd and K. A. Tolley, Calibrating the tree of vipers under the fossilized birth-death model. *Scientific Reports*, 9 (2019), 5510 https://doi.org/10.1038/s41598–019-41290-2

91. M. Ivanov, The first European pit viper from the Miocene of Ukraine. *Acta Palaeontologica Polonica*, 44/3 (1999), 327–334.

92. M. Ivanov, Fossil snake assemblages from the French Middle Miocene localities at La Grive (France). In Abstracts volume and excursions field guide, The 7th European workshop of vertebrate palaeontology, July 2-7, 2002, (Sibiu, 2002), pp. 26–27.

93. G. L. Georgalis, A. Villa, M. Ivanov, et al., Fossil amphibians and reptiles from the Neogene locality of Maramena (Greece), the most diverse European herpetofauna at the Miocene/Pliocene transition boundary. *Palaeontologia Electronica*, 22.3.68 (2019), 1–99.

94. R. Hoffstetter, Contribution à l'étude des Elapidae actuels et fossiles et de l'ostéologie des Ophidiens. *Archives du Muséum d'Histoire Naturelle de Lyon*, 15 (1939), 1–78.

95. Z. Szyndlar and G. A. Zerova, Neogene cobras of the genus *Naja* (Serpentes: Elapidae) of East Europe. *Annalen des Naturhistorischen Museums in Wien, A* 91 (1990), 53–61.

96. J.-C. Rage and Z. Szyndlar, *Natrix longivertebrata* from the European

Neogene, a snake with one of the longest known stratigraphic ranges. Neues Jahrbuch für Geologie und Paläontologie, Mh., 1 (1986), 56-64.

97. G. A. Zerova, Late Cainozoic localities of snakes and lizards of Ukraine. *Revue de Paléobiologie*, 7 (1993), 273-280.

98. M. Venczel and E. Ştiucă, Late middle Miocene amphibians and squamate reptiles from Tauţ, Romania. *Geodiversitas*, 30 (2008), 731-763.

99. M. Ivanov, *Hadi evropského kenozoika*. MS, PhD Thesis (Brno: Masaryk University, 1997), pp. 1-217. (in Czech)

100. M. Ivanov, The snake fauna of Devínska Nová Ves (Slovak Republic) in relation to the evolution of snake assemblages of the European Middle Miocene. *Acta Musei Moraviae, Scientiae geologicae*, 83 (1998), 159-172.

101. M. Böhme, A. Ilg, and M. Winklhofer, Late Miocene 'washhouse' climate in Europe. *Earth and Planetary Science Letters*, 275 (2008), 393-401.

102. J. Agustí and M. Antón, *Mammoths, Sabertooths, and Hominids - 65 million Years of Mammalian Evolution in Europe*, (New York: Columbia University Press, 2002).

103. P. Uetz, P. Freed, and J. Hošek, eds., The Reptile Database, www.reptile-database .org, (2020), accessed (20/04/2020).

104. M. Böhme, 3. Herpetofauna (Anura, Squamata) and palaeoclimatic implications: preliminary results. In G. Daxner-Höck, ed., Oligocene-Miocene vertebrates from the Valley of Lakes (Central Mongolia): morphology, phylogenetic and stratigraphic implications, *Annalen des Naturhistorischen Museums in Wien*, 108 (2007), 43-52.

105. G. L. Georgalis, J.-C. Rage, L. de Bonis, and G. D. Koufos, Lizards and snakes from the late Miocene hominoid locality of Ravin de la Pluie (Axios Valley, Greece). *Swiss Journal of Geosciences*, 111 (2018), 169-181.

106. F. Bachmayer and Z. Szyndlar, Ophidians (Reptilia: Serpentes) from the Kohfidisch fissures of Burgenland, Austria. *Annalen des Naturhistorischen Museums in Wien, A* 87 (1985), 79-100.

107. F. Bachmayer and Z. Szyndlar, A second contribution to the ophidian fauna (Reptilia, Serpentes) of Kohfidisch, Austria. *Annalen des Naturhistorischen Museums in Wien, A* 88 (1987), 25-39.

108. P. M. Tempfer, The Herpetofauna (Amphibia: Caudata, Anura; Reptilia: Scleroglossa) of the Upper Miocene Locality Kohfidisch (Burgerland, Austria). *Beiträge zur Paläontologie*, 29 (2005), 145-253.

109. S. Colombero, C. Angelone, E. Bonelli, et al., The Messinian vertebrate assemblages of Verduno (NW Italy): another brick for a latest Miocene bridge across the Mediterranean. *Neues Jahrbuch für Geologie und Paläontologie, A*, 272/3 (2014), 287-324.

110. V. Codrea, M. Venczel, L. Ursachi, and B. Răţoi, A large viper from the early Vallesian (MN 9) of Moldova (E-Romania) with notes on the palaeobiogeography of late Miocene 'Oriental vipers'. *Geobios*, 50 (2017), 401-411.

111. M. Venczel and G. Várdai, The genus *Elaphe* in the Carpathian Basin: Fossil record. *Nymphaea Folia naturae Bihariae*, 28 (2000), 65-82.

112. M. Venczel, Late Miocene snakes from Polgárdi (Hungary). *Acta zoologica cracoviensia*, 37/1 (1994), 1-29.

113. M. Venczel, Late Miocene snakes (Reptilia: Serpentes) from Polgárdi (Hungary): a second contribution. *Acta zoologica cracoviensia*, 41/1 (1998), 1-22.

114. S. Bailon, P. Bover, J. Quintana, and J. A. Alcover, First fossil record of Vipera Laurenti 1768 'Oriental vipers complex' (Serpentes: Viperidae) from the Early Pliocene of the western Mediterranean islands. *Comptes Rendus Palevol, 9 (2010)*, 147-154.

115. E. Torres, S. Bailon, P. Bover, and J. A. Alcover, Sobre la presencia de un vipérido de gran talla perteneciente al Complejo de Víboras Orientales en el yacimiento de Na Burguesa-1

(Mioceno Superior/Plioceno Inferior, Mallorca). *Jornadas de Paleontología SEP*, 30 (2014), 237–240.

116. H.-A. Blain, S. Bailon, and J. Agustí, The geographical and chronological pattern of herpetofaunal Pleistocene extinctions on the Iberian Peninsula. *Comptes Rendus Palevol*, 15 (2016), 731–744.

117. Z. Szyndlar, Fossil snakes from Poland. *Acta zoologica cracoviensia*, 28/1 (1984), 3–156.

118. F. Pokrant, C. Kindler, M. Ivanov, et al., Integrative taxonomy provides evidence for the species status of the Ibero-Maghrebian grass snake *Natrix astreptophora*. *Biological Journal of the Linnean Society*, 118 (2016), 873–888.

119. M. Delfino, T. Kotsakis, M. Arca, et al., Agamid lizards from the Plio-Pleistocene of Sardinia (Italy) and an overview of the European fossil record of the family. *Geodiversitas*, 30/3 (2008), 641–656.

120. G. L. Georgalis, A. Villa, E. Vlachos, and M. Delfino, Fossil amphibians and reptiles from Plakias, Crete: A glimpse into the earliest late Miocene herpetofaunas of southeastern Europe. *Geobios*, 49 (2016), 433–444.

121. A. Villa and M. Delfino, Fossil lizards and worm lizards (Reptilia, Squamata) from the Neogene and Quaternary of Europe: an overview. *Swiss Journal of Palaeontology*, 138/2 (2019), 177–211.

122. Z. Szyndlar, Ophidian fauna (Reptilia, Serpentes) from the Upper Miocene of Algora (Spain). *Estudios geológicos*, 41 (1985), 447–465.

123. Z. Szyndlar, Two new extinct especies of the genera *Malpolon* and *Vipera* (Reptilia, Serpentes) from the Pliocene of Layna (Spain). *Acta Zoologica Cracoviensia*, 31/27 (1988), 687–706.

124. S. Bailon and C. Verbeke, Reptiles escamosos del Mioceno final (MN13) de Salobreña (Granada, España). In B. Martínez-Navarro, P. Palmqvist, M. Patrocinio Espigares and S. Ros-Montoya, eds., *Libro de resúmenes XXXV Jornadas de la Sociedad Española de Paleontología*, (2019), pp. 37–38.

125. E. Syromyatnikova and A. Tesakov, Preliminary report on herpetofauna from the Solnechnodolsk locality (late Miocene), Russia. In A. MacKenzie, E. Maxwell and J. Miller-Camp, eds., *Abstracts of papers, 75th Annual Meeting of the Society of Vertebrate Paleontology*, 14–17 October 2015, (Dallas, 2015), p. 221.

126. A. M. Villa, G. Carnevale, M. Pavia, et al., An overview on the late Miocene vertebrates from the fissure fillings of Monticino Quarry (Brisighella, Italy), with new data on non-mammalian taxa. *Rivista Italiana di Paleontologia e Stratigrafia*, 127 (2) (2021) 297–354.

PART II

Palaeontology and the Marine-Origin Hypothesis

6

Sea-Serpentism

6.1 Introduction: The Genesis of a Myth

On 2 June 1849, the *Scientific American* (volume 4, p.290) reported news of a sea serpent sighting: '... the clipper ship Sophia Walker, Captain Wiswell, was chased around Cape Horn by an enormous sea serpent, half a mile long, and that Captain Wiswell was so terrified that his eyes stuck out far enough to hang a Quaker's hat upon.' A summary of sea serpent sightings dating back to 1895, published in the 12 October issue of the *Scientific American* of that year, specified that the creature 'always remains at the bottom of the sea except in July and August, when it makes its appearance at the surface of the water ... One will doubtless not fail to remark that the sea serpent has selected just those two months in which newspapers most suffer from a dearth of news!' Indeed, a modern statistical analysis confirmed: 'sea-serpent sightings do tend to be a summer phenomenon', although the idea that this was due to a dull season of the year was found to be 'difficult to test' [1: p.301].

Perhaps the earliest and most famous appearance of the sea serpent is on the *Carta marina* drawn by Olaus Magnus (1490–1557), a catholic priest who, following the conversion of his native Sweden to Lutheranism, sought exile in Rome. In 1527, Olaus began work on a map of northern Europe (mostly Scandinavia), which he completed in 1539 [2]. The *Carta marina* is the earliest, more or less correct map of northern Europe, which Olaus enriched and explained through a commentary book, *Historia de Gentibus Septentrionalibus* (History of the Northern People), first printed in Rome in 1555 [2]. The sea in Olaus's map was populated by all kinds of mythical creatures, amongst them two sea serpents located near the coast of Norway. The more northern one of these he identified, in the key to his map, as 'A worm 200 feet long wrapping itself around a big ship and destroying it'; the smaller southern one shown in a location near Bergen he identified as 'A sea snake, 30 or 40 feet long' [2: p.12]. A remarkable detail concerning the smaller southern sea serpent is that it is shown to propel itself forward through vertical undulations of its body, not sideward undulation as eels, or true snakes, would do. In his commentary book, Olaus was more explicit:

> They who in works of navigation on the coasts of Norway employ themselves in fishing or merchandise do all agree in this strange story, that there is a serpent there which is of a vast magnitude, namely 200 foot long, and moreover, 20 foot thick; and is wont to live in rocks and caves toward the sea-coast about Berge: which will go from his holes on a clear night in summer, and devour calves, lambs, and hogs, or else it goes into the sea to feed on polypus (octopus), locusts (lobsters), and all sorts of sea-crabs. He hath commonly hair hanging from his neck a cubit long, and sharp scales, and is black, and he has flaming, shining eyes. This snake disquites the shippers; and he puts up his head on high like a pillar, and catcheth away men, and he devours them . . . (translation from H. Lee [3: p.56]; also translated in Pontoppidan [4: p.208], who disputes the occurrence of the beast near Bergen; see also Oudemans [5: p.10], for a slightly more modern translation).

6.2 The Proliferation of a Myth

Olaus Magnus's report on sea serpents off the coast of Norway made it promptly into the *Historia Animalium*, published by the famous Swiss physician and naturalist Conrad Gesner (1516–1565) in Zurich in 1558 [6]. Gesner [6: p.1040] illustrates both sea serpents that appear on Olaus Magnus's *Carta marina* in mirrored woodcuts, the smaller one again displaying distinct vertical undulations of its body as it moves forward, the other one coiling its massive body around a ship while snatching a sailor off of it. Gesner writes of Olaus Magnus's sea serpent that it:

> . . . sometimes appears near Norway in fine weather, and is dangerous to sea-men, as it snatches away men from the ships. Mariners tell that it incloses ships, as large as our trading vessels . . . It sometimes makes such large coils above the water, that a ship can go through one of them . . . (translation from Oudemans [5: p.107]).

It has been argued that sea serpent sightings multiplied following the publication, in 1755, of the English translation of Erik Pontoppidan's (1698–1764) *The Natural History of Norway* [7]. There, Pontoppidan [4: p.195–196], Bishop of Bergen, wrote: 'The Soe Ormen, the Sea-Snake, *Serpens Marinus Magnus* . . . is a wonderful and terrible Sea-monster, which extremely deserves to be taken notice of by those who are curious to look into the extraordinary works of the great creator'. Most significantly, the sea serpent described and illustrated by Pontoppidan again propelled itself forward by vertical undulation of its body (Fig. 6.1). Pontoppidan [4: p.206] describes 'round lumps' or 'folds' that are visible when a sea serpent surfaces, and goes on to cite the German Scholar Adam Olearius (1599–1671), who in his '*Gottdorf Museum*' wrote: '. . . [in] the North sea, he saw in the water, which was very calm, a Snake, which appeared at that distance to be as thick as a pipe of wine, and had 25 folds' (Olearius [8: p.19]; translation from Pontoppidan [4: p.207]). Pontoppidan then goes on to surmise that the number of folds that 'may be counted on a Sea-snake . . . is not always the same, but depends on the various sizes of them' [4: p.207].

Figure 6.1 The Great Sea-Serpent, reproduced from Pontoppidan [4: plate 27].

6.3 The Scientific Approach to a Myth

The various sea serpent sightings reported throughout the nineteenth century have been collected and commented upon by Oudemans [5]. Here, only one of the most notorious cases will be recounted, involving a pathological specimen of the common black racer (*Coluber constrictor*) with a series of bumps along its back found near Gloucester, Massachusetts [9, 10]. The Gloucester Sea Serpent was first spotted near Cape Ann when it entered Gloucester Harbor on 6 August 1817, 'in pursuit of shoals of fishes, herrings, squids, &c. on which it feeds. Its motions are very quick; it was seen by great many, but all attempts to catch it have failed, although $5000 has been offered for its spoils' [11: p.433].

> In the month of August 1817, it was currently reported on various authorities, that an animal of very singular appearance had been recently and repeatedly seen in the harbor of Gloucester, Cape Ann, about thirty miles distant from Boston. It was said to resemble a serpent in its general form and motions, to be of immense size, and to move with wonderful rapidity; to appear on the surface of the water only in calm and bright weather; and to seem jointed or like a number of buoys or casks following each other in a line. In consequence of these reports, at a meeting of the Linnean Society of New England, holden at Boston on the 18th day of August, the Hon. John Davis, Jacob Bigelow, M. D., and Francis C. Gray, Esq.

were appointed a Committee to collect evidence with regard to the existence and appearance of any such animal.

Such opens the *Report of a Committee of the Linnean Society of New England, relative to a large marine animal, supposed to be a serpent, seen near Cape Ann, Massachusetts, in August 1817*, published in 1817 by Cummings and Hilliard in Boston (hereafter referred to as *Report* [12]). Eye-witnesses who claimed to have spotted the Sea Serpent were confronted with a list of questions drawn up by the Committee, and asked to testify under oath. The committee members were identified as 'men of fair and unblemished character in Gloucester' by one commentator [13: p.89], who also surmised that the classic summertime appearance of the Gloucester Sea Serpent may be explained by it being 'migratory, like the Coluber Natrix in Hungary, and may pass the winter season in Mexico or South America' [13: p.91]. The leading British geologist of the time, Charles Lyell (1797–1875), when commenting on the Gloucester Sea Serpent investigation, likewise emphasized: 'I am well acquainted with two of the three gentlemen' on the committee [14: p.111].

The various reports delivered to the Linnean Society committee were not entirely free of contradictions: 'These two points – the vertical or horizontal nature of the creature's undulations when moving, and the apparent 'bunches', or serrations exhibited by the outline of its back – seem to have formed rather a bone of contention among both the observers and the Committee' [15: p.39]. With much hindsight, Gould [15: p.40] wrote with reference to the putative vertical undulations of the Sea Serpent, 'the structure of its backbone must have been truly astonishing – it must have looked like a series of universal joints'. Because the Gloucester Sea Serpent was never caught or killed, it was at the time an unexpected opportunity to investigate the matter when on 27 September 1817, a small black snake was caught and killed 'in a field near Loblolly cove, about a mile and a half from Sandy Bay village' (*Report* [12: p.44]). The little black snake was brought to Boston for the Linnean Society committee members to inspect it: 'The animal had the general form and external characters of a serpent; but was remarkably distinguished from all others of that class known to your Committee, by a row of protuberances along the back, apparently formed by undulations of the spine' (*Report* [12: p.38]). The specimen was consequently named a new genus and species, *Scoliophis atlanticus* (the 'flexible Atlantic snake'). The committee members recognized the similarities the specimens shared with the common black racer (*Coluber constrictor*), a snake frequently seen in this area, but insisted that it differed 'strikingly ... by its undulating back' (*Report* [12: p.38]). Part of the specimen was eventually dissected in an attempt to clarify the functional anatomy of the vertebral musculoskeletal system in vertical undulation: 'The structure of the *spine* is very singular, and different from that of any serpent we have seen, or known to be described ... The structure of the different vertebrae varies to accommodate itself to this configuration [i.e., vertical undulation], so that the spine cannot be extended into a straight line without dislocation of its parts' (*Report* [12: p.40]). In a footnote to those observations, the committee added: 'This is the reverse of what Cuvier asserts of serpents in general' (*Report* [12: p.41]) – Georges Cuvier (1769–1832), of course, being the famous palaeontologist and functional and comparative anatomist at the National Museum of Natural History in

Paris. Cuvier's younger counterpart in Britain, Sir Richard Owen (1804–1892), would eventually side with the latter, however. In his account of the anatomy of the ophidian vertebral column, Owen noted: 'there is no natural undulation of the body upwards and downwards, it is permitted only from side to side...' [16: p.153].

Considering the snake with the bumpy back and its unique vertebral column to represent a new species, the committee members introduced it under the name *Scoliophis atlanticus*. Justice of the Peace Lonson Nash from Gloucester, who testified in front of the committee, was not alone in presuming the 'young' serpent 'to be the progeny of the remarkable Sea Serpent, that was so often seen in the harbour', and assured his audience: 'Should it thereafter be ascertained, that the great serpent has been seen on land here, I shall take the earliest opportunity to inform you of a fact, so important' (*Report* [12: p.46]). The assumption motivating this expectation was, of course, that the Sea Serpent would deposit its eggs on land in sandy soil near the shore, as modern sea turtles do.

In contrast, the French zoologist H. de Blainville (1777–1850) pointed out that *Scoliophis atlanticus* was identical with the common black racer (*Coluber constrictor*), except for its bumpy back. With respect to the conclusion that the small snake found on land should be an offspring of the animal seen in the sea – 'the probability diminishes to almost completely zero' [17: p.304]. This highly skeptical position was confirmed by the French naturalist and explorer Alexandre Lesueur (1778–1846), who inspected the little snake with the bumpy back from Loblolly cove, and compared it with *Coluber constrictor*. He found no characters in the external appearance of *Scoliophis* that would justify its recognition as a separate genus, and when it came to the vertebral column in a section of the body forming a bump, he observed: 'the vertebrae that constitute this part show a high degree of irregularity and indeed a notable deformation ... I have no doubt whatsoever that this represents a totally anomalous condition; perhaps the individual received a series of blows in its youth ...' [18: p.468]. Commenting on what became known as the *Scoliophis* 'hoax', the leading British Geologist Charles Lyell noted of the New England Linnean Society committee: 'Just as they were concluding their report, an unlucky accident raised a laugh at the expense of the Linnean Committee, and enabled the incredulous to turn the whole matter into ridicule' [14: p.112]. The 'accident' was the finding of the Loblolly cove serpent: 'It had a series of bumps on his back, caused by the individual happening to have a diseased spine' [14: p.112]. Indeed, during his travels in the United States, Lyell had realized what impact the Gloucester Sea Serpent saga had had on the broader public: 'After the year 1817, every marvelous tale was called in the United States a snake story ...' [14: p.113].

6.4 Names Change Meanings: The Myth Turns Real

Every incredible tale was called a 'snake story'; evidently, the term 'snake' no longer referred to snakes only in this context, but to whatever the 'marvelous tale' was about. A similar looseness of meaning became increasingly apparent with respect to the term 'sea serpent', which no longer referred to a giant marine snake only, but to any monstrous, unidentified animal spotted in the sea. The October 19, 1850 issue of the *Scientific American* reported: 'a large sea serpent has been seen in the Penobscot Bay . . . It is not known to what species of fish this monster belongs, neither is it improbable that in ancient times it was known as the 'leviathan' . . .' The 27 December 1879, issue of the *Scientific American* described how a giant octopus could create the appearance of a 'hydrophidian or marine serpent.' In the 19 March 1887 issue of the *Scientific American*, an anonymous letter writer professed: 'It has been my belief for some years that there is some fitful gigantic wanderer inhabiting the ocean . . . I think it is the sea serpent . . . I shall not attempt to classify it. Whether it belongs to the mammalia, reptilia, or pisces, whether it be ophidian, cetacean, or saurian, I must leave it to the naturalist to determine.' On 31 December 1892, the *Scientific American* reported a sea serpent sighting by the crew on the Liverpool mail steamer *Angola*, and concluded: 'As is usual in cases of the kind, the signatories have been subjected to a great deal of ridicule, but they are nevertheless convinced that the monster they saw was a marine mystery and a sea serpent.' The Inspector of Fisheries Frank Buckland concluded: 'It is, therefore, evident to my mind that the coast of Norway and the north of Scotland are occasionally visited by a living creature, which, for want of a better name, is called 'The Great Sea Snake'' [19: p.411].

In 1892, the Director of the Zoological Garden at the Hague, Anthonie Cornelius Oudemans (1858–1943) published his major treatise on *The Great Sea-Serpent* [5], of which an anonymous review in *Nature* stated: 'It is impossible, however, to treat this laborious work as a scientific treatise.' And yet: 'Enough has been written [on sea serpents] to prove that this volume is not without a certain amount of interest . . . to open its pages at random one is sure to be arrested by some startling phase of belief or by some marvelous narration . . .' [20]. Oudemans [5] chronologically listed and analysed reports of sea serpent sightings, carefully trying to discern phantasy from fact. Commenting on Conrad Gesner's [6] account of Olaus Magnus's sea serpent (see above), Oudemans [5: p.109] wrote: 'Here we meet with three other characteristics of the sea-serpent: it is harmless when not provoked, it encircles ships and turns them upside down, and its coils are so gigantic that a ship can go through one of them. The first characteristic is a real one: the sea-serpent is perfectly harmless. . .,' the other characterizations he deemed exaggerations. Of the *Report of the Committee* of the Linnean Society of New England, Oudemans [5: p.163] stated in support of the existence of the Gloucester Sea Serpent: 'We observe that the Linnean Society has exerted all its energies in the matter and has acted with greatest accuracy.' But what, in Oudemans's opinion, *is* the sea serpent, which he proposed to name – following Rafinesque [11] – *Megophias megophias*? 'It will be quite superfluous to tell my readers to which order of animals I think that this *Megophias megophias* belongs. It runs like a red thread through my whole volume, that I firmly believe that it belongs to the order of *Pinnipedia*' [5: p.546] – i.e., it resembles a sea-lion [5: p.568].

6.5 The Real Sea Serpent

In contrast to Oudemans's conclusion, the fossils of marine Late Cretaceous reptiles – the mosasaurs – had for some already become the material basis of real sea serpents. The

Philadelphia-based paleontologist Edward Drinker Cope (1840–1897) is perhaps best known for his engagement in the dinosaur 'Bone Wars' against Yale paleontologist Othniel Charles Marsh (1831–1899). These played out in the American West during the 'Great Dinosaur Rush' that lasted from 1877 through 1892. In the year 1869, Cope published in the *Proceedings of the Boston Society of Natural History* a paper titled 'On the reptilian orders Pythonomorpha and Streptosauria' [21]. What is of interest here is the name Pythonomorpha (see Polcyn et al., Chapter 7), which translates as 'reptiles that resemble pythons in their anatomy'. Cope [21: p.258] included two families in his Pythonomorpha, the Clidastidae and Mosasauridae, the latter two colloquially referred to as mosasaurs, marine reptiles from the Late Cretaceous. By the end of the Cretaceous, mosasaurs had ascended to the apex of the marine food-chain, many species exceeding four meters (13 feet) in length. Cope [21: p.262] himself described the fossil remains of a very large mosasaur as those of a new species, *Mosasaurus maximus* [see also 22: p.189]. Today, this species is treated as a junior synonym of *Mosasaurus hoffmannii* [23], famously the very first mosasaur to have ever been collected, at St. Pietersberg south of Maastrich (the Netherlands) between 1770 to 1774 [24]. The estimated total length reached by that species, based on new material of *Mosasaurus hoffmannii* from Penza (Russia), is 17 meters (55 feet) [25].

When reviewing the anatomy of mosasaurs, Cope [21] noted that they share a number of characteristics with lizards, but more interestingly they also share many characteristics with snakes. One of the most striking snake characters in mosasaurs concerns their lower jaw: 'The mandibular arch is very much like that of serpents' [21: p.255]. In lizards, as in all tetrapods, the dentary bones in the right and left mandible meet anteriorly in a syndesmotic suture (that may co-ossify in the adult, as in mammals). In mosasaurs and snakes, there is no mandibular symphysis, permitting the left and right mandible to move independently from one another. Another striking feature shared by snakes and mosasaurs is a joint in the middle of the mandible, between the dentary and the postdentary bones, allowing for further mobility and expansion. This led Cope [21: p.157] to conclude that mosasaurs' 'habit was to devour [their prey] whole', as snakes do.

> 'From the preceding evidence, we may look upon the mosasauroids and their allies as a race of gigantic, marine, serpent-like reptiles, with powers of swimming and running, like the modern Ophidia. Adding a pair of short anterior paddles, they are not badly represented by old Pontoppidan's figure of the sea serpent … Thus in the mosasauroids, we almost realize the fictions of snake-like dragons and sea serpents, in which men have been ever prone to indulge. On account of the ophidian part of their affinities, I have called this order the Pythonomorpha. ([21: p.257–258]; see also Fig. 6.1)

Cope's account of the Pythonomorpha was subsequently scrutinized by Owen, a critique which Cope [26] answered in a lengthy rebuttal. In it, Cope once again stressed his conclusion 'that the *Pythonomorpha* are nearer to the *Ophidia* than are the *Lacertilia*' [26: p.310]. Owen [27: p.714] criticized Cope for having claimed that the mosasaur represents 'a marine animal, scientifically entitled to be called 'serpent' … unless the 'great sea serpent' of our newspapers should establish its claims for admittance into the scientific catalogue.' Interestingly, in his reply Cope invoked the indeterminacy of meaning of the term 'sea serpent' again: 'As to the use of the term sea-serpent, since I have not referred these reptiles [i.e., the pythonomorphs] to the *Ophidia*, the term involves no error' [26: p.310].

6.6 The Sea Serpent's Legs

In a book review titled *The Serpent World*, Shine [28: p.1946] regretted: 'Unfortunately, Lee and Caldwell's work on the fabulous missing link (*Pachyrhachis*, 'the snake with legs') was published too recently for inclusion in the book.' What is meant here is the re-description by Caldwell and Lee [29] and Lee and Caldwell [30] of the fossil snake *Pachyrhachis problematicus* from marine sediments of mid-Cretaceous age, found near Ein Yabrud in the Judean Hills (West Bank, Middle East). The remarkable thing about this fossil snake is that it has small, but well-developed hindlimbs. On account of those limbs, it had already been praised as an 'almost ideal intermediate' between varanoids and snakes in what at that time was the standard textbook on Vertebrate Paleontology [31]. Caldwell and Lee's [29, 30] cladistic analysis supported this interpretation of *P. problematicus*, more specifically linking the 'snake with limbs' to mosasauroids, a clade of Cretaceous marine (anguimorphan or varanoid: Augusta et al., Chapter 8; but see [32]) squamates. Eventually, Cope's pythonomorph scenario, now cast in the light of the step-wise evolution of macrophagy in mosasaurs and the descendant snakes, was re-invented by claiming mosasaurs to 'represent a crucial intermediate stage' in the origin of snakes not from 'small burrowing lizards, as commonly assumed, but [from] large marine forms' instead [21, 33: p.655–656].

Doubts were soon cast on such identity of *Pachyrhachis* [34], later fueled by the discovery of two additional taxa of marine snakes with well-developed hindlimbs from the mid-Cretaceous of the Middle East (*Haasiophis* [35, 36] and *Eupodophis* [37, 38]). At the bottom of the dispute lies the question: are these marine 'snakes with limbs' the perfect missing link between snakes and mosasauroids, or are they more advanced snakes nested firmly among crown-group alethinophidians? The latter scenario implies, however, that limb reduction in basal snakes either occurred repeatedly [39], or alternatively, and more parsimoniously, that the hindlimbs in these mid-Cretaceous marine pachyophiid snakes would have re-developed from meager rudiments still present in their ancestors (see discussion in [35] and Zaher et al., Chapter 9).

One pioneering study of the developmental mechanisms underlying the evolution of the snake body plan is that of Cohn and Tickle [40]. Their study addressed the genetic basis of body regionalization, rather than body elongation per se, and how the development, or loss, of limbs is correlated with the axial regionalization (compartmentalization) of the body [41]. The postcranial body axis of a generalized tetrapod (think of a quadrupedal lizard) is compartmentalized into the neck, trunk and tail. On the boundary between neck and trunk are located the pectoral girdle and the forelimbs. On the boundary between trunk and tail are located the pelvic girdle and the hindlimbs. Cohn and Tickle [40] could show that the expression pattern of certain so called *Hox* genes sharply demarcates the neck from the trunk in tetrapods, thereby also specifying the identity of neck and trunk respectively. Gene expression patterns in a python indicated that in snakes, the identity of the trunk has been extended anteriorly up to the very first neck vertebra (atlas). That is to say, the ancestral neck in snakes has become thoracalized, although it is not entirely clear that the snake precloacal axial skeleton lost all signs of regionalization [42–44]. In the absence of the boundary zone signal between the ancestral neck and trunk there is no trace of a pectoral girdle or forelimb rudiments in any snake known, fossil and extant, nor do anterior limb buds sprout from the lateral body wall in any snake embryo (*Tetrapodophis*, an enigmatic fossil squamate from the Early Cretaceous of Brazil combines snake-like features with well-developed fore- and hindlimbs: [45]). The hindlimbs, on the other hand, are not affected through the thoracalization of the ancestral neck, and start to sprout from the python body wall, forming embryonic limb buds. But here again, certain molecular signals that structure limb development in a normal tetrapod (e.g., a quadrupedal lizard) are suppressed in the python limb bud, and the tissue in which these signals originate degenerates. Hindlimb development is thus truncated at an early stage in python. The result is that only a rudimentary femur develops. The chemical signals that fail to sustain the python limb bud growth and full-term morphogenesis are in large part mediated by Sonic hedgehog and fibroblast growth factors. When such growth factors are artificially inserted into the python limb bud, the latter was found to be thereby stimulated to grow to larger size than would be normal [40: p.476].

In a further follow-up study, even greater scrutiny was brought to bear on the analysis of limb reduction in pythons and boas [46]. Among a number of experiments the study also targeted the expression of a gene called *SOX9* in the python hindlimb bud. *SOX9* is a protein-coding gene that plays a regulatory role in skeleton formation (and also sex determination). As such, '*SOX9* expression in python hindlimbs delineates discrete skeletal condensations', initially marking out 'pre-chondrogenic skeletal condensations' of a 'Y-shaped domain' [46: p.2971]. This is, indeed, a classic stage of limb development in all tetrapods, where at an early stage both in the fore- and hindlimb the early cartilage condensation is in a Y-shaped form from which, in the hindlimb, the precursors of the femur, tibia and fibula differentiate [47]. Ultimately it proved possible to identify, in the python hindlimb bud, three discreet *SOX9* expression domains 'that resemble the three major segments … of the tetrapod limbs', and hence 'could be anlagen of the femur, tibia and fibula, and digital plate' [46: p.2971]. The upshot of this finding is 'that python embryos develop transitory cartilage condensations of the lower leg and foot', suggesting 'that re-acquisition of fully developed hindlimbs in extinct snakes may not have required de-novo re-evolution of lost structures, but could have resulted from persistence of the embryonic legs' [46: p.2971]. In summary, the 're-acquisition of complete limbs in some extinct snakes could have resulted from activation of a largely intact but dormant limb developmental program in snake embryos' [48: p.9]. The re-development of limbs from ancestral limb rudiments in squamates is thus a distinct possibility (see also [49]).

6.7 Conclusion

The snake is a powerful symbol of birth, death, and rebirth in mythology, alchemy, and psychology. In mythology, the snake resides either deep underground, or deeply submerged at the bottom of the sea or other bodies of water. In evolutionary biology, the classic scenarios concerning snake origins that have been identified by Lee et al. [33] in their sketch of the pythonomorph scenario are either from small, secretive or even burrowing lizard-like forms, or from large, marine macrophagous forms such as mosasaurs. The first scenario, the origin of snakes from small, burrowing lizard-like forms, was recently branded as a myth [50]. But sea serpentism is deeply steeped in mythology as well (see discussion in [51]). And either scenario is at odds with the squamate tree of life based on genomic data, where snakes come out as sister-group to a clade that comprises the sister-taxa Iguania and Anguimorpha [52–54]. Snakes, fossil and extant, have to this day not conclusively revealed the secret of their place of birth. To lift the veil that shrouds the evolutionary origin of snakes will require continued careful and unbiased morphological study, as well as a willingness to confront the issue why genomic data [53] yield a pattern of higher-level squamate relationships that is so fundamentally different from the tree based on morphological data [32], and counterintuitive as well (with dibamids at its base). It is fair to say that from a philosophical perspective we may never obtain certainty in our knowledge of snake origins, but we know one thing for certain: that the morphological and genomic data yield fundamentally incongruent trees of higher-level squamate relationships. It is time to ask the question why this should be so.

Acknowledgements

I thank Hussam Zaher and David Gower for inviting me to contribute to this volume. I am very grateful to Hussam Zaher, Aleta Quinn, Francisca Leal, and an anonymous reviewer who offered comments that greatly improved the content of the paper.

References

1. R. Westrum. Knowledge about sea-serpents. In R. Wallis, ed., *On the Margins of Science: The Social Construction of Rejected Knowledge* (Staffordshire: University of Keele, 1979), pp. 293–314.

2. J. Nigg, *Sea Monsters. A Voyage around the World's Most Beguiling Map* (Chicago: University of Chicago Press, 2013).

3. H. Lee, *Sea Monsters Unmasked* (London: William Clowes and Sons, 1883).

4. E. Pontoppidan, *The Natural History of Norway* (London: A. Linde, 1755).

5. A. C. Oudemans, *The Great Sea-Serpent. An Historical and Critical Treatise* (Leiden: J. Brill, 1892).

6. C. Gesner, *Historia Animalium, liber IV, qui est de piscium & aquatilium animantium natura* (Zürich: Christoph Froschauer, 1558).

7. D. Loxton and D. R. Prothero, *Abominable Science! Origins of the Yeti, Nessie, and Other Famous Cryptids* (New York: Columbia University Press, 2013).

8. A. Olearius, *Gottorffische Kunst-Cammer.* (Schleswig: J. Holwein)

9. L. Rieppel, Albert Koch's Hydrarchos craze. In C. Berkowitz, B. Lightman, eds., *Science Museums in Transition* (Pittsburgh: University of Pittsburgh Press, 2017), pp. 139–160.

10. L. Rieppel Hoaxes, Humbugs, and frauds. Distinguishing truth from untruth in early America. *Journal of the Early Republic*, 38 (2018), 501–529.

11. C. S. Rafinesque, Dissertation on Water Snakes, Sea Snakes, and Sea Serpents. *American Monthly Magazine and Critical Review*, 1 (1817), 431–435.

12. Anonymous, *Report of a Committee of the Linnean Society of New England, relative to a large marine animal, supposed to be a serpent, seen near Cape Ann, Massachusetts, in August 1817* (Boston: Cummings and Hilliard, 1817).

13. W. Peck, Some observations on the Sea Serpent. *Memoirs of the American Academy of Arts and Sciences*, 4 (1818), 86–91.

14. C. Lyell, *A Second Visit to the United States of North America* (New York: Harper & Brothers, 1849).

15. R. T. Gould, *The Case for the Sea-Serpent* (London: Philip Allan, 1930).

16. R. Owen, *A History of British Fossil Reptiles* (London: Cassell & Co, 1849-1884).

17. H. d. Blainville, Sur un nouveau genre de Serpent, Scoliophis, et le serpent de mer vu en Amérique en 1817. *Journal de Physique, de Chimie, et d'Histoire Naturelle*, 86 (1818), 297–304.

18. A. Lesueur, Sur le serpent nommé Scoliophis. *Journal de Physique, de Chimie, et d'Histoire Naturelle*, 86 (1818), 466–469.

19. F. Buckland, *Notes and Jottings from Animal Life* (London: Smith Elder, 1886).

20. Anonymous, The Great Sea-Serpent. *Nature*, 47 (1893), 506–507.

21. E. D. Cope, On the reptilian orders, Pythonomorpha and Streptosauria. *Proceedings of the Boston Society of Natural History*, 12 (1869), 250–266.

22. E. D. Cope, Synopsis of the extinct Batrachia, Reptilia and Aves of North America. *Transactions of the American Philosophical Society, ns*, 14 (1871), 1-252, i-xxxiii.

23. E. W. A. Mulder, Transatlantic latest Cretaceous mosasaurs (Reptilia Lacertilia) from the Maastrichtian type area and New Jersey. *Netherland Journal of Geosciences – Geologie en Mijnbouw*, 78 (1999), 281–300.

24. N. Bardet and J. W. M. Jagt, Mosasaurus hoffmanni, le 'Grand Animal fossile des Carrières de Maestricht': deux siècles d'histoire. *Bulletin du Muséum National d'Histoire Naturelle, 4ème série - section C – Sciences de La Terre, Paléontologie, Géologie, Minéralogie*, 18 (1996), 569–593.

25. D. V. Grigoriev, Giant *Mosasaurus hoffmanni* (Squamata, Mosasauridae) from the Late Cretaceous (Maastrichtian) of Penza, Russia. *Proceedings of the Zoological Institute RAS*, 318 (2014), 148–167.

26. E. D. Cope, Professor Owen on the Pythonomorpha. *Bulletin of the United States Geological and Geographical Survey of the Territories*, 4 (1878), 299–311.

27. R. Owen, On the rank and affinities in the reptilian class of the Mosasauridae, Gervais. *Quarterly Journal of the Geological Society of London*, 33 (1877), 682–715.

28. R. Shine, The serpent world. *Science*, 277 (1997), 1945–1946.

29. M. W. Caldwell and M. S. Y. Lee, A snake with legs from the marine Cretaceous of the Middle East. *Nature*, 386 (1997), 705–709.

30. M. S. Y. Lee and M. W. Caldwell, Anatomy and relationships of *Pachyrhachis problematicus*, a primitive snake with hindlimbs. *Philosophical Transactions of the Royal Society of London B*, 353 (1998), 1521–1552.

31. R. L. Carroll, *Vertebrate Paleontology and Evolution* (New York: W. H. Freeman, 1988).

32. J. Gauthier, M. Kearney, J. A. Maisano et al., Assembling the squamate tree of life: Perspectives from the phenotype and the fossil record. *Bulletin of the Peabody Museum of Natural History*, 53 (2012), 3–308.

33. M. S. Y. Lee, G. L. Bell, and M. W. Caldwell, The origin of snake feeding. *Nature*, 400 (1999), 655–659.

34. H. Zaher, The phylogenetic position of *Pachyrhachis* within snakes (Squamata, Lepidosauria). *Journal of Vertebrate Paleontology*, 18 (1998), 1–3.

35. O. Rieppel, H. Zaher, E. Tchernov, and M. J. Polcyn, The anatomy and relationships of *Haasiophis terrasanctus*, a fossil snake with well-developed hind limbs from the Mid-Cretaceous of the Middle East. *Journal of Paleontology*, 77 (2003), 536–558.

36. E. Tchernov, O. Rieppel, H. Zaher, et al., A fossil snake with limbs. *Science*, 287 (2000), 2010–2012.

37. J.-C. Rage and F. Escuillié, Un nouveau serpent bipède du Cénomanien (Crétacé). Implications phylétiques. *Comptes Rendus de l' Académie des Sciences Paris, Earth and Planetary Science* , 330 (2000), 513–520.

38. J.-C. Rage and F. Escuillié, The Cenomanian: stage of hindlimbed snakes. *Carnets Geologiques*, 2003/01 (2003), 1–11.

39. H. Zaher and O. Rieppel, The phylogenetic relationships of *Pachyrhachis problematicus*, and the evolution of limblessness in snakes (Lepidosauria, Squamata). *Comptes Rendus de l'Académie des Sciences, Paris (Série IIA), Earth and Planetary Science*, 329 (1999), 831–837.

40. M. J. Cohn and C. Tickle, Developmental basis of limblessness and axial patterning in snakes. *Nature*, 399 (1999), 474–479.

41. T. J. Sanger and J. J. Gibson-Brown, The developmental bases of limb reduction and body elongation in squamates. *Evolution*, 58 (2004), 2103–2106.

42. J. M. Woltering, F. J. Vonk, H. Muller, et al., Axial patterning in snakes and caecilians: evidence for an alternative interpretation of the Hox code. *Developmental Biology*, 332 (2009), 82–89.

43. T. Tsuihiji, M. Kearney, and O. Rieppel, Finding the neck-trunk boundary in snakes: anteoposterior dissociation of myological characteristics in snakes and its implications for their neck and trunk body regionalization. *Journal of Morphology*, 273 (2012), 992–1009.

44. J. J. Head and P. D. Polly, Evolution of the snake body form reveals homoplasy in amniote *Hox* gene function. *Nature*, 520 (2015), 86–89.

45. D. M. Martill, H. Tischlinger, and N. R. Longrich, A four-legged snake from the Early Cretaceous of Gondwana. *Science*, 349 (2015), 416–419.

46. F. Leal and M. J. Cohn, Loss and re-emergence of legs in snakes by modular evolution of sonic hedgehog and HOXD enhancers. *Current Biology*, 26 (2016), 2966–2973.

47. N. Shubin and P. Alberch, A morphogenetic approach to the origin and basic organization of the tetrapod limb. *Evolutionary Biology*, 20 (1986), 319–387.

48. F. Leal and M. J. Cohn, Developmental, genetic and genomic insights into the evolutionary loss of limbs in snakes. *The Journal of Genetics and Development*, 56 (2018), e23077.

49. J. J. Wiens, M. C. Brandley, and T. W. Reeder, Why does a trait evolve multiple times within a clade? Repeated evolution of snakelike body form in squamate reptiles. *Evolution*, 60 (2006), 123–141.

50. M. W. Caldwell, *The Origin of Snakes: Morphology and the Fossil Record* (Boca Raton: CRC Press, 2020).

51. O. Rieppel and H. Zaher, Re-building the bridge between mosasaurs and snakes. *Neues Jahrbuch für Geologie und Paläontologie, Abhandlungen*, 221 (2001), 111–132.

52. J. J. Wiens, C. R. Hutter, D. G. Mulcahy, et al., Resolving the phylogeny of lizards and snakes (Squamata) with extensive sampling of genes and species. *Biology Letters*, 8 (2012), 1043–1046.

53. J. W. Streicher and J. J. Wiens, Phylogenomic analyses of more than 4000 loci resolve the origin of snakes among lizard families. *Biology Letters*, 13 (2017), 20170393.

54. F. T. Burbrink, F. G. Grazziotin, R. A. Pyron, et al., Interrogating genomic-scale data for Squamata (lizards, snakes, and amphisbaenians) shows no support for key traditional morphological relationships. *Systematic Biology*, 69 (2020), 502–520.

7

Reassessing the Morphological Foundations of the Pythonomorph Hypothesis

Michael J. Polcyn, Bruno G. Augusta, and Hussam Zaher

7.1 Introduction

The search for the phylogenetic positions of snakes and mosasaurians within Squamata and their relationship with one another has a long and controversial history. Molecular analyses are increasingly providing insights into snake relationships with support for their inclusion in the clade Toxicofera (e.g., Ruane and Streicher, Chapter 10), but with alternate arrangements of Serpentes and Iguania as the sister to Anguimorpha [1, 2]. However, morphology-only analyses have failed to recover a well-supported position for snakes within Squamata [3–7]. Such lability has also plagued mosasaurian morphological systematics and likely for similar reasons as snakes; namely, their highly disparate morphology and imperfections in the fossil record of transitional forms. Our incomplete knowledge of the early evolution of the group has contributed to inconsistent phylogenetic hypotheses for mosasaurians recovered in recent squamate-wide analyses [5, 7].

Uncertainty of the phylogenetic position of mosasaurians is not a recent development. Shortly after their discovery in the eighteenth century, mosasaurs were considered fishes, crocodiles, and whales [8] but finally recognized as squamates [9, 10]. After Cuvier [10], mosasaurians were broadly accepted as being closely related to varanoid lizards, until Cope [11], confounded by what he considered a mosaic of snake, lizard, and sauropterygian characters presented by mosasaurs, rejected this. Cope instead erected a new order, Pythonomorpha, placing it in a trichotomy with his Ophidia and Lacertilia. Cope's arrangement led to protracted nineteenth century debates but no consensus (see [12, 13] and references therein).

With the discovery and growing knowledge of 'aigialosaurids', dolichosaurids, and other Cenomanian snake-like squamates, focus shifted, with aigialosaurids being considered by Nopcsa [14, 15] ancestral to snakes, dolichosaurids, and mosasaurs, and underpinning the hypothesis of a marine origin for snakes (see Zaher et al., Chapter 9). However, Cope's pythonomorph concept and any close relationship of snakes and mosasaurians was rejected by Camp [12] in favour of a varanid affinity, which heralded a long pause in the discussion (but see [16]).

Over 70 years later, Lee's [17] phylogenetic study of varanoid lizards led to recovery of a snake-mosasaur affinity and revival of the clade-name Pythonomorpha. Although used in a different sense from Cope [11], the revived Pythonomorph Hypothesis of snake origins was promoted and expanded [18–20] and critically rebutted [6, 21–24], culminating in what can respectfully be recognized as a stalemate [25]. The concept of Pythonomorpha has persisted and evolved with a renewed focus on dolichosaurids and a somewhat fluid diagnosis [26–33].

Table 7.1 presents a synoptic view of morphological characters used to diagnose Pythonomorpha. Understanding details of morphology from a mosasaurian perspective, that underlie purported pythonomorph synapomorphies, is a critical first step in assessing homology statements and character definitions. To that end, in this chapter we present new morphological data for exceptionally well-preserved early diverging (non-mosasaurid) mosasaurians (dolichosaurids and 'aigialosaurids') and the early diverging plioplatecarpine mosasaurid *Tethysaurus nopcsai*, and address issues that influenced prior analyses (e.g., narial retraction). Ultimately, our goals are to provide a succinct snapshot of the morphology of early diverging mosasaurians that purportedly unites mosasaurians with snakes, and to present new data clarifying morphology.

7.2 The Morphology of Early Diverging Mosasaurians

Morphological observations presented here are based on first-hand study of specimens, high-resolution computed tomography (HRCT) scans, and research casts. All specimens referred in the text are listed with their collection number in Supplementary Appendix 7.S1. We follow the phylogeny and nomenclature for Mosasauria given by Augusta et al. (Chapter 8). We use the term mosasaurians (= Mosasauria) to include dolichosaurids, 'aigialosaurids', and mosasaurids. For the purpose of this chapter, we use the term 'early diverging mosasaurians' for the mosasaurian assemblage, predominantly known from the Cenomanian–Turonian (~100–~90 Ma) and which includes dolichosaurids and 'aigialo-saurids'. We consider *Tethysaurus*, *Russellosaurus*, *Dallasaurus*, *Yaguarasaurus*, and Halisaurinae 'early diverging', generally morphologically plesiomorphic members of Mosasauridae. We further consider the remaining mosasaurians listed in this chapter as 'later diverging', generally morphologically more derived mosasaurids of the subfamilies Plioplatecarpinae, Mosasaurinae, and Tylosaurinae (Supplementary Figure 7.S1a). We present new data elucidating narial retraction in mosasaurians and thereafter generally follow the order outlined in Table 7.1. Due to space limitations, most morphological descriptions and supporting notes are in the Supplementary Appendix 7.S1, which generally follows the structure of this chapter.

Table 7.1 Synoptic listing of characters used to diagnose Pythonomorpha.

Letters in parentheses designate diagnosis citation: A = Cope [11]; B = Lee [17]; C = Lee and Caldwell [19]; D = Caldwell [38]; E = Lee [4]; F = Lee and Caldwell [40]; G = Pierce and Caldwell [30]; H = Palci and Caldwell [36]; I = Paparella et al. [31]. Minor differences in the wording of character descriptions were ignored, but where meaning was changed, we included both versions. Some entries were edited for clarity, designated by brackets around edited word or passage. Characters in italics indicate usage in most recent diagnosis [31]. Note 1: Cope's use of 'squamosal' is unclear to us in this context.

Premaxilla

Premaxillary teeth four or fewer (B,C,E,F,G)

Median premaxillary tooth absent (E,F,G)

Premaxilla contacts frontals (H)

Maxilla

There is merely a squamosal suture between the maxillary and premaxillary (A) [1]

Maxilla enters suborbital foramen (B,C)

Lacrimal foramen

Lacrimal foramen enclosed entirely by prefrontal (B,C)

Lacrimal foramen within prefrontal or absent (E,F,G)

Septomaxilla

Long posteromedial flange of septomaxilla (D,E,F,G,I)

Septomaxilla free of maxilla (E,F,G)

Palatine

Palatine with long anterior process (B)

Pterygoid

Pterygoids elongate and bear numerous teeth, and in one type are free, except at the extremities (A)

Anterior process of pterygoid distinct from lateral process (B,C,E,F,G)

Pterygoid teeth large and recurved (B,C)

Presence of pterygoid teeth (D)

Pterygoid teeth, when present, as large as maxillary teeth (E,F,G,H)

Ectopterygoid

Ectopterygoid-palatine contact absent, maxilla enters suborbital fenestra (H)

Frontal

Frontal is a single median element (H)

Table 7.1 (*cont.*)

Parietal

Parietal downgrowths prominent flanges (A,B,C,E,F,G,H,I)

Parietal-prootic contact extensive (E,F,G,H,I)

Basisphenoid

Basipterygoid processes do not project far from body of basisphenoid (B,C,E,F,G,I)

Basipterygoid processes expanded anteroposteriorly (B,C)

Body of basisphenoid extends far anteriorly beyond dorsum sellae (B)

Cultriform process straight in lateral view (B,C,E,F,G)

Anterior brain cavity floored by wide cultriform process (E,F,G)

Main body (wide portion) of parasphenoid extends anteriorly some distance in front of dorsum sella, before tapering into a narrow cultriform process (C)

Vidian canal is an open groove anteriorly (B,C)

Rear opening of vidian canal situated far posteriorly (B,C)

Otooccipital

Crista circumfenestralis encircling footplate of stapes (B,C)

Stapes footplate surrounded by flanges from prootic and opisthotic (E,F,G)

Supraoccipital

Supraoccipital sutured with parietals along its entire dorsal margin (B,C,H)

Supraoccipital sutured to parietal (E,F,G)

Supratemporal

Opisthotic projects free from the cranium, and is the suspensorium of the os quadratum (A)

Opisthotic [supratemporal] is supported by a pedestal projecting from the cranial walls, composed of the prolonged prootic in front, and the exoccipital [opisthotic] behind, which embraces the suspensorium for much of its length (A)

Squamosal [supratemporal] bone is present (A)

Supratemporal forms part of braincase and contacts prootic (C)

Supratemporal-prootic contact (B,E,F,G,H,I)

Supratemporal forms part of paroccipital process of braincase (E,F,G)

Supratemporal intercalated between quadrate and braincase (B)

Quadrate

There is no columella. [The stapes was unknown to Cope] (A)

Os quadratum is movably articulated to the opisthotic [supratemporal] (A)

Table 7.1 (*cont.*)

Os quadratum embraces and encloses the meatus auditorius externus (A)

Extracolumella with extensive contact with quadrate (B,C)

Quadrate with large, posteroventrally curved, suprastapedial process (H,I)

Quadrate suspended entirely by supratemporal (C)

Quadrate suspended mostly or entirely by supratemporal (E,F,G)

Dentary

Mandibular symphysis highly mobile (A,B,C,D,E,F,G,H,I)

Dentary straight in lateral view (E,F,G,I)

Meckelian groove confined to medial surface of lower jaw (B,C)

Anterior end of Meckel's canal on medial surface of dentary (E,F,G,I)

Development of a large subdental shelf (I)

Splenial-Dentary

Reduced splenial-dentary suture (D)

Anterior tip of splenial on medial surface of dentary (B,C,E,F,G)

Coronoid-Splenial

Posterior end of splenial does not overlap coronoid; convergent in *Lanthanotus* (B,C,E,F,G)

Coronoid does not contact splenial (B)

Splenial-Angular

Vertical articulation between splenial and angular (A,B,C,E,F,G,H)

[Splenial angular] greatly exposed in lateral view (H)

Surangular exposed in medial view of lower jaw (C)

Postdentary-dentary

Splenial does not substantially overlap postdentary elements (E,F,G,I)

Reduced overlap of the postdentary-dentary bones (D)

Coronoid

Coronoid with straight or convex ventral edge (C,E,F,G,I)

Posterior ramus of coronoid reduced or absent (B)

Subcoronoid fossa absent (B,E,F,G)

Anterior end of coronoid meets dentary directly (D)

Coronoid does not overlap dentary laterally (C)

Table 7.1 (*cont.*)

Adductor fossa

Adductor fossa faces dorsally (B,C,E,F,G)

Articular-prearticular

Articular and prearticular fused (E,F,G)

Compound bone

Angular [articular] bone is distinct (A)

Tooth implantation

Teeth have no fangs (A)

Discrete sockets (alveoli) under all marginal teeth (C)

Tooth replacement

Recumbent (horizontal) replacement teeth (B,C,E,F,G)

Vertebrae

Zygosphenes and zygantra present (A,B,C,D,E,F,G,H,I)

Absence of posterior process on atlas neural arch (B,C).

Condyles on centra circular in shape (B,C,E,F,G)

Five or more pygals (B)

Tail laterally compressed, i.e., transverse processes reduced anteriorly and absent posteriorly, chevrons and neural spines elongated (H)

Ribs

Abdominal cavity is long, and is surrounded by many short, curved ribs, which have a free anteroposterior movement on vertical, articulating surfaces, and which commence immediately behind the head (A)

Ribs begin from third cervical (B,C,E,F,G,I)

Girdles

Pelvic elements not suturally united (B,C,H,I)

Pubic plate oriented parasagittally, wide in lateral view (H)

Limbs

Femur stout (B)

Straight, short femur (C)

Table 7.1 (*cont.*)

Distal end of tibia gently convex for astragalocalcaneal articulation (I)
Astragalus and calcaneum separate (C,H,I)
Other postcranial
Cranial and axial epiphyses absent (B,C,I)
Osteoderms lost (B,C,H)

7.2.1 Snout, Palate, and Circumorbital Bones

Characters of the snout, dermal palate, and circumorbital series have played a prominent role in squamate systematics historically; however, due in part to a lack of data on that anatomy in early mosasaurians, phylogenetic analyses of Mosasauria using squamate-wide data sets have yielded varied and inconclusive results [5, 7, 34, 35]. Furthermore, the snouts of derived mosasaurids are highly modified by secondary marine and feeding adaptations, making direct comparisons with the plesiomorphic condition difficult. Cope interpreted the unique premaxillary–maxillary suture in mosasaurs as snake-like which is confounding, because mosasaur taxa available to him at that time were highly derived forms with the uniquely expanded premaxillary–maxillary contact described below, and bear no resemblance to the snake condition. The maxilla entering the suborbital fenestra was proposed by Lee [17, 19] to unite mosasaurs and snakes. This character is roughly equivalent to Palci and Caldwell's [36] character 59 'ectopterygoid-palatine contact absent, maxilla enters suborbital foramen [fenestra]' and is addressed below with the account of the ectopterygoid and its relationship with adjacent bones. No premaxillary or maxillary characters are included in the most recent diagnosis of Pythonomorpha [31].

Narial Retraction

Although narial retraction has long been recognized as a feature shared by varanoids and mosasaurians [5, 12, 13, 37], Caldwell et al. [34] criticized its use as a single character, arguing that topological relationships of constituent bones as opposed to the fenestration itself, should be considered instead. We generally agree with that, however, in doing so, Caldwell et al.'s recharacterization ensured the link between mosasaurians and varanids was weakened (p. 116 in [38]). Snout morphology in early diverging mosasaurians and early diverging plioplatecarpines clarifies narial retraction in Mosasauria because they preserve a morphology more comparable to the plesiomorphic varanoid condition and in some cases even more similar to varanids. We briefly summarize here and provide more details in Supplementary Appendix 7.S1.

All dolichosaurids that sufficiently preserve snout morphology allowing examination possess some degree of anteriorly retracted nares, in that the anterodorsal margin of the

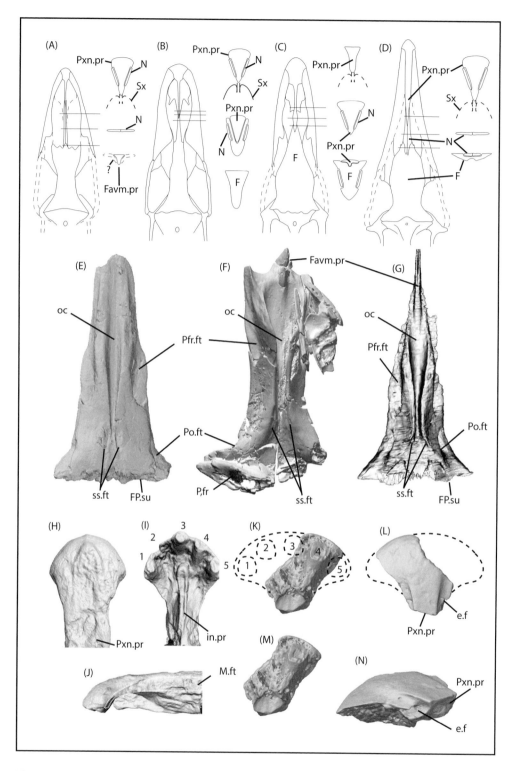

Figure 7.1 Narial retraction in Mosasauria, depicting schematic views of bony nares. Generalized dolichosaurid condition (A) based on *Pontosaurus lesinensis, P. kornhuberi,* and

maxilla is tall and possesses a long contact with the lateral edge of the premaxilla, forming a posteromedially trending premaxillary–maxillary suture. In some cases, the nasal process of the premaxilla is quite wide and forms a long posterior contact with the maxilla, as in *Coniasaurus* (Fig. 7.1H–J). All mosasaurians examined also possess posterior retraction of the nares, in that the posterior margin is formed largely by the frontal and prefrontal. In dolichosaurids, the anterior frontal is broad and the narial emargination largely transverse, whereas in 'aigialosaurids' and early diverging plioplatecarpines, the anterolateral parts of the frontal are greatly reduced, emarginating the posteromedial part of the nares with the anteromedial frontal projecting into the narial space (Fig. 7.1E–G). The relationship of the nasal process of the premaxilla to the frontal, and the variable morphology of the nasals, trends from being similar to the varanoid condition (in dolichosaurids and halisaurines) toward reduction and loss of the nasals in more derived mosasaurids. Within Mosasauria, contact between the premaxilla and frontal is present only in early diverging plioplatecarpines and derived mosasaurids, but was considered one of the defining characters of early pythonomorphs [36]. *Pontosaurus lesinensis* possesses narrow paired nasals, a condition shared with halisaurines (Supplementary Appendix 7.S1). In both cases the posterior nasals are much narrower than the frontal (Fig. 7.1A, D). The earliest instance of premaxillary–frontal contact is in the early Turonian *Tethysaurus*, which exhibits slight overlap. In the late Turonian plioplatecarpine *Yaguarasaurus* (Fig. 7.1C) the premaxilla overlies the frontal and forms a long spatulate posterior terminus, sutured to the anterodorsal surface, a condition common to all non-halisaurine derived mosasaurids. Both *Tethysaurus* and *Yaguarasaurus* (contra [39]) possess nasals ventrolateral to and spanning the anterior part of the frontal and posterior part of the nasal process of the premaxilla (not visible in dorsal view). In derived mosasaurids, there is a reduction in size and in some cases loss (or non-preservation) of the nasal bones.

Early diverging mosasaurians possess a typical varanid snout morphology with respect to narial retraction. The morphology associated with anterior narial retraction, and especially the increasing contribution of the premaxilla in the formation of the snout, stands in marked contrast to the regressed contact between premaxilla and maxillae in snakes, in which the loss of the premaxillary-maxillary suture is intimately related to increased snout kinesis. Equally important are details of the posteriorly retracted nares, in which early diverging mosasaurians share extremely narrow nasals separating the elongate nasal

Figure 7.1 (*cont.*) *Coniasaurus gracilodens*; generalized plesiomorphic mosasaurid condition (B) based on *Tethysaurus nopcsai*; generalized derived mosasaurid condition (C) based on *Yaguarasaurus columbianus*; halisaurine condition (D) based on YPM-40383 and *Eonatator coellensis*. Ventral view of frontals of (E) *Eonatator* cf. *sternbergi* (RMM-6890), (F) *Coniasaurus gracilodens* (NHMUK-R44141), and (G) *Tethysaurus nopcsai* (SMU-75486). Partial premaxilla of *Coniasaurus* sp. (DMNH-1602) in (H) dorsal, (I) ventral, and (J) left lateral views. Partial premaxilla of *Dallasaurus turneri* (SMU-76529) in (K,M) ventral, (L) dorsal, and (N) posterolateral and slightly dorsal oblique views. Abbreviations: e.f, foramen for ethmoid nerve; F, frontal; in.pr, incisive process; Favm.pr, frontal anterior ventral median process; M.ft, maxillary facet; N, nasal; P.fr, Parietal fragment; Pfr.ft, prefrontal facet; Po.ft, postorbitofrontal facet; Pxn.pr, nasal process of the premaxilla; ss.ft, supraseptale facet; Sx, septomaxilla.

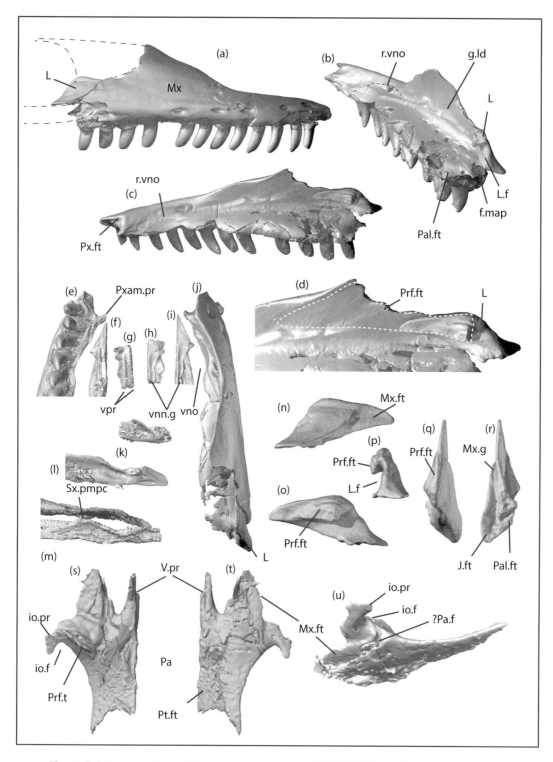

Figure 7.2 Snout region of *Coniasaurus gracilodens* (NHMUK-R44141) and the palatine of *Dolichosaurus longicollis* (NHMUK-R49002). Right maxilla and rearticulated lacrimal in (A) lateral,

process of the premaxilla from the frontal. The nasals are further reduced in more derived mosasaurids which, along with premaxillary–frontal suturing, contributes to loss of kinesis. This stands in contrast to the condition in stem snakes such as *Najash* [32] in which posteriorly broad nasals contact the frontal posteriorly and posterolaterally. Narial retraction in snakes appears to be intimately related to increasing snout kinesis, an evolutionary trend that is opposite to that in mosaraurians.

Premaxilla

Fewer teeth and absence of a median premaxillary tooth have been cited as pythonomorph characters (Table 7.1) [4, 17, 30, 40]. Although mosasaurs do possess fewer premaxillary teeth, these are pleurodont, symmetrically spaced, and a median tooth is present in some forms (Fig. 7.1 H–N; see also Supplementary Appendix 7.S1). These purported pythonomorph synapomorphies are not present in the most recent diagnosis of Pythonomorpha [31].

Septomaxillae and Vomers

Chemosensory adaptations of the dermal palate in squamates played a large role in their classification historically (e.g., [3, 12, 16]) but the morphology in mosasaurians, specifically of the septomaxilla, is only superficially known in a few taxa [41–43]. Camp [43] first reported the elongate posteromedial process of the septomaxilla in *Plotosaurus* and this character was hypothesized as a synapomorphy of Pythonomorpha by Caldwell [38] and others [4, 30, 31, 38, 40] and the character 'septomaxilla free of maxilla' was used by Lee [4], Lee and Caldwell [40], and Pierce and Caldwell [30], but not in more recent diagnoses.

HRCT scans of *Coniasaurus gracilodens* and *Tethysaurus nopcsai* provide new details of the septomaxilla and conjugate structures (Figs. 7.2 and 7.3; Supplementary Appendix 7.S1). The septomaxilla in *Tethysaurus* is composed of an anterior cupola, open medially, walled by the maxilla laterally, and a dorsolateral flange of the vomer posterolaterally

---

Figure 7.2 (*cont.*) (B) dorsal, posterolateral oblique, (C) medial, (D) medial, showing detail of maxilla-lacrimal contact, (E) ventral, and (J) dorsal views. Right vomer in (F) ventral, (I) dorsal, and (L) lateral views. Left vomer in (G) ventral, (H) dorsal, and (K) lateral views (K is reversed for comparison). Articulated anterior part of right vomer and septomaxilla of *Tethysaurus nopcsai* (SMU-75486) in (M) lateral view. Right lacrimal of *Coniasaurus gracilodens* in (N) lateral, (O) medial, (P) posterior, (Q) dorsal, and (R) ventral views. *Dolichosaurus longicollis* palatine in (S) dorsal, (T) ventral, and (U) posteroventral oblique views. Abbreviations: Ec, ectopterygoid; F, frontal; f.map, maxillary artery posterior; f.maa, foramen maxillary artery anterior; g.Ld, groove for lacrimal duct; in.pr, incisive process; io.f, infraorbital foramen; io.pr, infraorbital process; J, jugal; L, lacrimal; L.f, lacrimal foramen; L.f.d, dorsal lacrimal foramen; L.f.v, ventral lacrimal foramen; Mx, maxilla; Mx.ft, maxillary facet; Mx. g, groove to receive dorsal margin of maxilla; N, nasal; N.ft, nasal facet; P, parietal; Pa, palatine; Pa.f, palatine foramen; Pa.vp.ft, palatine vomer process facet; Pt.ft, pterygoid facet; Px, premaxilla; Pxam. pr, anteromedial process of premaxilla; Px.ft, premaxillary facet; Prf.ft, prefrontal facet; r.vno, recess for vomernasal organ; Sx, septomaxilla; Sx.pmpc, septomaxilla posteromedial process of cupola; Sx.ft, septomaxilla facet; V.pr, vomer process; vno, vomeronasal opening; vno.r, recess for vomernasal organ; vnn.g, groove for vomeronasal nerve; vpr, ventral parallel ridges; Scale bar = 5mm. See Supplementary Figure 7.S12 for colour version.

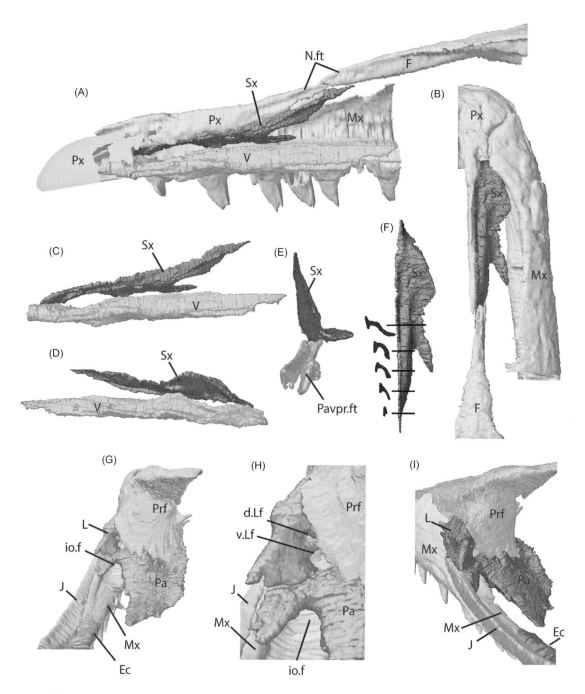

Figure 7.3 Morphology of the snout and antorbital series of *Tethysaurus nopcsai* from HRCT reconstructions. Parasagittally (slightly left of midline) sectioned μCT-based model of the skull of *Tethysaurus nopcsai* (SMU-75486) in (A) left medial and (B) dorsal views. Septomaxilla and vomer of same in (C) left lateral, (D) right lateral, and (E) posterior views. Right septomaxilla in (F), dorsal view with cross sections of posterior portion. Partial orbital series in (G) posterior, (H) detailed posterior, and (I) posterolateral oblique views. Abbreviations: Ec, ectopterygoid; d.lf, dorsal lacrimal foramen; F, frontal; io.f, infraorbital foramen; J, jugal; L, lacrimal; Mx, maxilla; N.ft, nasal facet; Pavpr.ft, facet for the vomer process of the palatine; Prf, prefrontal; Px, premaxilla; Sx, septomaxilla; V, vomer; v.lf, ventral lacrimal foramen. (A black and white version of this figure will appear in some formats. For the colour version, please refer to the plate section).

(Fig. 7.3B) yielding the incomplete neochoanate condition (sensu [44]). The short posterolateral process loosely overlies the maxilla and the longer posteromedial process (Fig. 7.3F) is ventrally concave and projects posterodorsally, contacting the ventrolateral nasal process of the premaxilla. Although the septomaxilla is not preserved in *C. gracilodens*, conjugate structures on the anteromedial maxilla and vomers leave no doubt its septomaxilla was similar (Fig. 7.2B, C, I, J).

Notwithstanding consequences of narial retraction and reduced anterior accommodation space, the morphology of the septomaxilla and vomer in *Tethysaurus* is similar in most respects to that of *Lanthanotus* and *Heloderma*. In *Lanthanotus*, the cupola is more domed, the posterolateral process elaborated, the palatine process of the vomer broad to receive the spatulate anterior palatine, and the dorsolateral process that meets the septomaxilla behind the incisura Jacobsoni more robust, and diagonally oriented. The relationship of the vomer with the blade-like vomer process of the palatine in *Tethysaurus* is most similar to, but more vertically oriented, than the sample of *Varanus* examined. The vomer process of the palatine in *Heloderma* is rod-like and articulates with a U-shaped trough in the posterior part of the vomer in that taxon. All varanoids have some form of a posteromedial process of septomaxilla (Supplementary Figure 7.S3a–e), although in *Varanus* the septomaxilla is structurally more complex. Extreme elongation of the posteromedial part of the septomaxilla occurs in *Varanus rudicollis*, for instance, though also laterally expanded and contacting the maxilla over an extended distance (Supplementary Figure 7.S3h). Interestingly, these taxa also possess retracted fleshy nares, suggesting septomaxilla elongation may be correlated with the opening of the dorsal wall of the nasal capsule.

Transverse reduction of the septomaxilla in snakes contributes to its high aspect ratio, giving the appearance of a long septomaxilla. However, when comparing the cupola length to the length of the posteromedial process (flange) of the septomaxilla of stem snakes (e.g., *Najash* [32]) to the one in varanoids and mosasaurs, it appears that the former is substantially shorter than the latter: only 24 per cent of the length of the cupola in *Najash* compared to 80 per cent in *Lanthanotus*, 43 per cent in *Heloderma*, 200+per cent in *Varanus dumerilii*, 145 per cent in *V. niloticus*, and 175 per cent in *Tethysaurus*. *Tethysaurus* falls within the range of *Varanus* ssp. and, along with the other varanoids measured, possesses a relatively longer posteromedial process than *Najash*. Thus, a 'long' posteromedial process of the septomaxilla in snakes cannot be considered a synapomorphy of Pythonomorpha [4, 30, 31, 38, 40].

In addition to comparisons of 'length', the posteromedial septomaxilla flange in *Najash* is blade-like and more similar to that of *Heloderma* (Supplementary Figure 7.S3b) than to the more complex, vaulted process in *Tethysaurus* (Fig. 7.3F). Furthermore, the septomaxilla in snakes lies primarily ventral to the nasal bones, whereas in mosasaurs (and varanoids) it lies primarily ventral to and contacts the nasal process of the premaxilla. The only similarity identified in *Najash* and *Tethysaurus* and inferred in *Coniasaurus gracilodens* is contact between the posterolateral margin of the cupola and a dorsal projection of the vomer, but this condition is also present in *Heloderma* and *Lanthanotus*, and further elaborated in *Varanus*.

Palatine

A single palatine character was proposed by Lee [17] – 'Palatine with long anterior process'. Although in mosasaurids the pterygoid process is markedly shorter and the anterior part more elongate, in dolichosaurids the pterygoid process is about the length of the anterior vomer process (Fig. 7.2S). Other aspects of the palatine in mosasaurians are markedly different than in snakes (Supplementary Appendix 7.S1). The edentulous palatine in mosasaurians has a long contact with the posteromedial surface of the maxilla. In dorsal view, between the maxillary contact and the vertically oriented blade-like vomer process, is a concave floor with intersecting sulci and its anterior part forms the concave posterior emargination of the internal choana (Fig. 7.2S–U; Supplementary Figure 7.S4a–c). Posterolaterally, the path for the maxillary artery and maxillary branch of the trigeminal nerve is not fully enclosed by the palatine, but merely roofed by a lateral projection that contacts the lacrimal, maxilla and jugal (Fig. 7.3G–I). In some mosasaurine mosasaurids, the lateral process is ventrolaterally elaborated, forming a structure superficially similar to the infraorbital foramen, but open ventrally (Supplementary Figure 7.S4d–g). In varanids, *Heloderma*, anguioids, *Shinisaurus*, and stem snakes like *Najash* [32], the path for the maxillary artery and maxillary branch of the trigeminal nerve passes through the palatine via the infraorbital foramen.

Lacrimal

A pythonomorph character 'lacrimal foramen within prefrontal or absent', was proposed by Lee [17], subsequently used by Lee [4], Lee and Caldwell [40], Pierce and Caldwell [30], but later dropped by Palci and Caldwell [36] and Paparella et al. [31]. Details of the lacrimal and its contacts with adjacent bones are poorly known in mosasaurians. We report here for the first time two well-preserved examples; the dolichosaurid *Coniasaurus gracilodens* (Fig. 7.2N–R) and *Tethysaurus nopcsai* (Fig. 7.3G–I; Supplementary Figure 7.S4d–i; Supplementary Appendix 7.S1). Notwithstanding changes in proportions, the general morphology and contacts of the lacrimal with adjacent bones in *C. gracilodens* and *T. nopcsai* is consistent with that in anguimorphs [45], but differs in the following respects. The lacrimal is approximately half the height of the prefrontal in contrast with the condition in *Varanus*. The articulation for the lacrimal is present as a facet on the ventrolateral part of the prefrontal in all mosasaurian specimens in which this region is preserved. Compared to the condition in *Varanus*, the ventromedial articulation of the lacrimal with the prefrontal in *T. nopcsai* and *C. gracilodens* is substantially reduced and the contact with the palatine is more extensive than its contact with the jugal in *T. nopcsai*. In both cases, the lacrimal foramen is open and delimited by both lacrimal and prefrontal walls and in no way resembles the condition in snakes, as previously suggested [4, 36, 38] (Table 7.1).

Pterygoid and Ectopterygoid

Five pterygoid and one ectopterygoid characters have been proposed as pythonomorph synapomorphies (Table 7.1). Cope [11] considered The pterygoids are elongate and bear numerous teeth, and in one type are free, except at the extremities, a snake character.

Lee [17] introduced 'Pterygoid teeth large and recurved' and 'Anterior process of pterygoid distinct from lateral process' (Table 7.1). Caldwell [38] included the mere 'presence of pterygoid teeth' in his diagnosis, noting it was equivocal. Most recently, 'pterygoid teeth, when present, as large as maxillary teeth' was used by Lee [4], Lee and Caldwell [40], Pierce and Caldwell [30], and Palci and Caldwell [36]. Paparella et al. [31] included no pterygoid characters in their diagnosis. Palci and Caldwell [36] also included 'ectopterygoid-palatine contact absent, maxilla enters suborbital fenestra' in their diagnosis, a character related to the previously abandoned 'maxilla enters suborbital foramen' [17].

The pterygoid has been described in the dolichosaurids *Coniasaurus gracilodens* [27] and *C. crassidens* [46]. Both possess teeth substantially smaller than the marginal teeth. The pterygoids are well known in mosasaurids [13] in which the pterygoid teeth are substantially smaller than the marginal teeth except in *Prognathodon* in which the pterygoid teeth approach the size of the marginal teeth. Additionally, the pterygoid in mosasaurs is further differentiated from both snakes and varanoids in the elaboration of a posteromedial process underlying the anterolateral parabasisphenoid. At the confluence of this process and the quadrate process, a dorsomedially facing fossa receives the short basipterygoid processes in a syndesmotic joint (Supplementary Figure 7.S4a, b). Similarly, a small medial process buttressing the basipterygoid process occurs in *Varanus komodoensis*, but not in some congeners (Supplementary Figure 7.S2g). The ectopterygoid in mosasaurians is gracile and generally L-shaped [47], its slender anterior ramus articulating with the jugal over an extended distance, clasping the posteromedial maxilla in some forms. Thus, the primary structural components bridging the pterygoid and the maxilla are the combined ectopterygoid and jugal (Fig. 7.3G–H). This is fundamentally different from both the varanoid and snake conditions in which a robust, rod-like ectopterygoid is the primary structural component and the jugal contacts it anteriorly and in a more dorsal position (as in *Varanus*) or barely contacts the posteromedial corner of the jugal as in *Najash* [32]. Loss of contact between ectopterygoid and palatine in snakes and mosasaurians [36] is fundamentally different. In snakes, loss of contact is associated with the elongate suborbital projection of the maxilla terminating in a typical relationship with the ectopterygoid, but well posterior to the palatine–maxillary articulation. In mosasaurians, as exemplified in *Tethysaurus*, the two elements are in close proximity but have lost contact due to the unique and significantly reduced mosasaurian ectopterygoid (Fig. 7.3G–I; Supplementary Figure 7.S4i). See Supplementary Appendix 7.S1 for more details on the mosasaurian ectopterygoid.

7.2.2 Skull Roof and Braincase

Fifteen pythonomorph skull roof and braincase synapomorphies have been previously proposed (Table 7.1), with only three included in the latest diagnosis of the group [31]: 'parietal downgrowths [as] prominent flanges', 'parietal-prootic contact extensive', and 'basipterygoid processes do not project far from body of basisphenoid'. Paparella et al. [31] scored *Pontosaurus kornhuberi*, *Adriosaurus*, *Mosasaurus hoffmannii*, and *Platecarpus tympaniticus* for the snake condition of parietal downgrowth and those same taxa (except *Adriosaurus)* for suturing of the parietal and prootic; however, Paparella et al.'s scoring for dolichosaurids cannot be confidently verified due to crushing and poor preservation

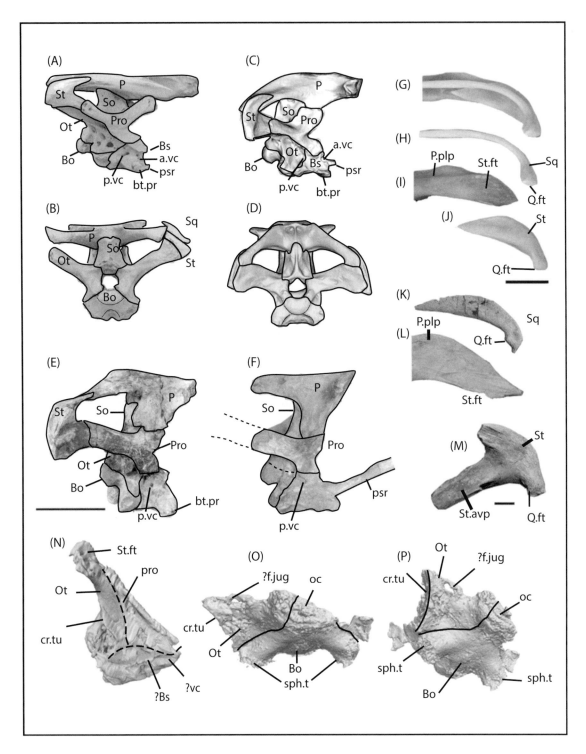

Figure 7.4 Braincase and suspensorium. Relationships of the braincase and suspensorium elements of *Tethysaurus nopcsai* (SMU-75486) in (A) right lateral, and (B) posterior views;

(Supplementary Figure 7.S1b–e). The mosasaurids in their matrix were scored for the short basipterygoid processes and dolichosaurids as unknown for this character.

Parietal Downgrowth

Pontosaurus kornhuberi appears to have some downgrowth, but no more than in *Varanus* (Supplementary Figure 7.S5a). Neither of the early diverging plioplatecarpines, *Tethysaurus* (Fig. 7.4A, B) and *Russellosaurus,* present any notable downgrowth (Supplementary Figure 7. S5b; see also Supplementary Appendix 7.S1). The alar process of the prootic in *Russellosaurus* shows evidence of ligamentous attachment to the parietal margin (fig. 3 in [47]). In halisaurines, there is moderate downgrowth of the lateral parietal margins which articulate with the distal part of the relatively short alar process of the prootic (Fig. 7.4C, D). However, downgrowth of the parietal descending processes is no greater than in adult *V. niloticus* (Supplementary Figure 7.S5a). The condition in mosasaurines such as *Clidastes* (Fig. 7.4E) is derived with substantial downgrowth of the parietal lateral margin articulating with the dorsal part of the extremely short alar processes of the prootic and posteriorly with the tall supraoccipital. This morphology is accentuated in *Prognathodon saturator* (Fig. 7.4F) [48]. Vertical expansion of the parietal facilitates increased area for muscle attachment, and together with the combination of sutural contacts with the anteriorly shifted braincase, forms a pillar-like akinetic structure, well suited to resist the increased bite forces in these large mosasaurians. There is no verifiable evidence of snake-like downgrowth of the lateral margins of the parietal nor suturing of the parietal and prootic in any early diverging mosasaurian. Development of these features in derived mosasaurids is the result of convergence.

Prootic-Supratemporal Contact

Contact between supratemporal and prootic is a long-standing putative pythonomorph character [4, 17, 36, 40] retained in the latest diagnosis [31] (Table 7.1). In stem snakes such as *Najash*, the supratemporal contacts the prootic but is superficial to it [32]. The contact is extensive and largely on the dorsal part of the prootic. As noted elsewhere [24], the supratemporal in mosasaurians is unique in the development of an anteroventral process that articulates medially with the paroccipital process, with its distal portion overlapped laterally by the prootic (Fig. 7.4A, C, M) (Supplementary Appendix 7.S1). In other words, the

---

Figure 7.4 (*cont.*) *Phosphorosaurus ponpetelegans* (HMG-1528) in (C) right lateral, and (D) posterior views. *Clidastes propython* (UAM-VP2001.2.1) in (E) right lateral view. *Prognathodon saturator* (NHMM-1998141) in (F) right lateral view. Details of temporal elements of *Varanus* sp. (unnumbered SMU specimen), in (G) articulated and (H–J) disarticulated left lateral views. Details of temporal elements in *Halisaurus* sp. (MGUAN-PA25) in (K–M) disarticulated left lateral views; Braincase elements of *Dolichosaurus longicollis* (NHMUK-R49002) in (N) right lateral oblique, (O) posterior, and (P) ventral posterolateral oblique views. Abbreviations: a.vc, anterior opening of vidian canal; Bo, basioccipital; Bs, basisphenoid; bt.pr, basipterygoid process of basisphenoid; cr.tu, crista tuberalis; f. jug, jugular foramen; Ot, otooccipital; oc, occipital condyle; P, parietal; P.plp, posterolateral process of parietal; Pro, prootic; psr, parasphenoid rostrum; p.vc, posterior opening of vidian canal; Q.ft, quadrate facet; So, supraoccipital; sph.t, spheno-occipital tubercle; St, supratemporal; St.avp, supratemporal anteroventral process; St.ft, supratemporal facet; Sq, squamosal; vc, vidian canal.

part of the supratemporal contacting the prootic is sandwiched between it and the paroccipital, fundamentally different from the condition in snakes. If mere contact between supratemporal and prootic is the criterion for scoring mosasaurians for the snake condition, that contact, albeit minor, is also present in some species of *Varanus* and in *Lanthanotus* (Supplementary Figure 7.S5g, h). Comparison of the condition of the temporal complexes in *Varanus* (Fig. 7.4G–J) and halisaurines (Fig. 7.4K–M) reveals striking similarities, with the exception of the novel anteroventral process of the supratemporal in mosasaurs.

Braincase

Braincase characters proposed as pythonomorph synapomorphies include: 'Basipterygoid processes do not project far from body of basisphenoid' [4, 17, 19, 30, 31, 40]; 'Basipterygoid processes expanded anteroposteriorly' [17, 19]; 'Body of basisphenoid extends far anteriorly beyond dorsum sellae' [17]; 'Cultriform process straight in lateral view' [4, 17, 19, 30, 40]; 'Anterior brain cavity floored by wide cultriform process' [4, 30, 40]; 'Main body (wide portion) of parasphenoid extends anteriorly some distance in front of the dorsum sellae, before tapering into a narrow cultriform process' [19]; 'Vidian canal is an open groove anteriorly' [17, 19]; 'Rear opening of vidian canal situated far posteriorly' [17, 19]; 'Crista circumfenestralis encircling footplate of stapes' [17, 19]; 'Stapes footplate surrounded by flanges from prootic and opisthotic' [17, 19, 30]; 'Supraoccipital sutured with parietals along its entire dorsal margin' [17, 19, 36], and 'Supraoccipital sutured to parietal' [4, 30, 40].

The majority of those characters have been addressed critically elsewhere [24, 49]. Although it is clear that the braincase of mosasaurids share no derived characters with snakes (Fig. 7.4A–F), details in dolichosaurids are largely unknown due to poor preservation with a single exception. *Dolichosaurus longicollis* was redescribed by Caldwell [28], which included some aspects of the braincase. Our HRCT exploration of the holotype revealed previously hidden elements and allowed reidentification of other elements that are briefly summarized here. The element described as the parietal [28] is the basioccipital and fragments of the otooccipitals, and the otooccipital and prootic are more extensive than previously described (Fig. 7.4N–P). An elongate, distally expanding paroccipital process is present, unlike the truncated element previously figured [28]. The basioccipital and articulated portions of the otooccipitals show a typical varanoid condition with a dorsolaterally sloped margin of the crista tuberalis, as in *Varanus*. Although not addressing any characters purportedly uniting snakes and mosasaurians, it confirms a typical varanid-like braincase in dolichosaurids. See additional comments on the braincase of *Dolichosaurus* from HRCT in the Supplementary Appendix 7.S1.

Basipterygoid Processes

The character 'basipterygoid processes do not project far from body of basisphenoid' [17, 31, 40] derives from the superficial similarity of the basipterygoid processes in some mosasaurs and snakes. This similarity led Lee [17] to consider the processes and their modifications as homologous, nothwithstanding that in snakes they arise from a different developmental pathway compared to other squamates [25]. Lee divided the morphology into two characters, one for overall length of the process and another for the nature of the distal articulating surface, using *Platecarpus* to illustrate the condition in mosasaurians. Broad fan-like basipterygoid processes (as Lee figured) are found only in tylosaurines and some plioplatecarpines. Halisaurines and mosasaurines have subrectangular, distally expanding basipterygoid processes [50, 51] (Supplementary Figure 7.S7b). Other than length and more anteroventral trajectory, their basipterygoid processes resemble the typical varanoid condition. The basipterygoid processes are short in all known mosasaurians compared with varanoids and the corresponding articulation with the pterygoids is syndesmotic within a deep fossa on the dorsomedian surface of the pterygoid at the confluence of the quadrate and basisphenoid processes (Supplementary Figure 7.S7c–d), buttressing the pterygoids from above.

7.2.3 Suspensorium and Quadrate

Dolichosaurids and halisaurines retain the plesiomorphic relationship of the supratemporal to the parietal and squamosal. Although knowledge of details in dolichosaurids is limited, portions of the suspensorium in *Coniasaurus gracilodens* and *Judeasaurus tchernovi* are preserved (Supplementary Appendix S7.1). The quadrate is supported from above by the posteroventral tip of the supratemporal and to a lesser extent, the posteroventral squamosal as in *Varanus* (Fig. 7.4G–I; Supplementary Figure 7.S5a). By the Turonian, early diverging plioplatecarpines show increased contribution of the squamosal to support of the quadrate (Supplementary Figure 7.S5b). Additionally, the squamosal and supratemporal develop an immobile sutural contact, accompanied by increased complexity in the supratemporal parietal contact [47].

The Quadrate

Three quadrate characters have been used in Pythonomorpha diagnoses. Cope (p. 253 in [11]) proposed 'The os quadratum embraces and encloses the meatus auditorius externus', which in his view was unique to his Pythonomorpha. Cope's character has been revived as 'quadrate with large, posteroventrally curved, suprastapedial process' [36] and, most recently, 'quadrate suprastapedial process recurved posteroventrally' [31].

Determining a 'large' suprastapedial process (SSP) is subjective unless that determination is made quantitatively. The SSP of the quadrate in mosasaurs is expanded compared to the *Varanus* condition and is also posteroventrally curved; however, a similar suprastapedial process is also present in some teiids (e.g., *Tupinambis* and *Dracaena*). Both elongation and curvature of the suprastapedial process is present in stem snakes, but appears exaggerated compared to the relatively short main shaft of the element. Comparing the length of the SSP with the distance from the dorsal coronoid-dentary junction to the centre of the mandibular glenoid, the posterior extent of the process in *Najash* is 22 per cent, versus 34 per cent in *Lanthanotus*, 23 per cent in *Varanus exanthematicus* and 16 per cent in *V. komodoensis*, and thus falls within the range of other varanoids. Comparing functional aspects of an elongate suprastapedial is revealing. In snakes such as *Najash* and *Cylindrophis*, the dorsomedial quadrate shaft and the entire length of the SSP contacts the supratemporal. In contrast, mosasaurians show the varanid condition, in which the SSP's contact with both the supratemporal and squamosal is restricted to a relatively small dorsomedial area, with all of the quadrate shaft and much of the SSP not participating at all in the support. A plausible explanation for the relatively large SSP in snakes is that it is purely a structural adaptation, increasing contact area medially with the supratemporal and thus increasing mechanical support for the quadrate. Additionally, there are other aspects of the quadrate in mosasaurs that suggest their specialized quadrate is an adaptation for underwater hearing (Supplementary Appendix S7.1), contrasting with snakes which have lost the typical impedance matching system of lizards and are instead optimized for transmission of vibrations from the substrate [52].

7.2.4 Mandibles

Several aspects of the mandibles have been cited as pythonomorph characters. These include a free dentary symphysis, geometry of the dorsal dentary margin, orientation of Meckel's groove, development of a medial ridge on the subdental shelf, a number of characters associated with reduction of overlap and increased mobility in the mid-mandible, reduction in the medial aspects of the coronoid, orientation of the adductor fossa, and fusion of the prearticular (Table 7.1).

Free Symphysis of Mandibles

Beginning with Cope [11], all Pythonomorpha diagnoses have included the free symphysis of the mandibles (Table 7.1). Although most dolichosaurids are too poorly preserved in this region to be scored, close examination and comparisons with previously published specimens show *Coniasaurus, Opetiosaurus* and *Haasiasaurus* have a symphysial morphology nearly identical to that in *Varanus* (Fig. 7.6H–J), and is also present but modified in derived mosasaurids (Supplementary Figure 7.S8c–i). This morphology is closely linked to another problematic character – 'anterior end of Meckel's canal on medial surface of dentary' [4, 30, 31, 40]. Meckel's canal in non-snake squamates is within the symphysial region [53] and therefore, by necessity is more or less medially facing. In *Coniasaurus, Opetiosaurus* and *Haasiasaurus* (Fig. 7.6I, J), the symphysial area dorsal and ventral to Meckel's canal are roughly the same as in *Varanus* (Fig. 7.6H) but in more derived mosasaurids, there is increased participation of the ventral part (Supplementary Figure 7.S8e, f) (Supplementary Appendix S7.1).

Dentary Margin

'Dentary straight in lateral view' has been proposed as a pythonomorph character [4, 41, 30, 31]. Paparella et al. [31] scored a straight dentary margin for *Dolichosaurus* and both *Pontosaurus* species as well as *Tetrapodophis*. The condition in available specimens of *Pontosaurus* and *Tetrapodophis* is unclear due to poor preservation. Broader comparisons reveal a curved margin in *Coniasaurus, Judeasaurus, Haasiasaurus,* but not *Opetiosaurus* (Supplementary Figure 7.S9c–e, g). Curvature is also present in some derived mosasasaurids [48] (Supplementary Appendix 7S.1).

Dentary Subdental Shelf Development

This character was initiated by Lee [17] but included only in the most recent diagnosis of Pythonomorpha [31]. As formulated [17] the character reads 'Subdental shelf medial to teeth. Present, 0. Absent, 1'. We take use of the term 'subdental shelf' to mean the dorsally projecting ridge of bone lingual to the teeth, hereafter referred to as the lingual ridge (sensu [22]). More recent use of the character [30, 31, 36] deviated from Lee's intent, with both snakes and mosasaurians scored using the form 'Subdental shelf: Weakly developed (0), large (1), absent (2)'. In *Coniasaurus*, there is no development of a lingual ridge of bone medial to the region containing pleurodont teeth (Fig. 7.6E, F). In *Haasiasaurus*, *Opetiosaurus*, and *Tethysaurus* (Supplementary Figure 7.S9e–h), there is a short lingual ridge of bone that forms the medial wall of a trough containing the pleurodont teeth, with *Tethysaurus* being the earliest mosasaurid known to possess poorly developed interdental ridges separating the bases of the teeth. The lingual ridge present in early diverging mosasaurians is derived for the group, not homologous to the basal plate described by Lessmann ([54]: discussed by [22]) and which provides direct support for the medial part of the tooth. The incipient lingual ridge of the latter three taxa presents an intermediate condition between *Dolichosaurus* and derived mosasaurids, in which the lingual ridge is substantially developed dorsally, in some cases reaching the level of the labial ridge (pleura) to form a deep alveolar groove [22]. This morphology consists of compact bone and has nothing to do with the type of tissues that provides attachment for the teeth and that have been intensely discussed in the literature [55]. In derived mosasaurids, the lingual ridge is pronounced but variable, and in some taxa (e.g., *Mosasaurus*, *Tylosaurus*, *Selmasaurus*) it is greatly developed and completely encloses the dorsomedial wall of the dentary, and along with development of interdental alveolar bone, forms discrete sockets. The relative height of the lingual ridge is quite variable among mosasaurians but there does seem to be a trend towards extreme development of the lingual ridge in some derived mosasaurids in separate subfamilies, as noted above. Other members of those same subfamilies retain low lingual ridges, suggesting this may be an adaptation to strengthen the jaw and/or provide additional lingual support for the tooth 'roots'. Notwithstanding the broad variation noted above, the scoring in recent work conflates alethinophidian tooth implantation [22] that results in the 'discrete sockets' of snakes with the clearly independently acquired derived condition seen in forms such as *Mosasaurus*.

Intramandibular Joint

Cope [11] noted the articulating surfaces of the angular and splenial in his concept of Pythonomorpha, considering it a snake character. Different aspects of the intramandibular region have been employed, including: 'Posterior end of splenial does not overlap coronoid' [4, 17, 19, 30, 40]; 'Coronoid does not contact splenial' [17]; 'Splenial does not substantially overlap postdentary elements' [4, 30, 31, 40]; 'Reduced overlap of the postdentary-dentary bones' [38]; 'Vertical articulation between the splenial and angular' [4, 11, 17, 19, 36, 40]; 'Splenial and angular greatly exposed in lateral view' [36]; and 'Surangular exposed in medial view of lower jaw' [19].

All of these characters are associated with increasing kinesis while maintaining structural support. Only one of these characters 'Splenial does not substantially overlap postdentary elements' is cited in the most recent diagnosis of Pythonomorpha [31]. Notwithstanding the non-quantitative nature of 'substantial', *Coniasaurus crassidens* (Fig. 7.5A), *Opetiosaurus*, and *Haasiasaurus* (Supplementary Figure 7.S9e) show contact of the posterodorsal splenial

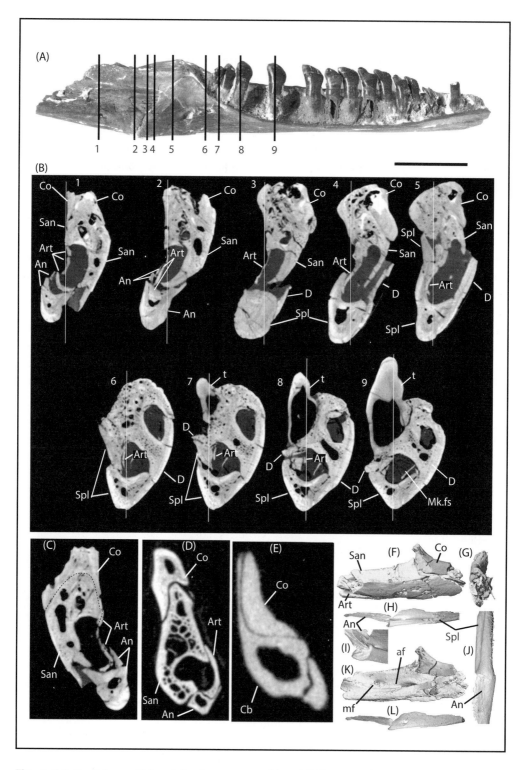

Figure 7.5 Partial mandibles of *Coniasaurus crassidens* (NHRI-R-3421) and *Dolichosaurus longicollis* (NHMUK-R49002). Partial left mandible of *C. crassidens* in (A) medial view with numbered vertical

and coronoid to the same or greater degree than seen in *Lanthanotus* in which the anteromedial coronoid lies adjacent to the posterior dentary, similar to the condition in *Najash*. However, that contact in mosasaurians, as minor as it is, is predominantly controlled by the posterodorsal extent of the splenial, because the coronoid medial processes (anterior and posterior) are highly reduced.

The morphology of the contact between the angular and splenial has historically played a prominent role in the Pythonomorph Hypothesis, included in most diagnoses [4, 11, 17, 19, 40], though not the most recent [31] in which the 'snake-mosasaur' condition was scored for *Lanthanotus*. Beyond the relatively transverse suture between the angular and the splenial (which is at least superficially similar in snakes and mosasaurs), there are substantial differences in the details of that contact, as there are in the morphology and contacts of the entire suite of bones involved in the intramandibular joint [6] (Supplementary Appendix S7.1).

The intramandibular regions of mosasaurs and snakes are fundamentally different, exhibiting only superficial similarities (as also interpreted by [6, 56]). The situation is best summarized by Gauthier [56]: 'Some similarity is to be expected, especially since there is but one place in a squamate mandible where a mobile joint could form between the dentary–splenial and postdentary bones.'

Posterior Mandible Characters

Characters of the postdentary mandible that have been used to support the Pythonomorph Hypothesis include: 'Posterior ramus of coronoid reduced or absent' [17]; 'Coronoid with straight or convex ventral edge' [4, 19, 30, 40]; 'Coronoid does not overlap dentary laterally' [19]; 'Anterior end of coronoid meets dentary directly' [38]; 'Subcoronoid fossa absent' [4, 17, 30, 40]; 'Adductor fossa faces dorsally' [4, 17, 19, 30, 40]; and 'Articular and prearticular fused' [4, 30, 40] (Table 7.1).

Reduction of the antero- and posteromedial processes of the coronoid is shared by mosasaurs and snakes; however, as was the case with the intramandibular complex, other major differences in the coronoid and its contacts suggest convergence rather than homology. Additionally, reduction of the medial processes render the character 'Subcoronoid fossa absent' [4, 17, 30, 40] redundant. Mosasaurians are quite variable in exposure of the surangular below the coronoid and above the articular.

Though the coronoid in early diverging mosasaurians possess reduced medial processes, it is otherwise comparable to *Varanus*, sharing only minor overlap of the posterodorsal

---

Figure 7.5 (*cont.*) lines to indicate position of (B) HRCT slices illustrating cross sections of intramandibular elements. Cross section (C) of *Coniasaurus* mandible between slices 1 and 2 of (B) compared with (D *Varanus komodoensis* (TNHC 95803; Digimorph HRCT data), and (E) *Cylindrophis ruffus* (FMNH 60958; Digimorph HRCT data). Partial right mandible (F-L) of *Dolichosaurus longicollis*. Articulated surangular and articular, with re-articulated coronoid in (F) lateral, (G) anterior, and (K) medial views. Associated left splenial, angular and fragmentary dentary in (H) medial, (I) posteromedial oblique, (J) ventral, and (L) lateral (reversed) views. Abbreviations: An, angular; af, adductor fossa; Art, articular; Cb, compound bone; Co, coronoid; D, dentary; mf, mandibular foramen; Mk.fs, Meckel's fossa; San, surangular; Spl, splenial; t, tooth. Scale in (a) = 5mm.

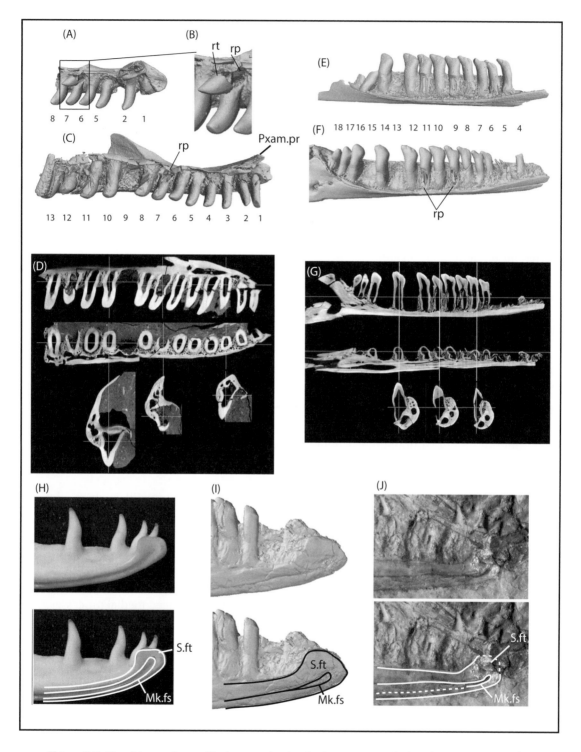

Figure 7.6 Dentition and mandibular symphysis. *Coniasaurus gracilodens* (NHMUK R44141) partial left maxilla in (A) medial view and (B) inset detailed view of the same, and partial right maxilla

dentary, a notched anterodorsal margin, and straddling the anterodorsal suranglar with most of the articulating surface on the lateral side of the surangular (Fig. 7.5C, d; Supplementary Figure 7.S10i–m). This condition is different from the that in the stem snake *Najash,* in which the coronoid substantially overlaps the posteromedial dentary and articulates on the medial side of the compound bone [32] as in the crown snake *Cylindrophis* (Fig. 7.5E).

The adductor fossa in all early diverging mosasaurians faces medially (e.g., *Dolichosaurus;* Fig. 7.5K; Supplementary Figure 7.S10b). In some derived mosasaurids, upgrowth of the dorsal articular margin obscures medial exposure. In all mosasaurians in which it can be examined, the articular and prearticular are fused with neither being fused to the surangular.

7.2.6 Marginal Dentition

The marginal dentition in derived mosasaurids possesses a highly modified implantation, that Cope [11] interpreted as snake-like. Lee and Caldwell [19] included 'teeth set in discrete sockets' in their list of 'pythonomorph characters of *Pachyrhachis'.* 'Recumbent (horizontal) replacement teeth' is included in all but the most recent diagnoses of Pythonomorpha [4, 17, 30, 40]. Though dolichosaurids and *Opetiosaurus* were reported to possess alveoli separated by interdental plates [40], this was subsequently corrected [31] based on recognition of pleurodont implantation in early diverging mosasaurians, as previously noted [57] in *Haasiasaurus.* Below, we provide a brief description of tooth implantation and replacement in dolichosaurids, 'aigialosaurids' and mosasaurids.

Tooth Implantation

In the dolichosaurids *Coniasaurus* and *Judeasaurus,* tooth implantation is pleurodont with minimally developed base of attachment [58] (Fig. 7.6A–E; Supplementary Figure 7.S9a–d), and without interdental and lingual ridges. Plicidentine is absent. All early diverging mosasaurians retain a fully pleurodont condition. *Opetiosaurus, Haasiasaurus,* and *Tethysaurus* have a low lingual ridge and in *Tethysaurus* thin interdental alveolar bone separates alveoli (Supplementary Figure 7.S9h). In early diverging and derived mosasaurids, there are varying degrees of interdental alveolar bone development and upgrowth of the lingual ridge of the subdental shelf, which along with the medial surface of the lingual ridge, surrounds the tooth bases and forms sockets in many taxa. Although Bertin et al. [55] illustrated the condition in what we consider derived mosasaurids (their figure 8f), early diverging mosasaurians do not have thecodont implantation, but instead all have a

---

Figure 7.6 (*cont.*) (reversed for comparison) in (C) medial view with inset of (D) select orthogonal HRCT slices. *Coniasaurus crassidens* (NHRI-R-3421) partial right mandible in (E) medial view (reversed) and left partial mandible in (F) medial view with inset (G) of select orthogonal HRCT slices of left mandible. Left dentary of (H) *Varanus* sp. (unnumbered SMU specimen), (I) *Coniasaurus* sp. (SMU 69019), and (J) *Opetiosaurus bucchichi* (NMW specimen, reversed for comparison). Abbreviations: Mk.fs, Meckel's fossa; Pxam.pr, anteromedial process of premaxilla; rp, resorption pit; rt, replacement tooth; S.ft, Symphysial facet.

pleurodont tooth implantation with upgrowth of the lingual ridge either absent (*Coniasaurus, Dolichosaurus*) or poorly developed (*Aigialosaurus, Pontosaurus, Haasiasaurus, Judeasaurus*). In that sense, *Coniasaurus* and *Dolichosaurus* retain a fully pleurodont implantation comparable to varanids and helodermatids, in which the teeth are ankylosed to the inclined labial wall of the tooth bearing bone, lacking a subdental shelf or lingual ridge (Supplementary Appendix 7.S1).

Tooth Replacement

The left maxilla of the holotype of *Coniasaurus gracilodens* preserves a single replacement tooth in a slightly recumbent orientation, posteromedial to the sixth tooth position which is partially resorbed as is its counterpart on the right (Fig. 7.6 A–C). Similar resorption is seen in *C. crassidens* although no replacement teeth are present (Fig. 7.6E, F). The resorption pits are more posteromedially oriented in *C. crassidens* than in *C. gracilodens,* but both species possess broadly open pulp cavities (Fig. 7.6D, G). Slightly recumbent replacement tooth crowns are present in *Haasiasaurus gittelmani* but, as in *C. gracilodens*, it is unclear whether this might result from postmortem displacement. The lack of plicidentine, development of resorption pits, and the posteromedial position of the replacement tooth in dolichosaurids is intermediate between iguanid and varanid replacement modes and consistent with an anguinoid condition [59]. A replacement tooth develops posteromedially while resorption initiates from the alveolar foramen medially or slightly posteromedially. The size of the resorption pit initially lags development of the replacement tooth (e.g., Fig. 7.6B), but later accelerates, with attachment tissues resorbed shortly before shedding the functional tooth, by which time the replacement tooth is nearly fully grown [16, 59]. This mode of replacement is seen in *Haasiasaurus* [57] which has numerous replacement teeth of various stages but no associated resorption pits. Resorption pits on the posteromedial base of attachment, likely associated with developing replacement teeth, is first seen in *Tethysaurus*. In derived mosasaurids resorption is more apparent and begins on or near the posteromedial dorsal surface of the bone of attachment, following the leading surface of the bell stage of the developing tooth, but ultimately creating deep crypts that accommodate the fully developed replacement tooth (see also Supplementary Figure 7.S9i and associated text).

Mosasaurians are more similar to anguioids than they are to snakes in terms of tooth implantation and replacement. In light of the other characters presented, recumbent replacement teeth are likely the result of convergence.

7.2.7 Postcrania

Many postcranial characters have been used to diagnose Pythonomorpha, including: 'Zygosphenes and zygantra present' [4, 11, 17, 19, 30, 31, 36, 38, 40]; 'Absence of posterior process on the atlas neural arch' [17, 19]; 'Condyles on centra circular in shape' [4, 17, 19, 30, 40]; 'Five or more pygals' [17]; 'Tail laterally compressed, i.e., transverse processes reduced anteriorly and absent posteriorly, chevrons and neural spines elongated' [36]; 'The abdominal cavity is long, and is surrounded by many short, curved ribs, which have a free antero-posterior movement on vertical, articulating surfaces, and which commence immediately behind the head' [11]; 'Ribs begin from third cervical' [4, 17, 19, 30, 40]; 'Pelvic elements not suturally united' [17, 19, 31, 36]; 'Pubic plate oriented parasagittally, wide in lateral view' [36]; 'Femur stout' [17]; 'Straight, short femur' [19]; 'Distal end of tibia gently convex for astragalocalcaneal articulation' [31]; 'Astragalus and calcaneum separate' [19, 31, 36]; 'Cranial and axial epiphyses absent' [17, 19, 31]; and 'Osteoderms lost' [17, 19, 36].

Most of these characters are secondary aquatic adaptations broadly distributed in marine amniotes that independently evolved limbs and hind parts optimized for swimming [60]. Changes in ossification patterns of the vertebral column, limb elements and bony epiphyses [61], and reduction in the mesopodials [62] are all common among secondarily derived marine amniotes, some of which may be the result of paedomorphosis [63]. As such, this suite of characters is prone to homoplasy as seen among subfamilies of mosasaurids [64] and thus appear to be of limited phylogenetic utility, especially at higher levels.

The most recent diagnosis of Pythonomorpha [31] cited five postcranial characters. Two of them (astragalus and calcaneum separate, and cranial and axial epiphyses absent) are subject to paedomorphism or ontogenetic variation. It is not uncommon for these elements to appear separate in early ontogeny and fuse later. The sub-adult condition in *Varanus exanthematicus* is illustrated by Sullivan [65]. A fused astragalus-calcaneum occurs in *Haasiasaurus* [57]. Ossification of epiphyses and other skeletal elements [66] generally tend to reduce during the evolution of mosasaurians.

A third character, distal end of tibia gently convex for astragalocalcaneal articulation, is difficult to judge. Some of the dolichosaurid specimens for which this was scored by Paparella et al. [31] are poorly preserved and badly crushed (Supplementary Figure 7.S11). A new specimen of *Coniasaurus* (Supplementary Figure 7.S11a), preserved 3-dimensionally, clearly illustrates a concave articulation on the distal tibia. It is not possible to judge if this was the case in *Adriosaurus* and *Acteosaurus* fossils given their badly crushed state (Supplementary Figure 7.S11b, c) and those taxa should be scored as unknown for this character. It is common in secondarily adapted marine amniotes to reduce ossification and retain only cartilaginous epiphyses, as in derived mosasaurids; however, halisaurines and 'aigialosaurids' present an intermediate step in retaining calcified but not fully fused epiphyses. Reduction and loss of ossification in carpal and tarsal elements occurs during the evolution of mosasaurids [62], but they otherwise retain or enhance a full complement of the other limb bones. Because they manifest so differently, homology of reduction in ossification of distal limb elements in stem snakes and in mosasaurians is unclear.

A fourth postcranial character, presence of zygosphenes and zygantra, which occurs in all dolichosaurids, 'aigialosaurids', and basal members of all mosasaurid subfamilies except Halisaurinae, is an unlikely pythonomorph character because zygosphenes and zygantra are broadly present in Squamata [3, 7]. Although not scored for these taxa by Gauthier et al. [7], zygosphenes and zygantra are present in *Saniwa ensidens*, *Estesia mongoliensis*, *Gobiderma pulchrum*, and *Telmasaurus grangeri*. In all non-snake squamates in which they are present, zygosphenes have an anterior medial notch. Given the differences in morphology in snakes and its distribution within varanoids, and other squamate groups, the phylogenetic utility of this character is questionable.

The fifth and final postcranial character we address is 'ribs begin third cervical vertebra or anterior'. As formulated, the scoring is binary; third vertebra and anterior versus fourth vertebra and posterior. There appears to be a cervical rib adjacent to the third cervical vertebra in the dolichosaurid *Pontosaurus kornhuberi* [29]; however, the cervical region in both *Pontosaurus lesinensis* and *Primitivus manduriensis* are too badly preserved to score this character (contra [31]). In the holotype of *Aigialosaurus dalmaticus* a short cervical rib lies predominantly on the axis, but is probably associated with the third cervical vertebra. Derived mosasaurids have cervical ribs beginning on the axis vertebra, optimized as an unambiguous synapomorphy of both Mosasauridae and Dibamidae according to [7]. The condition is unclear in *Najash* and *Dinilysia*. Ribs have been scored as present on the third or more anterior cervical vertebra [31] for the limbed Cenomanian snakes, *Pachyrhachis*, *Haasiophis*, and *Eopodophis*. However, ribs beginning on the 5th vertebra have been described for *Pachyrachis* [19]. X-rays of *Haasiophis* [67] show ribs beginning on the 5th or 6th vertebra. Gauthier et al. [7] used an ordered five-state scheme to formulate this character (one for each cervical vertebra 3–6 upon which the first rib appears). Their findings for pachyophiids are consistent with our own observations (contra [31]).

7.3 Conclusions and Closing Remarks

Most of the previously reported characters supporting the Pythonomorph Hypothesis are problematic, due to incomplete fossil preparation, artefacts of taphonomy, limited comparisons, misinterpretations of anatomy, incomplete taxon sampling, or inadequate character formulation and/or scoring. For many of the mosasaurian specimens used in prior work, taphonomic issues are substantial and need to be addressed and clearly communicated in future descriptive work and in those studies performing character scoring for phylogentic studies. Over-interpretation of fossils, influenced by prior phylogenetic hypotheses, may lead to confirmation bias in character formulation and scoring.

Augusta et al. (Chapter 8) present a cladistic analysis focused on inferring the position of mosasaurians within Squamata, including new data reported here. With the inclusion of those data, mosasaurians are found to lie within Anguimorpha, as the sister taxon to Varanoidea. The addition of data from well-preserved early branching mosasaurians provided new and corrected scoring for a handful of key characters in Augusta et al.'s study, including intermediate states that clarified homoplasy instances. The results include a dramatically reduced lability of Mosasauria compared to that found by Gauthier et al. [7] and subsequent work based on their matrix (see Augusta et al. and Zaher et al., Chapters 8 and 9). Nonetheless, even though parsimony analysis recovers Mosasauria as a proximal outgroup of Varanoidea, the morphology of mosasaurians presented herein raises some interesting questions. The apparently plesiomorphic state of the infraorbital foramen of the palatine (ventrally open), the unique reduction of the ventral part of the lacrimal and its relationship with the prefrontal, the reduction of the anterior ramus of the ectopterygoid and its reduced contact with the maxilla, the lack of plicidentine, and anguinoidean tooth replacement in early diverging mosasaurians, all deviate from the typical varanoid condition and allow alternate interpretations. However, the most parsimonious explanation based on the results of Augusta et al. (Chapter 8) is that mosasaurians are regressed varanoids, the palatine and infraorbital characters secondarily derived as marine adaptations associated with streamlining and narrowing of the skull along with narial retraction and reorganization of the nasal capsule for marine-adapted respiration and osmoregulation (see Supplementary Appendix 7.S1). The degree of narial retraction, posteriorly narrow nasals, and the unique lacrimal and palatine characters shared by dolichosaurids and other mosasaurians support the monophyly of this group to the exclusion of both varanoids and snakes. Nonetheless, two pythonomorph characters remain: the uniquely regressed medial coronoid and the splenial-angular contact. However, in light of the data presented here, it is increasingly difficult to interpret these as anything other than convergence.

Acknowledgements

Detailed acknowledgements are given in Supplementary Appendix 7.S1. The authors thank Olivier Rieppel, Nicholas Fraser, Louis Jacobs, and Anne Schulp for constructive criticism on the submitted and earlier versions of the manuscript.

References

1. T. M. Townsend, A. Larson, E. Louis, and J. R. Macey, Molecular phylogenetics of Squamata: the position of snakes, amphisbaenians, and dibamids, and the root of the squamate tree. *Systematic Biology*, 53 (2004), 735–757.

2. F. T. Burbrink, F. G. Grazziotin, R.A. Pyron, et al., Interrogating genomic-scale data for Squamata (lizards, snakes, and amphisbaenians) shows no support for key traditional morphological relationships. *Systematic Biology*, 69 (2020), 502–520.

3. R. Estes, K. de Queiroz, and J. A. Gauthier. Phylogenetic relationships within Squamata. In R. Estes and G. K. Pregill, eds., *Phylogenetic Relationships of the Lizard Families* (Stanford, California: Stanford University Press, 1988), pp. 119–281.

4. M. S. Y. Lee, Convergent evolution and character correlation in burrowing reptiles: towards a resolution of squamate relationships. *Biological Journal of the Linnean Society*, 65 (1998), 369–453.

5. J. L. Conrad, Phylogeny and systematics of Squamata (Reptilia) based on morphology.

Bulletin of the American Museum of Natural History, 310 (2008), 1–182.

6. O. Rieppel and H. Zaher, The intramandibular joint in squamates, and the phylogenetic relationships of the fossil snake *Pachyrhachis problematicus* Haas. *Fieldiana Geology*, 43 (2000), 1–69.

7. J. A. Gauthier, M. Kearney, J. A. Maisano, O. Rieppel, and A. Behlke, Assembling the squamate tree of life: Perspectives from the phenotype and the fossil record. *Bulletin of the Peabody Museum of Natural History*, 53 (2012), 3–308.

8. E. W. A. Mulder, Maastricht Cretaceous finds and Dutch pioneers in vertebrate palaeontology. In J. L. R. Touret and R. P. W. Visser, eds., *Dutch Pioneers of the Earth Sciences* (Amsterdam: Royal Netherlands Academy of Arts and Sciences, 2004), pp. 165–176.

9. A. G. Camper, Lettre de A.G. Camper à G. Cuvier sur les ossemens fossiles de la montagne de St. Pierre, à Maëstricht. *Journal de Physique, de Chimie, et d'Histoire Naturelle*, 51 (1800), 278–291.

10. G. Cuvier, Sur le grand animal fossile des carrières de Maestricht. *Annales du Museum National d'Histoire Naturelle*, 12 (1808), 145–176.

11. E. D. Cope, On the reptilian orders, Pythonomorpha and Streptosauria. *Proceedings of the Boston Society of Natural History*, 12 (1869), 250–266.

12. C. L. Camp, Classification of the lizards. *Bulletin of the American Museum of Natural History*, 48 (1923), 289–481.

13. D. A. Russell, Systematics and morphololgy of American mosasaurs (Reptilia, Sauria). *Bulletin of the Peabody Museum of Natural History*, 23 (1967), 1–241.

14. F. Nopcsa, Über die varanusartigen lacerten Istriens. *Beiträge zur Paläontologie Österreich-Ungarns und des Orients*, 15 (1903), 31–42.

15. F. Nopcsa, *Eidolosaurus* und *Pachyophis*: Zwei neue Neocom-Reptilien. *Palaeontographica*, 65 (1923), 99–154.

16. S. B. McDowell and C. M. Bogert, The systematic position of *Lanthanotus* and the affinities of the anguinomorphan lizards. *Bulletin of the American Museum of Natural History*, 105 (1954), 1–142.

17. M. S. Y. Lee, The phylogeny of varanoid lizards and the affinities of snakes. *Philosophical Transactions of the Royal Society of London*, B352 (1997), 53–91.

18. M. W. Caldwell and M. S. Y. Lee, A snake with legs from the marine Cretaceous of the Middle East. *Nature*, 386 (1997), 705–709.

19. M. S. Y. Lee and M. W. Caldwell, Anatomy and relationships of *Pachyrhachis problematicus*, a primitive snake with hindlimbs. *Philosophical Transactions of the Royal Society of London*. B353 (1998), 1521–1552.

20. M. S. Y. Lee, G. L. Bell, and M. W. Caldwell, The origin of snake feeding. *Nature*, 400 (1999), 655–659.

21. H. Zaher, The phylogenetic position of *Pachyrhachis* within snakes (Squamata, Lepidosauria). *Journal of Vertebrate Paleontology*, 18 (1998), 1–3.

22. H. Zaher and O. Rieppel, Tooth implantation and replacement in squamates, with special reference to mosasaur lizards and snakes. *American Museum Novitates*, 3271 (1999), 1–19.

23. H. Zaher and O. Rieppel, The phylogenetic relationships of *Pachyrhachis problematicus*, and the evolution of limblessness in snakes (Lepidosauria, Squamata). *Comptes Rendus de Séances de l'Académie des Sciences (Série IIA), Earth and Planetary Science*, 329 (1999), 831–837.

24. O. Rieppel and H. Zaher, The braincases of mosasaurs and *Varanus*, and the relationships of snakes. *Zoological Journal of the Linnean Society*, 129 (2000), 489–514.

25. O. Rieppel and M. Kearney, The origin of snakes: limits of a scientific debate. *Biologist (London)*, 48 (2001), 110–114.

26. M. W. Caldwell and J. A. Cooper, Redescription, palaeobiogeography and palaeoecology of *Coniasaurus crassidens* Owen, 1850 (Squamata) from the Lower Chalk (Cretaceous; Cenomanian) of SE England. *Zoological Journal of the Linnean Society*, 127 (1999), 423–452.

27. M. W. Caldwell, Description and phylogenetic relationships of a new species of *Coniasaurus* Owen, 1850 (Squamata). *Journal of Vertebrate Paleontology*, 19 (1999), 438–455.

28. M. W. Caldwell, On the aquatic squamate *Dolichosaurus longicollis* Owen, 1850 (Cenomanian, Upper Cretaceous), and the evolution of elongate necks in squamates. *Journal of Vertebrate Paleontology*, 20 (2000), 720–735.

29. M. W. Caldwell, A new species of *Pontosaurus* (Squamata, Pythonomorpha) from the Upper Cretaceous of Lebanon and a phylogenetic analysis of Pythonomorpha. *Memorie della Società Italiana di Scienze Naturali e del Museo Civico di Storia Naturale di Milano*, 34 (2006), 1–42.

30. S. E. Pierce and M. W. Caldwell, Redescription and phylogenetic position of the Adriatic (Upper Cretaceous; Cenomanian) dolichosaur *Pontosaurus*

lesinensis (Kornhuber, 1873). *Journal of Vertebrate Paleontology*, 24 (2004), 373–386.

31. I. Paparella, A. Palci, U. Nicosia, and M. W. Caldwell, A new fossil marine lizard with soft tissues from the Late Cretaceous of southern Italy. *Royal Society Open Science*, 5 (2018), 172411.

32. F. F. Garberoglio, S. Apesteguía, T. R. Simões, et al., New skulls and skeletons of the Cretaceous legged snake *Najash*, and the evolution of the modern snake body plan. *Science Advances*, 5 (2019), eaax5833.

33. M. C. Mekarski, D. Japundžić, K. Krizmanić, and M. W. Caldwell, Description of a new basal mosasauroid from the Late Cretaceous of Croatia, with comments on the evolution of the mosasauroid forelimb. *Journal of Vertebrate Paleontology*, 39 (2019), e1577872.

34. M. W. Caldwell, R. L. Carroll, and H. Kaiser, The pectoral girdle and forelimb of *Carsosaurus marchesetti* [sic] (Aigialosauridae), with a preliminary phylogenetic analysis of mosasauroids and varanoids. *Journal of Vertebrate Paleontology*, 15 (1995), 516–531.

35. G. L. Bell, Jr. A phylogenetic revision of North American and Adriatic Mosasauroidea. In J. M. Callaway and E. L. Nicholls, eds., *Ancient Marine Reptiles* (New York: Academic Press, 1997), pp. 293–332.

36. A. Palci and M. W. Caldwell, Redescription of *Acteosaurus tommasinii* von Meyer, 1860, and a discussion of evolutionary trends within the clade Ophidiomorpha. *Journal of Vertebrate Paleontology*, 30 (2010), 94–108.

37. M. deBraga and R. L. Carroll, The origin of mosasaurs as a model of macroevolutionary patterns and processes. *Evolutionary Biology*, 27 (1993), 245–322.

38. M. W. Caldwell, Squamate phylogeny and the relationships of snakes and mosasauroids. *Zoological Journal of the Linnean Society*, 125 (1999), 115–147.

39. M. E. Páramo-Fonseca, *Yaguarasaurus columbianus* (Reptilia, Mosasauridae), a primitive mosasaur from the Turonian (Upper Cretaceous) of Colombia. *Historical Biology*, 14 (2000), 121–131.

40. M. S. Y. Lee and M. W. Caldwell, *Adriosaurus* and the affinities of mosasaurs, dolichosaurs, and snakes. *Journal of Paleontology*, 74 (2000), 915–937.

41. T. Konishi and M. W. Caldwell, New specimens of *Platecarpus planifrons* (Cope, 1874) (Squamata: Mosasauridae) and a revised taxonomy of the genus. *Journal of Vertebrate Paleontology*, 27 (2007), 59–72.

42. N. Bardet, X. Pereda Suberbiola, and N.-E. Jalil, A new mosasauroid (Squamata) from the Late Cretaceous (Turonian) of Morocco. *Comptes Rendus Palevol*, 2 (2003), 607–616.

43. C. L. Camp, California mosasaurs. *Memoirs of the University of California*, 13 (1942), 1–68.

44. O. Rieppel, J. A. Gauthier, and J. A. Maisano, Comparative morphology of the dermal palate in squamate reptiles, with comments on phylogenetic implications. *Zoological Journal of the Linnean Society*, 152 (2008), 131–152.

45. J. L. Conrad, Skull, mandible, and hyoid of *Shinisaurus crocodilurus* Ahl (Squamata, Anguimorpha). *Zoological Journal of the Linnean Society*, 141 (2004), 399–434.

46. B. A. Bell, P. A. Murry, and L. W. Osten, *Coniasaurus* Owen, 1850 from North America. *Journal of Paleontology*, 56 (1982), 520–534.

47. M. J. Polcyn and G. L. Bell, Jr., *Russelosaurus coheni*, n. gen., n. sp., a 92 million-year-old mosasaur from Texas (USA), and the definition of the parafamily Russellosaurina. *Netherlands Journal of Geosciences*, 84 (2005), 321–334.

48. R. W. Dortangs, A. S. Schulp, E. W. A. Mulder, et al., A large new mosasaur from the Upper Cretaceous of The Netherlands. *Netherlands Journal of Geosciences*, 81 (2002), 1–8.

49. O. Rieppel and H. Zaher, Re-building the bridge between mosasaurs and snakes. *Neues Jahrbuch für Geologie und Paläontologie, Abhandlungen*, 221 (2001), 111–132.

50. T. Konishi, M. W. Caldwell, T. Nishimura, K. Sakurai, and K. Tanoue, A new

halisaurine mosasaur (Squamata: Halisaurinae) from Japan: the first record in the western Pacific realm and the first documented insights into binocular vision in mosasaurs. *Journal of Systematic Palaeontology*, 14 (2016), 809–839.

51. M. J. Polcyn, J. Lindgren, N. Bardet, et al., Description of new specimens of *Halisaurus arambourgi* Bardet & Pereda Suberbiola, 2005 and the relationships of Halisaurinae. *Bulletin de la Société Géologique de France*, 183 (2012), 123–136.

52. C. B. Christensen, J. Christensen-Dalsgaard, C. Brandt, and P. T. Madsen, Hearing with an atympanic ear: good vibration and poor sound-pressure detection in the royal python, *Python regius. Journal of Experimental Biology*, 215 (2012), 331–342.

53. C. M. Holliday, N. M. Gardner, S. M. Paesani, M. Douthitt, and J. L. Ratliff, Microanatomy of the mandibular symphysis in lizards: patterns in fiber orientation and Meckel's cartilage and their significance in cranial evolution. *Anatomical Record*, 293 (2010), 1350–1359.

54. M. H. Lessmann, Zur labialen Pleurodontie an Lacertilier-Gebissen. *Anatomischer Anzeiger*, 99 (1952), 35–67.

55. T. J. Bertin, B. Thivichon-Prince, A. R. LeBlanc, M. W. Caldwell, and L. Viriot, Current perspectives on tooth implantation, attachment, and replacement in Amniota. *Frontiers in Physiology*, 9 (2018), 1630.

56. J. A. Gauthier, Fossil xenosaurid and anguid lizards from the early Eocene Wasatch Formation, southeast Wyoming, and a revision of the Anguioidea. *Contributions to Geology, University of Wyoming*, 21 (1982), 7–54.

57. M. J. Polcyn, E. Tchernov, and L. L. Jacobs, The Cretaceous biogeography of the eastern Mediterranean with a description of a new basal mosasauroid from 'Ein Yabrud, Israel. In Y. Tomida, T. H. Rich and P. Vickers-Rich, eds., *Proceedings of the Second Gondwanan Dinosaur Symposium* (National Science Museum Monograph 15), (Tokyo: National Science Museum, 1999), pp. 259–290.

58. A. Haber and M. J. Polcyn, A new marine varanoid from the Cenomanian of the Middle East. *Netherlands Journal of Geosciences*, 84 (2005), 247–255.

59. O. Rieppel, Tooth replacement in anguinomorph lizards. *Zoomorphologie*, 91 (1978), 77–90.

60. N. P. Kelley and N. D. Pyenson, Evolutionary innovation and ecology in marine tetrapods from the Triassic to the Anthropocene. *Science*, 348 (2015), aaa3716.

61. A. Houssaye, Palaeoecological and morphofunctional interpretation of bone mass increase: an example in Late Cretaceous shallow marine squamates. *Biological Reviews*, 88 (2013), 117–139.

62. M. W. Caldwell, Ontogeny and phylogeny of the mesopodial skeleton in mosasauroid reptiles. *Zoological Journal of the Linnean Society*, 116 (1996), 407–436.

63. O. Rieppel, *Helveticosaurus zollingeri* Peyer (Reptilia, Diapsida) skeletal paedomorphosis, functional anatomy and systematic affinities. *Palaeontographica Abteilung A, Paläozoologie, Stratigraphie*, 208 (1989), 123–152.

64. G. L. Bell and M. J. Polcyn, *Dallasaurus turneri*, a new primitive mosasauroid from the Middle Turonian of Texas and comments on the phylogeny of Mosasauridae (Squamata). *Netherlands Journal of Geosciences*, 84 (2005), 177–194.

65. C. Sullivan, The role of calcaneal 'heel' as a propulsive lever in basal archosaurs and extant monitor lizards. *Journal of Vertebrate Paleontology*, 30 (2010), 1422–1432.

66. A. Houssaye, J. Lindgren, R. Pellegrini, et al., Microanatomical and histological features in the long bones of mosasaurine mosasaurs (Reptilia, Squamata) – implications for aquatic adaptation and growth rates. *PLoS ONE*, 8 (2013), e76741.

67. O. Rieppel, H. Zaher, E. Tchernov, and M. J. Polcyn, The anatomy and relationships of *Haasiophis terrasanctus*, a fossil snake with well-developed hind limbs from the Mid-Cretaceous of the Middle East. *Journal of Paleontology*, 77 (2003), 536–558.

A Review of Non-Mosasaurid (Dolichosaur and Aigialosaur) Mosasaurians and Their Relationships to Snakes

Bruno G. Augusta, Hussam Zaher, Michael J. Polcyn,
Anthony R. Fiorillo, and Louis L. Jacobs

8.1 Introduction

Mosasaurians are an extinct clade of aquatic and semiaquatic squamates whose fossils are found in upper Cretaceous marine rocks globally, with the earliest known forms from the lower Cretaceous (Berriasian–Hauterivian; 145–129 Ma) of Japan [1] and lower Cenomanian (upper Cretaceous; 98 Ma) of the Middle East [2]. They dominated global seas for over 30 million years until their demise at the end of the Maastrichtian (66 Ma) [3–5], at the K–Pg mass extinction event [6].

Mosasaurian fossils have been known since the late eighteenth century, with one of the first remains being an exceptionally large mosasaur skull, famously denominated at that time as 'Le Grand Animal de Maestricht' [7]. Currently, more than 100 species and thousands of specimens of mosasaurians are known, with their diversity broadly divided into three major lineages, with the likely paraphyletic early-branching lineages commonly referred to as dolichosaurs (or dolichosaurids; Dolichosauridae) and aigialosaurs (or aigialosaurids; Aigialosauridae), and the more deeply nested, specialized clade referred to as Mosasauridae. Mosasauridae includes four broadly accepted monophyletic subfamilies – Mosasaurinae, Plioplatecarpinae, Tylosaurinae, and Halisaurinae – with some exceptionally large forms such as the mosasaurine *Mosasaurus hoffmannii* (17.5 m) and the tylosaurine *Hainosaurus bernardi* (13m) [3, 8–18].

Non-mosasaurid mosasaurians have a confusing phylogenetic and taxonomic history. Recently, Madzia and Conrad [19] cleared some controversies surrounding the use of the higher-level names - such as Natantia Owen, 1851, Mosasauria Huxley, 1872, Mosasauri

Gadow, 1898, Mosasauromorpha Fürbringer, 1900, Mosasauroidea Camp, 1923, and Mosasauriformes Hay, 1930 – that are mostly junior synonyms of Mosasauridae, a name first proposed by Gervais [20]. Gervais [20] also coined Dolichosauridae, while Gorjanović-Kramberger [21] formulated Aigialosauridae, for the marine Tethyan forms known at that time (Zaher et al., Chapter 9). More recently, a consensus was reached to use the name Mosasauroidea for the minimally inclusive clade including aigialosaurids and mosasaurids [15, 19], i.e., non-dolichosaurid mosasaurians based on current understanding of phylogeny. Although a detailed nomenclatural revision of mosasaurians is still needed, especially regarding non-mosasaurid-mosasaurians, it is beyond the scope of this chapter, which follows Conrad's [15] and Madzia and Conrad's [19] usage of names for that purpose.

Dolichosaurids were relatively small and long-bodied lizards commonly interpreted as closely related to mosasauroids [15, 18, 22–24]. However, in the last 20 years, some authors have argued that dolichosaurids were instead more closely related to snakes [16, 25, 26]. Dolichosaurids are characterized primarily by an elongated neck (> 10 cervical vertebrae), reduced forelimbs, and zygosphene–zygantrum accessory articulation in dorsal vertebrae [1, 15, 16, 25, 26]. Dolichosaurids are found in Cenomanian-Turonian (100.5–89.8 Ma) marine sediments of North America and Eurasia [27–31]. Seventeen dolichosaurid taxa have been described, but *Acteosaurus crassicostatus* is probably a junior synonym of *Adriosaurus suessi* [32] and *Introrsisaurus pollicidens* was described on the basis of a single shed tooth from an entirely distinct geological context (Jurassic outcrop; [33]), so a more balanced position is to recognize 15 currently valid dolichosaurid taxa (listed in Supplementary Table 8.S1).

Aigialosaurs form a grade of semiaquatic marine squamates [10, 17, 23, 34], known from upper Cretaceous rocks of Europe, Asia, and North America [8, 35–37]. A close relationship between aigialosaurs and mosasaurids is supported by a suite of characters, such as the presence of a strongly curved suprastapedial process of the quadrate, expanded tooth bases, lateral compression, and dorsoventral deepening of the tail, and a scapula shorter than the coracoid [10, 15, 17]. McDowell and Bogert [38] considered the living lizard *Lanthanotus borneensis* to be a relict aigialosaur, but subsequent works rejected that hypothesis [3, 8, 39, 40]. Eight described taxa are here considered aigialosaurs (Supplementary Table 8.S2).

Cuvier [41] provided the first detailed monograph on a mosasaurian and concluded that it was related to extant monitor lizards (Varanidae), but also shared characters with Iguania. Subsequent studies [42–44] largely followed Cuvier's arrangement. However, Cope [45] suggested a closer relationship with snakes, arguing mosasaurs shared a mosaic of snake-like and lizard-like characters that precluded their subordination to either group, therefore erecting a new order, Pythonomorpha, to distinguish them from other lizards and snakes. Cope's hypothesis triggered an intense debate, with some authors in favour of a varanoid affinity [46–50], others in favour of a snake affinity [51, 52], and a third group suggesting a distinct subdivision for mosasaurs within lacertilians [53, 54]. Camp's [55] study of lizards reviewed this debate, concluding that mosasaurs were 'derived from a varanoid stock through the Aigialosauridae' ([55]; p. 418), suggesting a classification in which Cope's mosasaurs were allocated to the superfamily Mosasauroidea of platynotan anguimorphs.

Camp's conclusions were followed by most authors through the next decades [3, 33, 38, 39, 56–58], ending the debate on mosasaurian affinities in the pre-cladistic era. In the 1990s Lee [59] revisited mosasauroid affinities in a cladistic framework, reviving Cope's Pythonomorpha Hypothesis as a clade containing mosasauroids and snakes. Varying interpretations of the anatomy of mosasaurians and limbed snakes drove the discussion on clade affinities during the late 1990s and early 2000s (Zaher et al., Chapter 9), turning mosasaurian affinities into one of the most debated topics in squamate systematics. Because the Pythonomorpha Hypothesis is founded on the assumption that Mosasauria represents the sister group of snakes [16, 23, 24, 26], or that snakes are nested within a dolichosaurid assemblage [16, 25, 26, 59–62], investigating the phylogenetic signal retrieved in mosasaurs that supports this hypothesis represents an important step towards clarification of the origin of snakes and mosasaurs alike. More recently, Gauthier et al. [18] suggested an alternative hypothesis in which mosasaurians are distantly related to Anguimorpha, as sister to a monophyletic Scleroglossa.

Regardless of the large number of studies on mosasaurian anatomy and interrelationships, consensus has not been reached about their hypothesized affinities with snakes and other squamates. Four mainly conflicting hypotheses have been proposed: (1) the **Varanoid Hypothesis**, in which mosasaurians are more closely related to varanoid lizards than to other squamates ([3, 8, 10, 11, 15, 38, 55]; Zaher et al., this volume); (2) the **Pythonomorph Hypothesis**, in which mosasaurians are the sister group of snakes [23, 24, 59, 62, 63]; (3) the **Ophidiomorph Hypothesis**, in which Pythonomorpha is monophyletic but snakes are nested within dolichosaurids [25, 26, 28, 61, 62, 64]; and (4) the **Stem-scleroglossan Hypothesis**, in which mosasaurians are distantly related to varanoids and snakes [18]. A summary of the main hypotheses regarding dolichosaurid and aigialosaur phylogenetic affinities is provided in Figure 8.1.

Because both the Ophidiomorph and Pythonomorph Hypotheses are based on snakes being more closely related to mosasaurians than to any other squamate lineage (Fig. 8.1B, C, I, J, and K), and the former is also based on snakes being deeply nested within dolichosaurids (Fig. 8.1E and G), revising the anatomy and systematic relationships of dolichosaurids and aigialosaurs will also shed light on whether these taxa are related or not to the origin of snakes. Here we test these hypotheses and further discuss higher-level mosasaurian interrelationships with an expanded morphological dataset that incorporates new and previously known key specimens of non-mosasaurid mosasaurians. For our morphological observations, we benefitted from high-resolution μCT scan data of key specimens. Our primary focus is to examine mosasaurian cladistic relationships within squamates, with special reference to the putative relationships between mosasaurians and snakes, so we aimed to test their affinities with an expanded morphological dataset including an increased taxonomic sampling of well-preserved non-mosasaurid mosasaurians, and new characters and scorings (see Supplementary Appendix 8.S1 for details of taxon sampling and data partitions).

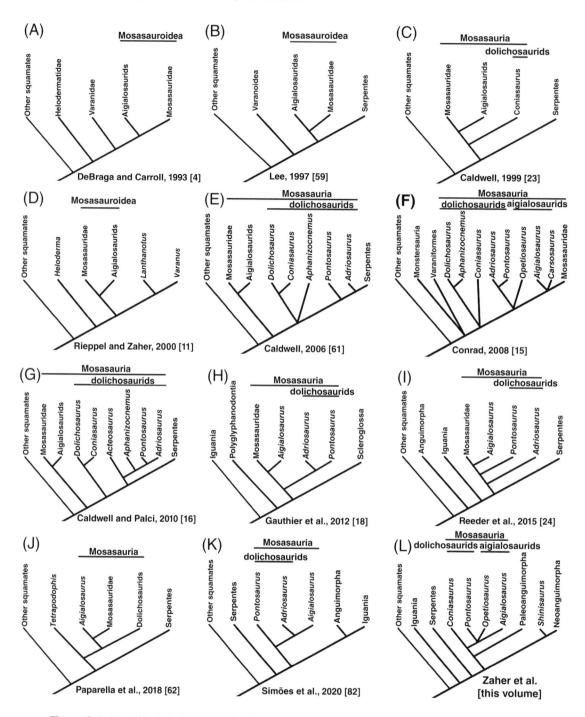

Figure 8.1 Simplified cladograms showing the main hypotheses of dolichosaurid and aigialosaur phylogenetic relationships. Horizontal bars highlight the position of dolichosaurids, aigialosaurs, Mosasauria, and Mosasauroidea as applied here.

8.2 An Historical Perspective of Dolichosaurids and Aigialosaurs

Owen [44] described the first dolichosaurids, *Coniasaurus crassidens* and *Dolichosaurus longicollis*, classifying them within his tribe Repentia, order Lacertilia, being separated from mosasaurs that were allocated into the tribe Natantia. Cornalia and Chiozza [65] described *Mesoleptos zendrinii* and recognized affinities with Owen's dolichosaurids. Von Meyer [66] named *Acteosaurus tommasinii* based on an articulated postcranial skeleton from Slovenia, recognizing affinities with Owen's dolichosaurids but classifying the taxon into his long-necked 'Macrotrachela', inclusive of plesiosaurs. Two additional dolichosaurids, *Pontosaurus* (*Hydrosaurus*) *lesinensis* Kornhuber, 1873 and *Adriosaurus suessi* Seeley, 1881, were named in the nineteenth century, although in their original descriptions they were considered related to varanoid lizards, and not particularly to any other known dolichosaurid. At the end of the nineteenth century, the first two aigialosaur taxa were described, *Aigialosaurus dalmaticus* Gorjanović-Kramberger, 1892 and *Carsosaurus marchesetti* Kornhuber, 1893. While Kornhuber [67] considered *Carsosaurus* a varanid, Gorjanović-Kramberger [21] reviewed the classification of all Cretaceous semiaquatic mosasaurians known at that time, allocating *Acteosaurus*, *Adriosaurus*, *Pontosaurus*, and *Aigialosaurus* to the newly erected Aigialosauridae, while *Mesoleptos* was placed in Varanidae and *Dolichosaurus* maintained in its own family Dolichosauridae. Kornhuber [68] described *Opetiosaurus bucchichi* and rejected the validity of Gorjanović-Kramberger's family Aigialosauridae, arguing for an inclusion of both *Opetiosaurus* and *Aigialosaurus* in Varanidae. Camp [55] considered Mosasauroidea to include only Mosasauridae, depicting aigialosaurids as direct ancestors of mosasaurids, and dolichosaurids as the sister group of Mosasauridae and Aigialosauridae. Nopcsa [50] named *Eidolosaurus trauthi* based on a slab containing a natural mould of an articulated postcranial skeleton and considered it, together with *Mesoleptos*, as an intermediate form between dolichosaurids and aigialosaurs.

After the prolific descriptions of dolichosaurids and aigialosaurs during the late nineteenth and early twentieth centuries, further new taxa were named at the end of the twentieth century. *Acteosaurus crassicostatus* was first described by Calligaris [69] from the Cenomanian–Turonian (100.5–89.8 Ma) of Slovenia, but Caldwell and Lee [32] considered it to be a junior synonym of *Adriosaurus suessi*. A series of early diverging Tethyan forms were described – including *Aphanizocnemus libanensis* Dal Sasso and Pinna, 1997 from the middle Cenomanian (97 Ma) of Lebanon; *Coniasaurus gracilodens* Caldwell, 1999 from the lower to middle Cenomanian (100.5–97 Ma) of England; *Haasiasaurus gittelmani* (Polcyn, Tchernov and Jacobs, 1999) from the lower Cenomanian (100.5 Ma) of Israel; *Carentonosaurus mineaui* Rage and Néraudeau, 2004 from the Cenomanian (100.5–93.9 Ma) of France; *Judeasaurus tchernovi* Haber and Polcyn, 2005 from the upper Cenomanian–lower Turonian (96–92 Ma) of Israel; *Pontosaurus kornhuberi* Caldwell, 2006 from the Cenomanian (100.5–93.9 Ma) of Lebanon; *Komensaurus carrolli* Caldwell and Palci, 2007 from the upper Cenomanian (93.9 Ma) of Slovenia (formerly referred in the literature to as the 'Trieste aigialosaur');

Adriosaurus microbrachis Palci and Caldwell, 2007 from the upper Cenomanian (93.9 Ma) of Slovenia; *Vallecillosaurus donrobertoi* Smith and Buchy, 2008 from the lower Turonian (92 Ma) of Mexico; *Adriosaurus skrbinensis* Caldwell and Palci, 2010 from the upper Cenomanian (93.9 Ma) of Slovenia; *Primitivus manduriensis* Paparella, Palci, Nicosia and Caldwell, 2018 from the upper Campanian–lower Maastrichtian (72.1 Ma) of Italy; and *Portunatasaurus krambergeri* Mekarski, Japundžić, Krizmanić and Caldwell, 2019 from the Cenomanian-Turonian (100.5-89.8 Ma) of Croatia – all loosely associated to either the dolichosaurid or aigialosaur lineages. Among those, *Pontosaurus*, *Aphanizocnemus*, and *Adriosaurus* were often found as supporting the Ophidiomorph Hypothesis, in which dolichosaurids form a paraphyletic assemblage with respect to snakes [25, 60, 61, 64].

8.3 Morphological versus Molecular Evidence in Squamate Systematics

Prior to discussing our phylogenetic results, we briefly review the two main competing hypotheses of squamate relationships supported by separate morphological and molecular evidence. Hypotheses of squamate relationships based on morphological data reconstruct the primary (basalmost) divergence as between Iguania and Scleroglossa [15, 18, 23, 25, 40, 70, 71]. An exception is the work of Simões et al. [63] in which Gekkonomorpha is sister to all other squamates. Within Scleroglossa, three major lineages are universally represented: Gekkota, Scincomorpha, and Anguimorpha, but their composition and proposed relationships with other scleroglossan lineages varies [15]. Most important for our work are the differences among in-group relationships of Anguimorpha. Major lineages of limbless squamates, such as snakes and dibamids, are variably found within Anguimorpha [18, 25, 40, 71] or not [15, 23, 72]. Varanoidea consistently comprises *Heloderma*, *Lanthanotus*, *Varanus,* and all descendants of their last common ancestor. Proposed anguimorph clades and taxa representing the closest relatives of varanoids vary in the literature, such as Anguidae [25], Xenosauridae [18], or a clade formed variously by both [15, 23, 40, 72]. In morphological analyses including fossil taxa, Mosasauria is consistently found either within Varanoidea [4, 11, 17, 73, 74], closely related to varanoids [15], or closely related to snakes [16, 25, 26, 59, 62, 63]. An exception to these scenarios is the global analysis of squamates provided by Gauthier et al. [18], in which mosasaurians are stem-scleroglossans.

In molecular analyses, the sister of all other squamates is either Dibamia [75–78] or Dibamia and Gekkota [24, 79]. Unidentata, a clade composed of squamates exclusive of Dibamia and Gekkota, is frequently found in molecular analyses [24, 75-77, 79]. Instead of recovering Iguania as sister to all other squamates, molecular analyses find a much less inclusive clade, Toxicofera, comprising iguanians, anguimorphs, and snakes [24, 75, 76, 79–81]. Within Anguimorpha, molecular analyses have found support for relationships distinct from those recovered by analyses of morphological data; instead of a monophyletic Varanoidea, they recover a dichotomy between Old (Paleoanguimorpha) and New World (Neoanguimorpha) anguimorphs. The position of mosasaurians in combined molecular and morphological trees varies, either within Anguimorpha and closely related to varanoids [76], within Toxicofera as the sister group of snakes [24], or within Toxicofera as the sister group of a clade comprising Anguimorpha and Iguania [82].

8.4 Character Scoring and the Quality of Phylogenetic Reconstructions of Mosasaurian Relationships

Recently, a series of studies discussed methods for phylogenetic reconstruction of mosasauroid relationships [83–86]. Although finding the best models for phylogenetic reconstructions represents an important step in any phylogenetic analysis, we address another problem often present when fossil taxa are included in datasets: character scorings based on poorly preserved specimens or on published interpretative drawings.

Simões et al. [84] applied various methods in an attempt to resolve in-group mosasauroid incongruences, including Maximum Likelihood, Bayesian inference, and implied weight maximum parsimony. Madzia and Cau [86] revised their dataset and criticized some of Simões et al.'s [84] choices, such as not integrating stratigraphic information into their Bayesian analysis, keeping all characters unordered without justification, and the reconstruction of a possibly artificial dolichosaurid grade by choosing *Adriosaurus suessi* as the outgroup. Madzia and Cau [86] also revised mosasaurian nomenclature and clade definitions. In a previous contribution, Simões et al. [83] questioned the quality of characters present in exceptionally large phylogenetic matrices, particularly regarding squamate relationships, concluded that many of the characters in larger matrices (such as in [18]), should be recoded or even excluded due to failure in tests of similarity, conjunction, independency, and due to the distinction between characters and character states. Laing et al. [85] criticized Simões et al.'s [83] deleting of 'problematic' characters, arguing that large matrices are inevitable and necessary for the development of systematics, because morphological data will only increase over time. Simões et al. [87] responded, arguing misinterpretation of their original ideas.

In assessing dolichosaurid character scorings in the morphological matrices of Conrad [15] and Gauthier et al. [18], especially regarding *Adriosaurus*, *Aphanizocnemus*, and *Eidolosaurus*, it is clear that many scored states are not observable in available specimens, suggesting that the authors relied on potentially misleading literature descriptions and interpretative drawings. For example, the recently redescribed *Adriosaurus suessi* [25] specimen NHMUK-R2867 is nearly complete but extremely poorly preserved, and we dispute many aspects of the redescribed and illustrated characters, especially in the skull. Figure 8.2 compares a recent photograph of the skull of NHMUK-R2867 with Lee and Caldwell's [25] interpretative drawing. In their discussion, Lee and Caldwell ([25]: p. 924–925) report four character states supporting the clade formed by *Adriosaurus* and their Ophidia: (1) premaxilla-maxilla contact mobile, (2) frontals paired, (3) postorbitofrontal ventral process large, (4) supratemporal superficial. However, none of these characters is clearly observable in the specimen. Breaks and erosion of the bony surfaces preclude precise, accurate assessment of *Adriosaurus* skull morphology. In particular, the posterior of the skull is an indistinguishable bony mass in our assessment, preventing confident interpretation of bones posterior to the frontal. A mediolateral break through the skull level with the orbits seems to have broken the frontal during taphonomic processes (Fig. 8.2C). We disagree with Lee and Caldwell's [25] interpretation of paired frontals in this specimen. The anterior portion of the frontal appears to be the best-preserved part of the skull, revealing two important features: a triradiate morphology anteriorly and lack of midline suture, consistent with a single bone. These features match more with the condition found in other anguimorphs, similar to the frontal of *Coniasaurus gracilodens* [27], and clearly distinct from that of snakes. Poor preservation is an issue in most dolichosaurid specimens from Europe, especially regarding the cranial region. The holotype of *Aphanizocnemus*, MSNM V783, has most of its skull badly crushed, and very few skull characters can be scored for this taxon (Fig. 8.2D). The *Eidolosaurus* holotype, GBA 1923-001-0001, lacks the skull and its postcranium is preserved as a mould (Fig. 8.2E), severely limiting the number of characters that can be scored.

Reliable character scoring can be aided by firsthand observations, better preserved specimens, appropriate visualization techniques, assessments of taphonomic processes, and conservative interpretation. Misinterpretation of morphology leads to inaccuracies in character scorings that can cause unreliable phylogenetic reconstructions. We are not suggesting exclusion of imperfect fossil specimens from systematic analysis, because they can also be very important in reconstructing phylogenetic relationships [88, 89], rather we stress particular care and diligence when scoring characters based on poorly preserved specimens or based only on the literature.

8.5 A Phylogenetic Reappraisal of Dolichosaurid and Aigialosaur Affinities

Our analysis (methods and additional discussion provided in Supplementary Appendix 8. S1) consistently recovered mosasaurians within Anguimorpha, as the sister of Varanoidea (Fig. 8.3A). This topology is notably distinct from that found by Gauthier et al. [18], which recovered Mosasauria outside Anguimorpha. Our results do not support the Pythonomorph Hypothesis because we did not find dolichosaurids, nor any other mosasaurian taxa, more closely related to snakes than to varanoid lizards. Our results are closer to those of Rieppel and Zaher [11], Rieppel et al. [73], Conrad [15], Wiens et al. [76], Conrad et al. [17], and Yi and Norell [74], although they differ in that mosasaurians are sister to Varanoidea instead of nested within the latter taxon. Synapomorphies for anguimorph clades are provided in Section 1.8 of Supplementary Appendix 8.S1.

Anguimorpha is supported by 14 unambiguous synapomorphies. Mosasauria could be scored for eight of them, including the presence of long and converging vomer ventral ridges, long slender palatine process clasped in a groove on the dorsal surface of vomer, and at least seven cervical vertebrae. Gauthier et al. [18] found Mosasauria to be outside of their 'Scleroglossa', as the sister group of that clade, presenting 24 synapomorphies supporting crown Scleroglossa, for which mosasaurians could be scored for only three.

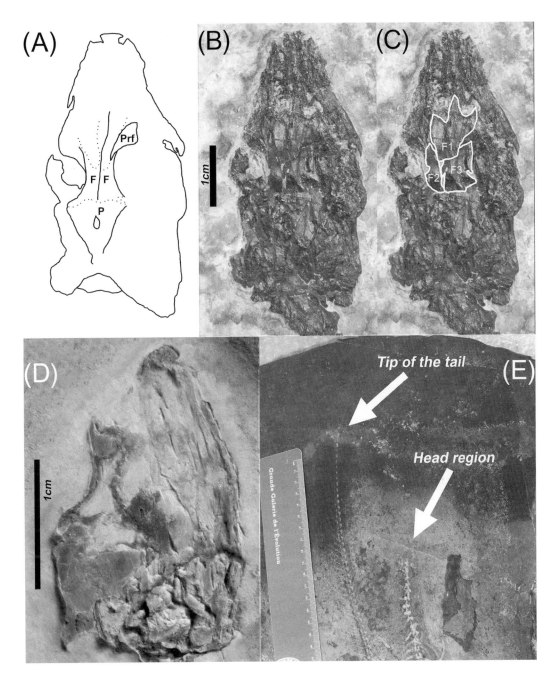

Figure 8.2 Dolichosaurids with poorly preserved skulls. (A) interpretation of the *Adriosaurus suessi* skull NHMUK-R2867, frontal morphology as presented by Lee and Caldwell [25] (redrawn from [25]); (B) recent photograph of the same specimen; (C) reinterpretation of the frontal morphology on the same specimen (white lines indicate the three main pieces of the frontal bone as interpreted here); (D) *Aphanizocnemus libanensis* skull MSNM V783; (E) *Eidolosaurus trauthi* holotype, GBA 1923-001-0001. Abbreviations: F, frontal; F1, F2 and F3, fragments of the frontal; P, parietal; Prf, prefrontal.

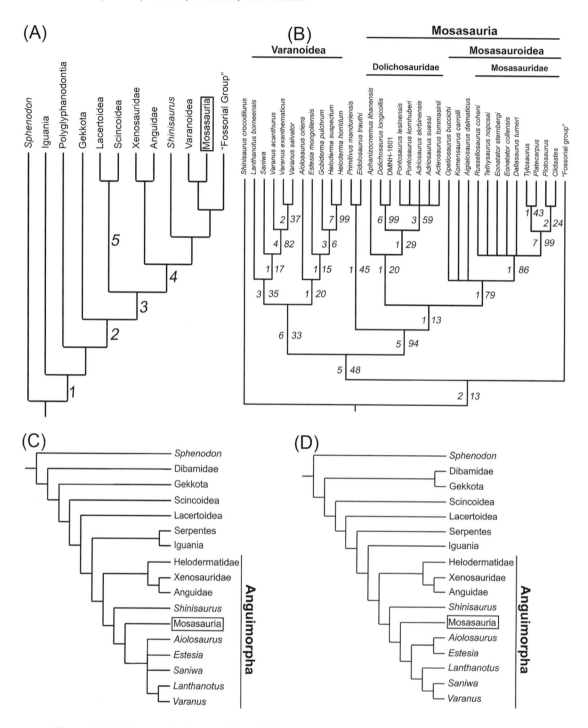

Figure 8.3 Phylogenetic relationships of dolichosaurids, aigialosaurs and mosasaurids, based on morphology, as recovered in this analysis. (A) simplified cladogram of the squamate topology recovered here, strict consensus of 63,040 MPTs. The "fossorial group" is the same recovered by Gauthier et al. [18], constituted by dibamids, amphisbaenians and snakes. Numbers indicate clades

By increasing taxonomic sampling of dolichosaurids, aigialosaurs, and plesiopedal (taxa retaining limbs for facultative terrestrial locomotion *sensu* Polcyn and Bell [34]) mosasaurids, we were able to better evaluate synapomorphies in Mosasauria. This analysis confirmed nine characters supporting 'Scleroglossa' [18], including base of squamosal temporal ramus lying against the parietal, absence of squamosal ascending process, and splenial extending anteriorly to nearly two-thirds of the mandible length. Of the six 'scleroglossan' synapomorphies of Gauthier et al. [18] that are certainly absent in mosasaurians, five are interpreted as reversed in some other anguimorphs:

(1)   temporal muscles originating dorsally on parietal (also reversed in Lanthanotidae + Varanidae)

(2)   prearticular and surangular unfused (also in *Heloderma* and some fossil varanoids)

(3)   simple, rod-like clavicle (also in *Varanus salvator* and *Shinisaurus*)

(4)   interclavicle anterior process <0.2 length of interclavicle (also in *Xenosaurus* and *Shinisaurus*)

(5)   absence of enlarged, subhemispherical distal epiphysis of ulna (also in *Xenosaurus*)

The remaining nine 'scleroglossan' characters could not be scored for dolichosaurids nor aigialosaurs due to lack of preservation. Running the analysis with unordered characters recovered mosasaurians as closely related to varanoids, and in a polytomy with varanids, *Heloderma*, *Gobiderma*, *Estesia,* and *Aiolosaurus.*

Origin of jaw musculature on parietal, narial retraction, expanded bases of marginal teeth, and lack of fusion between the articular-prearticular complex and surangular have supported the interpretation of mosasaurians being either closely related to [15] or nested within Varanoidea [4, 11, 17, 73, 74]. We recovered Mosasauria as the sister group of Varanoidea based on 13 unambiguous synapomorphies, including narial margin of maxilla arising at low angle, rod-like main portion of vomer, and prearticular and surangular unfused. These conditions are distinct from those found in Serpentes. *Primitivus manduriensis* and *Eidolosaurus trauthi*, originally described as dolichosaurids, are separated from other mosasaurians by not possessing characters such as a small carpal intermedium and V metatarsal not hooked.

A monophyletic Dolichosauridae was reconstructed, as in Paparella et al. [62], but in contrast to most other studies [15, 17, 25, 64]. *Aphanizocnemus*, recently proposed to be a scincogekkonomorph lizard instead of a dolichosaurid [90], was found to be sister of all other dolichosaurids. This taxon presents anatomical features distinct from those of typical mosasaurians (especially regarding limb morphology), but here it was recovered within

Figure 8.3 (*cont.*) as follows: (1) Squamata; (2) Scleroglossa; (3) Autarchoglossa; (4) Anguimorpha; (5) Scincomorpha; (B) phylogenetic relationships of anguimorphs, based on morphology, as recovered in this analysis. Numbers left of internal branches are Bremer support values, and numbers right of internal branches are Bootstrap values. (C) Topology recovered when using Wiens et al.'s [76] results as a molecular backbone constraint; (D) Topology recovered when using Reeder et al. [24] results as a molecular backbone constraint.

Dolichosauridae. A novel result of the present study is the recovery of two main dolichosaurid lineages: 'dolichosaurines' and 'acteosaurines'.

'Dolichosaurines' were relatively more elongate, with more cervical and dorsal vertebrae, and more specialized limbs and tail for swimming. They ranged from southern to southeastern portions of the Western Interior Seaway (coniasaur records reported from Germany [91] and Australia [92], are here considered too fragmentary for a reliable taxonomic identification). 'Acteosaurines' were slightly shorter, pachyostotic animals restricted to the palaeo-North Atlantic (around the present-day Adriatic Sea) [25, 26, 61].

The three aigialosaurs included here, *Opetiosaurus, Komensaurus,* and *Aigialosaurus,* were found in a polytomy with Mosasauridae, such that the monophyly of Aigialosauridae and its validity as a family was not supported. The mosaic anatomy of aigialosaurs, presenting intermediate characters between dolichosaurids, varanoids and mosasaurs, and the lack of more and better-preserved specimens, may be two important factors behind the poor resolution of this group's relationships in this analysis.

We were unable to resolve the earliest divergences within Mosasauridae, with *Russellosaurus, Tethysaurus, Dallasaurus, Eonatator,* and the hydropedal clade (those with fully developed aquatic hind limbs and girdles) forming a polytomy. This is perhaps because of inadequate representation of mosasaurids. Future contributions utilizing a larger taxonomic sample of mosasaurids in a more focused analysis should better reconstruct in-group relationships. The hydropedal mosasaur clade is supported by ten unambiguous synapomorphies, including frontal broadly overlapping prefrontal dorsally, a supratemporal longer than squamosal-parietal contact, and manual and pedal hyperphalangy.

To test alternative phylogenetic hypotheses for mosasaurian affinities, we performed two additional tests by applying two molecular phylogeny backbones in TNT [93] using the command 'force': (1) Wiens et al.'s [76] strict consensus (Fig. 8.3C), and (2) Reeder et al.'s [24] strict consensus (Fig. 8.3D). Both tests resulted in mosasaurians nesting within Paleoanguimorpha more closely related to varanids than to *Shinisaurus.* This result is remarkably similar to the outcome of Yi and Norell's [74] combined morphological and molecular analysis. Minor differences between the distinct backbones are: using Wiens et al.'s [76] strict consensus resulted in 199 additional steps required for finding the MPTs, while with Reeder et al.'s [24] strict consensus 213 additional steps were required. These tests highlight the stronger phylogenetic signal of mosasaurians as anguimorph lizards, especially close to varanoids, and not closely related to snakes (as in the Pythonomorph Hypothesis), nor being part of a more inclusive squamate clade (as in Gauthier et al. [18]). These results also argue against a close sister group relationship between mosasaurs and snakes.

8.6 The Pythonomorph and Ophidiomorph Hypotheses Revisited

The Pythonomorph and Ophidiomorph Hypotheses have changed fundamentally during the past two decades (Fig. 8.4). Multiple variations in Pythonomorpha topologies suggest

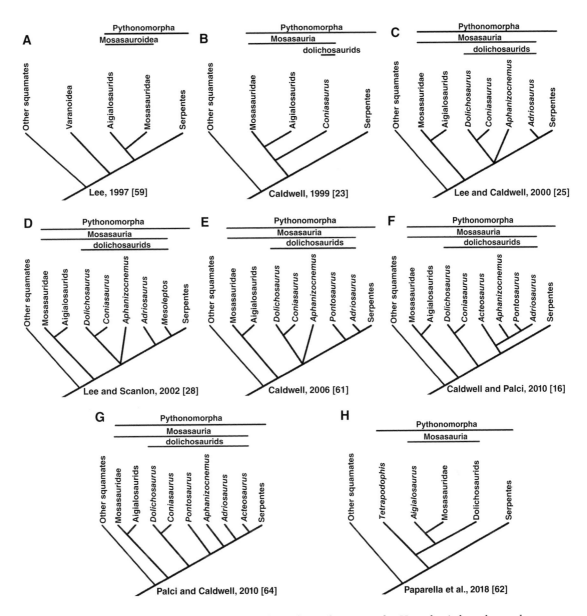

Figure 8.4 Simplified cladograms highlighting how the Pythonomorpha Hypothesis has changed through time. Black bars indicate the position of relevant higher taxa.

that new information on dolichosaurid anatomy and relationships will improve understanding of mosasauroid evolution and relationships within Squamata. For our dataset, constraining Mosasauria into a most-parsimonious sister-group relationship with snakes requires 5,287 steps, 11 more than in our unconstrained results. The more than 100,000 MPTs from this constrained analysis (Supplementary Figure 8.S9) presents different relationships within anguimorphs and also changes the relationships of non-snake radiations of Gauthier et al.'s [18] 'fossorial group'. Varanids were successively found as closest

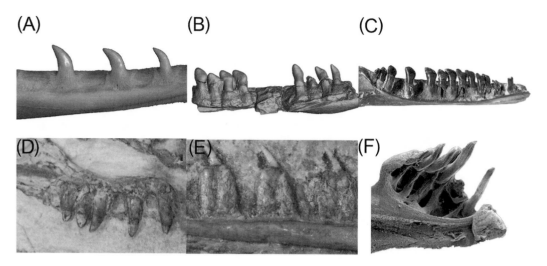

Figure 8.5 Tooth attachment and implantation. Tooth-bearing elements are depicted in medial view. (A) left dentary of the varanoid *Varanus rudicollis* (SMU-unnumbered); (B) left dentary of the dolichosaurid *Coniasaurus* sp. nov. (DMNH-1601); (C) left mandible of the dolichosaurid *Coniasaurus* cf. *C. crassidens* (NHMUK R 3421); (D) right maxilla of the dolichosaurid *Judeasaurus tchernovi* (HUJI P 4000); (E) right dentary of the mosasauroid *Opetiosaurus bucchichi* (GBA 1901/002/0001); (F) left dentary of the snake *Leptotyphlops koppesi* (MZSP 10992: courtesy of F. Curcio). Parts D and E flipped horizontally to facilitate comparison.

relatives of Pythonomorpha, with *Estesia mongoliensis* being the closest. Regarding the fossorial taxa, Dibamidae and Amphisbaenia were reconstructed as a clade as in Gauthier et al. [18], but within Scincidae, while *Sineoamphisbaena hexatabularis* and *Anniella pulchra* were within Anguidae.

A close relationship between mosasaurians and snakes in previous phylogenetic studies has typically been supported by characters related to limb and girdle reduction, tooth implantation and replacement, intramandibular joints, and the presence of zygantrum-zygosphene accessory articulations. Intense discussion about these sets of characters occurred during the late 1990s and early 2000s ([11, 25, 59, 94–100]; see also Zaher et. al., this volume), especially regarding tooth implantation and replacement patterns. However, most studies analysing tooth patterns in mosasaurians included only relatively deeply nested mosasaurids [96, 97, 100–102] and did not evaluate tooth attachment in less deeply nested forms that might be more likely to retain plesiomorphic conditions.

Overall patterns of tooth attachment and implantation in dolichosaurids are much more similar to those in varanoids than in snakes or deeply nested mosasaurids, presenting a plesiomorphic set of features defined as fully pleurodont (*sensu* Zaher and Rieppel [96]). In varanoids and dolichosaurids, tooth attachment sites are shallow, and the base of implantation is only slightly expanded (Fig. 8.5A–D), differing from the more expanded base in mosasauroids (Fig. 8.5E), and especially from the highly modified pleurodonty seen in 'higher' snakes (Fig. 8.5F; see also [96]). The cementum layer ankylosing teeth to bone is thin in both dolichosaurids and varanoids, but thicker independently in mosasauroids and 'higher' snakes. Finally, there are no interdental ridges in any dolichosaurid or early mosasauroid (Fig. 8.5B–E), whereas this is typical in snakes. Tooth attachment in dolichosaurids argues against the hypothesis of dolichosaurids being more closely related to snakes than to mosasauroids (see also Polcyn et al., Chapter 7). Similarities in the dentition of deeply nested mosasaurs and 'higher' snakes can be interpreted alternatively as convergence.

One of the main features of most works favoring the Ophidiomorph Hypothesis is dolichosaurids being more closely related to snakes than to mosasauroids [25, 26, 61, 64]. The hypothesis of snakes being most closely related to or even derived from dolichosaurids is based on proposed synapomorphies related to girdle and limb reduction, cervical vertebrae number, and cranial characters that are difficult to see in any dolichosaurid specimen. For example, none of the seven synapomorphies uniting dolichosaurids and snakes in a monophyletic Ophidiomorpha presented by Lee and Caldwell [25] can be adequately compared between the taxa: more than ten cervical vertebrae, scapulocoracoid reduced, interclavicle absent, forelimbs small, pelvis reduced, hindlimbs small, and pubis not expanded distally. The interpretation of snakes bearing elongate necks is based on the presence of hypapophyses along the axial series; however, developmental studies support the interpretation of snakes having short necks [103, 104]. Lack of girdle and limb elements is difficult to compare among limbless lineages. More recent works, such as Caldwell [61] and Palci and Caldwell [64], presented more synapomorphies supporting the monophyly of Pythonomorpha and Ophidiomorpha, but many of those are related to anatomical regions not quite visible in any dolichosaurid specimen, such as 'crista prootica (ridge on lateral surface of the prootic, overhanging the foramen pro nervi facialis) reduced to weak ridge, or absent' [61] and 'ectopterygoid-palatine contact absent, maxilla enters suborbital fenestra' [64]. The posterior process of the maxilla reaching or extending beyond the middle of the orbit is one of the synapomorphies uniting *Adriosaurus* and snakes according to Caldwell [61], but this character cannot be seen in available *Adriosaurus* material (Fig. 8.2), and it is not present in any mosasaurian taxon (Fig. 8.6A), which instead have short maxillary processes that do not extend past mid-orbit, as in varanoids and contrasting with the snake condition. Paired frontals were also presented as a character uniting *Adriosaurus* and snakes, but paired frontals are unknown in mosasaurians, with the paired frontals described by Lee and Caldwell [25] here interpreted as a single broken element, as discussed in Section 8.4 (Fig. 8.6B).

Our analysis inferred a grouping of dolichosaurids and mosasauroids in a monophyletic Mosasauria. A complete list of synapomorphies is provided in Supplementary Appendix 8. S1, but some are summarized here. The internasal process of the premaxilla is always exceptionally long, approaching the frontal (Fig. 8.6C). Mosasaurians present a typical and very particular quadrate morphology, with a highly developed suprastapedial process bending ventrally (Fig. 8.6D), and even touching the infrastapedial process in more deeply nested taxa [3]. Although some snakes have a somewhat expanded quadrate, it never approaches the particular condition seen in dolichosaurids and mosasauroids, in which the suprastapedial process is highly inclined ventrally. Mosasaurians also have a unique mandibular condition in which the surangular participates in the glenoid fossa (Fig. 8.6E),

Mosasauroidea · Dolichosauridae · Serpentes

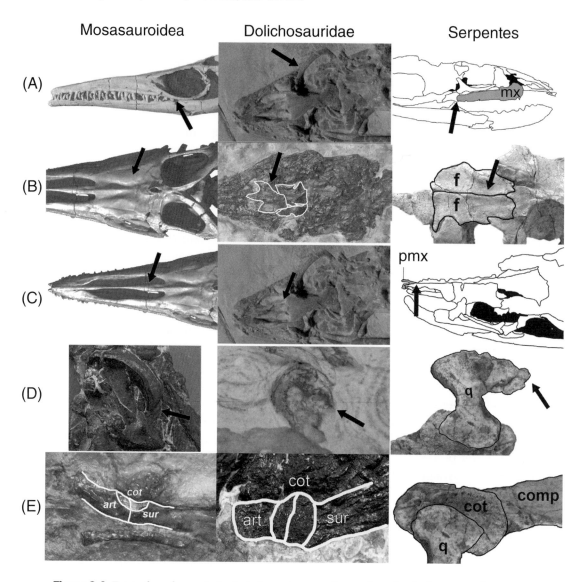

Figure 8.6 Examples of morphological characters arguing against the Pythonomorpha Hypothesis. (A) extension of the posterior process of the maxilla. From left to right: *Plotosaurus bennisoni* (UCMP 32778, courtesy of DigiMorph.org), *Pontosaurus kornhuberi* (MSNM V 3662) and *Najash rionegrina* (MPCA 500; line drawing based on Garberoglio et al., 2019 [107]). (B) fusion of the frontals; from left to right: *Plotosaurus bennisoni* (UCMP 32778, courtesy of DigiMorph.org), *Adriosaurus suessi* (NHMUK PR 2867; white lines represent a novel interpretation of the frontal morphology as discussed in section 8.4) and *Dinilysia patagonica* (MLP 26-410). (C) Extension of the internasal process of the premaxilla; from left to right: *Plotosaurus bennisoni* (UCMP 32778, courtesy of DigiMorph.org), *Pontosaurus kornhuberi* (MSNM V 3662) and *Najash rionegrina* (MPCA 500; line drawing based on Garberoglio et al., 2019, [107]). (D) Development of the quadrate suprastapedial process; from left to right: *Komensaurus carrolli* (MCSNT 11430-2), *Judeasaurus tchernovi* (HUJI P4000) and *Dinilysia patagonica* (MACN-RN 1013). (E) Participation of the surangular in the mandibular cotyle; from left to right: *Aigialosaurus dalmaticus* (BSP 1902 II 501), *Adriosaurus suessi* (NHMUK PR 2867) and *Dinilysia patagonica* (MACN-RN 1013). Abbreviations: art, articular; comp, compound bone; cot, cotyle; f, frontal; mx, maxilla; pmx, premaxilla; q, quadrate; sur, surangular.

which is not seen in snakes. In summary, we find clear morphological evidence to support a monophyletic Mosasauria, and we have not found evidence to support the hypothesis that dolichosaurids are more closely related to snakes than to mosasauroids.

8.7 Concluding Remarks and Future Prospects

This contribution emphasizes the importance of taxon and specimen selection, and the application of techniques that maximize the capture of previously missing data. Many mosasaurian characters are scored here for the first time through the application of micro-CT scanning (Chapter 7). These techniques, as applied here, and in the richly illustrated work of Gauthier et al. [18], are largely focused on osteology, but new techniques such as contrast staining [105, 106] are now setting the stage for deeper morphological analyses that can generate data for phylogenetic systematics.

Beyond these technologies, better and more complete data for mosasaurians will accrue through the discovery of new specimens and reanalysis of currently curated specimens. As discussed in Section 8.4, many characters, especially from the skull, cannot currently be scored for most dolichosaurids. Our proposed topology of relationships within Mosasauria and the monophyly of Dolichosauridae are tentative. Data from new and better-preserved fossils will test the relationships proposed here. However, we present what we consider to be a robust set of characters supporting the hypothesis of dolichosaurids, aigialosaurs and mosasaurids being anguimorphs in a sister-group relationship with Varanoidea.

The sampling of more dolichosaurids and plesiopedal mosasauroids than in previous studies likely contributed to our inference that Mosasauria is firmly nested within Anguimorpha, as the sister of varanoids. More complete taxonomic sampling with good-quality morphological data will hopefully resolve remaining conflicts in the systematics of the group. Based on currently available data, our interpretation is that a close phylogenetic relationship between mosasaurians and snakes can be rejected with confidence.

Acknowledgements

We thank Daniel Madzia, Anne Schulp and David Gower for providing constructive criticism of the submitted manuscript. We are grateful to the following colleagues for granting access to specimens under their care: Sandra Chapman and Michael Day (Natural History Museum, London); John Cooper (Booth Museum of Natural History); Ursula Göhlich (Natural History Museum, Vienna); Irene Zorn (Geological Survey of Austria, Vienna); Cristiano Dal Sasso (Museo di Storia Naturale di Milano), Deborah Arbulla (Museo Civico di Storia Naturale di Trieste), Ron Tykoski and Karen Morton (Perot Museum of Nature and Science). We are indebted to Jessica Maisano, Matthew Colbert and Alberto Carvalho for providing valuable help with CT-data treatment and analysis and Diana Vineyard for helping with funding and administrative documentation. We also want to thank Jessica Maisano and DigiMorph.org for providing 3D reconstructions of several squamate taxa scanned as part of the DigiMorph project funded by the National Science Foundation (DEB-0132227 and EF-0334961), and for allowing the reproduction of the *Plotosaurus bennisoni* 3D model in Figure 8.6. This research was financed in part by the Coordenação de Aperfeiçoamento de Pessoal de Nível Superior - Brasil (CAPES; Finance Code 001) and ISEM at SMU to B.A., and Fundação de Amparo à Pesquisa do Estado de São Paulo (BIOTA/ FAPESP, 11/50206-9 and 18/11902-9) and Conselho Nacional de Desenvolvimento Científico e Tecnológico (CNPq 306777/2014-2) to H.Z.

References

1. S. E. Evans, M. Manabe, M. Noro, S. Isajis, and M. Yamaguchi, A longbodied lizard from the Lower Cretaceous of Japan. *Paleontology*, 49 (2006), 1143–1165.

2. M. J. Polcyn, E. Tchernov, and L. L. Jacobs, The Cretaceous biogeography of the eastern Mediterranean with a description of a new basal mosasauroid from 'Ein Yabrud, Israel. In Y. Tomida, T.H. Rich and P. Vickers-Rich, eds., *Proceedings of the Second Gondwanan Dinosaur Symposium* (1999), pp. 259–290.

3. D. A. Russell, Systematics and morphology of American mosasaurs. *Bulletin of the Peabody Museum of Natural History*, 23 (1967), 1–241.

4. M. DeBraga and R. L. Carroll, The origin of mosasaurs as a model of macroevolutionary patterns and processes. *Evolutionary Biology*, 27 (1993), 245–322.

5. M. J. Polcyn, L. L. Jacobs, R. Araújo, A. S. Schulp, and O. Mateus, Physical drivers of mosasaur evolution. *Palaeogeography, Palaeoclimatology, Palaeoecology*, 400 (2014), 17–27.

6. W. B. Gallagher, K. G. Miller, R. M. Sherrell, et al., On the last mosasaurs: Late Maastrichtian mosasaurs and the Cretaceous-Paleogene boundary in New Jersey. *Bulletin de la Société Géologique de France*, 183 (2012), 145–150.

7. F. F. Pieters, P. G. Rompen, J. W. Jagt, and N. Bardet, A new look at Faujas de Saint-Fond's fantastic story on the provenance and acquisition of the type specimen of Mosasaurus hoffmanni Mantell, 1829.

Bulletin de la Société géologique de France, 183 (2012), 55–65.

8. R. L. Carroll and M. Debraga, Aigialosaurs: mid-Cretaceous varanoid lizards. *Journal of Vertebrate Paleontology*, 12 (1992), 66–86.

9. T. Lingham-Soliar, Anatomy and functional morphology of the largest marine reptile known, Mosasaurus hoffmanni (Mosasauridae, Reptilia) from the Upper Cretaceous, Upper Maastrichtian of the Netherlands. *Philosophical Transactions of the Royal Society of London B: Biological Sciences*, 347 (1995), 155–172.

10. G. L. Bell, A phylogenetic revision of North American and Adriatic Mosasauroidea. In J. M. Callaway and E. L. Nicholls, eds., *Ancient Marine Reptiles* (Academic Press: Cambridge, 1997), pp. 293–332.

11. O. Rieppel and H. Zaher, The intramandibular joint in squamates, and the phylogenetic relationships of the fossil snake *Pachyrhachis problematicus* Haas. *Fieldiana Geology*, 43 (2000), 1–69.

12. N. S. Bardet, X. P. Suberbiola, M. Iarochene, B. Bouya, and M. Amaghzaz, A new species of *Halisaurus* from the Late Cretaceous phosphates of Morocco, and the phylogenetical relationships of the Halisaurinae (Squamata: Mosasauridae). *Zoological Journal of the Linnean Society*, 143 (2005), 447–472.

13. G. L. Bell and M. J. Polcyn, *Dallasaurus turneri*, a new primitive mosasauroid from the Middle Turonian of Texas and comments on the phylogeny of Mosasauridae (Squamata). *Netherlands*

Journal of Geosciences, 84, Special Issue 3 (2005), 177–194.

14. J. Lindgren, The first record of *Hainosaurus* (Reptilia: Mosasauridae) from Sweden. *Journal of Paleontology*, 79 (2005), 1157–1165.

15. J. L. Conrad, Phylogeny and systematics of Squamata (Reptilia) based on morphology. *Bulletin of the American Museum of Natural History*, 310 (2008), 1–182.

16. M. W. Caldwell and A. Palci, A new species of marine ophidiomorph lizard, *Adriosaurus skrbinensis,* from the Upper Cretaceous of Slovenia. *Journal of Vertebrate Paleontology*, 30 (2010), 747–755.

17. J. L. Conrad, J. C. Ast, S. Montanari, and M. A. Norell, A combined evidence phylogenetic analysis of Anguimorpha (Reptilia: Squamata). *Cladistics*, 27 (2011), 230–277.

18. J. A. Gauthier, M. Kearney, J. A. Maisano, O. Rieppel, and A. Behlke, Assembling the squamate tree of life: Perspectives from the phenotype and the fossil record. *Bulletin of the Peabody Museum of Natural History*, 53 (2012), 3–308.

19. D. Madzia and J. Conrad, Mosasauridae. In K. C. de Queiroz, P. D. Cantino and J. A. Gauthier, eds., *Phylonyms: A Companion to the PhyloCode* (Boca Raton: CRC Press, 2020), pp. 1103–1108.

20. P. Gervais, *Zoologie et Paléontologie Generale* (Paris: A. Bertrand, 1852).

21. K. Gorjanović-Kramberger, *Aigialosaurus,* eine neue Eidechse aus den Kreideschiefern der Insel Lesina, mit Rücksicht auf die bereits beschriebenen Lacertiden von Comen und Lesina. *Glasnik hrvatskoga naravoslovnoga drustva (Societas historico-naturalis croatica) u Zagrebu*, 7 (1892), 74–106.

22. M. J. Polcyn and G. L. Bell, *Coniasaurus crassidens* and its bearing on varanoid-mosasauroid relationships. *Journal of Vertebrate Paleontology*, Supplement, 14 (1994), 42A.

23. M. W. Caldwell, Squamate phylogeny and the relationships of snakes and mosasauroids. *Zoological Journal of the Linnean Society*, 125 (1999), 115–147.

24. T. W.Reeder, T. M. Townsend, D. G. Mulcahy, et al., Integrated analyses resolve conflicts over squamate reptile phylogeny and reveal unexpected placements for fossil taxa. *PLoS One*, 10 (2015), e0118199.

25. M. S. Y. Lee and M. W. Caldwell, Adriosaurus and the affinities of mosasaurs, dolichosaurs, and snakes. *Journal of Paleontology*, 74 (2000), 915–937.

26. S. Pierce and M. W. Caldwell, Redescription and phylogenetic position of the Adriatic (Upper Cretaceous; Cenomanian) dolichosaur, *Pontosaurus lesinensis* (Kornhuber, 1873). *Journal of Vertebrate Paleontology*, 24 (2004), 376–389.

27. M. W. Caldwell, Description and phylogenetic relationships of a new species of *Coniasaurus* Owen, 1850 (Squamata). *Journal of Vertebrate Paleontology*, 19 (1999), 438–455.

28. M. S. Y. Lee and J. D. Scanlon, The Cretaceous marine squamate *Mesoleptos* and the origin of snakes. *Bulletin of the Natural History Museum London (Zoology Series)*, 68 (2002), 131–142.

29. A. Haber and M. J. Polcyn, A new marine varanoid from the Cenomanian of the Middle East. *Netherlands Journal of Geosciences*, 84, Special Issue 3 (2005), 247–255

30. L. L. Jacobs, K. Ferguson, M. J. Polcyn, and C. Rennison, Cretaceous $\delta[^{13}]$C stratigraphy and the age of dolichosaurs and early mosasaurs. *Netherlands Journal of Geosciences*, 84, Special Issue 3 (2005), 257–268.

31. M. C. Mekarski, D. Japundžić, K. Krizmanić, and M. W. Caldwell, Description of a new basal mosasauroid from the Late Cretaceous of Croatia, with comments on the evolution of the mosasauroid forelimb. *Journal of Vertebrate Paleontology*, 39 (2019), DOI: 10.1080/ 02724634.2019.1577872.

32. M. W. Caldwell and M. S. Y. Lee, Reevaluation of the Cretaceous marine lizard *Acteosaurus crassicostatus* Calligaris, 1993. *Journal of Paleontology*, 78 (2004), 617–619.

33. J. Seiffert, Upper Jurassic lizards from central Portugal. *Memórias do Serviço Geológico de Portugal (Nova Série)*, 22 (1973), 1–85.

34. M. J. Polcyn and G. L. Bell, *Russellosaurus coheni* n. gen., n. sp., a 92 million-year-old mosasaur from Texas (USA), and the definition of the parafamily Russellosaurina. *Netherlands Journal of Geosciences*, 84, Special Issue 3 (2005), 321–333.

35. A. R. Dutchak, A review of the taxonomy and systematic of aigialosaurs. *Netherlands Journal of Geosciences*, 84, Special Issue 3 (2005), 221–229.

36. Dutchak, A.R. and Caldwell, M.W., A redescription of *Aigialosaurus* (= *Opetiosaurus*) *bucchichi* (Kornhuber, 1901) (Squamata: Aigialosauridae) with comments on mosasauroid systematics. *Journal of Vertebrate Paleontology*, 29 (2009), 437–452.

37. K. T. Smith and M. L. Buchy, A new aigialosaur (Squamata: Anguimorpha) with soft tissues remains from the Upper Cretaceous of Nuevo León, Mexico. *Journal of Vertebrate Paleontology*, 28 (2008), 85–94.

38. S. B. McDowell and C. M. Bogert, The systematic position of *Lanthanotus* and the affinities of the anguinomorphan lizards. *Bulletin of the American Museum of Natural History*, 105 (1954), 1-142.

39. O. Rieppel, *The Phylogeny of Anguinomorph Lizards* (Basel: Birkhauser Verlag, 1980).

40. R. Estes, K. de Queiroz, and J. A. Gauthier, Phylogenetic relationships within Squamata. In R. P. Estes, ed., *Phylogenetic Relationships of the Lizard Families* (Stanford University Press: Stanford, 1988), pp. 119-282.

41. G. C. F. Cuvier, Sur le grand animal fossile des carriéres de Maestricht. *Annales du Muséum National d'Histoire Naturelle*, 12 (1808) 145-176.

42. G. Mantell, A tabular arrangement of the organic remains of the county of Sussex. *Transactions of the Geological Society of London*, 3 (1829), 201–216.

43. A. Goldfuss, The skull structure of the *Mosasaurus*, explained by means of a description of a new species of this genus. *Transactions of the Kansas Academy of Science*, 116 (2013), 27–46.

44. R. Owen, Description of the fossil reptiles of the Chalk Formation. In F. Dixon, ed., *The geology and fossils of the Tertiary and Cretaceous Formations of Sussex* (Longman, Brown, Green, and Longman: London, 1850), pp. 378-404.

45. E. D. Cope, On the reptilian orders, Pythonomorpha and Streptosauria. *Proceedings of the Boston Society of Natural History*, 12 (1869), 250–266.

46. R. Owen, On the rank and affinities in the reptilian class of the Mosasauridae, Gervais. *Quarterly Journal of the Geological Society of London*, 33 (1877), 682–715.

47. O. C. Marsh, New characters of mosasauroid reptiles. *American Journal of Science*, 19 (1880, 83–87.

48. G. H. C. L. Baur, On the characters and systematic position of the large sea-lizards, Mosasauridae. *Science*, 405 (1890), 262.

49. S. W. Williston, The relationships and habits of the mosasaurs. *The Journal of Geology*, 12 (1904), 43–51.

50. F. Nopcsa, *Eidolosaurus* und *Pachyophis*, zwei neue Neocom-Reptilien. *Palaeontographica*, 65 (1923), 99–154.

51. G. A. Boulenger, Notes on the osteology of *Heloderma horridum* and *H. suspectum*, with remarks on the systematic position of the Helodermatidæ and on the vertebræ of the Lacertilia. *Proceedings of the Zoological Society of London*, 59 (1891), 109–118.

52. E. D. Cope, Reply to Dr. Bauer's critique on my paper on the paroccipital bone of the scaled reptiles and the systematic position of the Pythonomorpha. *American Naturalist*, 29 (1895), 1003-1005.

53. H. F. Osborn, A complete mosasaur skeleton, osseous and cartilaginous. *Bulletin of the American Museum of Natural History*, 1 (1899), 167–188.

54. G. J. Fejérváry, Contributions to a monography on fossil Varanidae and on Megalanidae. *Annales Historico-Naturales Musei Nationalis Hungarici*, 16 (1918), 341–467.

55. C. L. Camp, Classification of the lizards. *Bulletin of the American Museum of Natural History*, 48 (1923), 289-481.

56. A. A. Bellairs, Observations on the snout of *Varanus*, and a comparison with that of other lizards and snakes. *Journal of Anatomy*, 83 (1949), 116.

57. G. Underwood, *Lanthanotus* and the anguinomorphan lizards: a critical review. *Copeia*, 1957 (1957), 20–30.

58. M. Borsuk-Białłynicka, Anguimorphans and related lizards from the Late Cretaceous of the Gobi Desert, Mongolia. *Palaeontologica Polonica*, 46 (1984), 5–105.

59. M. S. Y. Lee, The phylogeny of varanoid lizards and the affinities of snakes. *Philosophical Transactions of the Royal Society of London, Series B*, 352 (1997), 53–91.

60. M. W. Caldwell, *The Origin of Snakes: Morphology and the Fossil Record* (Boca Raton: CRC Press, 2020).

61. M. W. Caldwell, A New Species of 'Pontosaurus' (Squamata, Pythonomorpha) from the Upper Cretaceous of Lebanon and a phylogenetic analysis of Pythonomorpha. *Società Italiana di Scienze Naturali*, 34 (2006), 1–44.

62. I. Paparella, A. Palci, U. Nicosia and M. W. Caldwell, A new fossil marine lizard with soft tissues from the Late Cretaceous of southern Italy. *Royal Society Open Science*, 6 (2018), 172411.

63. T. R. Simões, M. W. Caldwell, M. Tałanda, et al., The origin of squamates revealed by a Middle Triassic lizard from the Italian Alps. *Nature*, 557 (2018), 706–709.

64. A. Palci and M. W. Caldwell, Redescription of *Acteosaurus tommasinii* von Meyer, 1860, and a discussion of evolutionary trends within the clade Ophidiomorpha. *Journal of Vertebrate Paleontology*, 30 (2010), 94-108.

65. E. Cornalia and L. Chiozza, Cenni geologici sull' Istria. *Giornale dell' I. R. Instituto Lombardo*, 3 (1852), 1–35.

66. H. von Meyer, *Acteosaurus tommasinii* aus dem schwarzen Kreide-Schiefer von Comen am Karste. *Palaeontographica*, 7 (1860), 223-231.

67. A. Kornhuber, Carsosaurus Marchesettii, ein neuer fossiler Lacertilier aus den Kreideschichten des Karstes bei Komen. *Abhandlungen der kaiserlich-königlichen geologischen Reichsanstalt zu Wien*, 17 (1893), 1–15.

68. A. Kornhuber, A., Opetiosaurus Bucchichi, eine neue fossile Eidechse aus der unteren Kreide von Lesina in Dalmatien. *Abhandlungen der kaiserlich-königlichen geologischen Reichsanstalt zu Wien*, 17 (1901), 1–24.

69. R. Calligaris, *Acteosaurus crassicostatus* nuova specie di Dolichosauridae negli Strati Calcarei Ittiolitici di Comeno. *Atti Museo Civico di Storia Naturale di Trieste*, 45 (1993), 29–34.

70. J. Hallermann, The ethmoidal region of *Dibamus taylori* (Squamata: Dibamidae), with a phylogenetic hypothesis on dibamid relationships within Squamata. *Zoological Journal of the Linnean Society*, 122 (1998), 385–426.

71. S. E. Evans and L. J. Barbadillo, A shortlimbed lizard from the Lower Cretaceous of Spain. *Special Papers in Palaeontology*, 60 (1999), 73–85.

72. K. Q. Gao and M. A. Norell, Taxonomic revision of *Carusia* (Reptilia: Squamata) from the Late Cretaceous of the Gobi Desert, and phylogenetic relationships of anguimorphan lizards. *American Museum Novitates*, 3230 (1998), 1–51.

73. O. Rieppel, J. L. Conrad, and J. A. Maisano, New morphological data for *Eosaniwa koehni* Haubold, 1977 and a revised phylogenetic analysis. *Journal of Paleontology*, 81 (2007), 760–769.

74. H. Y. Yi and M. A. Norell, New materials of *Estesia mongoliensis* (Squamata: Anguimorpha) and the evolution of venom grooves in lizards. *American Museum Novitates*, 3767 (2013), 1–31.

75. N. Vidal and S.B. Hedges, The phylogeny of squamate reptiles (lizards, snakes, and amphisbaenians) inferred from nine nuclear protein-coding genes. *Comptes Rendus Biologies*, 328 (2005), 1000–1008.

76. J. J. Wiens, C. A. Kuczynski, T. Townsend, et al., Combining phylogenomics and fossils in higher-level squamate reptile phylogeny: molecular data change the placement of fossil taxa. *Systematic Biology*, 59 (2010), 674–688.

77. R. A. Pyron, F.T. Burbrink, and J. J. Wiens, A phylogeny and revised classification of Squamata, including 4161 species of lizards and snakes. *BMC Evolutionary Biology*, 13 (2013), 1–53.

78. J. W. Streicher and J. J. Wiens, Phylogenomic analyses of more than 4000 nuclear loci resolve the origin of snakes among lizard families. *Biological Letters*, 13 (2017), e20170393.

79. F. T. Burbrink, F. G. Grazziotin, R. A. Pyron, et al., Interrogating genomic-scale data for Squamata (lizards, snakes, and amphisbaenians) shows no support for key traditional morphological relationships. *Systematic Biology*, 69 (2020), 502–520.

80. T. M. Townsend, A. Larson, E. Louis, and J. R. Macey, Molecular phylogenetics of Squamata: the position of snakes, amphisbaenians, and dibamids, and the root of the squamate tree. *Systematic Biology*, 53 (2004), 735–757.

81. N. Vidal and S.B. Hedges, The molecular evolutionary tree of lizards, snakes, and amphisbaenians. *Comptes Rendus Biologies*, 332 (2009), 129–139.

82. T. R. Simões, O. Vernygora, M. W. Caldwell, and S. E. Pierce, Megaevolutionary dynamics and the timing of evolutionary innovation in reptiles. *Nature Communications*, 11 (2020), 1–14.

83. T. R. Simões, M. W. Caldwell, A. Palci, and R. L. Nydam, Giant taxon-character matrices: quality of character constructions remains critical regardless of size. *Cladistics*, 33 (2017) 198–219.

84. T. R. Simões, O. Vernygora, I. Paparella, P. Jimenez-Huidobro, and M. W. Caldwell, Mosasauroid phylogeny under multiple phylogenetic methods provides new insights on the evolution of aquatic adaptations in the group. *PLoS ONE*, 12 (2017), e0176773.

85. A. M. Laing, S. Doyle, M. E. L. Gold, et al., Giant taxon-character matrices: The future of morphological systematics. *Cladistics*, 34 (2017), 333–335.

86. D. Madzia and A. Cau, Inferring 'weak spots' in phylogenetic trees: application to mosasauroid nomenclature. *PeerJ*, 5 (2017), e3782.

87. T. R. Simões, M. W. Caldwell, A. Palci, and R. L. Nydam, Giant taxon-character matrices II: a response to Laing et al. (2017). *Cladistics*, 34 (2017), 702–707.

88. J. A. Gauthier, A. G. Kluge, and T. Rowe, Amniote phylogeny and the importance of fossils. *Cladistics*, 4 (1988), 105–209.

89. M. J. Donoghue, J. A. Doyle, J. Gauthier, A. G. Kluge, and T. Rowe, The importance of fossils in phylogeny reconstruction. *Annual Review of Ecology and Systematics*. 20 (1989), 431–460.

90. M. M. Mekarski, The Origin and Evolution of Aquatic Adaptations in Cretaceous Squamates (Unpublished PhD Thesis: University of Alberta, 2017).

91. C. Diedrich, Ein dentale von *Coniosaurus crassidens* Owen (Varanoidea) aus dem Ober-Cenoman von Halle/Westf (NW-Deutschland). *Geologie und Paläontologie in Westfalen*, 47 (1997), 43–51.

92. J. D. Scanlon and S.A. Hocknull, A dolichosaurid lizard from the latest Albian (mid-Cretaceous) Winton Formation, Queensland, Australia. Transactions of the Kansas Academy of Science, Fort Hays Studies Special Issue –

Proceedings of the Second Mosasaur Meeting (2008), 131–136.

93. P. A. Goloboff and S. A. Catalano, TNT version 1.5, including a full implementation of phylogenetic morphometrics. *Cladistics*, 32 (2016), 221-238.

94. M. S. Y. Lee, Convergent evolution and character correlation in burrowing reptiles: towards a resolution of squamate relationships. *Biological Journal of the Linnean Society*, 65 (1998), 369-453.

95. H. Zaher, The phylogenetic position of *Pachyrhachis* within snakes (Squamata, Lepidosauria). *Journal of Vertebrate Paleontology*, 18 (1998), 1–3.

96. H. Zaher and O. Rieppel, Tooth implantation and replacement in squamates, with special reference to mosasaur lizards and snakes. *American Museum Novitates*, 3271 (1999), 1–19.

97. M. W. Caldwell, L. A. Budney, and D. O. Lamoureux, Histology of tooth attachment tissues in the Late Cretaceous mosasaurid Platecarpus. *Journal of Vertebrate Paleontology*, 23 (2003), 622-630.

98. M. W. Caldwell and A. Palci, A new basal mosasauroid from the Cenomanian (U. Cretaceous) of Slovenia with a review of mosasauroid phylogeny and evolution. *Journal of Vertebrate Paleontology*, 27 (2007), 863-883.

99. O. Rieppel and M. Kearney, Tooth replacement in the Late Cretaceous mosasaur *Clidastes*. *Journal of Herpetology*, 39 (2005), 688-692.

100. X. Luan, C. Walker, S. Dangaria, et al., The mosasaur tooth attachment apparatus as paradigm for the evolution of the gnathostome periodontium. *Evolution & Development*, 11 (2009), 247-259.

101. M. W. Caldwell, Ontogeny, anatomy and attachment of the dentition in mosasaurs (Mosasauridae: Squamata). *Zoological Journal of the Linnean Society*, 149 (2007), 687–700.

102. A. R. LeBlanc, D. O. Lamoureux, and M. W. Caldwell, Mosasaurs and snakes have a periodontal ligament: timing and extent of calcification, not tissue complexity, determines tooth attachment mode in reptiles. *Journal of Anatomy*, 231 (2017), 869–885.

103. C. Gomez, E. M. Özbudak, J. Wunderlich, et al., Control of segment number in vertebrate embryos. *Nature*, 454 (2008), 335–339.

104. M. J. Woltering, From lizard to snake; behind the evolution of an extreme body plan. *Current Genomics*, 13 (2012), 289–299.

105. R. Montero, J. D. Daza, A. M. Bauer, and V. Abdala, How common are cranial sesamoids among squamates? *Journal of Morphology*, 278 (2017), 1400-1411.

106. J. Ollonen, F. O. Silva, K. Mahlow, and N. Di-Poi, Skull development, ossification pattern, and adult shape in the emerging lizard model organism *Pogona vitticeps*: a comparative analysis with other squamates. *Frontiers in Physiology*, 9 (2018), 278.

107. F. F. Garberoglio, S. Apesteguía, T. R. Simões, et al., New skulls and skeletons of the Cretaceous legged snake *Najash*, and the evolution of the modern snake body plan. *Science Advances*, 5 (2019), eaax5833.

9

A Review of the Skull Anatomy and Phylogenetic Affinities of Marine Pachyophiid Snakes

Hussam Zaher, Bruno G. Augusta, Rivka Rabinovich,
Michael J. Polcyn, and Paul Tafforeau

9.1 Introduction

The Mesozoic saw an expansion of many terrestrial vertebrate lineages, including snakes, a highly specialized group of squamates with a controversial origin [1]. Although mainly terrestrial, snakes invaded the aquatic environment several times. The first known marine radiation occurred during the Cenomanian, on the northern Gondwana margins in the Mesogean Ocean and Neo-Tethys Sea [2–4; hereafter collectively referred to as Tethyan], and provided some of the most complete fossil snakes known, including the first known snakes with fully developed hindlimbs [5–7]. This radiation has in recent decades fueled an intense debate on the origin and early diversification of snakes [5–14]. Central to this debate are conflicting interpretations of the anatomy and phylogenetic affinities of these marine snakes, and more specifically of the limbed species *Eupodophis descouensi*, *Haasiophis terrasanctus*, and *Pachyrhachis problematicus*, known from sediments of the southern Tethyan margin in the Levant (Lebanon and Israel; [15]). Prior to the redescription of *Pachyrhachis* and the discovery of *Eupodophis* and *Haasiophis* in the 1990s, three other genera – *Simoliophis*, *Pachyophis*, and *Mesophis* – were already known from northern and southern Tethyan deposits of Egypt, France, and Boznia-Herzegovina [16–19]. However, the lack of well-preserved skulls limited their contribution to clarifying the anatomy of the group. Tethyan marine Cenomanian snakes are varyingly grouped together in the literature under two different family names – Pachyophiidae Nopcsa 1923 and Simoliophiidae Nopcsa, 1925 – the former having priority.

The first Tethyan marine snake to be described was *Simoliophis rochebrunei* (incorrectly spelled as *S. rochebruni* by Sauvage [16]). Sauvage [16] recognized *S. rochebrunei* as the oldest definitive snake known – until then, the oldest records were from the early Cenozoic ([20]; see also Chapter 4) – but with unclear affinities within the group. Rochebrune [21] suggested a possible association with extant typhlopid snakes but noted several conspicuous differences and especially the broad, high neural spine of *Simoliophis*.

In that same period, Cope [22] initiated the origin of snakes debate by suggesting that the mosasaurian families Mosasauridae and Clidastidae should be placed in a separate sub-order – Pythonomorpha – at the same rank as Lacertilia and Ophidia. This triggered a reaction in the scientific community of that time, which led to a series of exchanges on the merits of the new classification and validity of the morphological evidence supporting it (see Supplementary Appendix 9.S1 for a list of references; and [23, 24] for a review of that literature). However, the debate quickly shifted to focus on the findings of new aigialosaurids and dolichosaurids [25–29] – elongated marine mosasaurians with a suite of characteristics approaching the snake body-plan – and of the first articulated marine Cenomanian snakes [17, 19]. This collection of mid-Cretaceous aquatic forms from the Adriatic carbonate platform deposits seemed better suited than Cope's pythonomorphs to represent a series of intermediate forms linking lizards to extant snakes.

In a series of influential contributions, Nopcsa ([17, 18] and additional references in Supplementary Appendix 9.S1) further developed the hypothesis of a marine origin of snakes, proposing that snakes derived from a marine aigialosaurid ancestor. Nopcsa [17] revised Boulenger's classification of extant snakes, dividing them in Alethinophidia and Angiostomata, and creating the Cholophidia to accommodate the extinct forms *Simoliophis*, *Palaeophis*, *Archaeophis*, and his newly described *Pachyophis woodwardi*. According to Nopcsa [17], both Angiostomata and Alethinophidia descended from the Cholophidia. Similarly, dolichosaurids and mosasaurids would also derive independently from an aigialosaurid ancestor [17, 30], an opinion shared by Boulenger [31] and Dollo [32] (see also Supplementary Appendix 9.S1 for additional references).

Camp [23] did not accept a close affinity between snakes and any group of mosasaurian lizards, and classified Serpentes as a separate suborder at the same rank as Sauria, the latter including Mosasauroidea, Dolichosauridae, and Aigialosauridae. Apart from one notable exception [33], Camp's influential work was followed for most of the twentieth century, until Lee [34] revived the hypothesis of a marine origin of snakes by advancing lines of evidence similar to Cope, with *Pachyrhachis* representing the best-known morphological intermediate between mosasaurians and snakes [see 35 for a review]. Soon, dolichosaurids, aigialosaurids, and pachyophiids were viewed as successive stem taxa to crown Serpentes (e.g., [36]). However, in the last few years, these views were superseded in favor of a Jurassic origin of snakes [37–39], with 'parviraptorids' representing at least four stem-snake lineages. In that new scenario, the marine Cenomanian snakes are alethinophidians nested within the crown clade Serpentes [37], reconciling with previews views [7, 8]. However, 'parviraptorids' have had a complex taxonomic history, and have been attributed also to either Gekkonomorpha [40] and Anguimorpha [41]. New associated material currently under study (see Chapter 2) may offer additional information to resolve their affinities.

Despite the last twenty years of studies of marine Cenomanian snakes, and notwithstanding the recent shift away from the 'pythonomorph hypotheses' (see Chapters 7 and 8), several controversial aspects of their anatomy persist. Differences in observations and anatomical interpretations for the group have led to conflicting conclusions in broader combined analyses of snake relationships (e.g., [42, 43]). A recent review [39] did not attempt to clarify the inconsistencies created by the new evidence [37] and previous work on Cenomanian snakes (e.g., [44]), instead proposing another phylogenetic scenario that combines a pythonomorph clade and pachyophiids nested within derived alethinophidian snakes [45].

In this chapter, we revisit the available morphological evidence of pachyophiids based on new preparations, high resolution computed tomography (HRCT) scans, and Synchrotron images, to help formulate the basis for a consensus and clarify the peculiar morphology of this group. We also address the phylogenetic affinities and the challenges posed by taphonomy to the study and interpretation of these specimens.

9.2 Monophyly and Taxonomic Content of Pachyophiidae

Simoliophis, *Eupodophis*, *Pachyrhachis*, *Haasiophis*, *Pachyophis*, and *Mesophis* may form a clade of exclusively Tethyan marine snakes diagnosed by their highly pachyostotic mid- and posterior trunk vertebrae and ribs, with pachyostosis being characteristically concentrated in the vertebral centrum [46] (Supplementary Figure 9.S1). However, pachyostosis also occurs in the Tethyan marine dolichosaurids and aigialosaurids [15, 47] and has been reported in more than 35 secondarily marine amniote groups [48]. Pachyostosis might be a synapomorphy shared by early mid-Cretaceous Tethyan mosasaurians and snakes, secondarily lost within these groups, or might be a convergence favored by peculiar paleoenvironmental conditions of this period [47, 49]. Recently, the latter hypothesis received support through independent morphological and combined phylogenetic analyses [40, 42, 43, 50, 51], suggesting that pachyostosis is a pachyophiid instead of a pythonomorph synapomorphy. The recently described marine snake *Lunaophis aquaticus* [52] may represent a further example of convergence in which pachyostosis is concentrated in the zygapophyseal regions and does not affect the vertebral centra.

Apart from sharing mid- and posterior trunk vertebrae with pachyostotic centra, no other derived feature is known to be uniformly preserved in all six genera of pachyophiids. There are presently seven formally described species of pachyophiids (Supplementary Table 9.S1) and one possible undescribed species [53]. We summarized in Supplementary Tables 9.S1 and 9.S2 the bones preserved on each specimen of Pachyophiidae as well as their vertebral and tooth counts. Fossil completeness was assessed directly on each specimen, or through Synchrotron and HRCT images, and high-resolution photographs.

Simoliophis rochebrunei Sauvage, 1880 was originally described based on one disarticulated trunk vertebra from the Cenomanian of the Charentes region, France [16].

Disarticulated vertebrae of *Simoliophis* were later reported from Cenomanian marine localities throughout the northern and southern Tethyan margins [53]. This material includes a second valid species of *Simoliophis, S. libycus* [54], and at least one undescribed species from Bahariya, Egypt [10].

Eupodophis descouensi Rage and Escuillié, 2000 (Figs. 9.1 and 9.2) is known from the holotype (MIM F49; ex-Rh-E.F. 9001-9003) and three additional specimens (MSNM-V3660, V3661, V4014) referred to this species [46] (Supplementary Figures 9.S2–S9). The four nearly completely articulated specimens derive from two different localities in Lebanon [46]. The holotype preserves an almost complete skull and articulated body (Fig. 9.1; Supplementary Figures 9.S2–S4). However, preservation of the skull and vertebrae is poor, split across part and counterpart. The skull is divided in the horizontal plane, leaving only internal views of the braincase exposed on both slabs. Dorsal and ventral views of the skull are visible only through Synchrotron imaging reconstructions (Fig. 9.2). Preservation is similar in MSNM-V3661, but only one slab is known and available for study (Supplementary Figures 9.S5–S6). MSNM-V4014 is mostly complete, except for the ventrally exposed skull (Supplementary Figures 9.S7–S8). MSNM-V3660 preserves an articulated body, lacking skull and tail (the latter possibly hidden below mid-trunk vertebrae on the slab; Supplementary Figure 9.S9). Both the holotype and MSNM-V3660 are of similar size, and significantly larger than the other two specimens.

Rieppel and Head [46] suggested that the two smaller individuals might belong to a distinct species but conservatively assigned them to *Eupodophis descouensi*. Although distinct in size, they share with the holotype the presence of elongated chevron bones, posteromedially curved supratemporals (MSNM-V3661), anteromedially curved and distally tapering vomerine processes of the palatine (MSNM-V4014), anteroposteriorly narrow frontals, long and distally tapering, posterodorsally oriented coronoids, and straight distally

---

Box 9.1   Abbreviations used in figures and text

al.pa, alar process of the parietal ("lateral wing"); **amf**, anterior mylohyoid foramen; **ang**, angular; **asc**, anterior semicircular canal; **atl**, atlas; **ax**, axis; **bocc**, basioccipital; **bsph**, basisphenoid; **bspt**, basipterygoid process of the basisphenoid; **col**, stapes; **com**, compound bone; **co.pr**, notch for the articular contact of the prefrontal; **cor**, coronoid; **de**, dentary; **ect**, ectopterygoid; **ect.pt**, ectopterygoid process of the pterygoid; **f.col**, footplate of the columella; **fo**, fossa mandibularis; **f.o**, foramen ovalis; **fr**, frontal; **ju**, jugal; **lat.com**, lateral process of the compound bone; **l.du**, lacrimal duct; **l.lff**, left ventral subolfactory process ('lateral frontal flange'); **l.mfp**, left medial frontal pillar; **m.f**, mental foramen; **mx**, maxilla; **mx.pl**, maxillary process of the palatine; **na**, nasal; **n.prmx**, nasal process of the premaxilla; **of**, optic foramen; **oto**, otoccipital; **p.pr**, paroccipital process of the otoccipital; **pa**, parietal; **pf**, postfrontal; **pl**, palatine; **pl.mx**, palatine process of the maxilla; **post.mx**, posterior end of the maxilla; **prf**, prefrontal; **prmx**, premaxilla; **pro**, prootic; **prsph**, parasphenoid; **psc**, posterior semicircular canal; **prsph.r**, parasphenoid rostrum; **pt**, pterygoid; **qu**, quadrate; **set**, sella turcica; **s.cr**, sagittal crest; **so**, supraoccipital; **spl**, splenial; **st**, supratemporal; **sty**, quadrate stylohyal process; **tra**, trabecula; **v.pl**, vomerine process of the palatine; **V2**, **V3**, trigeminal foramen; **VII**, foramen for the hyomandibular ramus of the facial nerve; **vfn**, medial vertical lamina ('medial frontal flange') of the nasal.

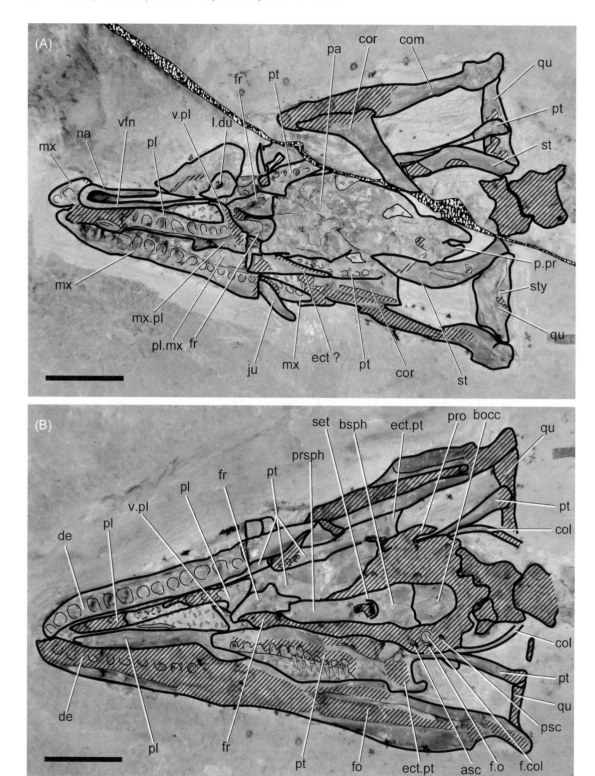

Figure 9.1 Photographs with superimposed interpretive line drawings of the skull of *Eupodophis descouensi* (holotype MIM F49); (A) ventral view of the dorsal surface; (B) dorsal view of the ventral surface. See Box 9.1 for explanation of abbreviations. See Supplementary Figures 9.S3 and S4 for colour versions of photographs without line drawings.

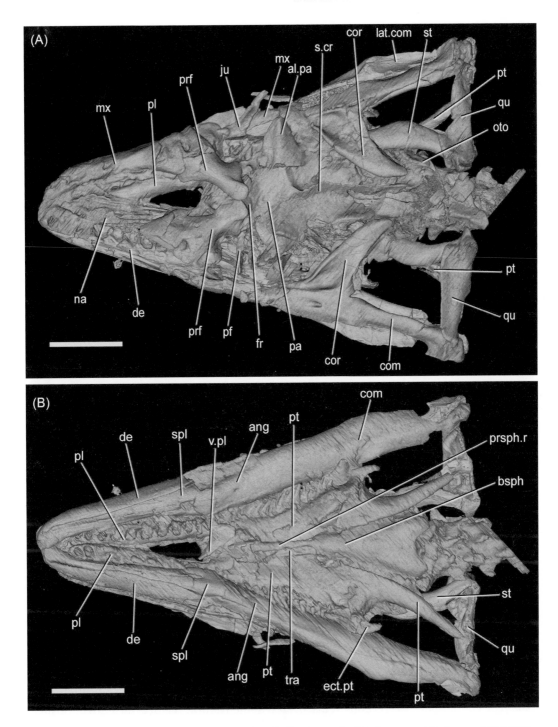

Figure 9.2 Synchrotron reconstructions of the skull of *Eupodophis descouensi* (holotype MIM F49); (A) dorsal surface; (B) ventral surface. See Box 9.1 for explanation of abbreviations.

tapering, flat jugals (MSNM-V3661). The combination of these characters is consistent with conspecificity.

Pachyrhachis problematicus Haas, 1979 (Fig. 9.3) is known from two mostly complete, articulated specimens from the limestone quarries of Ein Yabrud, Palestinian Territories. The holotype (HUJ-PAL EY692) comprises an almost complete skull and body [55, 56] (Supplementary Figures 9.S10–S11). Posterior trunk vertebrae, hindlimbs and tail are not preserved. The skull was removed from the slab and prepared on both sides, with the dorsal side embedded in resin. The referred specimen (HUJ-PAL EY691) retains an almost complete body with articulated hindlimbs and partially preserved skull and tail [57] (Supplementary Figures 9.S12–S13). The skull, also prepared on both sides, lies partially disarticulated below the vertebrae and ribs of mid-trunk vertebrae, on the right side of the ribcage. A string of articulated anteriormost vertebrae with developed hypapophyses is associated with the skull and also preserved below the ribcage. A second string of anterior trunk vertebrae is preserved slightly disconnected from the previous one.

Haasiophis terrasanctus Tchernov, Rieppel, Zaher, Polcyn & Jacobs, 2000 (Fig. 9.4) is known only from one complete and articulated specimen (HUJ-PAL EY695) (Supplementary Figures 9.S14–16). This was collected at the same quarries as the *Pachyrhachis* specimens and is prepared and mainly exposed in ventral view [35]. The dorsal side is embedded in resin. This is the best-preserved pachyophiid specimen known so far (Supplementary Table 9.S1).

Pachyophis woodwardi Nopcsa, 1923 (Fig. 9.5) is known from two specimens, the holotype (NHMW 1912-I8) and a referred specimen (NHMW 1912-I9) (Supplementary Figures 9.S17–S18). Both were collected in a quarry near Bileca, Bosnia-Herzegovina [58]. The holotype retains a well-preserved, articulated vertebral column exposed mostly in dorsal view, but no pelvic or hindlimb elements. Disarticulated cranial remains, including a left dentary and a possible posterior lower jaw, are also present. More cranial remains are preserved in the anteriormost region, but they are too badly crushed to be identified. The referred specimen, comprising three series of articulated but poorly preserved vertebrae, was recently redescribed [59]. Another specimen of marine snake from a quarry near Bileca, in the lower Turonian carbonates of Dubovac, was also recently described [60]. This specimen, composed of highly pachyostotic articulated vertebrae and ribs, is very similar to *Pachyophis woodwardi* in its neural arch and spine morphology, and highly pachyostotic ribs, and represents a third specimen of *P. woodwardi* or a congeneric species [60]. The Dubovac specimen reveals a centrum with a large, flat and square ventral surface, not visible in the first two specimens (Supplementary Figure 9.S1d), a characteristic unique to pachyophiids among snakes.

Mesophis nopcsai Bolkay, 1925 is known from a single specimen (NHM-BiH 2039) from a quarry in Selista, Bosnia-Herzegovina [19], and comprised originally part and counterpart slabs. The counterpart was recently rediscovered at the Natural History Museum of Bosnia-Herzegovina, while the main part is thought to be in a private collection [39]. It preserves a string of 79 articulated vertebrae exposed in ventral view. A cast of the holotype's counterpart is deposited at the Naturhistorisches Museum Wien, Austria (Supplementary Figure 9.S18).

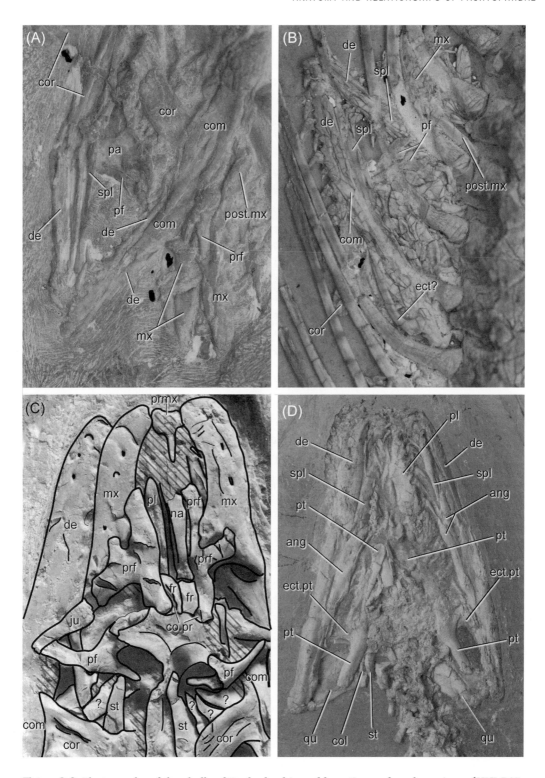

Figure 9.3 Photographs of the skulls of *Pachyrhachis problematicus*; referred specimen (HUJ-PAL EY691) in (A) dorsal, and (B) ventral views; holotype (HUJ-PAL EY692) in (C) dorsal (with superimposed interpretive line drawing), and (D) ventral views. Original photograph from Haas [55: fig.6] used in (C). See Box 9.1 for explanation of abbreviations. (A black and white version of this figure will appear in some formats. For the colour version, please refer to the plate section).

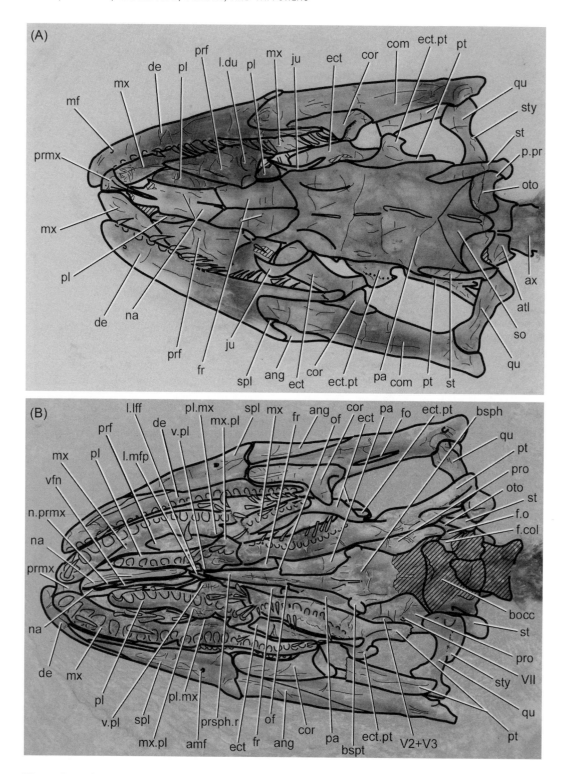

Figure 9.4 Photographs with superimposed interpretive line drawings of the skull of *Haasiophis terrasanctus* (holotype HUJ-PAL EY695); (A) dorsal view; (B) ventral view. See Box 9.1 for explanation of abbreviations. See Supplementary Figures 9.S15 and S16 for colour versions of photographs without line drawings.

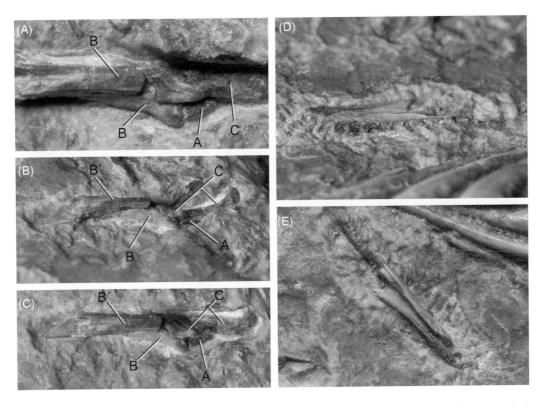

Figure 9.5 Skull fragments of *Pachyophis woodwardi* (holotype NHMW 1912-I8); (A–C) unidentified element; (D) dentary in dorsomedial view; (E) dentary in medial view. Letters A to C follow Lee et al. [58], who identified these structures as the angular (A), splenial (B), and compound bones (C). See Box 9.1 for explanation of abbreviations. (A black and white version of this figure will appear in some formats. For the colour version, please refer to the plate section).

9.3 A Review of the Morphological Evidence

Many contributions have dealt with pachyophiid anatomy. The most relevant, providing a strong platform from which we expand our observations, are as follow: *Pachyrhachis problematicus* [9, 55–57, 61–66]; *Haasiophis terrasanctus* [9, 35, 64, 66]; *Eupodophis descouensi* [6, 46, 66]; *Pachyophis woodwardi* [17, 58]; *Mesophis nopcsai* [19]; *Simoliophis rochebrunei* [18, 53]. These studies focused on a small number of key features that were discussed in detail, often resulting in contradictory interpretations [39]. On most of these issues (especially in *Haasiophis* and *Pachyrhachis*) there is now consensus. However, several aspects of pachyophiid anatomy remain controversial, and others have not been described adequately. We review these points below, following the anatomical divisions of Cundall and Irish [67].

Synchrotron and HRCT methods are presented in Supplementary Appendix 9.S2. Institutional collection acronyms and specimens examined are listed in Supplementary Appendix 9.S3. Phylogenetic methods and results are reported in Supplementary Appendix 9.S4.

9.3.1 Snout

Haas [55] and Lee and Caldwell [61] reported a **premaxilla** in *Pachyrhachis* (HUJ-PAL EY692) [contra 46]. The element is visible only in dorsal view, now under a thick layer of resin. A photograph of the dorsal surface of the skull before embedding [55] shows a small premaxilla with a concave anterior border and tapering nasal process (Fig. 9.3C). A premaxilla has been reported in *Eupodophis* (MSNM-V4014) [46], but we interpret that as the anterior tip of the right maxilla, still bearing the anteriormost maxillary tooth, and extending posteriorly below the right palatine (Supplementary Figure 9.S19). The premaxilla of *Haasiophis* has been described in detail [35], though the presence of ventral foramina and its toothless condition has been questioned [66]. A pair of long, large laminar vomerine processes have been reported [66], but available HRCT reconstructions do not support this interpretation (Fig. 9.4; Supplementary Figure 9.S20). The premaxilla in *Haasiophis* is similar to that of *Pachyrhachis* in being small, narrow and toothless, tightly (but not suturally) situated between the maxillae and anchored primarily to the nasals by a moderately long, distally tapering nasal process that projects between the descending laminae of the nasals (Fig. 9.4; Supplementary Figure 9.S20). Vomerine processes cannot be corroborated in any specimen.

Nasals are preserved in *Pachyrhachis*, *Haasiophis*, and *Eupodophis*. A left nasal is exposed in ventral view in the holotype of *Eupodophis* and the holotype of *Pachyrhachis* retains a similarly complete right nasal exposed in dorsal view (Figs. 9.1A and 9.3C). *Haasiophis* has both nasals preserved, exposed in both views (Fig. 9.4). The nasals in all three genera are long with a horizontal lamina and a moderately expanded medial vertical lamina. The latter is a typical alethinophidian feature. The holotype of *Eupodophis* offers the best view, the horizontal lamina tapers posteriorly towards and narrowly contacts the frontal (Fig. 9.1A). The morphology of the left nasal of *Eupodophis* holotype helps to identify a left fragmented nasal exposed in ventral view in MSNM-V4014, medial to the left dentary (Supplementary Figure 9.S19). The ventrally concave shape of the nasal is visible in the preserved surface of that bone, previously identified as a vomer [46]. Similarly, we identify in *Pachyrhachis* a long and narrow element that contacts the frontal posteriorly as a nasal (Fig. 9.3C). We therefore cannot corroborate a previous assertion [61] that the nasals in *Pachyrhachis* are short. The overall morphology of the nasal of *Haasiophis* differs from *Eupodophis* and *Pachyrhachis* by its posteriorly broader horizontal lamina and weak medial vertical lamina (probably due to dorso-ventral compression). Though the nasals are clearly separated posteriorly, the suture is not visible more anteriorly. It is difficult to determine the exact form of the nasals in dorsal view (Fig. 9.4A). In ventral view, the descending laminae clasp the nasal process of the premaxilla, suggesting that the nasals were separated throughout their length (Fig. 9.4B).

9.3.2 Braincase

Frontals are preserved in *Pachyrhachis*, *Haasiophis*, and *Eupodophis*, but their identification varies among authors. The holotypes of *Haasiophis* and *Eupodophis* have the best-preserved frontals. They are very distinct in size and shape. The frontals of *Haasiophis* were described [35] as a pair of broad and elongated bones that meet the parietal posteriorly in a U-shaped, anteriorly concave suture, and project anteriorly to meet the nasals beyond the frontal–prefrontal articulation. That study [35] also identified medial frontal pillars (flanges) exposed in ventral view, just in front of the parasphenoid rostrum. We confirm these observations, adding that only the left frontal retains its medial pillar, in contact with the left ventral subolfactory process, enclosing the olfactory tract as in alethinophidians (Fig. 9.4B; Supplementary Figure 9.S21). Although not lining up perfectly, due to postmortem displacement, the vertical laminae of the nasals in *Haasiophis* contact the corresponding medial frontal pillar ventrally, forming a prokinetic joint. The descending frontal flanges are preserved in contact with the lateral margin of the parasphenoid, closing the braincase ventrally (Fig. 9.4B; Supplementary Figure 9.S21). A small optic foramen is present at the triple junction of frontal–parietal–parasphenoid, formed in part by a weak notch in the anteroventral corner of the parietal (Fig. 9.4B). However, the size of the optic foramen cannot be ascertained because the frontal's posteroventral corner is poorly preserved. In dorsal view, the frontals meet the nasals in a V-shaped, anteriorly concave contact. HRCT reconstructions of the holotype of *Haasiophis* show that the two bones contact dorsally but do not seem to overlap each other as in *Dinilysia* and *Najash* (Supplementary Figure 9.S21).

The holotype of *Eupodophis* has short, narrow frontals, best exposed in ventral view (Fig. 9.1A). Dorsally, they are scarcely visible due to the overlap of the large prefrontal's articular knobs (Fig. 9.2A; Supplementary Figure 9.S22). The prefrontal precludes the lateral margin of the frontal from the orbit. Synchrotron images reveal poorly preserved medial frontal pillars pressed onto the medial wall of thick lateral frontal flanges (Supplementary Figure 9.S22). The latter extend ventrally, contacting the parasphenoid rostrum. The frontals contact the parietal posteriorly in a thick and narrow, slightly anteriorly W-shaped suture. The dorsal sagittal suture between the frontals is exposed in ventral (internal) view on the part (Fig. 9.1A).

A long, left frontal was identified in MSNM-V3661 with an anterior margin positioned at the level of the prefrontal-maxillary articulation and posterior end reaching the coronoid [46]. However, comparison with the holotype of *Eupodophis* suggests that the fronto-parietal suture is still preserved in MSNM-V3661, at the level of the anterior tip of the jugal (Supplementary Figure 9.S23). Therefore, the left frontal of MSNM-V3661 would be short, as in the holotype. The posterior two-thirds of the bone, previously identified as frontal [46], is the left portion of the parietal. Similarly, the bone previously identified as the right frontal [46] is a fragment of the right portion of the parietal that retains its contact with the prefrontal, as in the holotype of *Eupodophis*. Another striking similarity of MSNM-V3661 with the holotype of *Eupodophis* is the large knob-like proximal head of the prefrontal preserved on its right element (Supplementary Figure 9.S23).

Previously [55, 61], the holotype and the referred specimen of *Pachyrhachis* were interpreted as retaining frontals, mainly exposed in the dorsal and ventral sides of their respective slabs. Frontals were described as long and narrow, fitting posteriorly into a shallow notch of the anterior margin of the parietal and projecting anteriorly beyond the ascending processes of the maxillae. Anteriorly, the frontals would contact poorly preserved elements identified as nasals [61]. However, comparison with *Eupodophis* suggests that the frontals in the *Pachyrhachis* holotype are also covered laterally and partly dorsally by the long prefrontals which overlapped the lateral surface of the frontal table and contacted the parietal dorsally. The frontals are still visible in dorsal view, medial to the prefrontals (Fig. 9.3C); they are anteroposteriorly short and laterally narrow as in *Eupodophis*. The short length of the frontals is revealed by their midline sutural contact on the skull table. The dorsolateral margin of the frontals receives a large prefrontal knob, as in *Eupodophis*. According to this new interpretation of *Pachyrhachis*, the two elements labeled in the referred specimen as frontals by Lee and Caldwell [61: figs. 5-6] cannot be identified unequivocally. However, they may represent a pair of postfrontals ('postorbitofrontals' of [61]; see discussion in § 9.3.4).

The **parietal** is complete in *Haasiophis* and only partially preserved in *Pachyrhachis* and *Eupodophis*. It is at least twice as long as wide in all three taxa. *Pachyrhachis* and *Eupodophis* share a similar morphology of the parietal table, which bears well-developed 'postorbital' processes projecting laterally behind the orbit to form alar processes, or 'lateral wings', for the likely origin of hypertrophied adductor musculature thought to be associated with the massive coronoids (Figs. 9.2A and 9.3C). These alar processes are unique in not projecting from the anteroventral corner of the parietal, but rather only from the parietal's dorsolateral anterior margin. The sagittal crest is well developed throughout most of the posterior two-thirds of the parietal table, diverging anteriorly and delineating two broad, deep concavities on the dorsal surface, likely for further anchorage of the adductor muscles. As in *Pachyrhachis*, the anterior margin of the table of *Eupodophis* is thick with two shallow concavities in which the short, narrow frontals articulate (Figs. 9.2A and 9.3C). The posterior margin of the parietal and its exact sutural relations with the supraoccipital and prootic are not preserved in *Pachyrhachis* or *Eupodophis*. *Haasiophis* lacks parietal alar processes and the parietal table has straight lateral margins that broaden only slightly posteriorly (Fig. 9.4A). The absence of parietal alar processes is consistent with the plesiomorphically small coronoids. The sagittal crest is low and does not diverge anteriorly. The slightly curved facet projecting on the lateral aspect of the supraorbital process of the parietal is likely a ridge for the adductor muscles origin rather than a facet for a posterior orbital bone [35]. The element previously identified [35] as a postorbital is better interpreted as a jugal (see below). Therefore, *Haasiophis* lacks a posterior orbital element, an interpretation that better fits the morphology of the parietals, with smooth and straight lateral margins (Fig. 9.4A). Lateral descending flanges of the parietal are not preserved in *Pachyrhachis* and are fragmentary in the holotype of *Eupodophis*, but the left lateral descending flange is exposed on its internal view along with the left frontal in MSNM-V3661 (Supplementary Figure 9.S23). However, its connection with the parabasisphenoid, supraoccipital, and the prootic are lost. The lateral descending flanges in *Haasiophis* are well preserved and reach the parabasisphenoid in a sutural contact (Fig. 9.4B; Supplementary Figure 9.S21). A notch forming the optic foramen occurs on its anteroventral corner.

Prootics are well preserved in *Haasiophis*. They are partially visible ventrally, mainly covered by the supratemporals dorsally and pterygoids ventrally (Fig. 9.4B). In ventral view, the single trigeminal foramen is visible in the right prootic, posterior to the basisphenoid's small basipterygoid process and medial to the quadrate process of the pterygoid. A small foramen for the hyomandibular ramus of the facial nerve (VIIh) is exposed posterior to the trigeminal foramen (Fig. 9.4B). Unfortunately, the sutural contact between prootics and parietal is poorly preserved. Posteriorly, the left prootic is mostly covered laterally by the pterygoid but its posterior margin is partially exposed and reveals its contribution to the anterior and anterodorsal borders of a small fenestra ovalis in which a possible fragmentary **stapedial footplate** is still preserved in place (Fig. 9.4B; Supplementary Figure 9.S24). The footplate and fenestra are small when compared to the stem snakes *Dinilysia* and *Najash* [68, 69], resembling more the derived condition of the crown snake ear. The stapedial shaft is broken, with only the proximal part preserved. The posteroventral margin of the fenestra rotunda is badly eroded, along with most of the ventral surface of the posterior braincase. Thus, it is not possible to describe the contribution of the prootic, basioccipital, and otooccipital to the otic recess. The structure identified previously [35, 46] in *Haasiophis* and *Eupodophis* as a prootic flange was later [66] correctly identified as the ectopterygoid process of the pterygoid (Fig. 9.4B). The bone identified as a right exoccipital (otooccipital) in MSNM-V4014 [46] represents the ventral projection of the right prootic preserving the foramen for VIIh. We could not confirm the identification [61] of the prootic fragments in the holotype of *Pachyrhachis*.

Stapes are preserved in the holotypes of *Eupodophis* and *Pachyrhachis* [55] (Figs. 9.1B and 9.3B; Supplementary Figure 9.S22). In *Eupodophis*, the split of the braincase in two slabs exposes the fenestra ovalis on the broken surface of the fragmented prootic, with the small stapedial footplate visible in an internal view, flanked by the anterior and posterior semicircular canals (Fig. 9.1B). The stapedial shafts of the stapes are exposed on both sides of the posterior braincase, in a typical position dorsal to the pterygoids and ventral to the supratemporals (Supplementary Figures 9.S22, S24). They are long and project posteriorly, but instead of slender needle-like structures both shafts are robust elements that are bowed laterally, like the supratemporals. In *Pachyrhachis* the stapes is slender, but its ends are obscured (Fig. 9.3B).

Apparently, only *Haasiophis* has well-preserved **supraoccipital** and **otooccipitals** (Fig. 9.4A). Previously [35] the otooccipitals were interpreted as displaced laterally during dorsoventral compression, resulting in loss of their medial contact above the tectum of the foramen magnum. HRCT reconstructions of *Haasiophis* confirm these observations. However, we cannot confirm whether these elements were in sutural contact. A typically alethinophidian, reduced paroccipital process is visible on both sides and receives the supratemporals in a sutural contact laterally.

Both **parabasisphenoid** and **basioccipital** are preserved in *Pachyrhachis*, *Eupodophis*, and *Haasiophis*. Previously [46] the presence in *Eupodophis* of a basioccipital (MSNM-V3661) and a parabasisphenoid (MSNM-V3661 and V4014) was mentioned, but in MSNM-V3661 these elements are poorly preserved and might instead be parietal fragments in internal view. The possible anterior portion of the parasphenoid in MSNM-V4014 is also poorly preserved; it could as well be part of the right lateral descending flange of the parietal, aligned with the right palatine and prootic. The parabasisphenoid is badly crushed in *Haasiophis* and *Pachyrhachis* (Figs. 9.3D and 9.4B), but in *Eupodophis* (MIM F49) is mostly complete and revealed both dorsally (exposed on the slab) and ventrally (digitally segmented from HRCT scan) (Figs. 9.1 and 9.2). In dorsal view, the basisphenoid is bordered posteriorly by the basioccipital in a continuous transverse suture, forming a concave surface constituting part of the floor of the cranial cavity. A deep, narrowly U-shaped sella turcica is located in the anterior part of the basisphenoid, deeply recessed below the posterior dorsum sella (crista sellaris) (Fig. 9.2B; Supplementary Figure 9.S22). The parasphenoid rostrum extends from the anterior margin of the sella turcica into the orbital region as a narrowed structure, flanked laterally by deep trabecular grooves (Fig. 9.2B). Its constriction resembles more the alethinophidian condition, but the intertrabecular crest of the parasphenoid (sensu [70]) is expanded and rod-like, laterally roofing the trabecular grooves. In ventral view, the basisphenoid and basioccipital of MIM F49 form an anteriorly directed triangular shape with a smooth and ventrally convex surface (Fig. 9.2B). As in *Pachyrhachis*, basipterygoid processes are absent in MIM F49. In contrast, *Haasiophis* possesses small alethinophidian-like basipterygoid processes. Both basisphenoid and basioccipital show clear signs of pachyostosis (Fig. 9.2B; Supplementary Figure 9.S22). The sutural contacts with parietal and prootic are not preserved in MIM F49, and the main foramina and ducts in the parabasisphenoid are obscured due to intense internal breakage (Supplementary Figure 9.S22).

9.3.3 Palatomaxillary Complex

Long, needle-like recurved teeth lacking plicidentine are preserved in the maxillae, dentaries, palatines, and pterygoids of *Pachyrhachis*, *Eupodophis*, and *Haasiophis*, and in the dentary of *Pachyophis* (tooth counts in Supplementary Table 9.S2). Tooth implantation is of the 'modified alethinophidian type' [9] (Figs. 9.1–9.5). Teeth are set in distinct sockets formed by prominent interdental ridges extending onto the lingual wall of the subdental shelf. The pleura is low, and the lingual dental ridge (or basal plate of [71]) is more horizontally directed and leveled with the pleura. In *Haasiophis* and *Pachyrhachis* the enamel is distinctly [35] to only slightly [61] striated, respectively. In *Eupodophis* and *Pachyophis* the enamel of the preserved teeth appears smooth (Figs. 9.1B and 9.5D,E). Lingual and labial ridges are clearly visible in the teeth of *Haasiophis*, *Eupodophis* (MSNM-V4014), and *Pachyrhachis*, but are not apparent in *Pachyophis* (Fig. 9.5). Previously mentioned furrowed teeth [72] in *Pachyrhachis* could not be confirmed.

Although visible in both specimens of *Pachyrhachis*, the **maxillae** are poorly exposed and difficult to delineate in detail (Fig. 9.3). The same is true for the two Milan specimens of *Eupodophis* (MSNM-V3661 and V4014) (Supplementary Figures 9.S5–9.S8). *Haasiophis* has better-preserved maxillae in both views, and the holotype of *Eupodophis* reveals an almost complete right maxilla that helps clarify important features (Figs. 9.1, 9.2, and 9.4). The maxilla of *Haasiophis* and *Eupodophis* projects posteriorly beyond the orbit. The palatine process of the maxilla of *Eupodophis* is broad with an oblique anterior sloping margin and a straight posterior margin. It underlies a similarly broad maxillary process of the palatine at the level of the anterior margin of the frontals. The maxilla is mainly straight except for a weak medial curvature anteriorly, with a smooth bluntly rounded tip indicating no sutural contact with the premaxilla. *Eupodophis* and *Pachyrhachis* share a robust maxilla and high ascending process firmly clasped by two prefrontal processes. Posterior to the ascending process, the maxilla decreases abruptly in height and forms a dorsoventrally flattened lamina that receives the flattened jugal on its dorsal surface, and possibly the ectopterygoid

(Figs. 9.2 and 9.3). A flattened posterior end is also visible in both maxillae of the referred specimen of *Pachyrhachis* (Fig. 9.3A). As in *Pachyrhachis* and *Eupodophis*, the maxilla in *Haasiophis* articulates with the prefrontal on its medial surface at the level of its ascending process (Fig. 9.4). However, in contrast to the former two, it is not tightly clasped laterally by the prefrontal. Although not visible, the presence of an ascending process is indicated by the medial concave surface of the left maxilla in ventral view, which receives the anterior base of the medially clasping prefrontal process (Fig. 9.4B). A well-preserved palatine process of the maxilla is visible ventrally on both maxillae, and loosely articulating with the maxillary process of the palatine (Fig. 9.4B).

A complete **ectopterygoid** with contacts preserved is unknown in any specimen of *Eupodophis* or *Pachyrhachis*. Two elongated, slender J-shaped elements were previously [61] tentatively identified as ectopterygoids in the referred specimen of *Pachyrhachis*, and a large flattened element exposed in dorsal view on the left side of the skull of the holotype was also labeled. These two elements are very distinct from each other and cannot be identified with certainty as ectopterygoids. A fragment of a putative ectopterygoid is preserved along the medial surface of the right maxilla in the holotype of *Eupodophis* (Fig. 9.1A). However, this element is also broken on both extremities and seems to be intimately connected to an anterior fragment of pterygoid instead of its more posteriorly located ectopterygoid process. In any case, the putative ectopterygoids in both *Pachyrhachis* and *Eupodophis* are very distinct from the well-preserved ectopterygoids of *Haasiophis* (Fig. 9.4), which have long flat processes that articulate on the dorsal surface of the posterior end of the maxillae, as in typical alethinophidians. Although not visible on the right ectopterygoid, the posterior end of the left ectopterygoid of *Haasiophis* clasps the well-developed laterally projected ectopterygoid process of the pterygoid (Fig. 9.4B; Supplementary Figure 9.S24). A similarly large ectopterygoid process of the pterygoid is also preserved in *Eupodophis*, suggesting a similar clasping condition. However, ectopterygoids are not sufficiently preserved in *Eupodophis* and *Pachyrhachis* to confirm this. Although poorly preserved, the pterygoids of *Pachyrhachis* appear to have large laterally projecting ectopterygoid processes, visible in ventral view (Fig. 9.3D).

The **palatines** and **pterygoids** articulate via a narrow overlapping contact, forming a highly mobile unit. The palatines of *Eupodophis* and *Pachyrhachis* share a narrow antero-medially curved choanal process that tapers anteriorly to a pointed tip without articulating with the vomers (Figs. 9.1A and 9.3D). *Haasiophis* has a blunt, poorly developed, medially projected choanal process similar to that in some booids (sensu [1]) (Fig. 9.4B). In *Eupodophis*, *Pachyrhachis*, and *Haasiophis*, the anterior dentigerous process of the palatine projects freely between the maxilla and the vomeronasal complex. Posteriorly, the pterygoid bears a slender rod-like quadrate process that reaches the quadrate (Figs. 9.1–9.4). A basipterygoid articulation on the pterygoid is lacking in all three genera.

9.3.4 The Circumorbital Bones

Prefrontals are well preserved in all three genera and tightly clasp the ascending process of the maxillae in *Pachyrhachis* and *Eupodophis*, with a medial process fitting in a deep recess in the medial surface of the maxilla and a lateral process covering the dorsolateral surface of the ascending process (Figs. 9.2 and 9.3). In *Haasiophis*, the medial process also fits in a deep recess while the lateral process projects anteriorly on the dorsal surface of the maxilla covering its ascending process (Fig. 9.4). The prefrontal articulates on the lateral surface of the frontal via a salient knob-like process. In *Haasiophis*, the prefrontal knob articulates with the frontal via a notch at the middle of the lateral margin of a long frontal (Fig. 9.4A), whereas in *Pachyrhachis* and *Eupodophis* the prefrontal knob covers most of the lateral surface of a very short frontal and articulates with the anterior and anterolateral margins of the parietal (Figs. 9.2A and 9.3C; Supplementary Figure 9.S22). Unlike the slender knob-like process of *Haasiophis* and *Pachyrhachis*, the prefrontal knob of *Eupodophis* is massive and expanded mediolaterally. The preserved portion of prefrontal of MSNM-V3661 in ventral view closely resembles the expanded prefrontal knob of the holotype of the species (Supplementary Figure 9.S23). *Haasiophis* preserves an enclosed lacrimal duct on the posterior surface of the prefrontal (Fig. 9.4A), a condition that seems to be also present in *Eupodophis* (Fig. 9.1A). A preserved posteromedial (inner) surface of the prefrontal is not visible in any specimen.

The distinctly curved, boomerang-shaped elements preserved at the level of the orbits in *Haasiophis* (Fig. 9.4A) were identified previously [35] as postorbitals. However, these elements are better interpreted as **jugals**, associated with the posterodorsal surface of the maxilla, as in *Pachyrhachis* and *Eupodophis*. The absence of postfrontals or postorbitals in *Haasiophis* fits more accurately the morphology of the parietals, with smooth and straight lateral margins. The curved jugals of *Haasiophis* resemble the varanid condition, attached to the posterior end of the maxilla and directed posterodorsally without closing the posterior margin of the orbit. They are very distinct from the peculiar jugals in *Eupodophis* and *Pachyrhachis*, which are dorsoventrally flat, straight, club-like laminae with an expanded proximal extremity that tapers distally to a narrow tip (Figs. 9.1–9.3).

The identification of a **postfrontal** or **postorbitofrontal** bone in *Pachyrhachis* and *Eupodophis* is controversial. The triradiate element at the level of the posterior margin of the orbit and contacting the jugal in *Pachyrhachis* has been identified as a postorbitofrontal [61] or a postfrontal ([66]; sensu [68]), and a postorbital has been identified in *Haasiophis* and *Eupodophis* [35, 46]. As discussed in detail elsewhere [51] (see Supplementary Appendix 9.S4, characters 73, 74, 77, and 170), the posterodorsal orbital element of snakes clasping the fronto-parietal suture corresponds to a postfrontal in both extinct and extant forms. A postorbital has not been confirmed in any known snake [70]. The presence of a composite postorbitofrontal bone [61] cannot be ruled out in the extinct taxa, but embryological evidence does not support that hypothesis for extant taxa.

We consider that *Haasiophis* lacks a postfrontal (see above) and we could not identify in the holotype of *Eupodophis* an element similar to the long and slender postfrontal present in MSNM-V3661 ('postorbital 'of [46]). Instead, a large antero-posteriorly expanded laminar element is preserved ventral to the tip of the alar process of the parietal in the holotype (Fig. 9.2). Its dorsal end is excavated in a facet that articulates with the ventral surface of the tip of the alar process. This element appears to be a postfrontal whose peculiar shape differs significantly from the postfrontal of MSNM-V3661, the first evidence that these taxa may not be conspecific. However, the extremities are poorly preserved, and its morphology is difficult to delineate. The holotype of *Pachyrhachis*, on the other hand, has well-preserved postfrontals ('postorbitofrontals' of [61]) with the right element preserving a blunt rod-like distal extremity that reached the jugal, probably forming a closed posterior orbit [61], and a notched expanded proximal extremity with two divergent processes that probably clasped the alar process of the parietal and reached the frontoparietal suture below the prefrontal knob. However, the postfrontals are displaced posteriorly and the exact relation between the posterior orbital bones are not preserved. The elements identified previously as frontals [61] in the referred specimen of *Pachyrhachis* are here identified as postfrontals (*sensu* [51]) with their distal rod-like and peculiar proximal notched expanded extremities resembling the right postfrontal exposed in the holotype's dorsal surface. Nevertheless, this identification is provisional because the elements are displaced and poorly preserved (Supplementary Figure 9.S25).

9.3.5 Suspensorium and Mandible

The **supratemporal** was described in *Pachyrhachis*, *Eupodophis*, and *Haasiophis* as typically macrostomatan, being a slender and elongate element syndesmostically attached to the skull table and projecting posteriorly beyond the edge of the skull in a free-ending posterior process [6–8, 61]. However, important differences occur among these genera. A free-ending process of the supratemporal is present in *Pachyrhachis* and *Eupodophis* but not in *Haasiophis* where the supratemporal is mainly a straight, slender lamina resting on top of the skull table without projecting beyond the otooccipital posteriorly (Fig. 9.4; Supplementary Figure 9.S25). The holotype of *Eupodophis* possesses a surprisingly robust supratemporal, transversely thick, rod-like, and bowed laterally (Figs. 9.1 and 9.2; Supplementary Figure 9.S22). It tapers to a point anteriorly but retains its thickness posteriorly, terminating bluntly. The supratemporal rests on concave facets on the dorsal surface of the parietal (Supplementary Figure 9.S22). Although broken in three pieces, the right supratemporal of *Pachyrhachis* holotype appears similar, being thick and bowed laterally (Supplementary Figure 9.S26). A thick rod-like posterior extremity is exposed ventrally (Fig. 9.4D).

Eupodophis and *Pachyrhachis* share broad subrectangular **quadrates** that lack suprastapedial processes. The subrectangular shape is more accentuated in *Pachyrhachis*, where it is feebly bowed posteriorly (Fig. 9.3D; Supplementary Figure 9.S26) (see also [65]). The holotype of *Eupodophis* preserves an almost complete left quadrate that is very similar to the right quadrate in specimen MSNM-V4014 (Fig. 9.1A; Supplementary Figure 9.S8). The quadrate articulates with the supratemporal via a broad, oblique cephalic condyle (Fig. 9.1A). The broad rectangular shaft narrows abruptly at its distal end. The posterior (medial) margin of the quadrate shaft has a stylohyal at mid-height to receive the long stapedial shaft. The quadrate in *Pachyrhachis* bears a slightly more ventral stylohyal [65: fig. 6].

The quadrate of *Haasiophis* is a slender, mainly straight element with a slightly expanded cephalic condyle lacking a suprastapedial process (Fig. 9.4). A stylohyal is not preserved. The cephalic condyle articulates on the posterolateral surface of the supratemporal.

Dentaries are preserved in *Haasiophis, Eupodophis, Pachyrhachis*, and *Pachyophis*, being complete only in *Haasiophis* and in *Eupodophis* (specimen MSNM-V4014). Tooth counts are given in Supplementary Table 9.S2. The dentary of *Pachyophis* is exposed only in medial view. Our observations agree mostly with a previous description [58], except that a mobile symphysis cannot be inferred because the anterior tip is not preserved (Fig. 9.5). A smooth, rounded tip is preserved in *Haasiophis, Eupodophis*, and *Pachyrhachis*, confirming loose contact between the hemimandibles. Multiple mental foramina occur in *Pachyrhachis* [61], but only a single foramen is present in *Haasiophis* (Fig. 9.4A; Supplementary Figure 9.S20). The number of foramina in *Eupodophis* was not reported previously, but the holotype has at least four in its right dentary. Mental foramina are not exposed in *Pachyophis*. Dentary tooth implantation in all four taxa is of the 'modified alethinophidian type' [9].

The **splenial** forms a large, anteriorly tapered triangular lamina in *Eupodophis, Pachyrhachis*, and *Haasiophis*, (Figs. 9.2B, 9.3B and 9.4B). The anterior mylohyoid foramen is fully enclosed (Fig. 9.4B; Supplementary Figure 9.S20). Posteriorly, the splenial forms a vertically expanded margin articulating with the angular on its ventral extremity in *Haasiophis* and *Eupodophis*. However, in *Haasiophis, Pachyrhachis*, and *Eupodophis*, the more dorsal and medial margins of the articulation between splenial and **angular** seem to be interlocked in a tongue in groove contact, as seen in a ventral view of the holotype of *Pachyrhachis* (Fig. 9.3D) and in HRCT and Synchrotron images of the holotypes of *Haasiophis* and *Eupodophis*, respectively (Supplementary Figure 9.S20). Nonetheless, a precise configuration of the splenial-angular articulation cannot be described due to poor preservation of this region in all three taxa.

It was suggested previously [58] that the holotype of *Pachyophis* retains a fragmentary portion of the lower jaw that preserves in place a vertically oriented intramandibular joint formed by contact between splenial and angular. We were unable to confirm this (Fig. 9.5). The enigmatic element identified previously as the right splenial (element B of [58]) is a thin lamina that does not expand distally to form a typical articular surface for the angular, while the putative angular (element A) seems to represent a broken part of the larger putative compound bone (element C). Although identification of these elements as corresponding to parts of the lower jaw cannot be discarded, we reject the evidence for a splenial and angular forming an intramandibular joint. Further preparation is needed in order to provide more compelling basis for the identification of these elements.

The **coronoid** is triradiate in *Haasiophis, Pachyrhachis* and *Eupodophis*, a plesiomorphic condition within snakes. The hypertrophied coronoid process in *Pachyrhachis* and *Eupodophis* is markedly distinct from *Haasiophis*. Although smaller, the *Haasiophis* coronoid is robust and retains a large concave surface on the posterodorsal face of the coronoid process for a presumably well-developed bodenaponeurosis. The *Eupodophis* holotype and specimen MSNM-V3661 share a distally tapering posterodorsally directed coronoid (Fig. 9.2A; Supplementary Figure 9.S6) that is distinct from the distally broad coronoid present in *Pachyrhachis* holotype and referred specimen (Fig. 9.3).

The **articular** is preserved in *Haasiophis, Eupodophis*, and *Pachyrhachis* and, differently from mosasaurians [73], forms the entire surface of articulation for the quadrate, having a simple saddle-shaped form. A compound bone is present in these three genera.

9.4 Pachyophiids as Derived Alethinophidian Snakes

In the last 20 years, pachyophiids have been subject to conflicting interpretations of their skeletal anatomy. This is due to a number of factors, including different styles of preservation at various localities and, to a lesser degree, collecting procedures and preparation which in some cases unavoidably damages the specimen, removing or altering the preserved morphology. None of the western and northern Tethyan specimens preserves a skull or pelvic and hindlimb elements (at least visible on the slab). The southern Tethyan specimens have mostly complete, but badly crushed skulls. Microscopic examination of *Haasiophis* and *Pachyrhachis* demonstrates the individual bones are highly fractured and subject to plastic deformation, obfuscating many important details. Specimens of *Eupodophis*, though better preserved, were damaged during collection by slab splitting, causing loss of many structures. The two specimens with skulls acquired by the Museum of Milan are each known only by one slab, the counterparts missing.

The three best-preserved specimens also include well-developed pelves and hind limbs, not known to occur in any extant snake. These plesiomorphic attributes conflicted with seemingly derived skull morphology, leading to the conclusion that either macrostomy evolved early in the total-group Serpentes, or that hind limbs and pelves re-evolved within crown Serpentes. Currently, the scenario of multiple independent reductions of hindlimbs in snakes, first suggested in 1999 [62], is now accepted as the most plausible scenario [37, 42, 43, 50, 68, 74–76].

With the advent of large-scale molecular phylogenies, several long-standing concepts about extant snake interrelationships were challenged and ultimately revised in view of increasing molecular evidence. The term 'Macrostomata' represents a flagrant example, being traditionally applied to a morphologically supported clade of alethinophidian snakes that has never been retrieved in any recent molecular analysis (Chapter 10). Although more investigation is needed to rule out definitively the notion of a clade Macrostomata, functional morphology shows that macrostomy likely represents a much more complex condition that has no relation with the origin of snake feeding [77] (see Chapter 19).

Recently [51], the affinities of pachyophiids were assessed with an expanded morphological and molecular dataset for 90 extinct and extant species of the well-supported [1] crown clade Toxicofera, comprising anguimorphs, iguanians and snakes. The combined matrix incorporates a molecular partition of 523,686 aligned nucleotides and a morphological partition of 785 morphological characters, totaling 524,471 characters (see Supplementary Appendix 9.S4 and [51]). Parsimony analysis resulted in the strict consensus tree shown in Figure 9.6 (complete consensus tree and list of synapomorphies for relevant clades are provided in Supplementary Appendix 9.S4).

In this analysis, total-group Serpentes is resolved as the sister group of a well-supported clade formed by Iguania and Anguimorpha. Although mainly reflecting recent molecular phylogenetic analyses, a close affinity between varanids and iguanians was previously noticed by Northcutt [78] who drew attention to at least six shared derived forebrain and midbrain characters supporting the group he named Dracomorpha and which included varanids, iguanians and teiids. We use here the term Dracomorpha for the crown clade formed by the most recent common ancestor of *Iguana* and *Varanus* and all its descendants. Mosasaurians are retrieved as dracomorphans, nested within Anguimorpha as the sister group of varanids (Platynota *sensu* Camp [23]). These results reinforce previous conclusions that the mosasaurian 'wide-gaped' condition evolved independently from alethinophidian snakes [8, 9, 62, 64].

The Cretaceous *Tetrapodophis* and *Coniophis* are recovered as the earliest diverging members of the total-group Serpentes, corroborating previous analyses [76, 79]. The Cretaceous *Najash* and *Dinilysia* are resolved as the sister group of crown-clade Serpentes [68, 69]. Pachyophiids and the Australian Cenozoic 'madtsoiids' are retrieved as alethinophidians, robustly nested within crown Serpentes (Fig. 9.6) [35, 43, 50, 68, 79].

Despite preservational limitations, pachyophiids share a surprisingly large number of alethinophidian characters, including: long nasals with ventrally descending laminae contacting the frontals in a movable (prokinetic) joint; medial frontal pillars present; reduced paroccipital process of the otooccipital; deep, narrowly U-shaped sella turcica recessed below well-defined dorsum sella; basipterygoid process of basisphenoid absent or small and non-synovial; palatine process of maxilla and maxillary process of palatine overlap in a non-interlocking contact; ectopterygoid articulates broadly with dorsal surface of posterior end of maxilla (confirmed only in *Haasiophis*); palatine and pterygoid articulate via narrow, overlapping contact; anterior dentigerous process and short choanal process of palatine not tightly attached to vomer; supratemporal syndesmotically attached to skull table and substantially (only in *Pachyrhachis* and *Eupodophis*) projected beyond edge of skull in free-ending posterior process; quadrate shaft bearing a stylohyal at mid-height.

However, retention of multiple plesiomorphic features in pachyophiids might support a resolution of the polytomy at the base of Alethinophidia with them as stem rather than crown alethinophidians. These include: prefrontals articulating with lateral margin of frontals; prominent ascending process of the maxillae; ophidiosphenoid absent; broad, laterally projecting ectopterygoid process of pterygoid; jugal present; multiple mental foramina (only in *Pachyrhachis* and *Eupodophis*); coronoid triradiate; vertebrae without prezygapophyseal processes; well-developed hindlimb and pelvis. Nonetheless, despite the presence of these plesiomorphic features, pachyophiids are not retrieved as stem Serpentes (Fig. 9.6).

9.5 Concluding Remarks

Pachyophiids are an early-diverging lineage of the total-clade Alethinophidia, with a combination of primitive and derived cranial and postcranial features unknown in any living snake. Medial frontal pillars are preserved only in *Eupodophis* and *Haasiophis*, but their presence reinforces the view that pachyophiids had a movable prokinetic joint. Although they seem to have developed a prokinetic snout, the prefrontals of pachyophiids plesiomorphically articulate on the lateral surface of the frontals via a salient knob-like articulation that likely prevented the degree of movement known to occur in extant 'macrostomatan' alethinophidians (Chapter 19). On the other hand, even limited snout

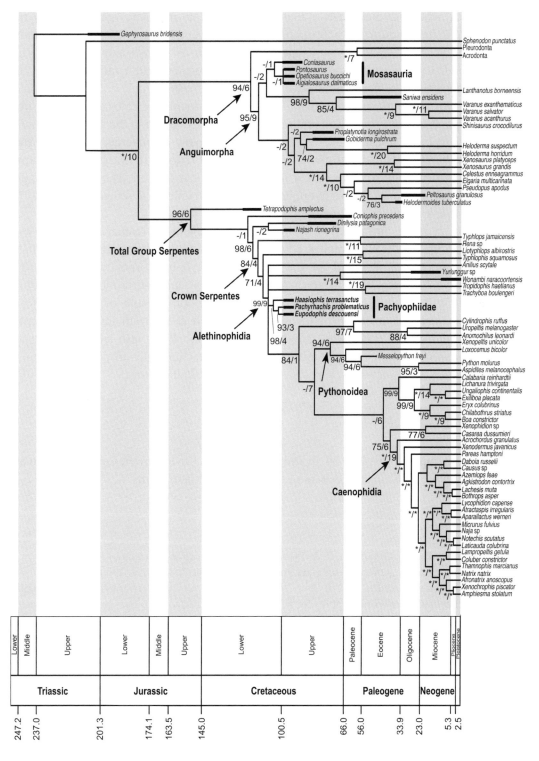

Figure 9.6 Temporally calibrated phylogenetic hypothesis of toxicoferan affinities, representing the consensus of four most parsimonious trees that resulted from Zaher and Smith's [51] analysis of the combined dataset (see Supplementary Appendix 9.S4). Timescale is in millions of years. Support values for internal branches are jackknife/Bremer, where values of 100% (jackknife) or >20 (Bremer) are indicated with an asterisk.

movements, when coupled with a highly kinetic palato-maxillary arch, suggests that pachyophiids had acquired synchronous unilateral upper jaw movements, a condition likely to be absent in macrophagous stem snakes such as *Najash* and *Dinilysia*, which retain broad, overlapping vomero-palatine and palato-pterygoid contacts, and in the microphagous extant scolecophidians with an immobile naso-frontal contact. The peculiar supratemporal and quadrate of *Pachyrhachis* and *Eupodophis* indicates a unique suspensorial morphology that provides an additional backward extension of the mandible, conferring a 'hyper-macrostomous' condition functionally convergent with extant alethinophidian booids and pythonoids [51]. *Haasiophis* lacks posteriorly projected supratemporals, which seem to have only a limited posterior extension. The peculiar 'hyper-macrostomous' nature of the pachyophiid suspensorial complex is not comparable to the condition in marine mosasaurians (contra [80]; see also Chapter 7).

Despite substantial advances in knowledge of pachyophiid cranial osteology, made possible through Synchrotron images, HRCT scans, and additional preparations of available specimens, important elements of the skull of pachyophiids remain unknown. The premaxilla, vomeronasal capsule and its connection with adjacent bones, ectopterygoid, postfrontal, and most of the posterior braincase and otic capsule of *Eupodophis* and *Pachyrhachis* are still unknown or incompletely described. Relevant information on the cranial morphology of *Pachyophis*, *Mesophis*, and *Simoliophis* is also lacking. We expect new fossil discoveries and rapidly advancing imaging technologies to provide additional information on these important aspects of pachyophiid skull anatomy.

Acknowledgements

See Supplementary Appendix 9.S6. In addition, we thank David Gower, Catherine Klein and Krister Smith for constructive reviews of the submitted manuscript.

References

1. F. T. Burbrink, F. G. Grazziotin, R. A. Pyron, et al., Interrogating genomic-scale data for Squamata (lizards, snakes, and amphisbaenians) shows no support for key traditional morphological relationships. *Systematic Biology*, 69 (2020), 502–520.

2. J. Dercourt, L. P. Zonenshain, L.- E. Ricou, et al., Geological evolution of the Tethys belt from the Atlantic to the Pamirs since the Lias. *Tectonophys*, 123 (1986), 241–315.

3. F. Berra and L. Angiolini, The evolution of the Tethys region throughout the Phanerozoic: A brief tectonic reconstruction. *American Association of Petroleum Geologists, Memoirs*, 106 (2014), 1–27.

4. M. J. Polcyn, E. Tchernov, and L. L. Jacobs, The Cretaceous biogeography of the eastern Mediterranean with a description of a new basal mosasauroid from 'Ein Yabrud, Israel. In Y. Tomida, T. H. Rich, and P. Vickers-Rich, eds., *Proceedings of the Second Gondwanan Dinosaur Symposium, National Science Museum Tokyo, Monographs* (1999), pp. 259–290.

5. M. W. Caldwell and M. S. Y. Lee, A snake with legs from the marine Cretaceous of the Middle East. *Nature*, 386 (1997), 705–709.

6. J.-C. Rage and F. Escuillié, Un nouveau serpent bipède du Cénomanien (Crétacé). Implications phylétiques. *Comptes Rendus de l'Academie des Sciences, Paris, Sciences de la Terre et des Planètes*, 330 (2000), 513–520.

7. E. Tchernov, O. Rieppel, H. Zaher, M. J. Polcyn, and L. L. Jacobs, A fossil snake with limbs. *Science*, 287 (2000), 2010–2012.

8. H. Zaher, The phylogenetic position of *Pachyrhachis* within snakes (Squamata, Lepidosauria). *Journal of Vertebrate Paleontology*, 18 (1998), 1–3.

9. H. Zaher and O. Rieppel, Tooth implantation and replacement in squamates, with special reference to mosasaur lizards and snakes. *American Museum Novitates*, 3271 (1999), 1–19.

10. J.-C. Rage and F. Escuillié, The Cenomanian: stage of hindlimbed snakes. *Carnets de Geologie*, 2003/01 (2003), 1–11.

11. J. D. Scanlon and M. S. Y. Lee, The Pleistocene serpent *Wonambi* and the early evolution of snakes. *Nature*, 403 (2000), 416–420.

12. J. D. Scanlon, Skull of the large non-macrostomatan snake *Yurlunggur* from the Australian Oligo-Miocene. *Nature*, 439 (2006), 839–842.

13. N. Vidal and S. B. Hedges, Molecular evidence for a terrestrial origin of snakes. *Proceedings of the Royal Society of London B*, 271 (2004), S226–S229.

14. S. Apesteguía and H. Zaher, A Cretaceous terrestrial snake with robust hindlimbs and a sacrum. *Nature*, 440 (2006), 1037–1040.

15. N. Bardet, A. Houssaye, J.-C. Rage, and X. Pereda Suberbiola, The Cenomanian-Turonian (late Cretaceous) radiation of marine squamates (Reptilia): the role of the Mediterranean Tethys. *Bulletin de la Société Géologique de France*, 179 (2008), 605–622.

16. H. E. Sauvage, Sur l'existence d'un reptile du type ophidien dans les couches à Ostrea columba des Charentes. *Comptes Rendus de la Société de l'Academie des Sciences, Paris*, 91 (1880), 671–672.

17. F. Nopcsa, *Eidolosaurus* und *Pachyophis*, zwei neue Neocom-Reptilien. *Palaeontograpjica*, 65 (1923), 99–154.

18. F. Nopcsa, Ergebnisse der Forschungsreisen Prof. E. Stromers in den Wüsten Ägyptens. II. Wirbeltier-Reste der Baharjie-Stufe (unterstes Cenoman). *Abhandlungen der Bayerische Akademie der Wissenschaftern, Mathematische-naturwissenschaftliche Abteilung*, 30 (1925), 5–27.

19. S. J. Bolkay, *Mesophis nopcsai* ngn sp. ein neues, schlangenähnliches reptil aus der unteren Kreide (Neocom) von Bilek-Selista (Ost-Hercegovina). *Glasnik zemaljskog Muzeja u Bosni i Hercegovini*, 37 (1925), 125–135.

20. R. Owen, *Palaeontology, or a systematic summary of extinct animals and their geological relations* (Edinburgh: Adam and Charlos Black, 1860).

21. A.-T. d. Rochebrune, Revision des ophidiens fossiles du Muséum d'Histoire naturelle. *Nouvelles Archives du Muséum d'Histoire Naturelle*, 2éme Série, 3 (1880), 271–296.

22. E. D. Cope, On the reptilian orders, Pythonomorpha and Streptosauria. *Proceedings of the Boston Society of Natural History*, 12 (1869), 250–266.

23. C. L. Camp, Classification of the lizards. *Bulletin of the American Museum of Natural History*, 48 (1923), 289–481.

24. D. A. Russell, Systematics and morphololgy of American mosasaurs (Reptilia, Sauria). *Peabody Museum of Natural History Bulletin*, 23 (1967), 1–241.

25. H. G. Seeley, On remains of a small lizard from the Neocomian rocks of Comén, near Trieste, preserved in the Geological Museum of the University of Vienna. *Quarterly Journal of the Geological Society of London*, 37 (1881), 52–56.

26. A. Kornhuber, Über einen neuen fossilen Saurier aus Lesina. *Abhandlungen der kaiserlich-königlichen geologischen Reichsanstalt*, 5 (1873), 75–90.

27. A. Kornhuber, *Carsosaurus Marchesettii*, ein neuer fossiler Lacertilier aus den Kreideschichten des Karstes bei Komen. *Abhandlungen der kaiserlich-königlichen geologischen Reichsanstalt*, 17 (1893), 1–15.

28. A. Kornhuber, *Opetiosaurus Bucchichi*, eine neue fossile Eidechse aus der unteren Kreide von Lesina in Dalmatien. *Abhandlungen der kaiserlich-königlichen geologischen Reichsanstalt*, 17 (1901), 1–24.

29. K. Gorjanović-Kramberger, *Aigialosaurus*, eine neue Eidechse aus den Kreideschiefern der Insel Lesina, mit Rücksicht auf die bereits beschriebenen Lacertiden von Comen und Lesina. *Glasnik hrvatskoga naravoslovnoga drustva u Zagrebu*, 7 (1892), 74–106.

30. F. Nopcsa, Über die varanusartigen lacerten Istriens. *Beiträge zur Paläontologie und Geologie Österreich-Ungarns und des Orients*, 15 (1903), 31–42.

31. G. A. Boulenger, Notes on the Osteology of *Heloderma horridum* and *H. suspectum*, with Remarks on the Systematic Position of the Helodermatidæ and on the Vertebræ of the Lacertilia. *Proceedings of the zoological Society of London*, 59 (1891), 109–118.

32. L. Dollo, Les ancêtres des mosasauriens. *Bulletin Scientifique de la France et de la Belgique*, 38 (1903), 137–139.

33. S. B. McDowell and C. M. Bogert, The systematic position of *Lanthanotus* and the affinities of the anguinomorphan lizards. *Bulletin of the American Museum of Natural History*, 105 (1954), 1–142.

34. M. S. Y. Lee, The phylogeny of varanoid lizards and the affinities of snakes. *Philosophical Transactions of the Royal Society of London B*, 352 (1997), 53–91.

35. O. Rieppel, H. Zaher, E. Tchernov, and M. J. Polcyn, The anatomy and relationships of *Haasiophis terrasanctus*, a fossil snake with well-developed hind limbs from the Mid-Cretaceous of the Middle East. *Journal of Paleontology*, 77 (2003), 536–558.

36. A. Palci and M. W. Caldwell, Redescription of *Acteosaurus tommasinii* von Meyer, 1860, and a discussion of evolutionary trends within the clade Ophidiomorpha. *Journal of Vertebrate Paleontology*, 30 (2010), 94–108.

37. F. F. Garberoglio, S. Apesteguia, T. Simoes, et al., New skulls and skeletons of the Cretaceous legged snake *Najash*, and the evolution of the modern snake body plan. *Science Advances*, 5 (2019), eaax5833.

38. F. F. Garberoglio, R. O. Gómez, T. R. Simões, M. W. Caldwell, and S. Apesteguía, The evolution of the axial skeleton intercentrum system in snakes revealed by new data from the Cretaceous snakes *Dinilysia* and *Najash*. *Scientific Reports*, 9 (2019), 1–10.

39. M. W. Caldwell, *The Origin of Snakes: Morphology and the Fossil Record* (Boca Raton: CRC Press, 2020).

40. J. L. Conrad, Phylogeny and systematics of Squamata (Reptilia) based on morphology. *Bulletin of the American Museum of Natural History*, 310 (2008), 1–182.

41. S. E. Evans, A new anguimorph lizard from the Jurassic and Lower Cretaceous of England. *Palaeontology*, 37 (1994), 33–49.

42. T. W. Reeder, T. M. Townsend, D. G. Mulcahy, et al., Integrated analyses resolve conflicts over squamate reptile phylogeny and reveal unexpected placements for fossil taxa. *PLoS ONE*, 10 (2015), e0118199.

43. A. Y. Hsiang, D. J. Field, T. H. Webster, et al., The origin of snakes: revealing the ecology, behavior, and evolutionary history of early snakes using genomics, phenomics, and the fossil record. *BMC Evolutionary Biology*, 15 (2015), 87.

44. M. W. Caldwell, Snake phylogeny, origins, and evolution: the role, impact, and importance of fossils (1869–2006). In J. S. Anderson and Sues, H.-D., ed., *Evolutionary Transitions and Origins of Major Groups of Vertebrates* (Bloomington, Indiana: Indiana University Press, 2007), pp. 253–302.

45. I. Paparella, A. Palci, U. Nicosia, and M. W. Caldwell, A new fossil marine lizard with soft tissues from the Late Cretaceous of southern Italy. *Royal Society Open Science*, 5 (2018), 172411.

46. O. Rieppel and J. J. Head, New specimens of the fossil snake genus *Eupodophis* Rage & Escuillié, from Cenomanian (Late Cretaceous) of Lebanon. *Memorie Soc Italiana Sci Nat*, 32 (2004), 1–26.

47. J.-C. Rage and D. Néraudeau, A new pachyostotic squamate reptile from the Cenomanian of France. *Palaeontology*, 47 (2004), 1195–1210.

48. A. Houssaye, 'Pachyostosis' in aquatic amniotes: a review. *Integrative Zoology*, 4 (2009), 325–340.

49. M. J. Polcyn, L. L. Jacobs, R. Araújo, A. S. Schulp, and O. Mateus, Physical drivers of mosasaur evolution. *Palaeogeography, Palaeoclimatology, Palaeoecology*, 400 (2014), 17–27.

50. J. Gauthier, M. Kearney, J. A. Maisano, O. Rieppel, and A. D. B. Behlke, Assembling the squamate tree of life: Perspectives from the phenotype and the fossil record. *Bulletin of the Peabody Museum of Natural History*, 53 (2012), 3–308.

51. H. Zaher and K. T. Smith, Pythons in the Eocene of Europe reveal a much older divergence of the group in sympatry with boas. *Biology Letters*, 16 (2020), 20200735.

52. A. Albino, J. D. Carrillo-Briceño, and J. M. Neenan, An enigmatic aquatic snake from the Cenomanian of Northern South America. *PeerJ*, 4 (2016), e2027.

53. J. C. Rage, R. Vullo, and D. Néraudeau, The mid-Cretaceous snake *Simoliophis rochebrunei* Sauvage, 1880 (Squamata: Ophidia) from its type area (Charentes, southwestern France): Redescription, distribution, and palaeoecology. *Cretaceous Research*, 58 (2016), 234–253.

54. L. A. Nessov, V. I. Zhegallo, and A. O. Averianov, A new locality of Late Cretaceous snakes, mammals and other vertebrates in Africa (western Libya). *Annales de Paléontologie*, 84 (1998), 265–274.

55. G. Haas, On a new snakelike reptile from the lower Cenomanian of 'Ein Yabrud, near Jerusalem. *Bulletin du Museum National d'Histoire Naturelle, Paris, Série* 4, 1 (1979), 51–64.

56. G. Haas, *Pachyrhachis problematicus* Haas, snakelike Reptile from the lower Cenomanian: ventral view of the skull.

Bulletin du Museum National d'Histoire Naturelle, Paris, Série 2 (1980), 87–104.

57. G. Haas, Remarks on a new ophiomorph reptile from the Lower Cenomanian of Ein Jabrud, Israel. In L. L. Jacobs, ed., *Aspects of Vertebrate History, in Honor of E H Colbert* (Flagstaff, Arizona: Museum of Northern Arizona Press, 1980), pp. 177–92.

58. M. S. Y. Lee, M. W. Caldwell, and J. D. Scanlon, A second primitive marine snake: *Pachyophis woodwardi* from the Cretaceous of Bosnia-Herzegovina. *Journal of Zoology*, 248 (1999), 509–520.

59. A. Houssaye, Rediscovery and description of the second specimen of the hind-limbed snake *Pachyophis woodwardi* Nopcsa, 1923 (Squamata, Ophidia) from the Cenomanian of Bosnia Herzegovina. *Journal of Vertebrate Paleontology*, 30 (2011), 276–279.

60. D. Đurić, D. Radosavljević, D. Petrović, M. Radonjić, and P. Vojnović, A new evidence for pachyostotic snake from Turonian of Bosnia-Herzegovina. *Geoloski anali Balkanskoga poluostrva*, (2017), 17–21.

61. M. S. Y. Lee and M. W. Caldwell, Anatomy and relationships of *Pachyrhachis problematicus*, a primitive snake with hindlimbs. *Philosophical Transactions of the Royal Society of London B*, 353 (1998), 1521–1552.

62. H. Zaher and O. Rieppel, The phylogenetic relationships of *Pachyrhachis problematicus*, and the evolution of limblessness in snakes (Lepidosauria, Squamata). Comptes Rendus de Séances de l'Académie des Sciences (Série IIA), *Earth and Planetary Science*, 329 (1999), 831–837.

63. H. Zaher and O. Rieppel, On the phylogenetic relationships of the Cretaceous snakes with legs, with special reference to *Pachyrhachis problematicus* (Squamata, Serpentes). *Journal of Vertebrate Paleontology*, 22 (2002), 104–109.

64. O. Rieppel and H. Zaher, The intramandibular joint in squamates, and the phylogenetic relationships of the fossil snake *Pachyrhachis problematicus* Haas. *Fieldiana Geology*, 43 (2000), 1–69.

65. M. J. Polcyn, L. L. Jacobs, and A. Haber, A morphological model and CT assessment of the skull of *Pachyrhachis problematicus* (Squamata, Serpentes), a 98 million year old snake with legs from the Middle East. *Palaeontologica Electronica*, 8 (2005), 1–24.

66. A. Palci, M. W. Caldwell, and R. L. Nydam, Reevaluation of the anatomy of the Cenomanian (Upper Cretaceous) hind-limbed marine fossil snakes *Pachyrhachis*, *Haasiophis*, and *Eupodophis*. *Journal of Vertebrate Paleontology*, 33 (2013), 1328–1342.

67. D. Cundall and F. J. Irish, The snake skull. In C. Gans, A. S. Gaunt and K. Adler, eds., *Biology of the Reptilia*, Vol. 20, *Morphology H, The Skull of Lepidosauria* (Ithaca, New York: Society for the Study of Amphibians and Reptiles, 2008), pp. 349–692.

68. H. Zaher and C. A. Scanferla, The skull of the Upper Cretaceous snake *Dinilysia patagonica* Smith-Woodward, 1901, and its phylogenetic position revisited. *Zoological Journal of the Linnean Society*, 164 (2012), 194–238.

69. H. Zaher, S. Apesteguía, and C. A. Scanferla, The anatomy of the Upper Cretaceous snake *Najash rionegrina* Apesteguía & Zaher, 2006, and the evolution of limblessness in snakes. *Zoological Journal of the Linnean Society*, 156 (2009), 801–826.

70. S. B. McDowell, The skull of Serpentes. In C. Gans, A. S. Gaunt, and K. Adler, eds., *Biology of the Reptilia*, Vol. 21, *Morphology I, The Skull and Appendicular Locomotor Apparatus of Lepidosauria* (Ithaca, New York: Society for the Study of Amphibians and Reptiles, 2008), pp. 467–620.

71. M. H. Lessmann, Zur labialen Pleurodontie an Lacertilier-Gebissen. *Anatomischer Anzeiger*, 99 (1952), 35–67.

72. E. Kochva, The origin of snakes and evolution of the venom apparatus. *Toxicon*, 25 (1987), 65–106.

73. M. deBraga and R. L. Carroll, The origin of mosasaurs as a model of macroevolutionary patterns and processes. *Evolutionary Biology*, 27 (1993), 245–322.

74. S. Apesteguia and H. Zaher, A Cretaceous terrestrial snake with robust hindlimbs and a sacrum. *Nature*, 440 (2006), 1037–1040.

75. S. M. Harrington and T. W. Reeder, Phylogenetic inference and divergence dating of snakes using molecules, morphology and fossils: new insights into convergent evolution of feeding morphology and limb reduction. *Biological Journal of the Linnean Society*, 121 (2017), 379–394.

76. D. M. Martill, H. Tischlinger, and N. R. Longrich, A four-legged snake from the Early Cretaceous of Gondwana. *Nature*, 349 (2015), 416–419.

77. N. J. Kley, Prey transport mechanisms in blindsnakes and the evolution of unilateral feeding systems in snakes. *American Zoologist*, 41 (2001), 1321–1337.

78. R. G. Northcutt. Forebrain and midbrain organization in lizards and its phylogenetic significance. In N. Greenberg and P. D. Maclean, eds., *Behavior and Neurology of Lizards* (Rockville: National Institute of Mental Health, 1978), pp. 11–64.

79. N. R. Longrich, B.-A. S. Bhullar, and J. A. Gauthier, A transitional snake from the Late Cretaceous Period of North America. *Nature*, 488 (2012), 205–208.

80. M. S. Y. Lee, G. L. Bell, and M. W. Caldwell, The origin of snake feeding. *Nature*, 400 (1999), 655–659.

PART III

Genomic Perspectives

Using Comparative Genomics to Resolve the Origin and Early Evolution of Snakes

SARA RUANE AND JEFFREY W. STREICHER

10.1 Introduction

The use of molecular characters for phylogenetic analysis revolutionized the way biologists reconstruct evolutionary histories. This began with biochemical comparisons of whole molecules and rapidly developed into the direct determination of nucleic acid sequences during the late twentieth century. Advancements in DNA sequencing technology during the early twenty-first century ushered in the reality of sequencing whole nuclear genomes from non-model groups, including snakes [1]. An exciting application of comparing whole genomes is the insight they provide into the origin and early evolution of ancient radiations. Despite this, many practical questions remain unanswered for how these comparisons are best made and to what extent whole nuclear genomes will be advantageous for phylogenetic inference.

The nuclear genomes of animals are often large, multidimensional, plastic and filled with evolutionary histories arising from both deterministic (e.g., selection) and stochastic (e.g., genetic drift, incomplete lineage sorting, duplication events) processes. Notably, localized genomic histories are often conflicting with respect to the overall species history, a quality referred to as gene tree–species tree discordance [2]. Analytical methods have been developed to deal with this issue but are exclusively for use with aligned, homologous DNA sequences (e.g., [3]). While these methods are presumed effective at resolving phylogenies despite the 'noise' of gene-tree discordance, emerging evidence suggests some relationships remain unclear, even when using hundreds or thousands of nuclear genes. This is particularly notable in snakes, where several ancient bifurcations remain uncompellingly resolved despite large amounts of nuclear data having been applied to phylogenetic inference [4, 5]. There are at least two possible explanations: First, the majority of these 'unresolved branches' are short and occur deep in the snake evolutionary tree, so it is possible that they correspond to ancient, rapid radiations that are best explained by polytomies and not the bifurcating model assumed by tree-based analyses [6]. Second, moderate rates of substitution may cause substantial homoplasy rendering molecular sequence data ineffective for resolving short bifurcations that occurred in deep time [7]. Given the second possibility, that sequence data may not be able to resolve short yet real bifurcations, there is the need to reexamine comparative frameworks for extracting phylogenetic signal from nuclear genomes outside of DNA, RNA, and amino acid sequences [8].

In this chapter we review the history of molecule-based phylogenetic analysis in snakes leading up to the contemporary genomics revolution. We then explore how future phylogenetic inference might be made via comparisons of genome characteristics in addition to, or instead of, aligned DNA, RNA, and amino acid sequences. First, we begin with a discussion of the types of characters that nuclear genomes offer systematists, and then speculate how existing analytical approaches from systematic biology might be applied to comparative genomics (sensu [9]). Using available snake genomes, we provide two examples of coding nuclear genome qualities as polymorphic multistate characters. We then discuss perceived opportunities and challenges associated with this approach for inferring phylogenies. Finally, we conclude with a review of contentious and unresolved relationships among snakes that are likely to be discussed during the comparative genomics era. Throughout, we follow the taxonomy presented in Burbrink et al. (2020) [5] unless otherwise noted. To aid with interpretation, bolded terms are further defined in the Glossary (Box 10.1).

10.2 The Evolution of Molecular Phylogenetics for Snakes

Although contemporary herpetologists may think of molecular methods for phylogenetic estimation as originating in the 1980s–1990s and becoming standard over the past two decades, the use of molecular data to infer taxonomic relationships for snakes dates back to at least the 1930s (Fig. 10.1; [10–12]). These early studies recognized that due to limitations of the fossil record and a 'sparsity of taxonomic characters' molecular methods could potentially provide an unbiased and quantitative method to inform snake systematics [13]. Unlike today, these studies did not directly use DNA sequences. For many of these early experiments, **serological reactions** (also known as **immunological correspondence**) were used. These procedures use blood sera from snakes, which contains antigens, substances that can create an immune response. The antiserum, a serum with known antibodies in it, is then prepared by injecting the snake sera into a rabbit, causing the rabbit to create antibodies to the antigens of that specific snake species. Blood is then drawn from the rabbit and spun down with the clear serum decanted; this is the antiserum. When a snake serum is then mixed with an antiserum, agglutination occurs if an antigen found in the snake blood is able to bind with antibodies found in the antiserum (meaning that if snake species A serum is mixed with the antiserum prepared using snake species A, a high degree of agglutination is expected because the antiserum should have the antibodies most

Box 10.1   Glossary of terms used in this chapter

Allozyme and Isozyme electrophoresis: An electrophoretic method for determining the degree of relatedness of taxa by separating out proteins (enzymes) of varying molecular weights and/or electrical charges by running them out on a gel; the different proteins (or alleles for a specific protein) have different rates of migration when subjected to electrical charges. This allows for the detection of genetic variation (allele frequencies) between species and/or within species and can be used to determine relatedness of taxa.

Cumulative and non-cumulative frequencies: Non-cumulative frequencies are essentially non-ordered character states (represented by frequencies in Matrix **a**). Cumulative frequencies treat multistate characters as ordered, so that the first category (= frequency bin) of Matrix **b** is equivalent to the frequency of individuals that have character values greater than the first frequency bin of Matrix **a**.

DIRS elements: DIRs1-like elements (*Dictyostelium* Intermediate Repeat sequence 1) are a category of tyrosinase recombinase retrotransposon TEs.

DNA-hybridization: A method for determining the degree of relatedness of taxa by taking the DNA from one species 'A' and mixing it with the DNA from another species 'B'; the double stranded DNA of each species is heated to allow the DNA strands to separate and then cooled together to subsequently anneal ('hybridize') between the two species when reforming the double stranded DNA. It is assumed that the more similar the sequences are to each other from the two species, the better they will bind together and thus require higher levels of energy (melting temperature) to be pulled apart when compared to less similar (less closely related) sequences.

DNA-transposons: These TEs differ from LTRs, LINEs and SINES in that they move using a DNA intermediate instead of an RNA intermediate.

Endogenous viral elements: DNA sequences that are derived from viruses. Can result from any virus that uses a host genome to replicate, including retroviruses, parvoviruses, bornaviruses, and circoviruses.

Gene ontology (GO) inference: Method for determining a gene's structure and function so that both paralogous (closely-related) genes can be identified within genomes, and homologous genes can be identified across species.

LINE elements: Long interspersed nucleotide elements are a class of retrotransposons that are fairly common in snake genomes (and other squamate reptile genomes).

LTR elements: Long terminal repeats (LTR) are sequences found at the terminus of retrotransposons. These sites are also used by some viruses to insert into host genomes.

PLE elements: Penelope-like elements (PLEs) are a category of retrotransposon TEs that are distinguished by specialized endonuclease domains and the ability to retain introns.

Restriction-site Associated (RAD) datasets: A DNA sequence dataset composed of hundreds or (usually) thousands of short loci (50–200 base pairs) with single nucleotide polyomorphisms (SNPs) that can be used for phylogenetic inference. Datasets are generated by using one or more enzymes to digest the nuclear genome of an organism at specific 'cut' sites, the resulting genomic fragments are sorted by size, and sequenced with the aim of generating a set of loci that contain SNPs for a particular species. Commonly used examples for snake phylogenetics include ddRAD-Seq, RAD-Seq, and genotyping-by-sequencing (GBS).

Serological reactions or Immunological correspondence: A method for determining the degree of relatedness of taxa by quantifying the degree of agglutination between the blood sera of a species 'A' when exposed to the antibodies (via an antiserum) of another species 'B'. It is assumed that the more closely related the taxa are to one another, the greater the level of agglutination observed, due to the ability for the blood sera antigens of species A to bind with the antisera antibodies of species B.

Box 10.1   (cont.)

Sex chromosome condition: Snakes have variable conditions of their sex chromosomes. In some species males, as in humans, are the heterogametic sex (= XY chromosomes); however, in most species females are the heterogametic sex (= ZW chromosomes). There are also some snakes with undifferentiated (in terms of size) sex chromosomes (= homomorphic).

Simple sequence repeats (SSRs): Also known as microsatellites, these are tandem repeats of di, tri- or tetra- nucleotides that appear throughout the nuclear genome. Often used in population genetics.

SINE elements: Short interspersed nuclear elements are a category of retrotransposon TEs that are non-autonomous and typically require an RNA intermediate to amplify. Non-autonomous TEs may be ideal phylogenetic markers.

Target-capture or Sequence-capture datasets: A DNA sequence dataset composed of hundreds to thousands of short (ca. 300 base pairs) to long (ca. 1,000 base pairs) loci generated by preparing libraries from fragmented genomic DNA and using probes to target specific regions of the genome for subsequent sequencing, often conserved portions such as exons. Commonly used examples for snake phylogenetics include ultra-conserved elements (UCEs), anchored-hybrid enrichment loci (AHEs), and the squamate conserved loci set (SqCL)

Transposable elements (TEs): Also known as 'jumping genes', this is an all-encompassing term for mobile genetic elements that can change their position in a genome. Sometimes the movement of TEs changes genome size or creates or reverses mutations.

specific for snake species A to bind to the antigens). When the blood of species A is mixed with antiserum made from species B, the degree of agglutination observed would be indicative of how similar the protein composition of the two snakes is and thus their degree of relatedness can be inferred.

Although these early studies typically lacked taxonomic breadth, they often inferred the same relationships as modern phylogenetic studies of snakes. For example, using antisera reactions, alligators and turtles were determined to be serologically closer to each other than they were to squamates [14], a finding that has been corroborated in recent studies that place testudines as the sister taxon to archosaurs [15, 16]. The same study [14] also found that within snakes, the elapids *Lapemis* and *Naja* were more closely related to each other than either is to *Coluber* or *Crotalus* and that within crotalines, *Sistrurus* and *Crotalus* were closer to one another than to *Agkistrodon*. In another study based on serological reactions [17], snakes in the tribe Lampropeltini (specifically *Pantherophis*, *Pituophis*, *Rhinocheilus*, and *Lampropeltis*) were found to be more closely related to each other than to a group of colubrines formed by *Coluber*, *Masticophis*, and *Opheodrys,* and these two groups of colubrids were more closely related to each other than to a group composed of the dipsadids *Diadophis*, *Heterodon*, and *Farancia* or to a natricid group of *Thamnophis* and *Nerodia.* Another early molecular method used for phylogenetic inference is subjecting body fluids (blood or fluids derived from blood) to electrophoresis and using migration rates and/or bands on the gel to determine relationships among taxa. An analysis of haemoglobin solutions from reptiles and amphibians, including 41 species of snakes [18],

Figure 10.1 Early contributors to the understanding that molecules extracted from snake tissues or cells have phylogenetic signal. (A) Fannie Eleanor Williams (1884–1963), Walter and Eliza Institute of Medical Research, Melbourne, Australia. Image reprinted from the public domain. (B) Charles Halliley Kellaway (1889–1952), Walter and Eliza Institute of Medical Research, Melbourne, Australia. Image reprinted from the Wellcome Collection under a Creative Commons Attribution license. (C) Shou-Hsian Mao (1922–2008), National Defense Medical Center, Taiwan, China. Image reprinted from [98] with permission of the Society for the Study of Amphibians and Reptiles (SSAR). (D) Garth L. Underwood (1919–2002), University of the West Indies, Jamaica and Trinidad; City of London Polytechnic, UK; The Natural History Museum, London, UK; Image reprinted from [99] with permission of the SSAR. (E) Herbert C. Dessauer (1921–2003), Louisiana State University, United States. Image reprinted from [100] with permission of the American Society of Ichthyologists and Herpetologists. (F) Robin Lawson (1933–2013), California Academy of Sciences, United States. Image courtesy of Robert Drewes.

showed that Boidae (*sensu latu*), Colubridae (*sensu latu*) and Crotalidae (*sensu latu*) formed groups that could be distinguished from each other. This early molecular work demonstrates that phylogenetic signal is clearly present at the scale of entire molecules and largely consistent with contemporary phylogenies.

In his paper *Molecular approaches to the taxonomy of colubrid snakes* Dessauer (Fig. 10.1) reviewed available molecular methods that could be applied to snake phylogenetics, including serological reactions as well as **DNA-hybridization** and high-resolution **electrophoresis** [19].

He highlighted that previous molecular studies that indicate a close relationship between snakes and some lizard families (iguanids specifically), that Colubridae (*sensu latu*) is closely related to elapids and crotalids, and that both Old and New World natricids form a group separate from a list of snakes currently classified as colubrids and dipsadids. George and Dessauer [13] went further, using these techniques to test three previously proposed specific taxonomies, based on morphological characters for 'colubrid snakes' [20, 21]. The molecular results closely aligned with Underwood's (Fig. 10.1; [21]) work which classified *Coluber* and *Pituophis* as colubrines within Colubridae, *Heterodon* as a xenodontine within Dipsadidae, and considered Homalopsidae, Natricidae and Acrochordidae as individual families separate from Colubridae, consistent with contemporary molecular phylogenies (e.g., [5]). As evidenced by intraspecific studies, early molecular methods were not limited to higher-level relationships but were also used to better understand taxonomy and biogeography (e.g., natricids [22]; sea snakes [23]; ratsnakes [24]).

Molecular work in snake systematics continued throughout the 1980s and early 1990s, with many researchers using immunological reactions and with **allozyme-based** studies that were popular in the early 1990s; these studies further corroborated earlier molecular studies on snake systematics and often included phylogenetic trees based on distance methods or parsimony (e.g., [25–27]). At this point, using molecules to infer snake phylogeny was no longer a rarity, as evidenced by Cadle's seminal monograph [28] *Phylogenetic relationships among advanced snakes, a molecular perspective*. This work reviewed prior studies and provided new molecular evidence for some now well-accepted relationships among snakes, namely that viperids and elapids are not sister taxa despite having front-fanged venom delivery systems, that *Atractaspis* is not a viperid and is more closely related to elapids, and that viperids are the sister taxon of the remaining Colubriformes (an assortment of dipsadids, colubrids, elapids, and *Atractaspis* in [28]).

Starting in the mid-1990s, direct DNA-sequencing began replacing immunological and allozyme-based studies for phylogenetic inference due to its relative ease and lowered costs. The first 'markers of choice' for DNA sequencing originated from the mitochondrial genome (mtDNA). Mitochondrial genes have several properties that make them useful for phylogenetics, namely high mutation rates, low effective population sizes via haploid inheritance, and an ease of sequencing due to high copy numbers in tissues such as muscle and liver. Additionally, as more studies used the same mtDNA markers, the ability to combine new sequences with previously published datasets increased samples sizes and taxonomic breadth without financial cost. This likely led to the self-perpetuation and continuance of these studies within herpetology, including population genetics, phylogeography, and deep-node resolution. Studies of mtDNA helped to identify cryptic species [29], determine familial and interfamilial relationships [30, 31], and test long-standing hypotheses pertaining to species limits [32]. By the mid-2000s it became clear that while extremely useful, mtDNA (and other single locus) studies were potentially limited in scope because single-gene trees may differ from the species tree. Thus, systematic studies would be better served by multilocus approaches including a variety of molecular markers [33, 34]. This led to an increase in snake studies that typically included mtDNA data along with anywhere from one to over a dozen nuclear loci [35–38].

Currently, sequencing and bioinformatics capabilities allow snake researchers to take genome-scale approaches to systematic studies (i.e., hundreds to thousands of nuclear loci; see [39] for a thorough review of these data types). These studies include **restriction-site associated DNA** datasets (RAD-seq), which result in thousands of single-nucleotide polymorphisms (SNPs) that can be used for population or phylogeographic studies. Although SNP datasets can also be used for evolutionary comparisons across taxonomic levels, **target capture** datasets comprised of longer markers (often ca. 300–1,000 base pairs in length) tend to be more useful for examining relationships at deeper levels; these types of datasets include ultraconserved elements (UCEs [40]), anchored hybrid enrichment loci (AHE [41]), squamate conserved loci (SqCL [42]), and rapidly evolving long exons (RELEC [43]). Encouragingly, studies using these genome-scale datasets have revealed that previous approaches, including electrophoretic studies (e.g., ratsnakes, [24] vs [44]) and mtDNA sequence studies (e.g., Western diamondback rattlesnakes, [45] vs [46]) are often consistent with recent and presumably improved methods of phylogenetic inference. However, genome-scale datasets have also revealed novel relationships including the monophyly of Pythonoidea + Booidea + Uropeltoidea + Bolyeriidae and several subgroups within this clade [4, 5, 47].

The natural and presumed next step in the analytical evolution of molecular phylogenetics for snakes is comparative genomics, comparing entire nuclear genomes among species. We suspect that existing approaches using DNA-sequence alignments (and models of nucleotide evolution) will remain relevant for comparing large subsets of the nuclear genome but that new methods will also be developed for phylogenetic inference. These broader-scale comparisons will require a framework for character coding, interpretation, and analysis that is infrequently discussed in the systematics literature [9, 48]. As such, a relevant first question is: what types of characters do snake genomes offer and how can they be coded for phylogenetic analysis?

10.3 Characteristics and Phylogenetic Coding of Snake Genomes

The nuclear genomes of snakes (ca. 1.5 Gb [1]) are typically smaller than mammal genomes (Fig. 10.2). At present, there have been 26 nearly complete nuclear genomes of snakes published (Table 10.1). Based on available data, the nuclear genomes of different snake species vary in several ways: overall size (number of nucleotides), GC-content and distribution (e.g., isochores [49]), number of protein-coding genes, gene family diversity [50], content and distribution of **transposable elements** (TEs [51]), content and distribution of **endogenous viral elements** (EVEs [52]), **sex chromosome condition** [53], and overall structural similarity (synteny level). Although similar in size to bird genomes, snake genomes have some qualities that are more mammal-like, such as TE content (Fig. 10.2). Thus, snake nuclear genomes are dynamic and offer a plethora of characteristics beyond nucleotide sequences to apply towards phylogenetic inference.

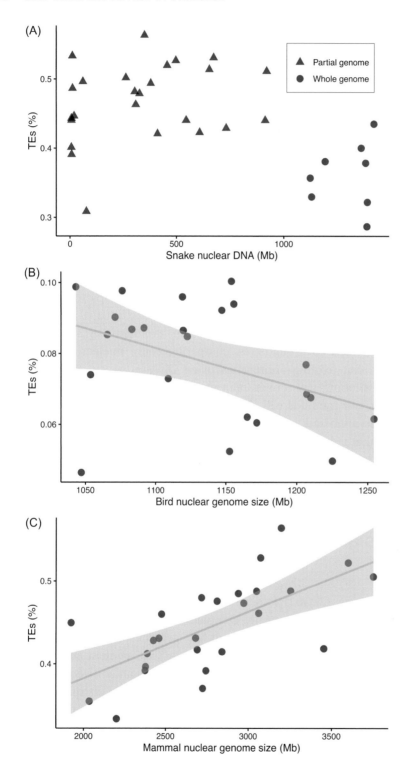

Figure 10.2 (A) Relationship between nuclear DNA sequencing effort and transposable element (TE) content in partial and whole snake genomes. (B) Relationship between genome size and TE content in birds. (C) Relationship between genome size and TE content in mammals. Data from [51] and NCBI Genome database.

Characters for phylogenetic analysis should have at least three qualities: (1) that homology can be reasonably assumed across operational taxonomic units, (2) that the character is heritable (thus being capable of accumulating phylogenetic signal), and (3) that the character can be coded for use in phylogenetic analysis. Many of the above nuclear genome qualities meet these criteria, so how might we analyse them? Unlike DNA-sequence alignment, which can be largely automated, preparing a comparative matrix of genome qualities is a nontrivial task. In many cases our ability to define the qualities we are interested in requires nuanced consideration. For example, the arrangement of genes is a proposed quality that might inform phylogenetic inference [9], however defining genes (and gene families) requires **gene ontology (GO) inference** before any scoring of a character matrix can take place. Given that some methods for establishing GO terms produce false positives and are controversial [54], analytical choices will need to be carefully considered and justified. Thus, coding genome-level characters for phylogenetic analysis involves decision making that is reminiscent of coding complex morphological structures [55, 56]. Though the considerations necessary for coding complex characters are far more bespoke than near-automated nucleic acid sequence alignment, the existing morphological and paleontological literature provides valuable insights into how we might use qualities of nuclear genomes to infer phylogenies (see [57]).

Without standardized assembly methods and character definitions, precisely coding genome qualities of snakes will be difficult. Furthermore, there is clearly variation in how the existing literature reports genome qualities; using the same raw data for *Boa constrictor,* there have been at least three different estimates for repetitive content level (e.g., LTR elements: 1.4%, 9.6%, and 2.3% in [58–60], respectively). Despite these current limitations, we provide two examples below that compare coding strategies using available snake genomes. We considered inter-assembly comparison valuable because detectable phylogenetic signal suggests character coding might be robust to substantial analytical 'noise'; therein providing a preliminary (albeit basic) test of the veracity of using genome-level characters for phylogenetic analysis of snakes.

Genome sizes of reptiles are known to have phylogenetic signal [61], so we selected qualities that might relate to genome size. As such, we coded two qualities using 26 snake genomes and select squamate reptile outgroups (Table 10.1): genome size (in Megabases, Mb) and genome-wide GC-content (%). These data were either collected from the original publications or the NCBI genome summary pages (https:// www.ncbi.nlm.nih.gov/genome/). Next, because transposable elements have been shown to have phylogenetic signal in a variety of taxa [62, 63] we reanalysed reptile genome data from Pasquesi et al. [51] for 14 categories of repetitive element content (including **SSRs, DNA transposons, LINEs, SINEs, LTRs**) from eight snakes and select squamate outgroups (Table 10.2). We only included taxa from Pasquesi et al. [51] with full genomes, because partial genomes included in this study had, on average, higher estimated TE levels (Fig. 10.2A). To transform percentage summaries into continuous data (GC-content, repetitive element content) we multiplied them by the total length (Mb) of each nuclear genome.

Table 10.1. Nuclear genome characteristics from published snake genomes and select reptile outgroups.

Taxonomy follows Burbrink et al. [5]. NCBI = National Centre for Biotechnology Information, Maryland, United States. Sec. Mil. Med. Uni. = Second Military Medical University, Shanghai, China.

Taxon	Family	Total length (Mb)	GC content (%)	Citations
Boidae + Pythonidae				
Boa constrictor	Boidae	1600.0	40.2	Bradnam et al. [101]
Python bivittatus	Pythonidae	1435.1	39.7	Castoe et al. [49]
Colubroidea				
Thermophis baileyi	Dipsadidae	1747.7	41.1	Li et al. [59]
Thamnophis elegans	Natricidae	1743.3	41.1	Rockefeller University (NCBI)
Thamnophis sirtalis	Natricidae	1424.9	41.8	McGlothlin et al. [102]
Pantherophis guttatus	Colubridae	1404.2	38.3	Ullate-Agote et al. [103]
Pantherophis obsoletus	Colubridae	1692.5	39.1	University of Geneva (NCBI)
Ptyas mucosa	Colubridae	1721.5	39.2	University of Geneva (NCBI)
Elapidae				
Pseudonaja textilis	Elapidae	1590.0	40.1	Earl et al. [104]
Laticauda colubrina	Elapidae	2024.7	35.6	Kishida et al. [105]
Laticauda laticauda	Elapidae	1558.7	40.1	Kishida et al. [105]
Hydrophis curtis	Elapidae	1910.0	39.7	Peng et al. [106]
Hydrophis melanocepahlus	Elapidae	1402.6	34.8	Kishida et al. [105]
Hydrophis hardwickii	Elapidae	1296.4	37.2	Sec. Mil. Med. Uni. (NCBI)
Hydophis cyanocinctus	Elapidae	1389.9	37.6	Sec. Mil. Med. Uni. (NCBI)
Emydocephalus ijimae	Elapidae	1625.2	40.3	Kishida et al. [105]
Naja naja	Elapidae	1768.5	40.4	Suryamohan et al. [107]

Table 10.1. (*cont.*)

Taxon	Family	Total length (Mb)	GC content (%)	Citations
Ophiophagus hannah	Elapidae	1594.1	40.6	Vonk et al. [108]
Notechis scutatus	Elapidae	1665.5	40.2	Univ. New South Wales (NCBI)
Viperidae				
Crotalus viridis	Viperidae	1340.2	39.4	Pasquesi et al. [51]
Crotalus pyrrhus	Viperidae	1126.5	38.5	Gilbert et al. [52]
Crotalus horridus	Viperidae	1520.3	34.3	University of Arkansas (NCBI)
Vipera berus	Viperidae	1532.4	41.3	Baylor College (NCBI)
Protobothrops flavoviridis	Viperidae	1413.2	38.2	Shibata et al. [97]
Protobothrops mucrosquamatus	Viperidae	1673.9	40.7	Aird et al. [109]
Deinagkistrodon acutus	Viperidae	1470.0	41.0	Yin et al. [57]
Other Toxicoferans				
Varanus komodoensis	Varanidae	1557.2	43.9	van Hoek et al. [110]
Pogona vitticeps	Agamidae	1716.7	42.1	Georges et al. [111]
Anolis carlinensis	Polychrotidae	1799.1	40.8	Alföldi et al. [112]
Outgroup				
Gekko japonicus	Gekkonidae	2550.0	45.5	Liu et al. [113]

The nuclear genome characters we collected can be considered polymorphic multistate characters that vary both within and among species. There are several methods for coding these characters including generalized frequency coding (GFC) and step-matrix-gap-weighting [64, 65]. Here, we used GFC on nuclear genome characters given that simulations suggest it outperforms step-matrix-gap-weighting [66]. Briefly, GFC requires four steps to code polymorphic multistate characters. First, for each quality of nuclear genome, observed variation is categorized as either counts (meristic data) or evenly-spaced ranges (continuous data), referred to hereafter as 'frequency bins'. In our two examples, we arbitrarily selected to divide genome qualities into eight frequency bins (though we acknowledge that exploring the impact of granularity, i.e., number of bins is worth undertaking at a later time). Second, an initial matrix (Matrix **a**) is made using **non-cumulative frequencies**. For

Table 10.2. Select repetitive element characteristics of snake nuclear genomes as reported by Pasquesi et al. [51].

Genome sizes and percentages of 14 repetitive elements are as reported in Supporting Data 4 and 6 in the original study. Taxonomy follows Burbrink et al. [5]. See Glossary in Box 10.1 for element definitions. *LINEs percentage is combined total from CR1–L3, L2, BovB, L1 and other elements; DNA transposons percentage is combined hAT, Tc1 and other elements. Each subcategory was treated as a separate set of eight phylogenetic characters, which produced character matrices of 98 and 112 characters for cumulative and non-cumulative analyses, respectively.

Taxon	Family	Total length (Mb)	SSRs	SINEs	LINEs*	PLEs	DIRS	LTRs	DNA*	Unclassified
Boidae + Pythonidae										
Boa constrictor	Boidae	1387.2	2.34	2.58	14.73	1.05	0.05	2.50	7.82	5.43
Python bivittatus	Pythonidae	1385.3	1.99	1.60	13.16	0.93	0.05	1.90	4.98	5.98
Colubroidea										
Thamnophis sirtalis	Natricidae	1122.6	2.83	3.90	10.35	1.11	0.87	2.60	10.48	6.37
Pantherophis guttatus	Colubridae	1358.4	2.60	3.53	14.04	1.66	0.74	3.38	12.43	3.99
Elapidae										
Ophiophagus hannah	Elapidae	1379.2	2.99	3.09	14.03	1.46	0.60	2.79	8.84	6.74
Viperidae										
Crotalus viridis	Viperidae	1129.3	2.26	1.83	14.22	1.49	1.11	4.25	9.92	5.03
Crotalus pyrrhus	Viperidae	1191.6	2.03	2.42	12.09	1.61	0.86	3.44	9.87	2.50
Deinagkistrodon acutus	Viperidae	1470.4	4.33	2.13	18.91	1.53	1.95	4.34	11.20	3.44
other toxicoferans										
Ophisaurus gracilis	Anguidae	1729.3	1.89	2.21	15.03	0.77	0.54	6.49	11.39	6.34
Pogona vitticeps	Agamidae	1592.7	2.57	2.49	14.84	0.67	0.83	3.32	11.38	4.32
Anolis carolinensis	Polychrotidae	1701.4	1.74	1.40	17.66	0.25	2.58	3.67	7.93	5.70
Outgroup										
Gekko japonicus	Gekkonidae	2402.0	1.07	6.90	15.05	0.69	1.98	3.31	2.97	10.53

each taxon, the observed frequencies of a character are placed into the frequency bins of Matrix **a**. Third, a second matrix (Matrix **b**) is constructed using **cumulative frequencies**, these summarize the frequency data obtained from Matrix **a**, and fill all columns to the right of an observed state (therein excluding the first bin). Finally, following Wiens' [67] frequency bins, the frequencies in Matrix **b** are replaced with letters to produce a new character matrix (Matrix **c1**). Non-cumulative frequencies can also be used from Matrix **a** to produce a lettered matrix (Matrix **c2**). Examples of these character matrices from our first dataset are available in Table 10.3. The maximum number of steps between each character state is limited to 25 (because of the Latin/Roman alphabet). However, this number of states is far more than has previously been used to code nuclear genome qualities in other vertebrate groups (e.g., [48]).

We constructed both cumulative and non-cumulative GFC matrices from snake genomes to compare their performance and ability to recover expected relationships (Fig. 10.3A). Specifically, we scored each method for its ability to recover the four branches in the expected topology. We used four ingroup taxa (Boidae + Pythonidae, Colubroidea, Elapidae, and Viperidae) and two outgroup taxa (other toxicoferans and *Gekko japonicus*). We grouped Boidae and Pythonidae together (given their supported monophyly [4, 5]). The outgroup taxon 'other toxicoferans' included *Varanus komodoensis*, *Pogona vitticeps*, and *Anolis carolinensis* in the first analysis and *Ophisaurus gracilis*, *Pogona vitticeps*, and *Anolis carolinensis* in the second analysis. Once coded using GFC, we used the phangorn package [68] in R statistical software [69] to load GFC matrices and conduct UPGMA analyses. Because of program limitations, all analyses in R recognized our letter matrix as 'amino acid' data. To ensure that inappropriate distance corrections were not performed, we used the 'JC69' model [70] which treats all character state changes as equally probable. We rooted all trees with *Gekko japonicus*, given that it was the most divergent outgroup included in these analyses.

In the first analysis (genome size, GC-content), GFC coding resulted in a character matrix of 14 characters using cumulative frequencies (Fig. 10.3C) and 16 characters for non-cumulative frequencies. The cumulative frequency coding tree recovered 75 per cent of the expected relationships (Fig. 10.3B). These 'correctly' inferred relationships included the monophyly of snakes and the sister relationship between Colubroidea and Elapidae. Only the placement of Viperidae and Boidae + Pythonidae differed from the expected tree. In contrast, the non-cumulative tree did not recover snakes as monophyletic and recovered only one of the expected relationships (25 per cent), that between Colubroidea and Elapidae (Fig. 10.3D). Results from our second analysis, the coding of content levels from 14 repetitive elements, were based on larger character matrices (cumulative, 98 characters; non-cumulative, 112 characters). The results were overall similar to the first analysis, with the cumulative frequency tree recovering 75 per cent of expected relationships, however, the non-cumulative frequency tree failed to recover any of the expected relationships (Fig. 10.4). Interestingly, the placement of Viperidae differed between cumulative analyses in our examples. In the first analysis, Viperidae was the sister taxon of Boidae + Pythonidae whereas in the second it was the sister taxon of a clade containing Boidae + Pythonidae, Elapidae and Colubroidea.

Table 10.3. Character matrices derived by generalized frequency coding (after [63]) for Nuclear genome size (Mb) and GC-content (eight characters each).

Genome size characters are abbreviated as total length (TL) and GC-content characters as (GC). N = sample sizes upon which frequencies are based. Matrix A is non-cumulative frequencies. Matrix B is cumulative frequencies. Matrix C1 is coded cumulative frequencies. Matrix C2 is coded non-cumulative frequencies. Coding in both C matrices was based on table from Wiens [67]. See Table 10.1 for underlying data. B + P = Boidae + Pythonidae; COL = Colubroidea; ELAP = Elapidae; VIP = Viperidae; TOX = other toxicoferans; GEK = *Gekko japonicus*.

Taxon	N	TL1	TL2	TL3	TL4	TL5	TL6	TL7	TL8	GC1	GC2	GC3	GC4	GC5	GC6	GC7	GC8
Matrix A																	
B + P	2	0.5	0.5	0	0	0	0	0	0	0.5	0.5	0	0	0	0	0	0
COL	6	0.33	0.67	0	0	0	0	0	0	0.33	0.66	0	0	0	0	0	0
ELAP	11	0.28	0.63	0.09	0	0	0	0	0	0.36	0.64	0	0	0	0	0	0
VIP	7	0.71	0.29	0	0	0	0	0	0	0.57	0.43	0	0	0	0	0	0
TOX	3	0	1	0	0	0	0	0	0	0	1	0	0	0	0	0	0
GEK	1	0	0	0	1	1	0	0	0	0	0	0	1	1	0	0	0
Matrix B																	
B + P	2	0.5	0	0	0	0	0	0	N/A	0.5	0	0	0	0	0	0	N/A
COL	6	0.67	0	0	0	0	0	0	N/A	0.66	0	0	0	0	0	0	N/A
ELAP	11	0.72	0.09	0	0	0	0	0	N/A	0.64	0	0	0	0	0	0	N/A
VIP	7	0.29	0	0	0	0	0	0	N/A	0.43	0	0	0	0	0	0	N/A
TOX	3	1	1	0	0	0	0	0	N/A	1	1	0	0	0	0	0	N/A
GEK	1	1	1	1	1	0	0	0	N/A	1	1	1	1	0	0	0	N/A

Matrix C1

B + P	2	M	A	A	A	A	A	N/A	M	A	A	A	A	A	N/A
COL	6	Q	A	A	A	A	A	N/A	Q	A	A	A	A	A	N/A
ELAP	11	S	C	A	A	A	A	N/A	Q	A	A	A	A	A	N/A
VIP	7	H	A	A	A	A	A	N/A	K	A	A	A	A	A	N/A
TOX	3	Y	Y	A	A	A	A	N/A	Y	Y	A	A	A	A	N/A
GEK	1	Y	Y	Y	A	A	A	N/A	Y	Y	A	A	A	A	N/A

Matrix C2

B + P	2	M	A	M	A	A	A	A	M	A	A	A	A	A	A
COL	6	I	Q	I	A	A	A	A	Q	A	A	A	A	A	A
ELAP	11	H	P	J	A	A	A	A	Q	A	A	A	A	A	A
VIP	7	R	H	O	A	A	A	A	K	A	A	A	A	A	A
TOX	3	A	Y	A	A	A	A	A	Y	Y	A	A	A	A	A
GEK	1	A	A	A	Y	A	A	A	A	A	A	A	A	A	A

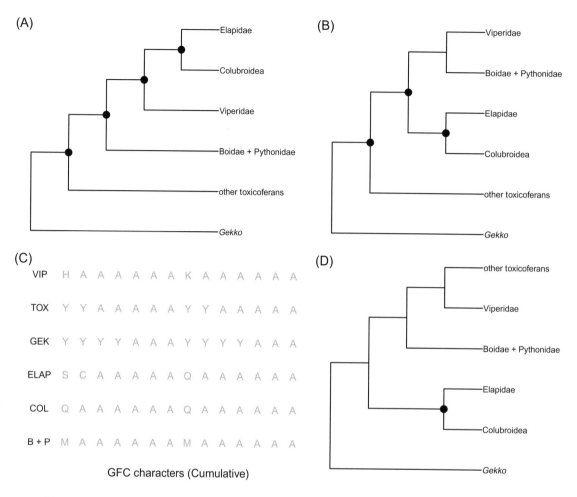

Figure 10.3 Test case of using generalized frequency coding (GFC) to code two qualities of nuclear genomes (assembly size and GC content) as characters for phylogenetic inference. (A) Expected phylogeny based on available literature. (B) Cumulative frequency derived UPGMA tree based on 14 coded characters (C: data from Matrix C1 in Table 10.3.) that is mostly consistent with the expected example, except for the positions of Viperidae and Boidae + Pythonidae. (D) Non-cumulative frequency derived UPGMA tree that is very different from the expected tree, with some outgroups (other toxicoferans) nested within the ingroup (Serpentes). Solid circles indicate nodes consistent with *a priori* expectations from (A). All trees have been rooted with *Gekko*. See Table 10.3 for explanation of taxon abbreviations in matrix in part (C).

These basic examples of coding nuclear genome qualities as phylogenetic characters allow for two conclusions: (1) phylogenetic signal can likely be extracted using methods such as GFC and (2) the method of character coding matters because we found that cumulative frequency far outperformed non-cumulative frequency coding (which in one case produced a tree devoid of 'correct' relationships; Fig. 10.4B). This suggests that decisions about character coding will be important when comparing snake nuclear genomes in a phylogenetic context.

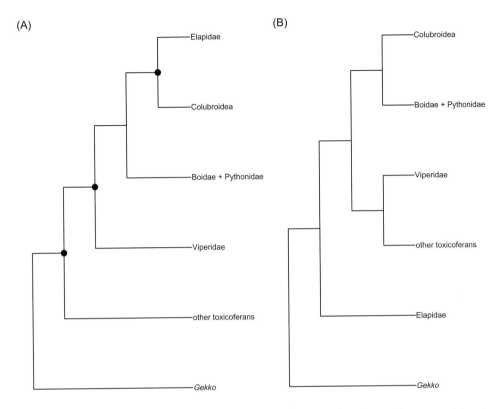

Figure 10.4 Test case of using generalized frequency coding (GFC) to code 14 repetitive element categories of snake nuclear genomes as characters for phylogenetic analysis. (A) UPGMA tree resulting from cumulative GFC of 98 characters. (B) UPGMA tree resulting from non-cumulative analysis of 112 coded characters. Solid circles indicate nodes consistent with *a priori* expectations from Fig. 10.3A. All trees have been rooted with *Gekko*.

10.4 Unresolved and Contentious Branches for Snakes

As research on snake origins and relationships continues, systematists are likely to be most interested in using technical advancements to study unresolved relationships. Modern molecular work has clearly done much to advance our understanding of snake systematics, even resolving the taxonomic placement of some enigmatic taxa including *Brachyorrhos* [71], *Xenophidion* [72], and *Xylophis* [73]. However, despite the plethora of data now available for a broad taxonomic sampling of snakes, some branches within Serpentes remain contentious and/or poorly supported, with different molecular datasets and tree-inference methods producing varying results.

10.4.1 Scolecophidia

Large and broadly sampled familial or generic-level phylogenies for snakes began appearing in the 1990s and 2000s and continue to the present day [4, 5, 35, 42, 74–81].

Across these phylogenies there are several poorly supported and/or inconsistent place-ments that persist regardless of the datasets or methodologies used. For example, although molecular phylogenetic studies challenge the monophyly of Scolecophidia (*sensu latu*), with most studies over the past decade indicating non-monophyly, the placement of major scolecophidian lineages varies [4, 5, 42, 78–80, 82, 83]. Specifically, these studies all find that Anomalepididae are, at minimum, an independent lineage (not sister to a lineage compris-ing only other extant scolecophidians), although this may be dependent on the methods used when estimating the phylogeny (see [4]). However, even among the phylogenies that support a lack of scolecophidian monophyly, the position of Anomalepididae remains inconsistent (and with varying support), with the sister to all other living snakes (Alethinophidia) found to be either Anomalepididae (e.g., [5, 42, 47, 78, 84]) or all non-anomalepidid scolecophidians [4, 79, 82]. This uncertainty is particularly noteworthy because it has a profound impact on inferring ancestral states of the earliest divergences within Serpentes. Furthermore, not all scolecophidian families (e.g., Xenotyphlopidae) or major subfamilial lineages have been included in genome-scale analyses, leaving the non-monophyly of scolecophidians a hypothesis that requires further testing.

10.4.2 Colubroidea

Even when using well-sampled, large datasets and high-powered computational analyses, the relationships of families within Colubroidea (*sensu stricto*) (e.g., Natricidae and Grayiidae) and the relationships of some families within Elapoidea (e.g., Elapidae and Lamprophiidae) remain uncertain. This is indicated by low-support values and/or incon-sistent placement in recent studies [5, 47], as well as in studies with fewer nuclear loci [79–82, 84]. A commonality in these cases is relatively short internal branches [5], something that, as described in the Introduction (§ 10.1), remain a challenge in inferring phylogeny across the tree of life [2, 85] including for squamate systematic studies in general [36, 79].

10.4.3 Toxicofera and Macrostomata

Comparative genomics will likely be applied to understanding molecular snake phylogenies with particular relationships that greatly contrast with phylogenetic inferences based on traditional morphological characters. It was long agreed that within squamates, snakes were part of the Scleroglossa [86], a group consisting of gekkotans and autarchoglossans (i.e., all squamates exclusive of Iguania). However, molecular work began challenging this hypothesis, with phylogenies indicating that Iguania + Anguimorpha shared a more recent common ancestor with snakes [87–89]. This group (including snakes) was named Toxicofera [89]. Molecular evidence repeatedly supports the validity of Toxicofera to a degree that we would argue it is not contentious at this point [5, 47, 79, 82, 84, 90]. Given this, comparative genomics may offer little additional information beyond further testing of the sister taxon for Serpentes, i.e., the Serpentes (Iguania + Anguimorpha) relationship suggested by most genome-scale datasets [5, 47, 90].

Similarly, the lack of support that the 'macrostomatan' snakes form a clade within Serpentes has been repeatedly found to be a poorly supported hypothesis when using molecular data, specifically due to the evidence for Amerophidia (Aniliidae +

Tropidophiidae). Although found in recent genomic-scale phylogenies (e.g., [4, 5, 47]), this relationship was also frequently recovered in older, single locus studies (e.g., [76]), across various multilocus studies [36, 77–80, 82], and from one morphological study [91].

10.5 Conclusions

In this chapter we summarized almost 100 years of molecular studies that have informed our understanding of how the snake tree of life came to exist. Given the large amount of concordance between early and modern molecular phylogenetic analyses (§ 10.2), one may wonder if comparing whole genomes among snakes will provide any truly novel insights into snake phylogeny. As outlined previously (§ 10.4), despite genome-scale datasets often corroborating previous phylogenetic hypotheses, there are still plenty of uncertainties. It is these uncertainties that we think comparative nuclear genomics will be most relevant for addressing as we continue to study the origins of snakes. However, it remains unclear if we are broadly ready in terms of our analytical toolkits and resources to infer phylogenies using comparative genomics.

Although there is clearly phylogenetic signal in genome-level characters of snakes, we see several challenges to address before the potential of systematic comparative genomics can be fully explored. As discussed above (§ 10.3), current snake genome assemblies vary substantially in their quality, completeness, and the methodologies used to annotate them. As such, automated pipelines for the assembly and annotation of snake genomes from raw data would add precision to nuclear genome comparisons. Though these bioinformatics pipelines are currently available, they are computationally intensive, making them inaccessible to most snake systematists. Genomes are lacking from several major lineages of Serpentes including many of the early diverging radiations (e.g., Scolecophidia, Amerophidia, and Uropeltoidea). We recommend sequencing (near) complete genomes from representatives of these groups prior to exploring comparative genomics. Using GFC or similar methods may be helpful for extracting some phylogenetic signal, but may not be appropriate for coding all characteristics of nuclear genomes (e.g., sex chromosome condition). Performing more localized comparisons with particular nuclear genome qualities may also be beneficial because global summaries may miss subtle phylogenetic signal. For example, GC content may differ between macro- and micro-chromosomes [92]. Finally, establishing pragmatic methods for coding different qualities of snake genomes will likely require collaboration and validation from genomics researchers, much as developmental biologists have improved the use of morphological characters for phylogenetic inference. Key research areas that will aid approaches to comparative genomics include patterns of gene-family and transposable-element evolution [62, 63, 93]. The development of probabilistic models for how genomic qualities change through time would aid in leveraging likelihood-based approaches to phylogenetic inference.

Advances in molecular methodologies, in particular sequencing technologies and computational power, have opened the door to integrating whole genomes with evolutionary research. Furthermore, some of these sequencing technologies enable us to generate sequence data from historical, formalin-fixed museum specimens, allowing them to be placed in a molecular phylogeny [94, 95]. This also creates the possibility of eventually sequencing partial or whole nuclear genomes from extinct or rare species (*sensu* [96]). Comparative genomics of snakes has already informed an array of research disciplines including in the areas of toxicology [97] and genomic-landscape dynamics [51]. Although we emphasize that comparative genomics (whether applied as described in this chapter or with yet to be developed methods) is unlikely to be a panacea to all systematic problems or questions, we are confident that it will provide further insight into the early evolution of snakes. Clarifying the origin of snakes, however, will remain at least partly a palaeontological endeavour because additional fossils are needed to better place snake origins in the broader context of squamate reptile evolution.

Acknowledgements

We sincerely thank David Gower and Hussam Zaher for the invitation to present our joint talk 'Conflicting histories: assessing phylogenetic signal across the genomes of snakes' at the Linnean Society in London on 24 June 2019. We also thank them for editing this chapter. We thank Giulia Pasquesi and Davide Pisani for reviews that greatly enhanced the quality of this chapter. We are grateful to Henrique Costa, Rayna Bell, Bob Drewes, Bob Hansen (SSAR), Leo Smith (ASIH) and Kraig Adler (SSAR) for assistance with locating photographs for Figure 10.1 and the other participants of the 'Contribution to the Origin and Early Evolution of Snakes' symposium for an inspirational set of talks and interactions.

References

1. H. M. I. Kerkkamp, R. Manjunatha Kini, A. S. Pospelov, et al., Snake genome sequencing: Results and future prospects. *Toxins (Basel)*, 8 (2016), 360.
2. J. H. Degnan and N. A. Rosenberg, Gene tree discordance, phylogenetic inference and the multispecies coalescent. *Trends in Ecology and Evolution*, 24 (2009), 332–40.
3. C. Zhang, M. Rabiee, E. Sayyari, and S. Mirarab, ASTRAL-III: polynomial time species tree reconstruction from partially resolved gene trees. *BMC Bioinformatics* 19 (2018), 153.
4. J. W. Streicher and J. J. Wiens, Phylogenomic analyses reveal novel relationships among snake families. *Molecular Phylogenetics and Evolution*, 100 (2016), 160–169.
5. F. T. Burbrink, F. G. Grazziotin, R. A. Pyron, et al., Interrogating genomic-scale data for Squamata (lizards, snakes, and amphisbaenians) shows no support for key traditional morphological relationships. *Systematic Biology*, 69 (2020), 502–520.
6. D. H. Huson, R. Rupp, and C. Scornavacca, *Phylogenetic Networks: Concepts, Algorithms and Applications* (Cambridge: Cambridge University Press, 2010).
7. M. C. Brandley, D. L. Warren, A. D. Leaché, and J. A. McGuire, Homoplasy and clade support. *Systematic Biology*, 58 (2009), 184–198.
8. J. L. Boore and S. I. Fuerstenberg, Beyond linear sequence comparisons: the use of genome-level characters for phylogenetic reconstruction. *Philosophical Transactions*

of the Royal Society London B, 363 (2008), 1445-1451.

9. J. L. Boore, The use of genome-level characteristics for phylogenetic reconstruction. *Trends in Ecology and Evolution*, 21 (2006), 439-446.

10. C. H. Kellaway and F. E. Williams, The serological and blood relationships of some common Australian snakes. *Australian Journal of Experimental Biology and Medical Science*, 8 (1931), 123-132.

11. G. C. Bond, Serological Studies of the Reptilia: I. Hemagglutinins and hemagglutinogens of snake blood. *The Journal of Immunology*, 36 (1939), 1-9.

12. G. C. Bond and N. P. Sherwood, Serological studies of the Reptilia: II. The hemolytic property of snake serum. *The Journal of Immunology*, 36 (1939), 11-16.

13. D. W. George and H. C. Dessauer, Immunological correspondence of transferrins and the relationships of colubrid snakes. *Comparative Biochemistry and Physiology*, 33 (1970), 617-627.

14. E. Cohen, Immunological studies of the serum proteins of some reptiles. *The Biological Bulletin*, 109(1955), 394-403.

15. N. G. Crawford, B. C. Faircloth, J. E. McCormack, et al., More than 1000 ultraconserved elements provide evidence that turtles are the sister group of archosaurs. *Biology Letters*, 8 (2012), 783-6.

16. D. J. Field, J. A. Gauthier, B. L. King, et al., Toward consilience in reptile phylogeny: miRNAs support an archosaur, not lepidosaur, affinity for turtles. *Evolution & Development*, 16 (2014), 189-196.

17. D. D. Pearson, Serological and immuno-electrophoretic comparisons among species of snakes. *Bulletin of the Serological Museum* 36 (1966), 8.

18. H. C. Dessauer, W. Fox, and J. R. Ramírez, Preliminary attempt to correlate paper-electrophoretic migration of hemoglobins with phylogeny in Amphibia and Reptilia. *Archives of Biochemistry and Biophysics*, 71 (1957), 11-16.

19. H. C. Dessauer, Molecular approach to the taxonomy of colubrid snakes. *Herpetologica*, 23 (1967), 148-155.

20. H. G. Dowling, Hemipenes and other characters in colubrid classification. *Herpetologica*, 2(1967), 138-142.

21. G. L. Underwood, *A Contribution to the Classification of Snakes* (London: British Museum (Natural History), 1967).

22. S. -H. Mao and H. C. Dessauer, Selectively neutral mutations, transferrins and the evolution of natricine snakes. *Comparative Biochemistry and Physiology Part A: Physiology*, 40 (1971), 669-680.

23. S. -H. Mao, B. -Y. Chen, and H. -M. Chang, The evolutionary relationships of sea snakes suggested by immunological cross-reactivity of transferrins. *Comparative Biochemistry and Physiology Part A: Physiology*, 57 (1977), 403-406.

24. R. Lawson and H. C. Dessauer, Electrophoretic evaluation of the colubrid genus *Elaphe* (Fitzinger). *Isozyme Bulletin*, 14 (1981), 83.

25. H. G. Dowlings, R. Highton, G. C. Maha, and L. R. Maxson, Biochemical evaluation of colubrid snake phylogeny. *Journal of Zoology*, 201 (1983), 309-329.

26. J. E. Cadle, H. C. Dessauer, C. Gans, and D. F. Gartside, Phylogenetic relationships and molecular evolution in uropeltid snakes (Serpentes: Uropeltidae): allozymes and albumin immunology. *Biological Journal of the Linnean Society*, 40 (1990), 293-320.

27. J. B. Slowinski, A phylogenetic analysis of the New World coral snakes (Elapidae: *Leptomicrurus*, *Micruroides*, and *Micrurus*) based on allozymic and morphological characters. *Journal of Herpetology*, 29 (1995), 325-338.

28. J. E. Cadle (1988). Phylogenetic relationships among advanced snakes: a molecular perspective. *University of California Publications in Zoology*, 119 (1988), 1-77.

29. C. R. Feldman and G. S. Spicer (2002). Mitochondrial variation in sharp-tailed

snakes (*Contia tenuis*): Evidence of a cryptic species. *Journal of Herpetology*, 36 (2002), 648.

30. F. Kraus and W. M. Brown, Phylogenetic relationships of colubroid snakes based on mitochondrial DNA sequences. *Zoological Journal of the Linnean Society*, 122 (1998), 455–487.

31. J. B. Slowinski and J. S. Keogh, Phylogenetic relationships of elapid snakes based on cytochrome b mtDNA sequences. *Molecular Phylogenetics and Evolution*, 15 (2000), 157–164.

32. F. T. Burbrink, R. Lawson, and J. B. Slowinski, Mitochondrial DNA phylogeography of the polytypic North American rat snake (*Elaphe obsoleta*): a critique of the subspecies concept. *Evolution*, 54 (2000), 2107–2118.

33. D. J. Funk and K. E. Omland, Species-level paraphyly and polyphyly: Frequency, causes, and consequences, with insights from animal mitochondrial DNA. *Annual Review of Ecology, Evolution, and Systematics*, 34 (2003), 397–423.

34. D. Rubinoff and B. S. Holland, Between two extremes: mitochondrial DNA is neither the panacea nor the nemesis of phylogenetic and taxonomic inference. *Systematic Biology*, 54 (2005), 952–961.

35. R. Lawson, J. B. Slowinski, B. I. Crother, and F. T. Burbrink, Phylogeny of the Colubroidea (Serpentes): new evidence from mitochondrial and nuclear genes. *Molecular Phylogenetics and Evolution*, 37 (2005), 581–601.

36. J. J. Wiens, C. A. Kuczynski, S. A. Smith, et al., Branch lengths, support, and congruence: Testing the phylogenomic approach with 20 nuclear loci in snakes. *Systematic Biology*, 57 (2008), 420–431.

37. L. S. Kubatko, H. L. Gibbs, and E. W. Bloomquist, Inferring species-level phylogenies and taxonomic distinctiveness using multilocus data in *Sistrurus* rattlesnakes. *Systematic Biology*, 60 (2011), 393–409.

38. K. L. Sanders, M. S. Y. Lee, Mumpuni, T. Bertozzi, and A. R. Rasmussen, Multilocus phylogeny and recent rapid radiation of the viviparous sea snakes (Elapidae: Hydrophiinae). *Molecular Phylogenetics and Evolution*, 66 (2013), 575–591.

39. J. W. Streicher and S. Ruane, *Phylogenomics of snakes. In* eLS. Chichester: John Wiley & Sons Ltd, 2018, DOI: 10.1002/9780470015902.a0027476).

40. B. C. Faircloth, J. E. McCormack, N. G. Crawford, et al., Ultraconserved elements anchor thousands of genetic markers spanning multiple evolutionary timescales. *Systematic Biology*, 61 (2012), 717–726.

41. A. R. Lemmon, S. A. Emme, and E. M. Lemmon, Anchored hybrid enrichment for massively high-throughput phylogenomics. *Systematic Biology*, 61 (2012), 727–44.

42. S. Singhal, M. Grundler, G. Colli, and D. L. Rabosky, Squamate conserved loci (SqCL): A unified set of conserved loci for phylogenomics and population genetics of squamate reptiles. *Molecular Ecology Resources*, 17 (2017), e12–e24.

43. B. R. Karin, T. Gamble, and T. R. Jackman, Optimizing phylogenomics with rapidly evolving long exons: Comparison with anchored hybrid enrichment and ultraconserved element. *Molecular Biology and Evolution*, 37 (2020), 904–922.

44. X. Chen, A. R. Lemmon, E. M. Lemmon, R. A. Pyron, and F. T. Burbrink, Using phylogenomics to understand the link between biogeographic origins and regional diversification in ratsnakes. *Molecular Phylogenetics and Evolution*, 111 (2017), 206–218.

45. T. A. Castoe, C. L. Spencer, and C. L. Parkinson, Phylogeographic structure and historical demography of the western diamondback rattlesnake (*Crotalus atrox*): A perspective on North American desert biogeography. *Molecular Phylogenetics and Evolution*, 42 (2007), 193–212.

46. D. R. Schield, D. C. Card, N. R. Hales, et al., The origins and evolution of chromosomes, dosage compensation, and mechanisms underlying venom regulation in snakes. *Genome Research*, 29 (2019), 590–601.

47. S. Singhal, T. J. Colston, M. R. Grundler, et al., Congruence and conflict in the higher-level phylogenetics of squamate reptiles: An expanded phylogenomic perspective. *Systematic Biology*, 70 (2021), 542–557.

48. G. E. Sims, S. -E. Jun, G. A. Wu, and S. -H. Kim, Whole-genome phylogeny of mammals: Evolutionary information in genic and nongenic regions. *Proceedings of the National Academy of Sciences USA*, 106 (2009), 1777–17082.

49. T. A. Castoe, A. P. J. de Koning, K. T. Hall, et al., The Burmese python genome reveals the molecular basis for extreme adaptation in snakes. *Proceedings of the National Academy of Sciences USA*, 110 (2013), 20645–20650.

50. M. W. Giorgianni, N. L. Dowell, S. Griffin, et al., The origin and diversification of a novel protein family in venomous snakes. *Proceedings of the National Academy of Sciences USA*, 117 (2020), 10911–10920.

51. G. I. M. Pasquesi, R. H. Adams, D. C. Card, et al., Squamate reptiles challenge paradigms of genomic repeat element evolution set by birds and mammals. *Nature Communications*, 9 (2018), 2774.

52. C. Gilbert, J. M. Meik, D. Dashevsky, et al., Endogenous hepadaviruses, bornaviruses and circoviruses in snakes. *Proceedings of the Royal Society B*, 281 (2014), 1122.

53. B. Augstenová, M. Johnson Pokorná, M. Altmanová, et al., ZW, XY, and yet ZW: Sex chromosome evolution in snakes even more complicated. *Evolution*, 72 (2018), 1701–1707.

54. P. Palvildis, J. D. Jensen, W. Stephan, and A. Stamatakis, A critical assessment of storytelling: gene ontology categories and the importance of validating genome scans. *Molecular Biology and Evolution*, 29 (2012), 3237–3248.

55. J. W. Archie, Methods for coding variable morphological features for numerical taxonomic analysis. *Systematic Zoology*, 34 (1985), 326–345.

56. R. S. Thorpe, Coding morphometric characters for constructing distance Wagner networks. *Evolution*, 38 (1984), 244–255.

57. J. V. Freudenstein, Characters, states and homology. *Systematic Biology*, 54 (2005), 965–973.

58. W. Yin, Z-J. Wang, Q-Y. Li, et al., Evolutionary trajectories of snake genes and genomes revealed by comparative analyses of five-pacer viper. *Nature Communications*, 7 (2016), 13107.

59. J. -T. Li, Y. -D. Gao, L. Xie , et al., Comparative genomics investigation of high elevation adaptation in ectothermic snakes. *Proceedings of the National Academy of Sciences USA*, 115 (2018), 8406–8411.

60. D. C. Card, R. H. Adams, D. R. Schield, et al., Genomic basis of convergent island phenotypes in boa constrictors. *Genome Biology and Evolution*, 11 (2019), 3123–3143.

61. C. L. Organ, R. Godínez Moreno, and S. V. Edwards, Three tiers of genome evolution in reptiles. *Integrative and Comparative Biology*, 48 (2008), 494–504.

62. J. O. Kriegs, G. Churalov, M. Kiefmann, et al., Retrotransposed elements as archives for the evolutionary history of placental mammals. *PLoS Biology*, 4 (2006), e91.

63. D. Vitales, S. Garcia, and S. Dodsworth, Reconstructing phylogenetic relationships based on repeat sequence similarities. *Molecular Phylogenetics and Evolution*, 147 (2020), 106766.

64. E. N. Smith and R. L. Gutberlet Jr., Generalized frequency coding: A method of preparing polymorphic multistate characters for phylogenetic analysis. *Systematic Biology*, 50 (2001), 156–169.

65. J. J. Wiens, Character analysis in morphological phylogenetics: problems and solutions. *Systematic Biology*, 50 (2001), 689–699.

66. A. M. Lawing, J. M. Meik, and W. E. Schargel, Coding meristic characters for phylogenetic analysis: A comparison of step-matrix gap weighting and generalized frequency coding. *Systematic Biology*, 57 (2008), 167–173.

67. J. J. Wiens, Polymorphic characters in phylogenetic systematics. *Systematic Biology*, 44 (1995), 482–500.

68. K. P. Schliep, Phangorn: phylogenetic analysis in R. *Bioinformatics*, 27 (2011), 592–593.

69. R Development Core Team, R: *A Language and Environment for Statistical Computing* (Vienna: R Foundation for Statistical Computing, 2019).

70. T. H. Jukes and C. R. Cantor, Evolution of protein molecules. In H. N. Munro, ed. *Mammalian protein metabolism. Volume 3* (New York: Academic Press, 1969), pp. 21–132.

71. J. C. Murphy and K. L. Sanders, First molecular evidence for the phylogenetic placement of the enigmatic snake genus *Brachyorrhos* (Serpentes: Caenophidia). *Molecular Phylogenetics and Evolution*, 61 (2011), 953–957.

72. R. Lawson, J. B. Slowinski, and F. T. Burbrink, A molecular approach to discerning the phylogenetic placement of the enigmatic snake *Xenophidion schaeferi* among the Alethinophidia. *Journal of Zoology*, 263 (2004), 285–294.

73. V. Deepak, S. Ruane, and D. J. Gower, A new subfamily of fossorial colubroid snakes from the Western Ghats of peninsular India. *Journal of Natural History*, 52 (2018), 2919–2934.

74. P. J. Heise, L. R. Maxson, H. G. Dowling, and S. B. Hedges, Higher-level snake phylogeny inferred from mitochondrial DNA sequences of 12S rRNA and 16S rRNA genes. *Molecular Biology and Evolution*, 12 (1995), 259–265.

75. H. G. Dowling, C. A. Hass, S. B. Hedges, and R. Highton, Snake relationships revealed by slow evolving proteins: a preliminary survey. *Journal of Zoology*, 240 (1996), 1–28.

76. J. B. Slowinski and R. Lawson, Snake phylogeny: evidence from nuclear and mitochondrial genes. *Molecular Phylogenetics and Evolution*, 24 (2002), 194–202.

77. R. A. Pyron, F. T. Burbrink, G. R. Colli, et al., The phylogeny of advanced snakes (Colubroidea), with discovery of a new subfamily and comparison of support methods for likelihood trees. *Molecular Phylogenetics and Evolution*, 58 (2011), 329–342.

78. R. A. Pyron, H. K. D. Kandambi, C. R. Hendry, et al., Genus-level phylogeny of snakes reveals the origins of species richness in Sri Lanka. *Molecular Phylogenetics and Evolution*, 66 (2013), 969–978.

79. J. J. Wiens, C. R. Hutter, D. G. Mulcahy, et al., Resolving the phylogeny of lizards and snakes (Squamata) with extensive sampling of genes and species. *Biology Letters*, 8 (2012), 1043–1046.

80. A. Figueroa, A. D. McKelvy, L. L. Grismer, C. D. Bell, and S. P. Lailvaux, A species-level phylogeny of extant snakes with description of a new colubrid subfamily and genus. *PloS ONE*, 11 (2016), e0161070.

81. H. Zaher, R. W. Murphy, J. C. Arredondo, et al., Large-scale molecular phylogeny, morphology, divergence-time estimation, and the fossil record of advanced caenophidian snakes (Squamata: Serpentes). *PloS ONE*, 14 (2019), e0216148.

82. Y. Zheng and J. J. Wiens, Combining phylogenomic and supermatrix approaches, and a time-calibrated phylogeny for squamate reptiles (lizards and snakes) based on 52 genes and 4162 species. *Molecular Phylogenetics and Evolution*, 94 (2016), 537–547.

83. A. Miralles, L. Marin, D. Markus, et al., Molecular evidence for the paraphyly of Scolecophidia and its evolutionary

implications. *Journal of Evolutionary Biology*, 31 (2018), 1782–1793.

84. R. A. Pyron, R. A., F. T. Burbrink, and J. J. Wiens, A phylogeny and revised classification of Squamata, including 4161 species of lizards and snakes. *BMC Evolutionary Biology*, 13 (2013), 93.

85. A. Rokas and S. B. Carroll, Bushes in the tree of life. *PLoS Biology*, 4 (2006), e352.

86. R. K. Estes, K. de Queiroz, and J. Gauthier, Phylogenetic relationships within Squamata. In R. Estes and G. Pregill, eds., *Phylogenetic Relationships of the Lizard Families* (Stanford: Stanford University Press 1988), pp.119–281.

87. K. M. Saint, C. C. Austin, S. C. Donnellan, and M. N. Hutchinson, C-mos, a nuclear marker useful for squamate phylogenetic analysis. *Molecular Phylogenetics and Evolution*, 10 (1998), 259–263.

88. T. M. Townsend, A. Larson, E. Louis, and J. R. Macey, Molecular phylogenetics of Squamata: the position of snakes, amphisbaenians, and dibamids, and the root of the squamate tree. *Systematic Biology*, 53 (2004), 735–757.

89. N. Vidal and S. B. Hedges, The phylogeny of squamate reptiles (lizards, snakes, and amphisbaenians) inferred from nine nuclear protein-coding genes. *Comptes Rendus Biologies*, 328 (2005), 1000–1008.

90. J. W. Streicher and J. J. Wiens, Phylogenomic analyses of more than 4000 nuclear loci resolve the origin of snakes among lizard families. *Biology Letters*, 13 (2017), 20170393.

91. D. S. Siegel, A. Miralles, and R. D. Aldridge, R. D., Controversial snake relationships supported by reproductive anatomy. *Journal of Anatomy*, 218 (2011), 342–348.

92. D. R. Schield, D. C. Card, R. H. Adams, et al., Incipient speciation with biased gene flow between two lineages of the Western Diamondback Rattlesnake (*Crotalus atrox*). *Molecular Phylogenetics and Evolution*, 83 (2015), 213–223.

93. T. A. Williams, G. J. Szöllosi, A. Spang, et al., Integrative modeling of gene and genome evolution roots the archaeal tree of life. *Proceedings of the National Academy of Sciences USA*, 114 (2017), E4602–4611.

94. S. Ruane and C. C. Austin, Phylogenomics using formalin-fixed and 100+ year-old intractable natural history specimens. *Molecular Ecology Resources*, 17 (2017), 1003–1008.

95. S. Ruane, New data from old specimens. *Copeia* (in press).

96. C. Y. Feigin, A. H. Newton, L. Doronina, et al., Genome of the Tasmanian tiger provides insights into the evolution and demography of an extinct marsupial carnivore. *Nature Ecology and Evolution*, 2 (2018), 182–192.

97. H. Shibata, T. Chijiwa, N. Oda-Ueda, et al., The habu genome reveals accelerated evolution of venom protein genes. *Scientific Reports*, 8 (2018), 11300.

98. K. Adler, *Contributions to the History of Herpetology*, Volume 3 (Vancouver: Society for the Study of Amphibians and Reptiles, 1012), 564 pp.

99. R. S. Thorpe, Garth Underwood (1919–2002): A vision of reptile systematics. *Herpetological Review*, 34 (2003), 6–7.

100. E. A. Liner and C. J. Cole, Herbert C. Dessauer. *Copeia*, 2003 (2003), 195–199.

101. K. R. Bradnam, J. N. Fass, A. Alexandrov, et al., Assemblathon 2: evaluating de novo methods of genome assembly in three vertebrate species. *GigaScience*, 2 (2013), 2047-217x-2-10.

102. J. W. McGlothlin, J. P. Chuckalovcak, D. E. Janes, et al., Parallel evolution of tetrodotoxin resistance in three voltage-gated sodium channel genes in the garter snake *Thamnophis sirtalis*. *Molecular Biology and Evolution*, 31 (2014), 2836–2846.

103. A. Ullate-Agote, M. C. Milinkovitch, and A. C. Tzikia, The genome sequence of the corn snake (*Pantherophis guttatus*), a valuable resource for EvoDevo studies in squamates. *International Journal of Developmental Biology*, 58 (2014), 881–888.

104. S. T. H. Earl, G. W. Birrell, T. P. Wallis, et al., Post-translational modification accounts for the presence of varied forms of nerve growth factor in Australian elapid snake venoms. *Proteomics*, 6 (2006), 6554–6665.

105. T. Kishida, Y. Go, S. Tatsumoto, et al., Loss of olfaction in sea snakes provides new perspectives on the aquatic adaptation of amniotes. *Proceedings of the Royal Society B*, 286 (2019), 2019.1828.

106. C. Peng, J-L. Ren, C. Deng, et al., The genome of Shaw's seasnake (*Hydrophis curtus*) reveals secondary adaptation to its marine environment. *Molecular Biology and Evolution*, 37 (2020), 1744–1760.

107. K. Suryamohan, S. P. Krishnankutty, J. Guillory, et al., The Indian cobra reference genome and transcriptome enables comprehensive identification of venom toxins. *Nature Genetics*, 52 (2020), 106–117.

108. F. J. Vonk, N. R. Casewell, C. V. Henkel, et al., The king cobra genome reveals dynamic gene evolution and adaptation in the snake venom system. *Proceedings of the National Academy of Sciences USA*, 110 (2013), 20651–20656.

109. S. D. Aird, J. Arora, A. Barua, et al., Population genomic analysis of a pitviper reveals microevolutionary forces underlying venom chemistry. *Genome Biology and Evolution*, 9 (2018), 2640–2649.

110. M. L. van Hoek, M. D. Prickett, R. E. Settlage, et al., The Komodo dragon (*Varanus komodoensis*) genome and identification of innate immunity genes and clusters. *BMC Genomics*, 20 (2019), 684.

111. A. Georges, Q. Li, J. Lian, et al., High-coverage sequencing and annotate assembly of the genome of the Australian dragon lizard *Pogona vitticeps*. *GigaScience*, 4 (2015), 45.

112. J. Alföldi, F. Di Palma, M. Grabherr, et al., The genome of the green anole lizards and a comparative analysis with birds and mammals. *Nature*, 477 (2011), 587–591.

113. Y. Liu, Q. Zhou, Y. Wang, et al., *Gekko japonicus* genome reveals evolution of adhesive toe pads and tail regeneration. *Nature Communications*, 6 (2015), 10033.

The Evolution of Squamate Chitinase Genes (*CHIAs*) Supports an Insectivory–Carnivory Transition during the Early History of Snakes

Christopher A. Emerling

11.1 Introduction

As genome sequencing and assembly continues to improve in terms of feasibility, an increasing wealth of genomic resources have become available for research in the public domain. In regards to snakes, 2013 saw the first published serpent genomes [1, 2], and the National Center for Biotechnology Information (NCBI) hosts 27 snake assemblies available for research at the time of this writing. Although these are but a tiny fraction of the >3,800 extant species of Serpentes, they have provided evidence for the genomic basis of a number of major evolutionary events during the early history of snakes, including limb loss [2–6], the expansion of olfactory and vomeronasal sensory repertoires [1, 7], the minimization of gustation in conjunction with the forking of the tongue [4, 8], and the reduction of light-dependent physiological processes [1, 4, 9], including colour vision.

One topic that has been understudied in snake history concerns evolutionary shifts in diet. Although most of the predominantly faunivorous non-serpent squamates include arthropods in their diet [10], most extant snakes can be described as carnivorous, largely feeding on vertebrate prey [11]. Certain modifications in snake anatomy and predation strategies likely reflect such a change in diet, including the presence of recurved teeth [12], and various prey-disabling traits, such as constriction and envenomation (see Chapters 12, 18, and 19). However, there are some notable exceptions to this dietary trend in serpents: Dipsadidae, Pareidae, and Uropeltidae are or include annelid and/or mollusc specialists, Homalopsidae and Natricidae include crustacean specialists, and some of the earliest diverging lineages ('Scolecophidia') are today represented by myrmecophagous species, primarily specializing in eating ants and termites [11]. Furthermore, consumption of arthropods (= 'insectivory', as applied here; sometimes termed arthopodophagy) has been recorded in some boids, viperids, elapids, lamprophiids, and colubroids [11], making it unclear whether a major dietary shift from generalized faunivory to a more strict carnivory was associated with the early evolution of snakes.

If such an evolutionary transition in diet did indeed occur, it may be reflected in extant serpent genomes. One set of genes that may help clarify snake history encodes chitinase enzymes. Chitinases degrade chitin, a polysaccharide that makes up much of the exoskeletons of arthropods, including insects. Chitinases are found across disparate vertebrates [13–15], although genetically they have been best characterized in placental mammals [16–18]. Notably, the acidic mammalian chitinases (*CHIA*s), which are found in the digestive tracts of mammals [19–21], positively correlate with the proportion of invertebrates, primarily arthropods, in the diet [16]. Specifically, up to five functional *CHIA*s, inherited from a probable insectivorous common ancestor, are found in placental mammalian lineages that likely have had a long history of sustained insectivory (e.g., afrotherian insectivores [Afroinsectiphilia], tree shrews [Scandentia], armadillos [Cingulata], anteaters [Vermilingua], tarsiers [Tarsiidae]), whereas committed herbivores and carnivores have inactivated and/or deleted four or all five of these *CHIA*s (e.g., elephants, manatees, and hyraxes [Paenungulata], sloths [Folivora], Old World fruit bats [Pteropodidae], toothed whales [Odontoceti], rhinos and horses [Perissodactyla], carnivorans). These observations are consistent with predictions from regressive evolution, in which adapting to a new niche renders a formerly useful trait superfluous [22], frequently leading to the loss and/or inactivation of genes underlying such traits [23–26].

Given the widespread consumption of arthropods by many squamates [10], it seems plausible that chitinases may be widespread in this clade. Chitinases have been reported in the digestive tracts of some lizards [15, 27, 28], suggesting that these species may also possess *CHIA*s. However, one question is whether squamate *CHIA*s vary in number as in placental mammals, with the number of genes having expanded and/or contracted over evolutionary history. By extension, if the lineage leading to Serpentes inherited chitinases from an ancestral squamate, did snakes lose *CHIA*s *en masse,* as expected when adapting to carnivory? This chapter seeks to address these questions by examining *CHIA* gene content in serpent and other squamate genomes.

11.2 Materials and Methods

11.2.1 Data Collection

I began by BLASTing (discontiguous megablast) [29] a reference human *CHIA* (NM_021797.4) against the nucleotide collection on NCBI, restricting the search to the taxon Squamata. NCBI's Eukaryotic Genome Annotation pipeline creates gene models for many of their stored genome assemblies, providing access to annotated gene sequences. I downloaded all gene models with annotations that suggested homology with *CHIA*, such as acidic mammalian chitinase-like, chitinase-3-like protein 1 and chitotriosidase-1, and imported them into Geneious R9 [30]. I then performed additional BLASTs using the newly obtained gene models and also searched NCBI's nucleotide collection directly based on the annotation name to ensure completeness. I recorded whether the sequence was predicted to be a pseudogene, as indicated by the 'low quality' annotation, designated when a predicted frameshift and/or premature stop codon has been 'corrected' during the creation of the model. Gene models for the following squamates were available at the time of this project: *Gekko japonicus* (Schlegel's Japanese gecko; Gekkonidae), *Lacerta agilis* (sand lizard; Lacertidae), *Podarcis muralis* (common wall lizard; Lacertidae), *Zootoca vivipara* (viviparous lizard; Lacertidae), *Pogona vitticeps* (central bearded dragon; Agamidae*), Anolis carolinensis* (green anole; Dactyloidae*), Python bivittatus* (Burmese python; Pythonidae), *Protobothrops mucrosquamatus* (brown-spotted pit viper; Viperidae), *Notechis scutatus* (tiger snake; Elapidae), *Pseudonaja textilis* (eastern brown snake; Elapidae), *Pantherophis guttatus* (corn snake; Colubridae), *Thamnophis elegans* (western terrestrial garter snake; Natricidae), and *Thamnophis sirtalis* (common garter snake; Natricidae). Family-level classification follows that employed by Burbrink et al. [31].

After obtaining the gene models, to determine synteny, I BLASTed (megablast) each species' sequences against their own respective genome assemblies in NCBI's whole-genome shotgun contigs database (WGS). I recorded the contigs and scaffolds with the best hits (nearly always 100 per cent identity) and the coordinates encompassing each predicted gene. Obtaining further synteny information (i.e., any non-chitinase genes) involved BLASTing (megablast) the intergenic and flanking (15,000+ bp) regions of representative species against the nucleotide collection, restricting the search to the identical respective species. I then recorded the annotation and accession number for each non-chitinase gene.

Within Geneious, I aligned whole gene sequences from WGS contigs and scaffolds to their respective gene models using default settings in MUSCLE [32] to understand exon structure. I also used this information to obtain additional snake sequences that do not yet have gene models. I took the whole *CHIA* gene for *Python bivittatus*, encompassing the exons, introns and flanking regions, and BLASTed it (discontiguous megablast) against the WGS for the following species: *Crotalus viridis* (prairie rattlesnake; Viperidae), *Laticauda laticaudata* (blue-lipped sea krait; Elapidae), *Naja naja* (Indian cobra; Elapidae), *Ophiophagus hannah* (king cobra; Elapidae), and *Ptyas mucosa* (oriental ratsnake; Colubridae). These taxa were chosen to maximize the number of phylogenetically distinct lineages available for reconstructing *CHIA* history in early snakes, rather than to character-ize the *CHIA* complements for all available snake species. Furthermore, given that the *Gekko japonicus* assembly is relatively fragmentary, with many closely positioned genes being found on separate contigs and scaffolds (Supplementary Table 11.S1), I BLASTed (discontiguous megablast) the gene models of this species against the WGS for the gekkonid *Paroedura picta* (ocelot gecko), which has a far more contiguous assembly. All sequences were imported into Geneious for analysis.

11.2.2 Analyses

Within Geneious, I took the coding sequences of all putative squamate chitinase sequences and performed several rounds of automated sequence alignments (MUSCLE, nucleotide and translation align), plus manual adjustments. This was further guided by gene-tree estimations using the GTR-CAT model as implemented in RAxML v. 8.2.11 [33] with default parameters, until arriving on a final master alignment. Using this alignment, I translated all squamate sequences and examined them for the chitinolytic domain characteristic of functional chitinases (DXXDXDXE) [34], and the six cysteines towards the C-terminus associated with chitin-binding function [35].

After characterizing the functionality of each chitinase gene, I aligned this final batch of sequences (MUSCLE, translation align) to a *CHIA* alignment of placental mammals [16] and human *CHIT1* (chitotriosidase; NM_003465), *CHI3L1* (chitinase 3 like 1; NM_001276), *CHI3L2* (chitinase 3 like 2; NM_004000) and *OVGP1* (oviductal glycoprotein 1; NM_002557) sequences (Supplementary Appendix 11.S1), due to their being part of the wider chitinase gene family [36]. I then executed a RAxML v. 8.2.12 analysis on CIPRES (RAxML-HPC2 on XSEDE) [37] to estimate the gene tree for the squamate and mammalian chitinase genes. I used 500 bootstrap replications, but otherwise implemented default parameters. For comparison purposes, I have included dietary information for all of the focal taxa in Supplementary Table 11.S2.

11.3 Results

Despite the description of CHIAs as acidic 'mammalian' chitinases, these proteins are clearly not restricted to mammals, consistent with previous analyses [13, 36]. All examined squamates have at least one *CHIA* and as many as eight (Fig. 11.1, Supplementary Table 11. S1, Supplementary Appendix 11.S2). Notably, the examined snakes only possess a single *CHIA*, strongly contrasting with the other examined squamates: the next lowest count belongs to *Pogona vitticeps* with five *CHIA*s, seven are present in the gekkonids and *Anolis carolinensis*, and eight were found in the lacertids. Furthermore, all squamates examined have at least one chitotriosidase (*CHIT1*) and at least one chitinase-like-lectin (*CHI3L1*). Of the examined assemblies, the snake *Thamnophis elegans* is the only species with two of the former, and the lizard *A. carolinensis* is the sole species that has two of the latter.

The gene tree and synteny analyses provided the identities of some of the *CHIA* paralogs (Figs. 11.1 and 11.2, Supplementary Appendix 11.S2), but some appear to be distinct to squamates, or particular clades of squamates, and are not found in placental mammals. Placental mammal *CHIA*s have been named *CHIA1*, *CHIA2*, *CHIA3*, *CHIA4* and *CHIA5*, forming distinct clades and consistent syntenic positions in association with the genes *OVGP1*, *PIFO* (protein pitchfork) and *DENND2D* (DENN domain containing 2D) [16]:

$$OVGP1 \rightarrow CHIA1 \rightarrow CHIA2 \rightarrow\leftarrow PIFO \leftarrow CHIA5 \leftarrow CHIA4 \leftarrow CHIA3 \leftarrow DENND2D$$

One squamate *CHIA* seems to clearly be an ortholog of *CHIA3*, supported by its phylogenetic clustering with mammalian *CHIA3* (100 per cent bootstrap support [BS]) and its syntenic orientation (Figs. 11.1 and 11.2, Supplementary Appendix 11.S2). *CHIA4* and *CHIA5* also appear to be present in squamates. Although they do not group with

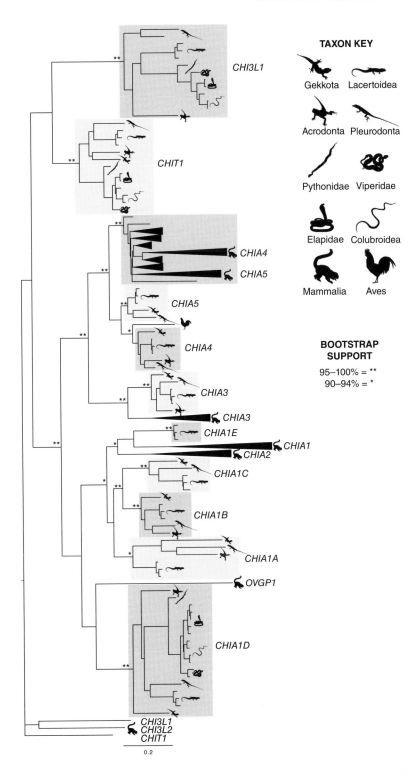

Figure 11.1 Phylogram showing the estimated relationships of squamate chitinase genes to those of placental mammals, and a bird (*Gallus gallus*). This figure is a simplification of Supplementary Appendix 11.S2. Distinct clades are indicated by boxes and names, and taxonomic representation is shown by silhouettes (obtained from phylopic.org). Bootstrap support values are indicated for internal branches at the base of distinct paralog clades, and internal branches that are more basal than these.

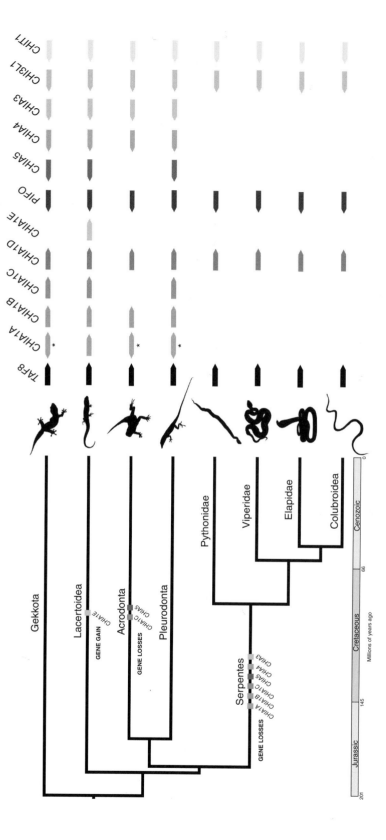

Figure 11.2 Chronogram of relevant squamate lineages, with branch lengths derived from [40], showing the syntenic pattern of *CHIA* genes in the taxa studied and the inferred patterns of *CHIA* gains and losses. Pointed ends of gene symbols indicate direction of gene relative to other genes on the same chromosome. Exceptions are *CHIA1A* sequences indicated by an asterisk and showing two pointed ends: these are located on separate contigs or scaffolds. Whether this is due to the gene truly being located elsewhere in the genome or is an artifact of the assembly will require further research. Note that although indicated as Serpentes, the clade represented is Afrophidia, due to the absence of data for amerophidian and scolecophidian taxa. Silhouettes obtained from phylopic.org. (A black and white version of this figure will appear in some formats. For the colour version, please refer to the plate section).

mammalian *CHIA4* and *CHIA5*, respectively, they are found immediately upstream of *CHIA3*, and form distinct squamate clades (*CHIA4* 86 per cent BS, *CHIA5* 99 per cent BS). Although the topology may reflect separate genes for mammalian and squamate *CHIA4* and *CHIA5*, it is highly plausible that this pattern could reflect a distortion of the topology due to gene conversion.

To elaborate, a previous study provided evidence of gene conversion in *CHIA4* and *CHIA5* in numerous major lineages of mammals (Chiroptera [bats], Afrosoricida [African insectivores], Macroscelidea [elephant shrews], Tubulidentata [aardvark], Eulipotyphla ['true' insectivores], Scandentia [tree shrews], Cingulata [armadillos]), to the point that it appeared to greatly distort the 'true' gene-tree topology [16]. Specifically, when estimating the phylogeny for placental mammal *CHIA*s, *CHIA4* became nested within *CHIA5*, and all other *CHIA4* sequences were nested within a paraphyletic arrangement of chiropteran *CHIA4* sequences. Furthermore, the sister group to *CHIA4* was estimated to be chiropteran *CHIA5*s. However, upon removing all such bat sequences from the alignment, the resulting topology yielded *CHIA4* and *CHIA5* as monophyletic sister groups. Given the evidence for phylogenetically widespread gene conversion in placental mammals, and the importance of tandem gene orientation in gene conversion [38], gene conversion may similarly explain the lack of *CHIA4* and *CHIA5* monophyly between mammals and squamates.

Beyond *CHIA3*, *CHIA4,* and *CHIA5*, the five remaining squamate *CHIA*s appear to phylogenetically group with *CHIA1*, *CHIA2,* and human *OVGP1*, each representing a distinct clade (Fig. 11.1, Supplementary Appendix 11.S2). At least four of the five, when present, are in the same syntenic sequence, inverted in orientation relative to the *CHIA3*–*CHIA5* grouping (Fig. 11.2), similar to *CHIA1*, *CHIA2* and *OVGP1* in mammals. Given that the topology and gene order prevent confident identification of any of the squamate genes as *CHIA1* or *CHIA2*, here I tentatively describe them as *CHIA1A*, *CHIA1B*, *CHIA1C*, *CHIA1D,* and *CHIA1E*, which are oriented in that order when present on the same contig or scaffold. *CHIA1A*, however, is typically found on a separate contig or scaffold (Supplementary Table 11.S1), suggesting a separate genomic location for this gene. Furthermore, each of these *CHIA* clades has consistently strong bootstrap support: 92 per cent (*CHIA1A*), 99 per cent (*CHIA1E*) and 100 per cent (*CHIA1B*, *CHIA1C*, *CHIA1D*). Whereas *CHIA1A*, *CHIA1B*, *CHIA1C*, *CHIA1D*, *CHIA3*, *CHIA4,* and *CHIA5* are all represented in the gekkonids, *Anolis carolinensis,* and the three lacertids, *CHIA1E* is found only in the lacertids, the agamid *Pogona vitticeps* lacks *CHIA1C* and *CHIA5*, and the snakes possess only *CHIA1D*. Possible pseudogenes include *CHIA1C* in *A. carolinensis* and *CHIA1D* in *Gekko japonicus* and *Notechis scutatus*.

Exon structure varies across the *CHIA* paralogs (Supplementary Table 11.S1): *CHIA1A* appears to range from 10 exons in *Gekko japonicus*, 15 in *Anolis carolinensis* and 20 in *Zootoca vivipara*; *CHIA1B* has either nine (*Paroedura picta*) or 11 (*Z. vivipara*); *CHIA1C* has 5 (*Z. vivipara*), 10 (*A. carolinensis*) or 11 (*P. picta*); *CHIA1D* has 10 exons across all 12 examined species; *CHIA1E* has 11 in *Z. vivipara*; *CHIA3* has nine (*A. carolinensis*) or 11 (*P. picta, Z. vivipara*); *CHIA4* has 11 (*A. carolinensis, P. picta, Z. vivipara*); and *CHIA5* has seven (*A. carolinensis*) or 11 (*P. picta, Z. vivipara*). For comparison, all five mammalian *CHIA*s are reported to have 11 exons [16]. Although some deviations from this number in squamates may be the result of assembly errors and/or poor gene model reconstruction, at least some are plausibly real, such as the consistent recovery of 10 exons in *CHIA1D* in 12 species. Future studies will benefit from mRNA sequence data to validate the exon structure suggested by these gene models.

Among the CHIAs, there is variability in the presence of functional chitinolytic and chitin-binding domains (Supplementary Table 11.S1). Whereas CHIA1B, CHIA1E, CHIA3, CHIA4 and CHIA5 all have both domains intact in all species examined, CHIA1A, CHIA1C and CHIA1D appear to vary in domain functionality. In CHIA1A, *Gekko japonicus* lacks both domains, *Pogona vitticeps* and *Anolis carolinensis* lack a chitin-binding domain, whereas all three lacertids retain both intact. Furthermore, *Zootoca vivipara* appears to lack a chitinolytic domain in CHIA1C, *A. carolinensis* seemingly has no chitin-binding domain, and the other three squamates have both intact. Finally, CHIA1D has a canonical chitinolytic domain in 16 of 18 sampled squamates, the exceptions being both *Thamnophis* spp., but the exon containing the chitin-binding domain is absent in all squamate species. As expected for chitinase-like-lectins [39], the CHI3L1 sequences appear to lack the potential for chitinase activity, given the absence of both the chitinolytic and chitin-binding domains. By contrast, CHIT1 varies, with most sequences possessing a chitinolytic domain and lacking a chitin-binding domain. The exceptions to the former include one of the two *Thamnophis elegans* CHIT1s, and those of *Pogona vitticeps* and all three lacertids. The only *CHIT1* gene models retaining chitin-binding domains are, again, *Pogona vitticeps* and the lacertids.

11.4 Discussion

This is the first detailed analysis of the evolution of vertebrate chitinase genes in a taxon other than placental mammals, and it reveals that multiple *CHIAs* have a history at least as old as the origin of crown amniotes, roughly 320 million years ago [40]. Given uncertainties in the exact relationships of the paralogs, it is unclear how many *CHIAs* would have been present in the earliest amniotes, but it is plausible that there were at least five and perhaps more. The phylogenetic distribution of *CHIA* genes, however, more clearly suggests that the earliest crown squamates, some 200 million years ago [40, 41], had seven enzymatically active *CHIAs* (Fig. 11.2). Terrestrial ecosystems had been colonized repeatedly by chitinous arthropods, perhaps as early as 300 million years prior to the first squamates, with tremendous diversification during this intervening period [42–45]. Presumably, as vertebrates adapted on land, such chitinous prey provided an invaluable energetic resource [46]. Being able to hydrolyze the primary component of arthropod exoskeletons would liberate other tissues for digestion, making chitinases an indispensable part of the earliest squamates' genomic repertoire. The high frequency with which arthropods remain a part of typical lizards' diets [10] supports this inference.

By contrast, the snakes sampled for this study appear to have lost nearly all of their *CHIAs*, suggesting the retention of only one, at least by the common ancestor of Afrophidia (all snakes except Amerophidia and 'Scolecophidia'). The absence of additional chitinases in afrophidian snakes is supported not only by the absence of gene models for the examined species, but the retention of the same syntenic orientation with flanking genes (Fig. 11.2), typically on the same contigs and scaffolds (Supplementary Table 11.S1). Further interrogations of the intergenic regions failed to yield evidence of *CHIA* pseudogenes in snakes, though more sensitive analyses may uncover some in future studies. Together, these results suggest that the absence of other *CHIAs* is not an issue of genome incompleteness, but rather reflects a major loss of these genes between when the serpent lineage diverged from other extant toxicoferans and prior to the origin of afrophidian snakes.

The extensive loss of *CHIAs* in placental mammals is typically associated with inferred transitions from insectivory to either herbivory or carnivory [16], the latter of which plausibly explains the pattern seen in snakes. Although variations in diet exist, snakes largely consume vertebrate prey [11], and a shift to such a diet is expected to lead to regression of the chitinase complement that was probably present in their arthropod-consuming ancestors. Whereas most cases of *CHIA* gene loss have resulted in pseudogenes in placental mammals [16], the genes appear to be completely deleted, at least in afrophidian snakes.

Underlying this largely carnivorous behaviour is the presence of recurved teeth in most major lineages of serpents, a trait notably also found in carnivorous helodermatids and varanids [12, 47]. Importantly, early snake and putative snake fossils typically possess recurved teeth [48–51], further suggesting a transition to carnivory occurred during the early evolution of Serpentes. Cretaceous snake fossils, including vertebrate gut contents in *Pachyrhachis problematicus* and the putative stem serpent *Tetrapodophis amplectus* [48, 52], plus suggestive predatory behaviour on sauropod hatchlings in *Sanajeh indicus* [53], further bolster this interpretation.

A limitation to this study, as well as to all genomic research on snakes to this point, is the sparse taxonomic representation for assembled genomes. Critically, although a handful of afrophidian genomes have been sequenced and assembled (e.g., Pythonidae, Viperidae, Elapidae, Colubroidea), as of this writing, there are currently no genome assemblies available for the sister group to afrophidians (Amerophidia), nor the more basally-diverging 'scolecophidian' clades (Anomalepididae, Typhlopoidea, Leptotyphlopidae). Notably, the scolecophidian lineages predominantly, or exclusively, consume arthropods [11]. Therefore, it remains to be seen whether they retain the typical squamate complement of chitinase genes, or if they represent a secondary specialization in insectivory that inherited a depauperate set of *CHIAs* from a lineage of carnivorous stem snakes. In placental mammals, both patterns have seemingly been uncovered, with multiple lineages of insectivorous mammals retaining the ancestral set of five *CHIAs*, presumably reflecting a continuation of the dietary habits that they inherited, whereas some chitinophagous lineages such as mysticetes (baleen whales) and pholidotans (pangolins) appear to have inherited a single *CHIA* from non-insectivorous ancestors [16].

The possibility of a secondary adaptation to insectivory applies not only to scolecophidians but also to several lineages of afrophidian snakes. Such examples include the colubrids *Scolecophis atrocinctus* (black-banded snake), *Tantilla* spp. (centipede snakes),

Eirenis spp. (dwarf racers), *Stenorrhina* spp., and *Opheodrys* spp. (green snakes) [11]. To be better adapted at consuming chitin, these species may employ novel duplicated *CHIA*s, amplify the expression of any remaining *CHIA*s, or perhaps utilize chitinophagous microbiota. Indeed, the latter two have been suggested as possible means by which myticetes and pholidotans have compensated for their diminished *CHIA* paralog counts [16, 54], and the discovery here of an apparently lacertid-specific chitinase (CHIA1E) suggests gene duplication remains a viable option. Future researchers would do well to explore these poorly studied taxa to better understand the genomic evolution underpinning the anatomical and physiological traits associated with the origin and diversification of snakes, including the inferred major transitions in dietary habits.

Acknowledgements

I thank Jeff Streicher and Mareike Janiak for taking the time to review the manuscript, and Jason Head, Hussam Zaher and David Gower for additional helpful comments, all of which improved the manuscript.

References

1. T. A. Castoe, A. P. J. De Koning, K. T. Hall, et al., The Burmese python genome reveals the molecular basis for extreme adaptation in snakes. *Proceedings of the National Academy of Sciences USA*, 110 (2013), 20645–20650.
2. F. J. Vonk, N. R. Casewell, C. V. Henkel, et al., The king cobra genome reveals dynamic gene evolution and adaptation in the snake venom system. *Proceedings of the National Academy of Sciences USA*, 110 (2013), 20651–20656.
3. A. Ullate-Agote, M. C. Milinkovitch, and A. C. Tzikia, The genome sequence of the corn snake (*Pantherophis guttatus*), a valuable resource for EvoDevo studies in squamates. *International Journal of Developmental Biology*, 58 (2014), 881–888.
4. C. A. Emerling, Genomic regression of claw keratin, taste receptor and light-associated genes provides insights into biology and evolutionary origins of snakes. *Molecular Phylogenetics and Evolution*, 115 (2017), 40–49.
5. C. R. Infante, A. G. Mihala, S. Park, et al., Shared enhancer activity in the limbs and phallus and functional divergence of a limb-genital cis-regulatory element in snakes. *Developmental Cell*, 35 (2015), 107–119.
6. E. Z. Kvon, O. K. Kamneva, U. S. Melo, et al., Progressive Loss of function in a limb enhancer during snake evolution. *Cell*, 167 (2016), 633-642.e11.
7. T. Kishida, Y. Go, S. Tatsumoto, et al., Loss of olfaction in sea snakes provides new perspectives on the aquatic adaptation of amniotes. *Proceedings of the Royal Society B*, 286 (2019), 2019.1828.
8. H. Zhong, S. Shang, X. Wu, et al., Genomic evidence of bitter taste in snakes and phylogenetic analysis of bitter taste receptor genes in reptiles. *PeerJ*, 5 (2017), e3708.
9. B. W. Perry, D. C. Card, J. W. McGlothlin, et al., Molecular adaptations for sensing and securing prey and insight into amniote genome diversity from the garter snake genome. *Genome Biology and Evolution*, 10 (2018), 2110–2129.
10. S. Meiri, Traits of lizards of the world: Variation around a successful evolutionary

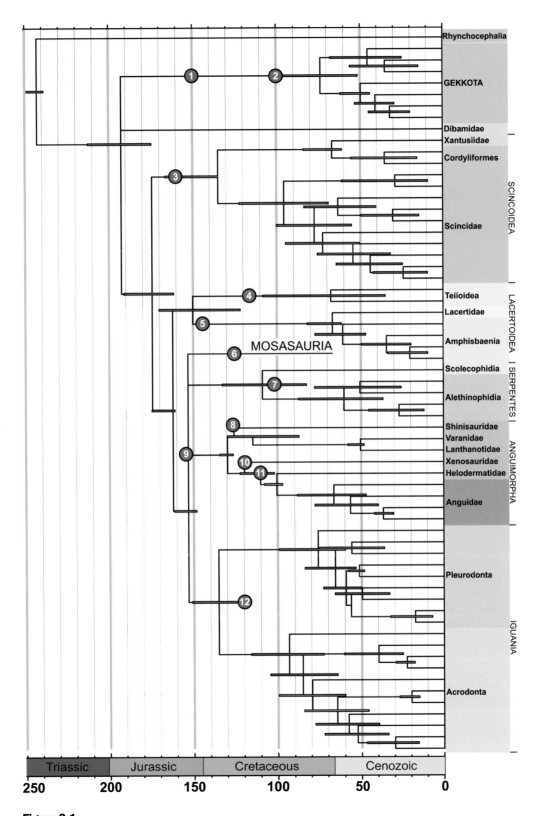

Figure 2.1

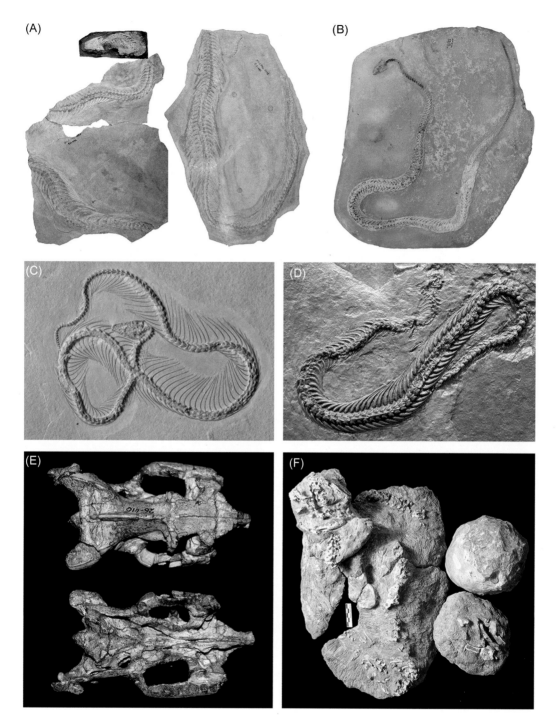

Figure 3.1

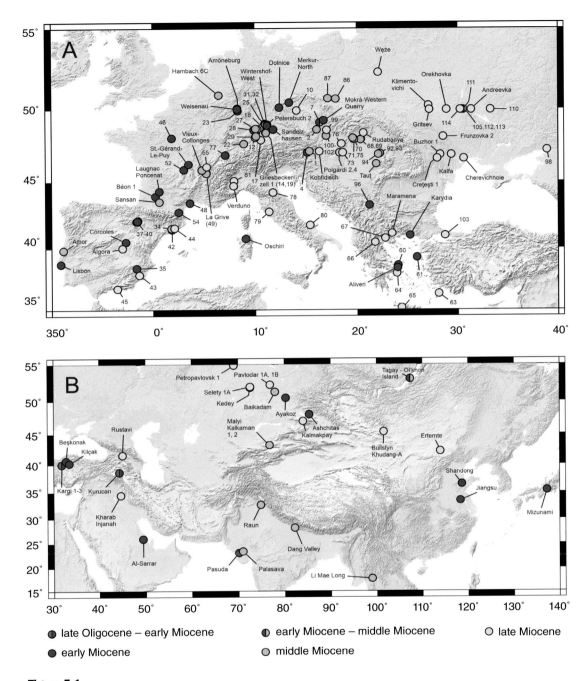

Figure 5.1

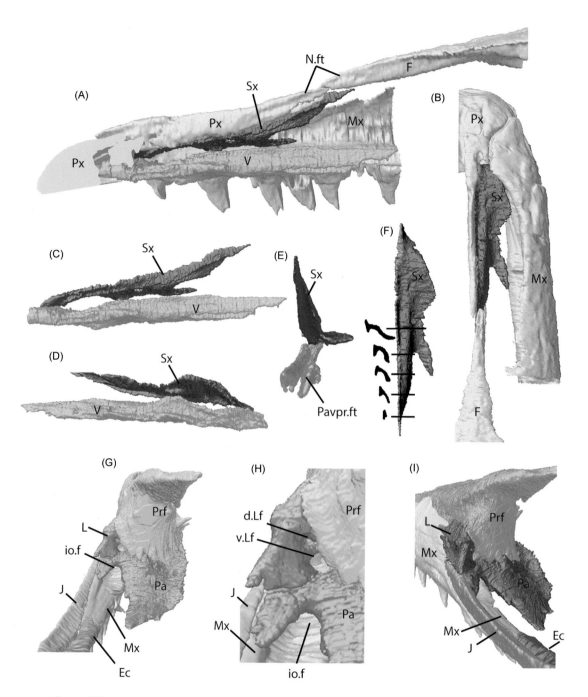

Figure 7.3

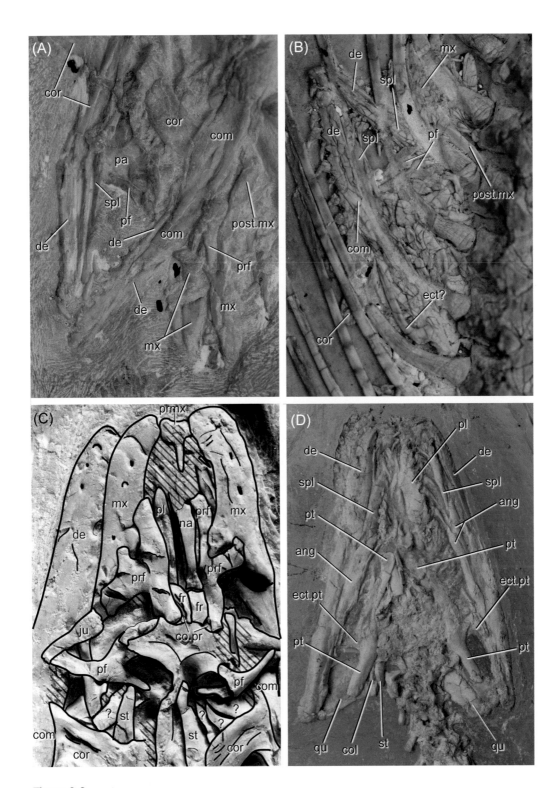

Figure 9.3

Figure 9.5

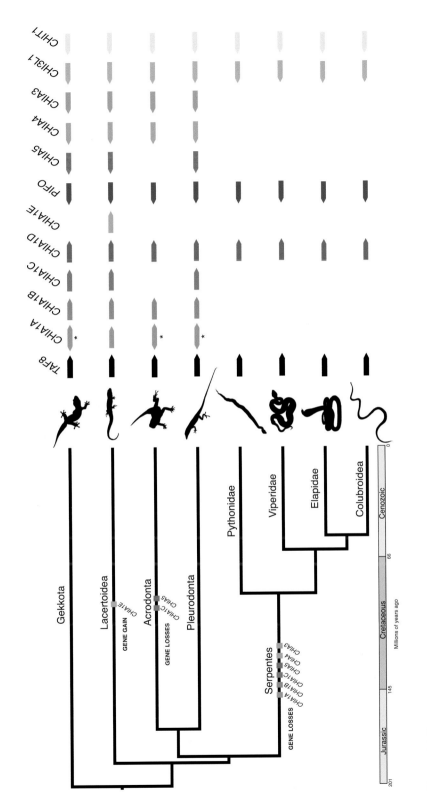

Figure 11.2

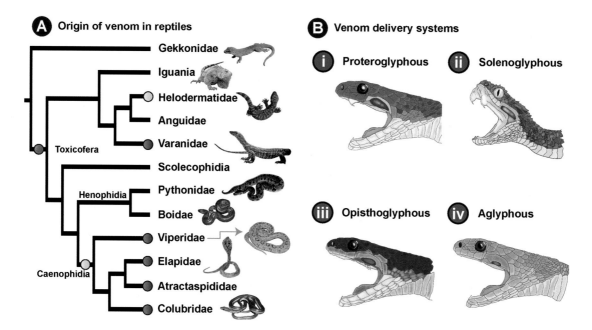

A Origin of venom in reptiles

- Gekkonidae
- Iguania
- Helodermatidae
- Anguidae
- Varanidae
- Scolecophidia
- Pythonidae
- Boidae
- Viperidae
- Elapidae
- Atractaspididae
- Colubridae

Toxicofera

Henophidia

Caenophidia

B Venom delivery systems

i Proteroglyphous

ii Solenoglyphous

iii Opisthoglyphous

iv Aglyphous

Figure 12.1

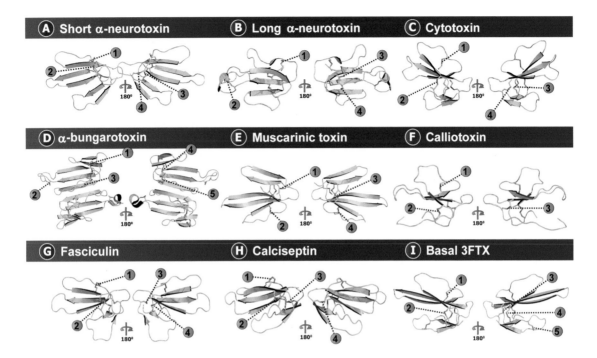

Figure 12.2

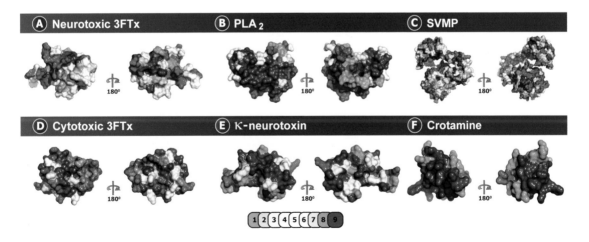

Figure 12.3

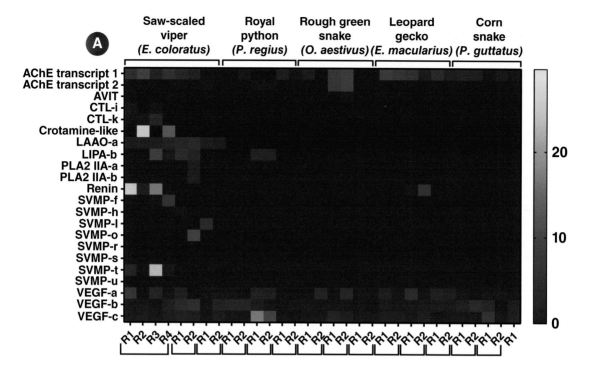

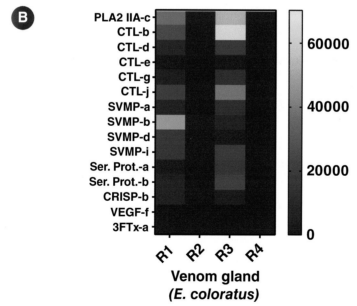

Figure 12.4

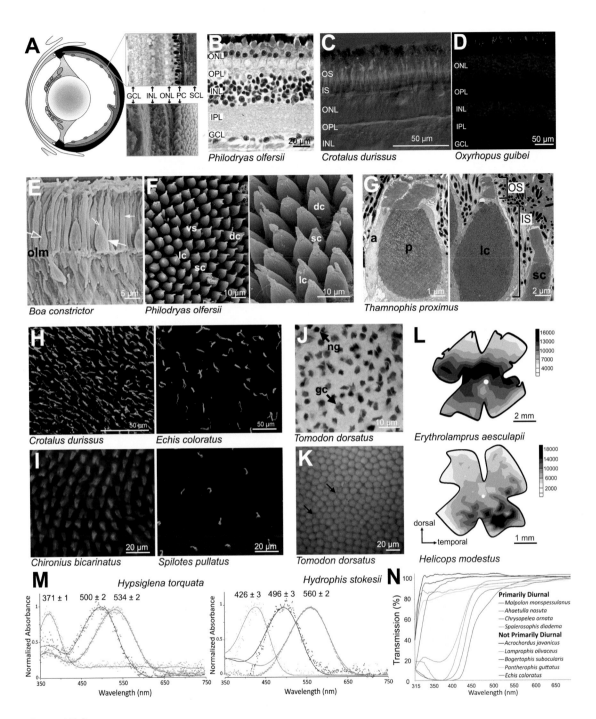

A

GCL INL ONL PC SCL

B ONL / OPL / INL / IPL / GCL — 20 µm
Philodryas olfersii

C OS / IS / ONL / OPL / INL — 50 µm
Crotalus durissus

D ONL / OPL / INL / IPL / GCL — 50 µm
Oxyrhopus guibei

E olm — 5 µm
Boa constrictor

F vs / dc / lc / sc — 10 µm / dc / sc / lc — 10 µm
Philodryas olfersii

G a / p — 1 µm / OS / IS / lc / sc — 2 µm
Thamnophis proximus

H 50 µm
Crotalus durissus   *Echis coloratus* — 50 µm

J ng / gc — 10 µm
Tomodon dorsatus

L 16000 / 13000 / 10000 / 7000 / 4000 — 2 mm
Erythrolamprus aesculapii

18000 / 14000 / 10000 / 6000 / 2000 — 1 mm
dorsal / temporal
Helicops modestus

I 20 µm
Chironius bicarinatus   *Spilotes pullatus* — 20 µm

K 20 µm
Tomodon dorsatus

M *Hypsiglena torquata*
371 ± 1   500 ± 2   534 ± 2
Normalized Absorbance / Wavelength (nm)

Hydrophis stokesii
426 ± 3   496 ± 3   560 ± 2
Normalized Absorbance / Wavelength (nm)

N Transmission (%) / Wavelength (nm)
Primarily Diurnal
— *Malpolon monspessulanus*
— *Ahaetulla nasuta*
— *Chrysopelea ornata*
— *Spalerosophis diadema*
Not Primarily Diurnal
— *Acrochordus javanicus*
— *Lamprophis olivaceus*
— *Bogertophis subocularis*
— *Pantherophis guttatus*
— *Echis coloratus*

Figure 15.4

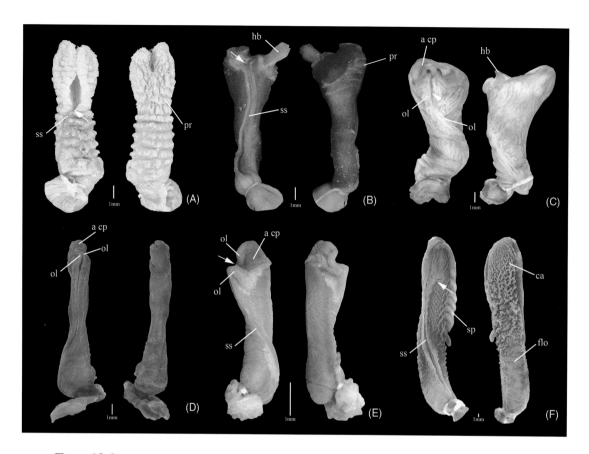

Figure 16.4

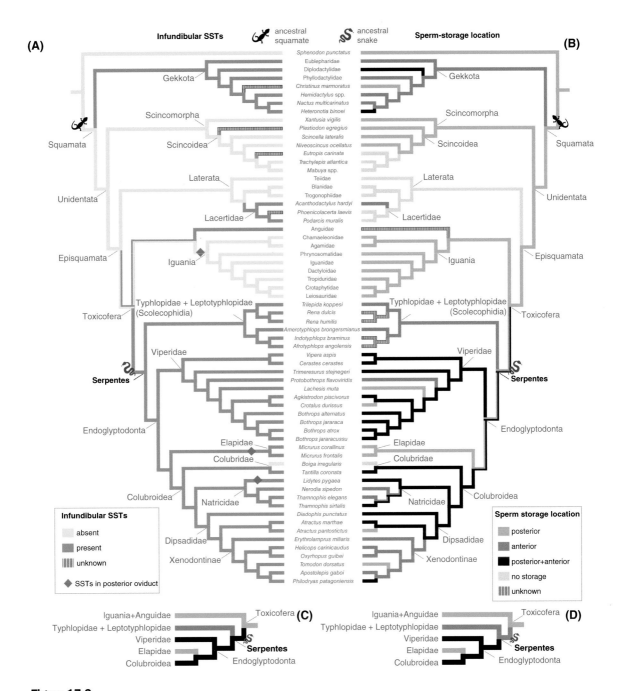

Figure 17.2

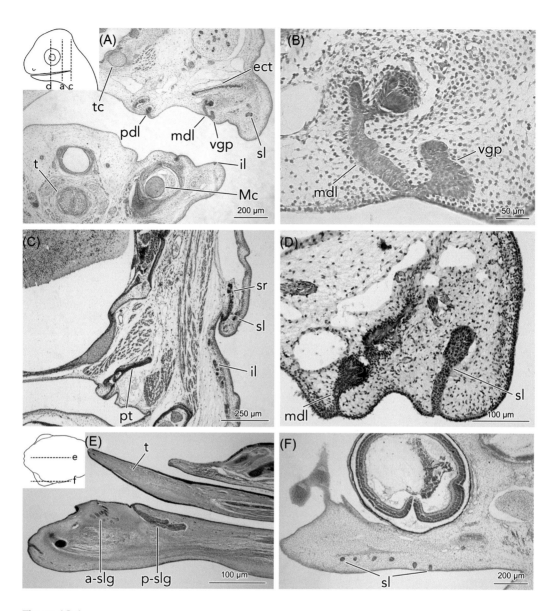

Figure 18.1

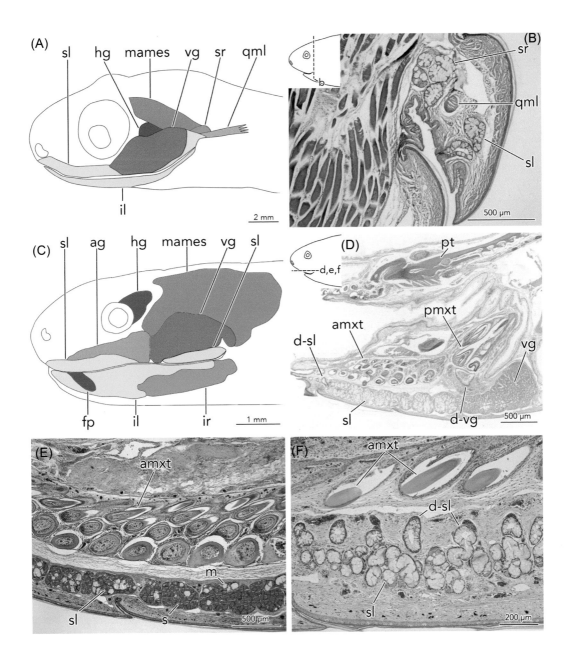

Figure 18.4

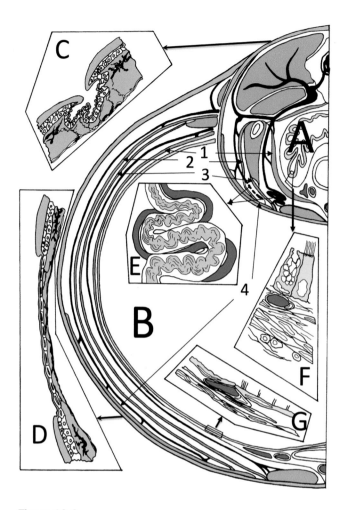

Figure 19.2

design. *Global Ecology and Biogeography*, 27 (2018), 1168-1172.

11. M. C. Grundler, SquamataBase: A natural history database and R package for comparative biology of snake feeding habits. *Biodiversity Data Journal*, 8 (2020), e49943.

12. T. Rowe, *DigiMorph*. www.DigiMorph.org (2020)

13. Y. H. Chenand, and H. Zhao, Evolution of digestive enzymes and dietary diversification in birds. *PeerJ*, 7 (2019), e6840.

14. D. P. German, B. C. Nagle, J. M. Villeda, et al., Evolution of herbivory in a carnivorous clade of minnows (Teleostei: Cyprinidae): Effects on gut size and digestive physiology. *Physiological and Biochemical Zoology*, 83 (2010), 1-18.

15. C. Jeuniaux, Chitinase: an addition to the list of hydrolases in the digestive tract of vertebrates. *Nature*, 192 (1961), 135-136.

16. C. A. Emerling, F. Delsuc, and M. W. Nachman, Chitinase genes (CHIAs) provide genomic footprints of a post-Cretaceous dietary radiation in placental mammals. *Science Advances*, 4 (2018), eaar6478.

17. M. C. Janiak, M. E. Chaney, and A. J. Tosi, Evolution of acidic mammalian chitinase genes (*CHIA*) is related to body mass and insectivory in primates. *Molecular Biology and Evolution*, 35 (2018), 607-622.

18. K. Wang, S. Tian, J. Galindo-González, et al., Molecular adaptation and convergent evolution of frugivory in Old World and neotropical fruit bats. *Molecular Ecology*, 29 (2020), 4366-4381.

19. R. G. Boot, E. F. C. Blommaart, E. Swart, et al., Identification of a novel acidic mammalian chitinase distinct from chitotriosidase. *Journal of Biological Chemistry*, 276 (2001), 6770-6778.

20. S. Strobel, A. Roswag, N. I. Becker, T. E. Trenczek, and J. A. Encarnação, Insectivorous bats digest chitin in the stomach using acidic mammalian chitinase. *PLoS One*, 8 (2013), e72770.

21. M. Ohno, M. Kimura, H. Miyakazi, et al. Acidic mammalian chitinase is a proteases-resistant glycosidase in mouse digestive system. *Scientific Reports*, 6 (2016), 37756.

22. D. Fong, T. Kane, and D. Culver, Vestigialization and loss of nonfunctional characters. *Annual Review of Ecology and Systematics*, 26 (1995), 249-268.

23. C. A. Emerling and M. S. Springer, Eyes underground: Regression of visual protein networks in subterranean mammals. *Molecular Phylogenetics and Evolution*, 78 (2014), 260-270.

24. R. W. Meredith, G. Zhang, M. T. P. Gilbert, E. D. Jarvis, and M. S. Springer, Evidence for a single loss of mineralized teeth in the common avian ancestor. *Science*, 346 (2014), 1254390.

25. R. Albalat and C. Cañestro, Evolution by gene loss. *Nature Reviews Genetics*, 17 (2016), 379-391.

26. P. Feng, J. Zheng, S. J. Rossiter, D. Wang, and H. Zhao, Massive losses of taste receptor genes in toothed and baleen whales. *Genome Biology and Evolution*, 6 (2014), 1254-1265.

27. R. S. Marsh, C. Moe, R. B. Lomneth, J. D. Fawcett, and A. Place, Characterization of gastrointestinal chitinase in the lizard *Sceloporus undulatus garmani* (Reptilia: Phrynosomatidae). *Comparative Biochemistry and Physiology Part B: Biochemistry and Molecular Biology*, 128 (2001), 675-682.

28. I. Koludarov, T. N. W. Jackson, B. op den Brouw, et al., Enter the dragon: The dynamic and multifunctional evolution of Anguimorpha lizard venoms. *Toxins*, 9 (2017), 242.

29. S. F. Altschul, W. Gish, W. Miller, E. W. Myers, and D. Lipman, Basic local alignment search tool. *Journal of Molecular Biology*, 215 (1990), 403-410.

30. M. Kearse, R. Moir, A. Wilson, et al., Geneious Basic: An integrated and extendable desktop software platform for the organization and analysis of sequence data. *Bioinformatics.* 28 (2012), 1647-1649.

31. F. T. Burbrink, F. G. Grazziotin, R. A. Pyron, et al., Interrogating genomic-scale data for

Squamata (lizards, snakes, and amphisbaenians) shows no support for key traditional morphological relationships. *Systematic Biology*, 69 (2020), 502–520.

32. R. C. Edgar, MUSCLE: Multiple sequence alignment with high accuracy and high throughput. *Nucleic Acids Research*, 32 (2004), 1792–1797.

33. A. Stamatakis, RAxML version 8: A tool for phylogenetic analysis and post-analysis of large phylogenies. *Bioinformatics*, 30 (2014), 1312–1313.

34. A. M. Olland, J. Strand, E. Presman, et al., Triad of polar residues implicated in pH specificity of acidic mammalian chitinase. *Protein Science*, 18 (2009), 569–578.

35. L. W. Tjoelker, L. Gosting, S. Frey, et al., Structural and functional definition of the human chitinase chitin-binding domain. *Journal of Biological Chemistry*, 275 (2000), 514–520.

36. M. Hussain and J. B. Wilson, New paralogues and revised time line in the expansion of the vertebrate GH18 family. *Journal of Molecular Evolution*, 76 (2013), 240–260.

37. M. A. Miller, W. Pfeiffer, and T. Schwartz, *Creating the CIPRES Science Gateway for inference of large phylogenetic trees. 2010 Gateway Computing Environments Workshop (GCE)* (2010), 1–8.

38. J. M. Chen, D. N. Cooper, N. Chuzhanova, C. Férec, and G. P. Patrinos, Gene conversion: Mechanisms, evolution and human disease. *Nature Reviews Genetics*, 8 (2007), 762–775.

39. G. H. Renkema, R. G. Boot, F. L. Au, et al., Chitotriosidase a chitinase, and the 39-kDa human cartilage glycoprotein, a chitin-binding lectin, are homologues of family 18 glycosyl hydrolases secreted by human macrophages. *European Journal of Biochemistry*, 251 (1998), 504–509.

40. I. Irisarri, D. Baurain, H. Brinkmann, et al., Phylotranscriptomic consolidation of the jawed vertebrate timetree. *Nature Ecology and Evolution*, 1 (2017), 1370–1378.

41. Y. Zheng and J. J. Wiens, Combining phylogenomic and supermatrix approaches, and a time-calibrated phylogeny for squamate reptiles (lizards and snakes) based on 52 genes and 4162 species. *Molecular Phylogenetics and Evolution*, 94 (2016), 537–547.

42. J. Lozano-Fernandez, A. R. Tanner, M. N. Puttick, et al., A Cambrian–Ordovician terrestrialization of arachnids. *Frontiers in Genetics*, 11 (2020), 182.

43. L. S. F. Lins, S. Y. W. Ho, and N. Lo, An evolutionary timescale for terrestrial isopods and a lack of molecular support for the monophyly of Oniscidea (Crustacea: Isopoda). *Organisms, Diversity and Evolution*, 17 (2017), 813–820.

44. B. Misof, S. Liu, K. Meusemann, et al., Phylogenomics resolves the timing and pattern of insect evolution. *Science*, 346 (2014), 763–767.

45. R. Fernández, G. D. Edgecombe, and G. Giribet, Phylogenomics illuminates the backbone of the Myriapoda Tree of Life and reconciles morphological and molecular phylogenies. *Scientific Reports*, 8 (2018), 83

46. S. P. Modesto, D. M. Scott, and R. R. Reisz, Arthropod remains in the oral cavities of fossil reptiles support inference of early insectivory. *Biology Letters*, 5 (2009), 838–840.

47. K. M. Melstrom, The relationship between diet and tooth complexity in living dentigerous saurians. *Journal of Morphology*, 278 (2017), 500–522.

48. D. M. Martill, H. Tischlinger, and N. R. Longrich, A four-legged snake from the Early Cretaceous of Gondwana. *Science*, 349 (2015), 416–419.

49. M. W. Caldwell, R. L. Nydam, A. Palci, and S. Apesteguía, The oldest known snakes from the Middle Jurassic-Lower Cretaceous provide insights on snake evolution. *Nature Communications*, 6 (2015), 5996.

50. N. R. Longrich, B. A. S. Bhullar, and J. A. Gauthier, A transitional snake from the Late Cretaceous period of North America. *Nature*, 488 (2012), 205–208.

51. M. W. Caldwell and M. S. Y. Lee, A snake with legs from the marine Cretaceous of the Middle East. *Nature*, 386 (1997), 705–708.

52. G. Haas, On a new snakelike reptile from the Lower Cenomanian of Ein Jabrud, near Jerusalem. *Bulletin du Muséum National d'Histoire Naturelle Paris*, 1 (1979), 51–64.

53. J. A. Wilson, D. M. Mohabey, S. E. Peters, and J. J. Head, Predation upon hatchling dinosaurs by a new snake from the Late Cretaceous of India. *PLoS Biology*, 8 (2010), e1000322.

54. J. G. Sanders, A. C. Beichman, J. Roman, et al., Baleen whales host a unique gut microbiome with similarities to both carnivores and herbivores. *Nature Communications*, 6 (2015), 8285.

Origin and Early Diversification of the Enigmatic Squamate Venom Cocktail

Vivek Suranse*, Ashwin Iyer*, Timothy N. W. Jackson*, and Kartik Sunagar*

12.1 Introduction

Nature is replete with examples of strategies so effective that they evolve time and again, independently in divergent lineages. These strategies are represented by functional traits associated with them. Eyes are one such trait, and venom is another. Venom has evolved independently in over 100 lineages across the animal kingdom, including cnidarians, scorpions, spiders, mammals, and squamate reptiles [1]. Venom is a secretory concoction of biochemically disparate components that is deployed by one organism (the venomous organism) to disrupt the normal physiological and biochemical functions of another (the target organism) in ways that facilitate feeding, defence, or competitor deterrence. Venoms are produced by specialized tissues (either a compartmentalized gland or specialized cells), and actively inoculated to the target animal through the infliction of wounds [2]. Thus, in contrast to the historical definition, the modern definition of venom is inclusive of haematophagous (blood-feeding) animals, such as vampire bats, lampreys, and leeches that possess adaptations to facilitate their unique diet [2]. During the course of evolution, animal venoms have diversified under the influence of natural selection to attack myriad regulatory systems responsible for homeostasis. To efficiently deliver this complex bioactive cocktail, venomous organisms have evolved diverse delivery systems such as fangs, forcipules, harpoons, spines, stingers, and stinging cells (the nematocytes of cnidarians).

Though over a hundred convergent origins of venom have been documented in the animal kingdom, the emergence of this trait in squamate reptiles has been a matter of contentious debate. Our understanding of the origin and evolution of squamate venoms

---

* These authors contributed equally

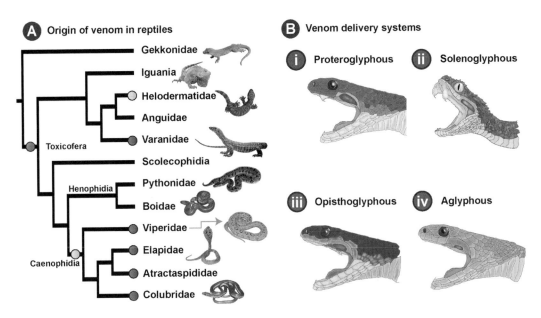

Figure 12.1 (A) Alternative hypotheses of the evolution of venom in squamate reptiles, showing single (red), dual (yellow) and multiple (purple) origins. The phylogeny shown here is based on [10]. (B) The position of fangs and venom glands (red) in front-fanged (i: proteroglyphous and ii: solenoglyphous), rear-fanged (iii: opisthoglyphous) and fangless (iv: aglyphous) snakes, highlighting the diversity of venom-delivery systems in this lineage. (A black and white version of this figure will appear in some formats. For the colour version, please refer to the plate section).

has been somewhat hampered by a historical focus of venom research on species that are medically important to humans. Traditionally, reptiles that did not pose a medical threat to humans were considered non-venomous, and venom was perceived to be restricted to deeply nested clades. In the recent past, venom research from an ecological and evolutionary perspective has argued against this popular belief. Nonetheless, multiple distinct scenarios have been described in attempts to explain the evolutionary origins of venom in squamates (Fig. 12.1A). The so-called 'Toxicofera Hypothesis', for example, argues for a single early origin of venom in the common ancestor of the Toxicofera clade, which comprises snakes (Serpentes) and anguimorph and iguanian lizards [3–6]. In contrast, others have favoured a hypothesis with at least two independent origins of venom in squamates, one in 'advanced snakes' (Caenophidia) and another in helodermatid anguimorphs [7, 8]. Furthermore, multiple origins of venom in snakes have also been suggested [9]. Though our current understanding may point to a single origin of venom in squamates, unequivocal evidence to resolve this long-standing debate is lacking. Although tracing the precise origin of venom in squamates is crucial for understanding the evolution and early diversification of snakes and lizards, the complex and dynamic nature of venom evolution and its possible existence as a 'marginal trait' in many taxa may result in this knowledge remaining elusive. In this chapter, we summarize current understanding of the phylogenetic origins of squamate venoms and the mechanisms that have underpinned their evolutionary diversification. We highlight the major limitations of various hypotheses that have been proposed to explain the origin(s) of venom in this enigmatic clade.

12.2 Venom from an Ecological Perspective

Venom composition is influenced by numerous ecological and environmental factors, such as diet, ontogeny, predator pressure, intraspecific competition, and geographic distribution [11–14]. Animal venoms often evolve remarkable specificity towards the natural prey and predators of the organisms that produce them, which may render them relatively ineffective against mammals, including humans. However, a natural anthropocentric bias in our thinking has historically seen only animals that are capable of inflicting clinically significant envenomings on humans labelled 'venomous'. Having been predominantly assessed within this anthropocentric framework, the ecological roles of venom have largely been ignored. Recently, the importance of understanding venom as an ecological trait has begun to be more widely recognized, although 'venom ecology' remains a neglected area within toxinology [15, 16].

Considerable differences in snake venom composition and activity are observable at various phylogenetic levels. Snake venoms can vary starkly among families [17], closely related species [18], geographically disparate populations of the same species [11, 13, 19–21], and even developmental stages in the life of a single individual [22–24]. Although the majority of venom components are shared by all snakes, some toxin types in the venoms of Elapidae, Viperidae, and 'non-front-fanged' lineages within Caenophidia, have become the defining features of each of these clades. While differences in venom composition at higher taxonomic levels are explained by the dynamic recruitment and diversification of particular toxin types, prominent differences at lower taxonomic levels are attributed to differing ecology and environmental conditions.

Diet is a crucial ecological factor governing venom composition. Stark variations in venom profiles associated with dietary shifts are frequently encountered in widely distributed snake species [11, 25]. Because snake venoms are evolutionary innovations that primarily facilitate subduing of prey and defense against predators, they often evolve remarkable specificity and potency towards natural prey or predators. For example, some saw-scaled vipers (*Echis* spp.) that chiefly prey upon scorpions attain a venom composition that is highly potent towards arthropods [26], while tree snake (e.g., *Boiga irregularis*) venoms contain toxins that are minimally toxic to mammals but extremely toxic to the lizards and birds that they primarily feed upon [27]. Likely as a consequence of the energetic costs associated with the production of complex venom cocktails, complete degeneration of toxin-encoding genes and associated venom-delivery apparatus has been observed in oophagous sea snakes that no longer depend on venom for predation [28]. A switch in the relative expression of venom components has also been reported across the developmental stages of snake species, strongly correlating with dietary shifts [22, 23]. The molecular mechanisms underpinning such dramatic shifts are not fully understood, although small RNAs (e.g., microRNA) have been implicated in shaping ontogenetic changes in venom profiles [29].

The reciprocal evolution of increased toxin potency and toxin resistance is another important process in venom evolution. Several lineages have convergently evolved resistance by adopting similar molecular strategies, such as altering receptors targeted by toxins, overexpression of decoy receptors, or evolving toxin-specific enzyme inhibitors [30–32]. Adaptive venom resistance in ground squirrels (*Spermophilus beecheyi*) against rattlesnake (*Crotalus viridis oreganus*) venoms, and in opossums (Marsupialia: Didelphidae) that feed on pit vipers, are among the best-studied examples of such 'chemical arms races' [33–35]. Ground squirrels and opossums have independently acquired toxin inhibitors in their blood, while convergent mutations in the acetylcholine receptors of the Egyptian mongoose (*Herpestes ichneumon*) and honey badger (*Mellivora capensis*) enable these carnivores to prey on various snake species [36, 37].

12.3 Composition of Venom Cocktails

Venoms are complex cocktails of bioactive molecules, such as proteins, carbohydrates, salts, and other organic components [38]. Despite the enormous diversity of toxic molecules in venoms, several components have been independently recruited by many animal lineages. Such remarkable convergence may be the result of evolutionary contingencies, such as the existence of ubiquitous molecular targets in prey and predators [39]. Despite being constituted by diverse toxin types from distinct gene families (Table 12.1), many components exert toxicities by functioning in concert. For instance, hyaluronidase in snake venoms acts as a 'spreading factor' by degrading hyaluronan of the extracellular matrix and facilitating the diffusion of other venom components [40, 41]. Similarly, the DNase enzyme in some snake venoms can prevent the ensnaring of toxins by Neutrophil Extracellular Traps (NETs). These NETs are formed during a process termed 'NETosis', where neutrophils extrude their nuclear DNA as traps to limit diffusion of venom proteins [42]. Snake venoms may also contain a variety of proteases that can act synergistically to perturb the blood coagulation cascade [43]. Likewise, various pre- and post-synaptic elapid neurotoxins can simultaneously antagonize a diversity of receptors involved in neurotransmission, resulting in the impairment of signal transduction and rapid paralysis [44].

Most snakes capable of delivering clinically severe bites to humans belong to Elapidae (cobras, kraits, coral snakes, taipans, mambas, and sea snakes), Viperidae (vipers, adders, and rattlesnakes), and Atractaspidinae (mole vipers and stiletto snakes). Many non-front-fanged caenophidian snakes do not pose a medical threat to humans; however, boomslangs (*Dispholidus typus*), twig snakes (genus *Thelotornis*) and some keelbacks (genus *Rhabdophis*) can inflict medically significant envenomings [45]. Elapid venoms are chiefly constituted by type I PLA_2 and 3FTx, while viperid venoms are enriched with type II PLA_2, SVMP, and SVSP [46]. Non-front-fanged caenophidian venoms split the difference, typically being dominated either by 3FTx or SVMP [45].

In contrast to Caenophidia, which includes the venomous lineages discussed above, members of Henophidia (including pythons, boas, sunbeam snakes, Asian pipe snakes, and shieldtail snakes), have historically been considered non-venomous. This is in part

Table 12.1. The distribution of toxins in distinct lineages of toxicoferan squamates.

Only those toxins whose roles in envenomation have been validated experimentally are shown. NFF: non-front-fanged; 3FTx: three-finger toxin; SVMP: snake venom metalloproteinase; SVSP: snake venom serine protease; CRiSP: cysteine-rich secretory proteins; KUN: Kunitz peptides; CVF: cobra venom factor; LAAO: L-amino acid oxidase; NP: natriuretic peptides; VEGF: vascular endothelial growth factor; NGF: nerve growth factor; DNase: deoxyribonuclease; HYAL: hyaluronidase; AChE: acetylcholinesterase; NTD: 5′-nucleotidase; PDE: phosphodiesterase; LEC: lectins; CYS: cystatins; AVIT: AVIT-domain peptides; DEF: defensins; VEF: veficolins; HEL: helokinestatins; EXE: exendins; GTx: goannatyrotoxin; CTx: cholecystoxin; ESP: epididymal secretory protein; VESP: vespryn. Enzymatic toxins are marked with asterisks.

	3FTx	SVMP*	SVSP*	PLA*	CRiSP	KUN	CVF	LAAO*	NP
Elapidae	✓	✓	✓	✓ (I)	✓	✓	✓	✓	✓
Viperidae	✓	✓	✓	✓ (II)	✓	✓	✗	✓	✓
NFF Caenophidia	✓	✓	✓	✓	✓	✓	✓	✓	✓
Henophidia	✓	✓	✗	✓	✓	✗	✓	✓	✗
Varanidae	✗	✗	✗	✓ (III)	✓	✓	✗	✗	✓
Helodermatidae	✗	✓	✓	✓ (III)	✓	✓	✗	✓	✓
Iguanidae	✗	✗	✗	✗	✓	✓	✓	✓	✗

	VEGF	NGF	DNase*	HYAL*	AChE*	NTD*	PDE*	LEC	CYS
Elapidae	✓	✓	✓	✓	✓	✓	✓	✓	✓
Viperidae	✓	✓	✓	✓	✓	✓	✓	✓	✓
NFF Caenophidia	✓	✓	✗	✓	✓	✓	✓	✓	✓
Henophidia	✓	✓	✗	✓	✓	✓	✓	✓	✓
Varanidae	✗	✗	✗	✗	✗	✗	✗	✓	✓
Helodermatidae	✗	✗	✗	✓	✓	✓	✓	✓	✓
Iguanidae	✗	✗	✗	✗	✗	✗	✗	✓	✓

	AVIT	DEF	VEF	HEL	EXE	GTx	CTx	ESP	VESP
Elapidae	✓	✓	✗	✗	✗	✗	✗	✓	✓
Viperidae	✗	✓	✗	✗	✗	✗	✗	✓	✓
NFF Caenophidia	✗	✓	✗	✗	✗	✗	✗	✗	✓
Henophidia	✗	✗	✓	✗	✗	✗	✗	✓	✓
Varanidae	✓	✓	✓	✗	✗	✓	✓	✓	✓
Helodermatidae	✗	✓	✓	✓	✓	✗	✗	✗	✓
Iguanidae	✗	✓	✓	✗	✗	✗	✗	✗	✓

because the most well-known henophidians, the pythons (Pythonoidea) and boas (Boidae), typically employ constriction to subdue prey. However, some henophidians, including Uropeltoidea (Asian pipe snakes and shieldtail snakes), simply swallow prey after biting it repeatedly, without reliance on constriction. Homologues of venom proteins, including 3FTx, lectin, CRiSP, CVF, and vespryn, have been identified in the oral-gland transcriptomes of both pythonoids and uropeltoids [47]. 3FTx and lectin, in particular, were not only identified as dominant transcripts in various gland types of these snakes but were also detected in the predominantly mucosal oral secretions of several species. In complete contrast to the rapid diversification of 3FTxs in Caenophidia [48], 3FTxs from distinct families of henophidian snakes are extremely conserved, despite being separated by >90 million years of evolutionary history [47]. In the absence of selection pressures, protein-encoding genes often follow a neutral evolutionary path that leads to degeneration (e.g., [28]). Evolutionary conservation of 3FTx homologues in henophidians suggests that their evolution is strongly constrained, which may be indicative of an important yet unidentified function [47]. Given their low abundance in the oral secretions of henophidians, it seems unlikely that 3FTxs play a role in prey subjugation, but further investigation of the oral secretions of non-caenophidians is a priority for researchers interested in unravelling the early evolutionary history of snake venom [49].

Given their medical relevance, the Gila monster (*Heloderma suspectum*) and beaded lizard (*H. horridum*) are the only lizards to have typically been considered 'venomous'. Helodermatid lizard venoms are predominantly composed of kallikrein-like serine proteases and type III PLA_2s, along with defensins, helokinestatins, and exendins in lower amounts [50]. Exendin-4 from *H. suspectum* venom has been a lead compound for the manufacture of an anti-diabetic drug [51]. Considering their ability to inflict clinically significant envenoming in humans and the biodiscovery potential of their venoms, Helodermatidae remains the most studied group of toxiferan lizards. Their close relatives, the monitor lizards (Varanidae and Lanthanotidae) are typically regarded as non-venomous despite varanid dental glands, homologous to the venom glands of snakes and helodermatid lizards, having been identified (and christened 'the gland of Gabe') decades ago [52]. Manifestations of Komodo dragon (*V. komodoensis*) bites including rapid swelling, localized disruption in blood clotting and shooting pain, were considered to result from pathogenic bacteria in the lizards' saliva [53]. These assumptions were later challenged by studies that argued for the plausibility of Komodo dragons utilizing bioactive secretions (i.e., venom) rather than bacteria in prey capture [54]. Monitor lizard salivary secretions were subsequently shown to vary greatly in composition with CRiSPs, kallikrein-like serine proteases, and type III PLA_2 predominantly present in all varanid species (Table 12.1; [55]).

Similarly, again despite the evidence of homologous dental glands [52], iguanian lizards were ignored by toxinologists until recently, when transcripts belonging to several protein families that are homologous to snake venom toxins, including LAAO, β-defensins, and cystatins, were identified in iguanian oral-gland transcriptomes [47]. Comparisons of the snake and toxicoferan lizard oral secretions reveal several other shared protein families, including CRiSP, natriuretic peptide, hyaluronidase, and kallikrein, which have been proposed to be part of the ancestral toxicoferan 'venom arsenal' [56].

12.4 Evolutionary Origin and Diversification of Squamate Venom

Toxin-encoding genes are theorized to have evolved from endophysiological genes that perform various biochemical and regulatory functions in the producing animal [3]. Over the course of evolution, these genes have diversified into a plethora of structural forms with diverse biochemical functions. This astounding toxin diversity has been explained by various molecular and evolutionary mechanisms.

Gene duplication has played a major role in the expansion and diversification of venom. Toxin-encoding genes are hypothesized to follow a 'birth and death' model of gene evolution where, post-duplication, one of the copies is relieved of pre-existing selection pressures and has increased propensity to acquire changes [57, 58]. Owing to the accelerated rate of changes, it may undergo degeneration (pseudogenization) or acquire novel functions (neo-functionalization), resulting in the emergence of multi-locus gene families with remarkable structural and functional diversity (Fig. 12.2; [59, 60]). However, this redundancy may not necessarily underpin gene diversification and instead could potentially translate to elevated expression levels. Concerted evolution, where a high degree of sequence similarity is maintained between duplicates through homologous recombination [61], leads to the overexpression of identical toxins [62]. Overexpression of specific toxins can drastically alter venom biochemistry and potency, thereby facilitating evolutionary adaptations. Stochastic gene and domain degeneration have also influenced the snake venom repertoire. Domain loss in viperid SVMP has led to the origin of remarkable structural–functional disparity [59] and has potentiated the origin of novel neurotoxins in non-front-fanged caenophidians [63].

Several theories have been proposed to explain the origin of venom proteins. According to the recruitment hypothesis, following duplication of an endophysiological gene with expression across a broad range of tissue types, one of the duplicates is recruited and selectively expressed in the venom gland. This is followed by the accumulation of variations and the origination of novel toxin functions [57, 58, 64, 65]. In contrast, the restriction hypothesis proposes that genes that are expressed in multiple physiological tissues, including the venom gland, may undergo duplication followed by subfunctionalization: the division of ancestral role among duplicates. Over time, the expression of one of the duplicates is restricted to the venom gland, while the other retains its multi-tissue expression profile [66]. A stepwise intermediate nearly-neutral evolutionary recruitment (SINNER) model has recently been proposed [67], based on the transcriptional profiling of toxin-encoding gene homologues in various tissues of the Burmese python (*Python bivittatus*). According to this model, which shares similarities with the restriction hypothesis, toxin-encoding genes originate when a recently duplicated precursor follows three crucial steps for recruitment into the venom gland: (1) a near-constitutive low expression of this gene in the venom gland, (2) venom gland specific over-expression, and (3) an overall decrease in expression across physiological tissue types. Thus, these models explain how endophysiological genes that are typically expressed in the oral glands or other physiological tissues, may have served as the primal stock for the origin of squamate venoms [3].

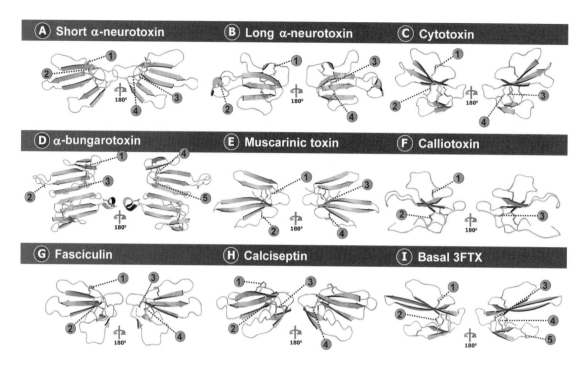

Figure 12.2 The remarkable structural and functional diversity in 3FTxs, illustrated by homology models of 3FTxs and their disulfide bonds. Short- (A) and long-chain (B) α-neurotoxins target various nicotinic acetylcholine receptor (nAChR) subtypes, while cytotoxins (C) bind to anionic lipids in the plasma membrane leading to cell lysis. Heterodimeric α-bungarotoxins (D) antagonize nAChRs at the neuromuscular junctions, while muscarinic 3FTxs (E) antagonize muscarinic AChRs. Calliotoxins (F) are known to activate the voltage-gated sodium channels (Nav1.4), whereas fasciculins (G) inhibit AChE. Calciseptin (H) selectively blocks L-type calcium channels, while basal 3FTxs (I) remain uncharacterized. (A black and white version of this figure will appear in some formats. For the colour version, please refer to the plate section).

Various molecular processes could underpin the diversification of toxin-encoding genes. Among these, alternative splicing and trans-splicing, the inclusion of exons from the same or different transcripts, respectively, have been shown to produce diverse protein isoforms in vertebrates [68]. Genome-wide searches and transcriptomic profiling of toxin-encoding genes in the habu pit viper (*Protobothrops flavoviridis*) has revealed the contribution of these processes in generating toxin diversity [69]. The role of alternative splicing in generating physiological and venom isoforms of AChE in the banded krait (the elapid *Bungarus fasciatus*) [70] and helokinestatin in the Gila monster [71] have also been well-documented.

Positive Darwinian selection, which is characterized by an increased non-synonymous (nucleotide changes that alter the encoded amino acid) to synonymous (nucleotide changes that do not alter the resultant amino acid) substitution rate, has played a significant role in snake venom diversification (Fig. 12.3; [48, 59, 72, 73]). The rapid accumulation of variation in exposed residues (RAVER) model proposes that changes introduced by positive selection mostly manifest in regions that do not perturb a molecule's structural or functional integrity, such as loops

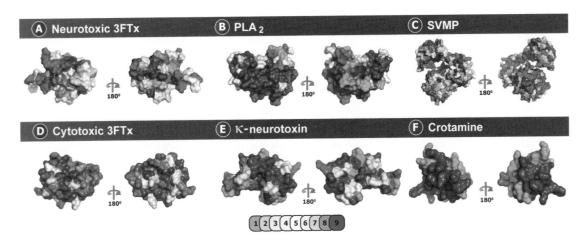

Figure 12.3 Molecular evolution of snake venom proteins. Homology models, shown here, highlight the degree of evolutionary conservation on a gradient scale of blue to maroon (1–9). Amino acid sites under positive selection are shown in red. Although many toxins rapidly diversify under positive selection (A–C), those that target conserved cellular features (D–E) or experience structural constraints (F), undergo negative selection. (A black and white version of this figure will appear in some formats. For the colour version, please refer to the plate section).

and surface-exposed residues [48]. Historically, positive selection has been predominantly invoked to explain the diversification of venoms, while the effects of negative (or purifying) selection, which is responsible for sequence conservation, were largely unrecognized. Recently, a two-speed mode for venom evolution in evolutionarily old and young lineages has been suggested [39]. Toxin-encoding genes in evolutionarily ancient lineages with origins between 700–400 millions of years ago (Mya; e.g., cnidarians, scorpions, centipedes, spiders, octopus, cuttlefish, and squid) were found to be under the influence of strong negative selection, while evolutionarily younger lineages originating between 55–30 Mya (e.g., cone snails and caenophidian snakes) were typified by toxin-encoding genes diversifying rapidly under positive selection. It has been postulated that during adaptation to substantial shifts in ecology and/or the environment, toxin-encoding genes may experience strong positive selection and generate the necessary venom variation for targeting novel prey and predators. However, once potent toxins are generated, and the venom cocktail is evolutionarily tinkered to impart effective potency, the rate of evolution slows down, and the toxin-encoding genes evolve under negative selection that preserves structural and functional integrity [39]. A strong influence of negative selection can also be noted in toxins that serve a defensive role, exert toxicities by targeting conserved features, or experience structural constraints (Fig. 12.3; [21, 39, 48]).

12.5 Venom Delivery Systems

The snake venom system comprises the venom glands and fangs. Snakes can be grouped morphologically on the presence, absence, and position of fangs into rear-fanged (opisthoglyphous), front-fanged (solenoglyphous and proteroglyphous), and those lacking fangs (aglyphous) (Fig. 12.1B). In solenoglyphous snakes, the hollow hypodermic needle-like fangs are the sole teeth on the shortened and highly mobile maxillae and point posteriorly. The maxillae are hinged with the braincase, allowing the fangs to be folded away or to swing forward into an injecting position. Proteroglyphous elapid snakes are equipped with short hollow fangs at the front of less mobile maxillae, which they share with additional teeth. The grooved fangs of opisthoglyphous snakes are found towards the back of the maxillae and are preceded on that bone by numerous teeth that function to secure prey [74].

The ophidian venom gland is an extensively modified dental gland (Chapter 18). In front-fanged snakes, the *adductor externus profundus* (Viperidae) or *superficialis* (Elapidae) muscles overlap the venom glands to facilitate a high-pressure injection of venom. The evolution of this efficient delivery apparatus, ca. 50 Mya, has had a substantial impact on the evolution of venom itself, likely driving the rapid diversification of many venom toxin superfamilies. This fascinating coevolution of venom and venom-delivery apparatus underscores the fact that changes in gross anatomy can greatly influence molecular evolution [48, 75]. In contrast to their elapid and viperid counterparts, most non-front-fanged caenophidians lack such musculature. This prevents rapid and efficient injection of venom, and these snakes often chew on prey to attempt inoculation. Interestingly, however, rudimentary compressor musculature has evolved several times independently in non-front-fanged snakes [4]. The toxin-secreting dental gland of non-front-fanged species is sometimes referred to as Duvernoy's gland [76], but these glands are developmentally homologous to the venom glands in front-fanged snakes [74], and appear to fulfil the same primary function (venom production and secretion) rendering this term redundant ([49]). Fascinatingly, despite the diversity of venom-delivery apparatuses, the venom glands and associated venom-injecting teeth of all caenophidian snakes share common evolutionary and developmental origins [77].

Non-caenophidian snakes, which do not depend on venom for prey incapacitation, are aglyphous (Fig. 12.1B). In addition to the venom glands, snakes possess a diversity of oral glands, including labial and rictal glands (Oliveira and Zaher, Chapter 18). Labial glands are responsible for mucus secretion, but the precise role of rictal glands remains unclear. Although rictal gland secretions of some henophidians were reported in the earlier part of the twentieth century to be toxic to birds [78], transcripts homologous to caenophidian toxins were retrieved from rictal glands of these snakes only recently [47]. This perhaps suggests a common evolutionary origin of ophidian oral glands. In contrast to caenophidians, the mandibular and maxillary glands of henophidians are composed chiefly of mucous- rather than protein-secreting (serous) cells, and they provide lubrication to facilitate swallowing feathered avian or furred mammalian prey [47].

In contrast to snakes, toxicoferan lizards have relatively simple venom-delivery systems [71]. Iguanians have retained plesiomorphic seromucous mandibular and maxillary dental glands [56], while a loss of maxillary glands and supporting musculature has been observed in some of their anguimorphan relatives [47]. Though few species have been investigated to date, dental glands of iguanians that feed on vertebrates appear to have a greater proportion of serous cells than those of insectivorous and herbivorous species, which are largely composed of mucoid cells. In contrast, all non-toxiceran lizards investigated possess a largely mucoid mandibular gland, with or without a maxillary gland. Both Anguimorpha and Iguania have grooved mandibular teeth associated with their dental glands, which may serve as an apparatus for venom delivery in some anguimorphs [47]. Interestingly, the glands of the Gila monster and some monitor lizards of the genera *Varanus* and *Lanthanotus* are enlarged and have convergently evolved well-structured lumens and segregation of the protein- and mucus-secreting regions [71].

12.6 The Long-Standing Debate on the Origin of Venom in Squamate Reptiles

Given the highly dynamic evolution of both venom and venom-delivery systems within Squamata, tracing the evolutionary origins of the venom system has been challenging, but several hypotheses have been proposed. An early hypothesis proposed two independent origins: in the common ancestor of Caenophidia, and in helodermatid lizards (Fig. 12.1A; [7]). On the other hand, it was conjectured that these venom systems arose as derivations of homologous dental glands [52], although the phylogenetic relationships between snakes and anguimorph lizards remained unclear.

More recently, molecular phylogenetics has recovered strong support for a clade comprising Serpentes, Anguimorpha, and Iguania [10], and because all venomous squamates are members of this assemblage, the clade was christened Toxicofera [6]. Transcriptomic profiling of the oral glands of toxicoferans resulted in the identification of a set of common transcripts shared by most members [56]. Many of these genes exhibited either high sequence similarity to known toxins or were assigned a putative toxic role given their enzymatic function. Phylogenetic reconstruction of the evolutionary histories of these genes inferred reciprocal monophyly of such putative toxins and their endophysiological homologues. Furthermore, consideration of the anatomy of toxicoferan lizard oral glands revealed that dental glands (i.e., modified labial glands with ducts opening at the base of the teeth) are a synapomorphy of the clade [79], a finding anticipated by previous work [52]. Given the lack of evidence of an ecological role in feeding or defence for the secretions of these glands in iguanians, they were referred to as an incipient venom system [80]. Iguanians and the anguimorph *Pseudopus apodus* have both mandibular and maxillary dental glands, whereas other anguimorphs and snakes were argued to have lost the maxillary and mandibular glands, respectively [79]. The presence of protein-secreting dental glands in both the upper (maxillary) and lower jaw (mandibular), therefore, is an inferred trait of the most recent common ancestor (MRCA) of all toxicoferans. Taken together, these findings were interpreted as providing support for a single early origin of the venom system in Toxicofera [56]. Following the proposal of this hypothesis, transcriptomic profiling of salivary glands of a diversity of toxicoferans provided evidence for the recruitment of several additional putatively toxic genes in the toxicoferan MRCA [47, 54, 71].

However, the Toxicofera Hypothesis of venom origins, which relied heavily on the inference of ancestral recruitment and reciprocal monophyly of putative toxin-encoding genes, has been challenged. It has been argued that failing to compare the expression profiles of oral glands with physiological tissues may result in the incorrect identification of endophysiological proteins as venom toxins. Indeed, comparative transcriptomics involving multiple tissues from various toxicoferans revealed a broad tissue-expression profile for many genes that were initially hypothesized as ancestrally recruited toxins, and a lack of consistent venom or salivary gland-specific overexpression was noted [81]. Furthermore, reinvestigation of phylogenetic histories of toxin-encoding genes and their endophysiological homologues resulted in paraphyletic groupings. These discrepancies were leveraged as arguments against the Toxicofera Hypothesis and, in concordance with Kochva's two-origin theory, to conjecture at least two origins of venom in squamates, with a possible third independent origin in monitor lizards [81]. Moreover, some authors have also argued for multiple independent origins of venom systems within Caenophidia, with Elapidae, Viperidae, and some medically important non-front-fanged species having evolved venom and associated delivery systems independently after divergence from their common ancestor [9].

12.7 The Verdict: Evidence for and against the Toxicofera Hypothesis of Squamate Venoms

Understanding the phylogeny of snakes and lizards is crucial for unravelling the origin of venom in squamates. Had lizards and snakes formed reciprocally monophyletic clades, a single early origin of venom would have been highly unlikely. Molecular phylogenetics has overturned previous hypotheses of squamate relationships and provides strong support for a clade (Toxicofera) comprising Iguania, Anguimorpha, and Serpentes [6, 10, 82].

Anatomical investigations of toxicoferans have uncovered several shared morphological features. The dental glands primarily responsible for toxin secretion are associated with the upper jaw of snakes and the lower jaw of most anguimorph lizards, while iguanian dental glands are associated with both jaws. Dental glands appear to be a synapomorphy of Toxicofera and, thus, the inferred MRCA of this clade likely possessed mixed seromucous dental glands associated with both jaws; an arrangement retained in Iguania, while snakes and anguimorphs evolved specialized glands with discrete serous and mucous-secreting regions, associated with the upper and lower jaws, respectively [56]. Utilization of venom for predation and/or defence may have influenced the diversification of snake and anguimorph lizard venom systems, while those of many iguanians remained incipient due to their herbivorous and insectivorous diets.

Transcriptome sequencing of dental glands from diverse toxicoferans revealed several shared, putative toxin-encoding genes, leading to the proposition of 'basal (sic) recruitment' of such toxins in the MRCA of toxicoferans and a single early origin of venom in this clade [47, 54, 56, 71]. In the absence of functional validation, the oral-gland expression of genes homologous to those that encode toxins does not necessarily provide evidence that the products of these genes possess a toxic function. It is important to sample diverse tissues to distinguish endophysiological proteins with increased expression across broad tissue types, including venom glands. Furthermore, it must be noted that whether or not a gene product is a functional toxin is a question about the deployment of that gene product in the ecology of the producing organism, and not simply about its homology, site of expression, or *in vitro* activity [80]. Comparative transcriptomics of various tissues from a few toxicoferans have revealed a broad tissue-expression profile for many of the genes that were presumed to be ancestrally recruited as toxins, and this has been interpreted as evidence against the Toxicofera Hypothesis [81]. However, for clear hypothesis testing, it is imperative to comprehensively and comparatively analyse multiple tissue types from a diversity of both toxicoferan and non-toxicoferan reptiles, which studies thus far have failed to do.

Although many of these proteins are unlikely to serve a venom function, at least in most toxicoferans, the toxic role of proteolytic enzymes with venom-gland overexpression cannot be ruled out. Proteases can not only participate in the processing of venom proteins but can also induce non-specific pathologies in the victim when injected in large amounts. Although it has been stated otherwise [81], proteins with broad tissue expression could, therefore, serve multiple functions, and do not necessarily require post-transcriptional processing, such as alternative splicing. Therefore, broad tissue-expression profiles alone cannot be used to discount the role of venom gland overexpressed proteins in envenomation. Moreover, given the necessity for increased secretion, many venom genes may possess strong promoters, resulting in a 'leaky' expression profile in non-venom gland tissues. Thus, the direct comparison of expression values, which are extremely low for many toxin-encoding genes (Fig. 12.4A; [81]), without the usage of appropriate statistics can be misleading [83]. Furthermore, Hargreaves et al.'s [81] expression data exhibits unusually large variability between venom gland tissue replicates (Fig. 12.4B), possibly because they source samples at various time points (0, 16, 24, and 48 hours post-milking). As the authors leverage these expression values to support their findings and argue against the Toxicofera Hypothesis, sourcing venom gland tissues from the same time point would have been an appropriate strategy. Further, the authors did not establish orthology between toxin-encoding genes and their endophysiological homologues, which could lead to erroneous interpretations. Thus, the claims made by Hargreaves et al., [81] based on their transcriptomic data to refute the Toxicofera Hypothesis need reassessment.

The Toxicofera Hypothesis was also challenged on the grounds of non-reciprocal monophyly of toxins and endophysiological proteins [81]. However, most gene trees presented by Hargreaves et al. [81] included low branch support and/or unresolved relationships. Therefore, conclusions drawn regarding the role of many proteins in envenomation, based exclusively on potentially unreliable tree topologies, are questionable. Moreover, non-reciprocal monophyly could also stem from the limited divergence of genes from their plesiotypic endophysiological homologues, as seen in co-opted genes that serve multiple functions and those that are recently recruited into venom.

Based on the interpretation of the word 'venomous', multiple origins of venom have also been proposed in squamates, including at least three independent origins in Caenophidia

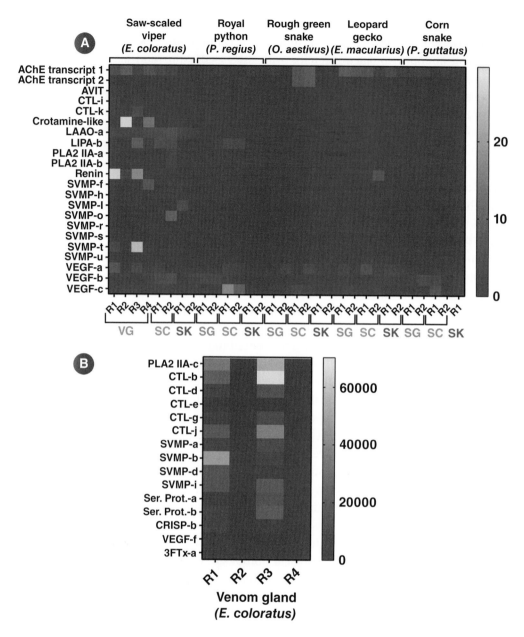

Figure 12.4 The estimated transcript abundance values for many toxin-encoding genes sequenced from the scent gland (SC), skin tissue (SK) and venom gland (VG) or salivary gland (SG) of painted saw-scaled viper (*Echis coloratus*), royal python (*Python regius*), rough green snake (*Opheodrys aestivus*), leopard gecko (*Eublepharis macularius)* and corn snake (*Pantherophis guttatus*) by Hargreaves et al. [81] are either extremely low (A) or exhibit large deviations across replicates (R1–R4; B). These heatmaps were generated using the supplementary data from [81]. (A black and white version of this figure will appear in some formats. For the colour version, please refer to the plate section).

and another in helodermatid lizards [9]. This interpretation is based on the argument that the pharmacological or biochemical testing of 'toxic substances' in the saliva of particular medically unimportant squamates (i.e., most non-front-fanged snakes) is insufficient, and their actual biological roles should be investigated [9]. Although we agree with this suggestion, it should be noted that biological roles of 'toxic substances' have nothing to do with toxicity to humans. Such anthropocentric assumptions are invalidated by many studies that clearly demonstrate the potent prey-specific activities of snake venoms [26, 27]. These salivary secretions evolve to facilitate predation or defense of their producers, the ability of venoms to inflict severe clinical symptoms in humans should not be a criterion for their inclusion into the definition of venom. Thus, considering the modern definition of venom and the phylogenetic history of Caenophidia, we consider it highly unlikely that venom has multiple independent origins within this lineage. On the other hand, high-pressure venom delivery systems have undoubtedly evolved multiple times within the group, i.e., there has been independent specialization towards the use of venom as a primary trophic trait within Caenophidia.

Though currently available evidence provides space for multiple interpretations, the picture it paints suggests that dental glands are a synapomorphy of Toxicofera and that this glandular anatomy represents an exaptation that paved the way for the subsequent evolution of specialized venom systems. Thus, whether there was a single origin of the venom system in the Toxicofera MRCA or multiple origins in separate lineages depends largely on the usage of the term 'venom system'. Although an incipient squamate venom system may have had a single early origin, specializations in venom systems have very likely originated convergently in distinct toxicoferan clades. It is interesting to note that the research on embryonic development and morphogenesis of snake dentition has shown that both front and rear fangs have a common evolutionary origin, and that the extant diversity of snakes may have been underpinned by the convergent specializations of the venom glands and their associated dental structures [77]. Current evidence is not sufficient to draw conclusions regarding the precise origin of venom in reptiles, and we may never be in possession of such evidence. It seems probable, however, that the MRCA of Caenophidia possessed a functional venom system. Given the presence of large toxin-secreting oral glands in plesiomorphic non-caenophidian genera (e.g., *Cylindrophis* and *Anilius*), as well as in highly divergent scolecophidian lineages, venom may even be an ancestral snake character, the occurrence of numerous reversals within that clade notwithstanding (see [80] and [49] for further discussion).

12.8 Conclusion

The origin of venom in squamates has been a topic of never-ending debate. The rapid diversification of venom systems as an adaptation to differing ecologies and environments makes them among the most dynamic features of squamate reptiles. With the paucity of whole-genome and ecological data available for toxicoferans and their outgroups, the current understanding of venom origins is primarily based on anatomical, phylogenetic, and transcriptomic studies. A comparative genomics approach to examine the synteny (i.e., colocalization) of toxin-encoding genes and their endophysiological homologues in the genomes of both toxicoferan reptiles and their close, non-venomous relatives may further our understanding of the evolutionary origin of squamate venom. However, given the complexity of venom systems and their dynamic nature, elucidating the evolutionary history of this trait could remain an arduous challenge.

Acknowledgements

The authors are thankful to editor David Gower (National History Museum, London) and reviewers Giulia Zancolli (University of Lausanne) and Nicholas Casewell (Liverpool School of Tropical Medicine) for their invaluable inputs and suggestions which greatly helped to improve the quality of the manuscript. We thank Genevieve Jackson for contributing snake venom-gland illustrations in Figure 12.1B.

References

1. V. Schendel, L. D. Rash, R. A. Jenner, and E. A. B. Undheim, The diversity of venom: the importance of behavior and venom system morphology in understanding its ecology and evolution. *Toxins (Basel)*, 11 (2019), 666.

2. B. G. Fry, K. Roelants, D. E. Champagne, et al., The toxicogenomic multiverse: convergent recruitment of proteins into animal venoms. *Annual Review of Genomics and Human Genetics*, 10 (2009), 483–511.

3. B. G. Fry, From genome to 'venome': molecular origin and evolution of the snake venom proteome inferred from phylogenetic analysis of toxin sequences and related body proteins. *Genome Research*, 15 (2005), 403–420.

4. B. G. Fry, H. Scheib, L. van der Weerd, et al., Evolution of an arsenal: structural and functional diversification of the venom system in the advanced snakes (Caenophidia). *Molecular and Cellular Proteomics*, 7 (2008), 215–46.

5. B. G. Fry, N. Vidal, L. van der Weerd, E. Kochva, and C. Renjifo, Evolution and diversification of the Toxicofera reptile venom system. *Journal of Proteomics*, 72 (2009), 127–136.

6. N. Vidal and S. B. Hedges. The phylogeny of squamate reptiles (lizards, snakes, and amphisbaenians) inferred from nine nuclear protein-coding genes. *Comptes Rendus Biologies*, 328 (2005), 1000–1008.

7. E. Kochva, Oral glands of the Reptilia. Biology of the Reptilia. 8 (1978), 43–162.

8. F. H. Pough, R. M. Andrews, J. E. Cadle, et al., *Herpetology* 3rd edn. (New Jersey: Prentice Hall, 2004).

9. K. V. Kardong, S. A. Weinstein, and T. L. Smith, Reptile venom glands: form, function, and future. In S. P. Mackessy, ed., *Handbook of Venoms and Toxins of Reptiles* (Boca Raton, FL: CRC Press, 2009), pp. 65–91.

10. F. T. Burbrink, F. G. Grazziotin, R. A. Pyron, et al. Interrogating genomic-scale data for Squamata (lizards, snakes, and amphisbaenians) shows no support for key traditional morphological relationships. *Systematic Biology*, 69 (2020), 502–520.

11. J. C. Daltry, W. Wüster, and R. S. Thorpe, Diet and snake venom evolution. *Nature*, 379 (1996), 537–40.

12. H. L. Gibbs, L. Sanz, J. E. Chiucchi, T. M. Farrell, and J. J. Calvete, Proteomic

analysis of ontogenetic and diet-related changes in venom composition of juvenile and adult Dusky Pigmy rattlesnakes (*Sistrurus miliarius barbouri*). *Journal of Proteomics*, 74 (2011), 2169–2179.

13. G. Zancolli, J. J. Calvete, M. D. Cardwell, et al., When one phenotype is not enough: divergent evolutionary trajectories govern venom variation in a widespread rattlesnake species. *Proceedings of the Royal Society B*, 286 (2019), 20182735.

14. N. R. Casewell, T. N. W. Jackson, A. H. Laustsen, and K. Sunagar, Causes and Consequences of Snake Venom Variation. *Trends in Pharmacological Sciences*, 41 (2020), 570–581.

15. T. N. W. Jackson, H. Jouanne, and N. Vidal, Snake venom in context: Neglected clades and concepts. *Frontiers in Ecology and Evolution*, 7 (2019), 332.

16. M. V. Modica, K. Sunagar, M. Holford, and S. Dutertre,. Diversity and evolution of animal venoms: Neglected targets, ecological iInteractions, future perspectives. *Frontiers in Ecology and Evolution*, 8 (2020), 65.

17. F. C. Cardoso, C. R. Ferraz, A. Arrahman, et al., Multifunctional toxins in snake venoms and therapeutic implications: from pain to hemorrhage and necrosis. *Frontiers in Ecology and Evolution*, 7 (2019), 218.

18. N. R. Casewell, R. A. Harrison, W. Wüster, and S. C. Wagstaff, Comparative venom gland transcriptome surveys of the saw-scaled vipers (Viperidae: *Echis*) reveal substantial intra-family gene diversity and novel venom transcripts. *BMC Genomics*, 10 (2009), 564.

19. R. B. Currier, R. A. Harrison, P. D. Rowley, G. D. Laing, and S. C. Wagstaff, Intra-specific variation in venom of the African Puff Adder (*Bitis arietans*): Differential expression and activity of snake venom metalloproteinases (SVMPs). *Toxicon*, 55 (2010), 864–873.

20. R. R. Senji Laxme, S. Khochare, H. F. de Souza, et al., Beyond the 'big four': Venom profiling of the medically important yet neglected Indian snakes reveals disturbing antivenom deficiencies. *PLoS Neglected Tropical Diseases*, 13 (2019), e0007899.

21. K. Sunagar, E. A. Undheim, H. Scheib, et al., Intraspecific venom variation in the medically significant Southern Pacific Rattlesnake (*Crotalus oreganus helleri*): biodiscovery, clinical and evolutionary implications. *Journal of Proteomics*, 99 (2014), 68–83.

22. T. N. Jackson, I. Koludarov, S. A. Ali, et al., Rapid radiations and the race to redundancy: An investigation of the evolution of Australian elapid snake venoms. *Toxins (Basel)*, 8 (2016), 309.

23. S. P. Mackessy, K. Williams, and K. G. Ashton, Ontogenetic variation in venom composition and diet of *Crotalus oreganus concolor*: a case of venom paedomorphosis? *Copeia*, 2003 (2003), 769–782.

24. D. R. Rokyta, M. J. Margres, M. J. Ward, and E. E. Sanchez, The genetics of venom ontogeny in the eastern diamondback rattlesnake (*Crotalus adamanteus*). *PeerJ*, 5 (2017), e3249.

25. L. Sanz, H. L. Gibbs, S. P. Mackessy, and J. J. Calvete, Venom proteomes of closely related *Sistrurus rattlesnakes* with divergent diets. *Journal of Proteome Research.* 5 (2006), 2098–2112.

26. A. Barlow, C. E. Pook, R. A. Harrison, and W. Wüster, Coevolution of diet and prey-specific venom activity supports the role of selection in snake venom evolution. *Proceedings of the Royal Society B*, 276 (2009), 2443–2449.

27. S. P. Mackessy, N. M. Sixberry, W. H. Heyborne, and T. Fritts, Venom of the Brown Treesnake, *Boiga irregularis*: ontogenetic shifts and taxa-specific toxicity. *Toxicon*, 47 (2006), 537–548.

28. M. Li, B. G. Fry, and R. M. Kini, Eggs-only diet: its implications for the toxin profile changes and ecology of the marbled sea snake (*Aipysurus eydouxii*). *Journal of Molecular Evolution*, 60 (2005), 81–89.

29. J. Durban, A. Perez, L. Sanz, et al., Integrated 'omics' profiling indicates that miRNAs are

modulators of the ontogenetic venom composition shift in the Central American rattlesnake, *Crotalus simus simus. BMC Genomics*, 14 (2013), 234.

30. B. Ujvari, N. R. Casewell, K. Sunagar, et al., Widespread convergence in toxin resistance by predictable molecular evolution. *Proceedings of the National Academy of Sciences of the USA*, 112 (2015), 11911–11916.

31. A. H. Rowe, Y. Xiao, M. P. Rowe, T. R. Cummins, and H. H. Zakon, Voltage-gated sodium channel in grasshopper mice defends against bark scorpion toxin. *Science*, 342 (2013), 441–446.

32. M. L. Holding, D. H. Drabeck, S. A. Jansa, and H. L. Gibbs, Venom resistance as a model for understanding the molecular basis of complex coevolutionary adaptations. *Integrative and Comparative Biology*, 56 (2016), 1032–1043.

33. J. E. Biardi, D. C. Chien, and R. G. Coss. California ground squirrel (*Spermophilus beecheyi*) defenses against rattlesnake venom digestive and hemostatic toxins. *Journal of Chemical Ecology*, 32 (2006), 137–154.

34. S. A. Jansa and R. S. Voss, Adaptive evolution of the venom-targeted vWF protein in opossums that eat pitvipers. *PLoS One*, 6 (2011), e20997.

35. M. L. Holding, J. E. Biardi, and H. L. Gibbs, Coevolution of venom function and venom resistance in a rattlesnake predator and its squirrel prey. *Proceedings of the Royal Society B*, 283 (2016), 20152841.

36. D. Barchan, S. Kachalsky, D. Neumann, et al., How the mongoose can fight the snake: the binding site of the mongoose acetylcholine receptor. *Proceedings of the National Academy of Sciences of the USA*, 89 (1992), 7717–7721.

37. D. H. Drabeck, A. M. Dean, and S. A. Jansa. Why the honey badger don't care: Convergent evolution of venom-targeted nicotinic acetylcholine receptors in mammals that survive venomous snake bites. *Toxicon*, 99 (2015), 68–72.

38. K. Sunagar, N. Casewell, S. Varma, et al., Deadly innovations: unraveling the molecular evolution of animal venoms. In P. Gopalakrishnakone and J. J. Calvete, eds., *Venom Genomics and Proteomics* (Dordrecht: Springer, 2014), pp. 1–23.

39. K. Sunagar and Y. Moran. The rise and fall of an evolutionary innovation: Contrasting strategies of venom evolution in ancient and young animals. *PLoS Genetics*, 11 (2015), e1005596.

40. K. S. Girish, D. K. Jagadeesha, K. B. Rajeev, and K. Kemparaju. Snake venom hyaluronidase: an evidence for isoforms and extracellular matrix degradation. *Molecular and Cellular Biochemistry*, 240 (2002), 105–110.

41. A. T. Tu and R. R. Hendon, Characterization of lizard venom hyaluronidase and evidence for its action as a spreading factor. *Comparative Biochemistry and Physiology B*, 76 (1983), 377–383.

42. G. D. Katkar, M. S. Sundaram, S. K. NaveenKumar, et al., NETosis and lack of DNase activity are key factors in *Echis carinatus* venom-induced tissue destruction. *Nature Communications*, 7 (2016), 11361.

43. Q. Lu, J. M. Clemetson, and K. J. Clemetson. Snake venoms and hemostasis. *Journal of Thrombosis and Haemostasis*, 3 (2005), 1791–1799.

44. S. Xiong and C. Huang. Synergistic strategies of predominant toxins in snake venoms. *Toxicology Letters*, 287 (2018), 142–154.

45. C. M. Modahl and S. P. Mackessy, Venoms of rear-fanged snakes: New proteins and novel activities. *Frontiers in Ecology and Evolution*, 7 (2019), 279.

46. T. Tasoulis and G. K. Isbister, A review and database of snake venom proteomes. *Toxins (Basel)*, 9 (2017), 9.

47. B. G. Fry, E. A. Undheim, S. A. Ali, et al., Squeezers and leaf-cutters: differential diversification and degeneration of the venom system in toxicoferan reptiles. *Molecular and Cellular Proteomics*, 12 (2013), 1881–1899.

48. K. Sunagar, T. N. Jackson, E. A. Undheim, et al., Three-fingered RAVERs: Rapid

Accumulation of Variations in Exposed Residues of snake venom toxins. *Toxins (Basel)*, 5 (2013), 2172–2208.

49. T. N. Jackson, B. Young, G. Underwood, et al., Endless forms most beautiful: The evolution of ophidian oral glands, including the venom system, and the use of appropriate terminology for homologous structures. *Zoomorphology*, 136 (2017), 107–130.

50. K. W. Sanggaard, T. F. Dyrlund, L. R. Thomsen, et al., Characterization of the gila monster (*Heloderma suspectum suspectum*) venom proteome. *Journal of Proteomics*, 117 (2015), 1–11.

51. M. K. K. Yap and N. Misuan, Exendin-4 from *Heloderma suspectum* venom: From discovery to its latest application as type II diabetes combatant. *Basic & Clinical Pharmacology & Toxicology*, 124 (2019), 513–527.

52. E. Kochva. The origin of snakes and evolution of the venom apparatus. *Toxicon*, 25 (1987), 65–106.

53. W. Auffenberg. *The Behavioral Ecology of the Komodo Monitor* (Gainsville, FL: University Presses of Florida, 1981).

54. B. G. Fry, S. Wroe, W. Teeuwisse, et al., A central role for venom in predation by *Varanus komodoensis* (Komodo Dragon) and the extinct giant *Varanus* (*Megalania*) *priscus*. *Proceedings of the National Academy of Sciences of the USA*, 106 (2009), 8969–8974.

55. I. Koludarov, T. N. Jackson, B. op den Brouw, et al., Enter the dragon: The dynamic and multifunctional evolution of Anguimorpha lizard venoms. *Toxins (Basel)*, 9 (2017), 242.

56. B. G. Fry, N. Vidal, J. A. Norman, et al., Early evolution of the venom system in lizards and snakes. *Nature*, 439 (2006), 584–588.

57. M. Nei, X. Gu, and T. Sitnikova, Evolution by the birth-and-death process in multigene families of the vertebrate immune system. *Proceedings of the National Academy of Sciences of the USA*, 94 (1997), 7799–7806.

58. B. G. Fry, W. Wüster, R. M. Kini, et al., Molecular evolution and phylogeny of elapid snake venom three-finger toxins. *Journal of Molecular Evolution*, 57 (2003), 110–129.

59. N. R. Casewell, S. C. Wagstaff, R. A. Harrison, C. Renjifo, and W. Wüster, Domain loss facilitates accelerated evolution and neofunctionalization of duplicate snake venom metalloproteinase toxin genes. *Molecular Biology and Evolution*, 28 (2011), 2637–2649.

60. K. Suryamohan, S. P. Krishnankutty, J. Guillory, et al., The Indian cobra reference genome and transcriptome enables comprehensive identification of venom toxins. *Nature Genetics*, 52 (2020), 106–117.

61. D. D. Brown, P. C. Wensink, and E. Jordan, A comparison of the ribosomal DNA's of *Xenopus laevis* and *Xenopus mulleri*: the evolution of tandem genes. *Journal of Molecular Biology*, 63 (1972), 57–73.

62. Y. Moran, H. Weinberger, J. C. Sullivan, et al., Concerted evolution of sea anemone neurotoxin genes is revealed through analysis of the *Nematostella vectensis* genome. *Molecular Biology and Evolution*, 25 (2008), 737–747.

63. A. Brust, K. Sunagar, E. A. B. Undheim, et al., Differential evolution and neofunctionalization of snake venom metalloprotease domains. *Molecular and Cellular Proteomics*, 12 (2013), 651–663.

64. D. Kordis and F. Gubensek, Adaptive evolution of animal toxin multigene families. *Gene*, 261 (2000), 43–52.

65. N. R. Casewell, W. Wüster, F. J. Vonk, R. A. Harrison, and B. G. Fry. Complex cocktails: the evolutionary novelty of venoms. *Trends in Ecology and Evolution*, 28 (2013), 219–229.

66. A. D. Hargreaves, M. T. Swain, M. J. Hegarty, D. W. Logan, and J. F. Mulley, Restriction and recruitment-gene duplication and the origin and evolution of snake venom toxins. *Genome Biology and Evolution*, 6 (2014), 2088–2095.

67. J. Reyes-Velasco, D. C. Card, A. L. Andrew, et al., Expression of venom gene homologs in diverse python tissues suggests a new model for the evolution of snake venom.

Molecular Biology and Evolution, 32 (2015), 173–183.

68. Q. Lei, C. Li, Z. Zuo, et al., Evolutionary Insights into RNA trans-Splicing in Vertebrates. *Genome Biology and Evolution*, 8 (2016), 562–577.

69. T. Ogawa, N. Oda-Ueda, K. Hisata, et al., Alternative mRNA splicing in three venom families underlying a possible production of divergent venom proteins of the Habu Snake, Protobothrops flavoviridis. *Toxins (Basel)*, 11 (2019), 581.

70. X. Cousin, S. Bon, J. Massoulie, and C. Bon, Identification of a novel type of alternatively spliced exon from the acetylcholinesterase gene of *Bungarus fasciatus*. Molecular forms of acetylcholinesterase in the snake liver and muscle. *Journal of Biological Chemistry*, 273 (1998), 9812–9820.

71. B. G. Fry, K. Winter, J. A. Norman, et al., Functional and structural diversification of the Anguimorpha lizard venom system. *Molecular and Cellular Proteomics*, 9 (2010), 2369–2390.

72. V. J. Lynch, Inventing an arsenal: adaptive evolution and neofunctionalization of snake venom phospholipase A2 genes. *BMC Evolutionary Biology*, 7 (2007), 2.

73. K. Sunagar, W. E. Johnson, S. J. O'Brien, V. Vasconcelos, and A. Antunes, Evolution of CRISPs associated with toxicoferan-reptilian venom and mammalian reproduction. *Molecular Biology and Evolution*, 29 (2012), 1807–1822.

74. K. Jackson, The evolution of venom-delivery systems in snakes. *Zoological Journal of the Linnean Society*, 137 (2003), 337–354.

75. I. Koludarov, T. N. Jackson, A. Pozzi, and A. S. Mikheyev, Family saga: reconstructing the evolutionary history of a functionally diverse gene family reveals complexity at the genetic origins of novelty. *bioRxiv*, (2019), 583344.

76. A. M. Taub. Ophidian cephalic glands. *Journal of Morphology*, 118 (1966), 529–542.

77. F. J. Vonk, J. F. Admiraal, K. Jackson, et al., Evolutionary origin and development of snake fangs. *Nature*, 454 (2008), 630–633.

78. M. Phisalix and R. Caius. L'extension de la fonction venimeuse dans l'ordre entière des ophidiens et son existence chez des familles ou elle n'avait pas été soupçonnée jusqu'içi. *Journal de Physiologie et de Pathologie Générale*, 17 (1918), 923–964.

79. B. G. Fry, N. R. Casewell, W. Wüster, et al., The structural and functional diversification of the Toxicofera reptile venom system. *Toxicon*, 60 (2012), 434–448.

80. T. N. Jackson and B. G. Fry, A tricky trait: Applying the fruits of the 'function debate' in the philosophy of biology to the 'venom debate' in the science of toxinology. *Toxins*, 8 (2016), 263.

81. A. D. Hargreaves, M. T. Swain, D. W. Logan, and J. F. Mulley, Testing the Toxicofera: comparative transcriptomics casts doubt on the single, early evolution of the reptile venom system. *Toxicon*, 92 (2014), 140–156.

82. T. Townsend, A. Larson, E. Louis, and J. R. Macey. Molecular phylogenetics of squamata: the position of snakes, amphisbaenians, and dibamids, and the root of the squamate tree. *Systematic Biology* 53 (2004), 735–757.

83. A. Conesa, P. Madrigal, S. Tarazona, et al., A survey of best practices for RNA-seq data analysis. *Genome Biology*, 17 (2016), 13.

PART IV

Neurobiological Perspectives

13

Using Adaptive Traits in the Ear to Estimate Ecology of Early Snakes

HONGYU YI

13.1 Introduction

Snakes losing limbs marks a major shift of body plan in tetrapod evolution, like whales gaining flippers and birds gaining wings. It remains unsettled whether body elongation and limb reduction occurred in ancestral snakes as an adaptive response to living in subsurface habitats or in the ocean [1–4]. In the past decade, new approaches have been applied to identify fossorial and aquatic specialists in the early evolution of snakes, using virtual models of the ear of fossils [4–7]. High-resolution X-ray computed tomography (HRXCT) can image internal structures of extant and fossil snakes without damaging specimens, and morphological datasets of snake endocranial sensory systems are developing rapidly [8, 9]. This chapter reviews the osseous part of the inner ear of major snake lineages, comparing shape of the bony labyrinth and stapes among habitat generalists and (fossorial or aquatic) specialists. This chapter also reviews quantitative approaches to analysing bony labyrinth geometry and to estimating habitat use of extinct snakes.

13.2 The Snake Inner Ear: General Anatomy and Function

The snake inner ear comprises the membranous labyrinth and the bony labyrinth. The membranous labyrinth is formed of sensory cells and membranous tubes filled with endolymph fluid. Surrounding the membranous labyrinth is the bony labyrinth, formed of ossified canals and cavities (Fig. 13.1). The bony labyrinth includes three structures: the cochlea, vestibule, and semicircular canals (Fig. 13.1). The cochlea houses the cochlear duct that processes air-borne sound. The vestibule and semicircular canals house sensory organs to register movement and maintain positional equilibrium: the saccule, utricle, and

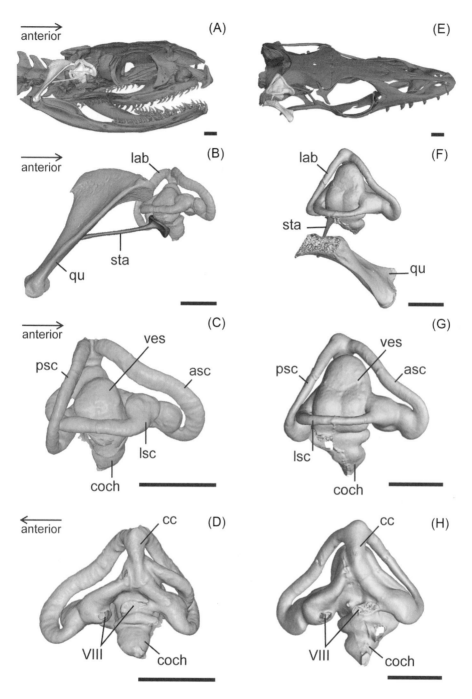

Figure 13.1 Anatomy of the squamate ear. (A–D) *Ptyas mucosa*, a terrestrial snake. (E–F) *Varanus indicus*, a terrestrial lizard. (A) Skull of *P. mucosa* showing bony elements related with sound perception. (B) Isolated sound-transmitting and processing apparatuses of *P. mucosa*. (C–D) The bony labyrinth of *P. mucosa* in lateral and medial view, respectively. (E) Skull of *V. indicus* showing bony elements related with sound perception. (F) Isolated sound-transmitting and processing apparatuses of *V. indicus*. (G–H) The bony labyrinth of *V. indicus* in lateral and medial view, respectively. Anatomical abbreviations: asc, anterior semicircular canal; lab, bony labyrinth; cc, common crus; coch, cochlea; lsc, lateral semicircular canal; psc, posterior semicircular canal; stp, stapes; qu, quadrate; ves, vestibule; VIII, foramina for vestibulocochlear nerve. Scale bar = 3 mm.

semicircular ducts [10]. A branch of the vestibulocochlear nerve penetrates the medial wall of the vestibule (Fig. 13.1D, H).

In lizards, environmental sound is transmitted into the inner ear via the external ear and the middle ear [11]. Snakes differ from most lizards in lacking an external ear, but they maintained the stapes bone in the middle ear to transmit sound vibrations (Fig. 13.1). Additional environmental vibrations are transmitted via muscles and bones in the skull [12].

The osseous inner ear, or the bony labyrinth, is associated with the perception of sound and equilibrium in vertebrates [10, 13]. As with other ossified elements in the skeleton, the bony labyrinth can be readily imaged using HRXCT. A series of CT images can be stacked virtually to form a three-dimensional virtual model of the inner ear (Fig. 13.1C–D, G–H). As CT models of the snake ear accumulate, particular morphological specializations have been shown to occur repeatedly in clades with similar habits, including burrowing (fossorial) or swimming [4, 6]. In the following section, morphology of the ossified inner and middle ear is reviewed for all major families of Serpentes (Fig. 13.2). The phylogeny and classification of extant squamates considered here follows that employed in two recent large-scale studies [14, 15].

13.3 Morphological Variation of the Bony Labyrinth in Habitat Generalists and Specialists

Many snakes are habitat generalists. They are good climbers, swimmers, and racers in various terrestrial terrains [16]. However, some snakes live predominantly in one type of habitat. Burrowing snakes spend more or less of their lives in litter, soil and/or sand, with more dedicated burrowers feeding on soil vertebrates and invertebrates. Most marine snakes are pelagic, spending their entire life cycle in the ocean [16].

Soil habitats are typically characterized by tight space, dim light, and environmental sound transmitted via partly solid substrate. In contrast, the habitats of marine snakes typically have open spaces, more intense light, and sound vibrations transmitted by air and water. In response to these different environmental conditions, sensory organs (e.g., eyes and ears) evolve adaptively [4, 6, 17–20]. Variation in the morphology of the snake inner ear correlates to some extent with the primary habitat of a species [4, 7, 17, 21, 22]. In the following section, the morphology of the inner ear of snake families is reviewed, including the morphology of fossilized inner ears of some stem snakes.

13.3.1 Typhlopidae, Leptotyphlopidae, and Anomalepididae

The three families comprise small, possibly miniaturized burrowers generally known as blind snakes or worm snakes [16]. They form a monophyletic Scolecophidia in snake phylogenies generated from morphological and total evidence (morphology + molecule) data [3, 9, 23, 24], but recent large-scale molecular phylogenetic studies find Scolecophidia to be paraphyletic, with Anomalepididae more closely related to Alethinophidia than to other blind snakes [2, 14, 25–28].

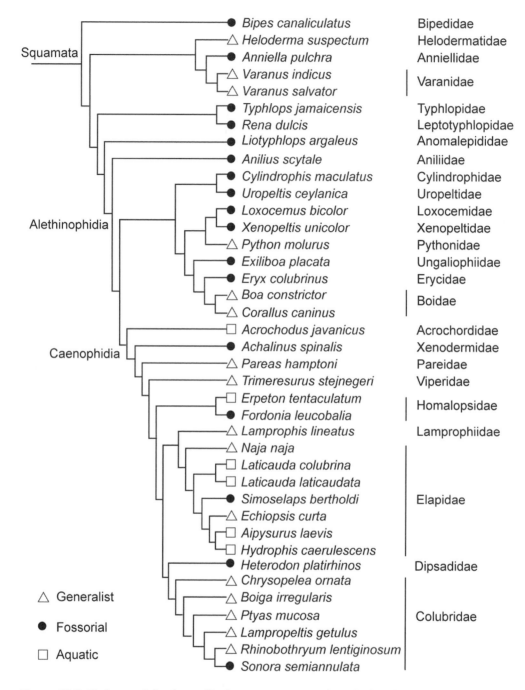

Figure 13.2 Phylogeny (after [14, 15]) of extant squamates described in this chapter, showing convergent evolution of fossorial and aquatic habits.

The bony labyrinths of the three families as sampled in this study (Fig. 13.3B–D) share short and stout semicircular canals that have not been found in alethinophidians (Figs. 13.3–13.5). Wide semicircular canal diameters occur in *Typhlops jamaicensis* (Typhlopidae), *Rena dulcis* (Leptotyphlopidae), and *Liotyphlops argaleus* (Anomalepididae) (Fig. 13.3B–D). Given current comparative data [4, 6, 29], cross-sectional expansion of the semicircular canals cannot be linked with miniaturization or fossoriality. It remains unclear whether the membranous ducts inside the semicircular canals are also stout. In small fossorial snakes and lizards, including *Cylindrophis* (Fig. 13.3F), *Anniella* (Fig. 13.5K), and the miniaturized *Dibamus* (fig. S1 in [4]), the semicircular canals are slender. In fossorial alethinophidians, including *Xenopeltis* and *Loxocemus*, slender semicircular canals are more common (Fig. 13.3H–I).

Vestibule shape varies considerably among the three sampled scolecophidian families. In *T. jamaicensis* (Typhlopidae), the vestibule is enlarged ventral to the lateral semicircular canal (Fig. 13.3B–C). This differs from the morphology in other burrowing squamates, whose vestibule is most expanded in the part dorsal to the lateral semicircular canal. Unlike *T. jamaicensis*, *R. dulcis* (Leptotyphlopidae) has a small vestibule with no expansion (Fig. 13.3D). Although commonly considered a fossorial species, *R. dulcis* (the Texas blind snake) and other leptotyphlopids have been observed climbing trees [16]. It remains unclear whether *R. dulcis* differs from *T. jamaicensis* and *L. argaleus* in habitat preferences.

Despite the varying degree of vestibule expansions, the three species sampled here share a large fenestra ovalis that opens to a large stapedial footplate, which is similar to the condition in other burrowing squamates (Fig. 13.3B–D). The stapes of *T. jamaicensis* has a short, cylindrical shaft. In *R. dulcis* and *L. argaleus*, the stapedial shaft bifurcates distally.

13.3.2 Alethinophidia: Aniliidae, Cylindrophiidae, and Uropeltidae

The majority of extant snakes constitute the Alethinophidia to the exclusion of Typhlopidae, Gerrhopilidae, Leptotyphlopidae, and Anomalepididae [14]. Most extant alethinophidians comprise the clade Caenophidia ('higher' snakes). Among non-caenophidian alethinophidians, there are several clades of small to medium, fossorial snakes, including Aniliidae, Cylindrophiidae, Uropeltidae, and Anomochilidae (Fig. 13.2). These burrowing specialists display morphological adaptations in the ear corresponding to their habitat use [30, 31].

The bony labyrinth of *Anilius scytale* has slender semicircular canals surrounding a large vestibule, which is shared by *Cylindrophis maculatus* and *Uropeltis ceylanica* (Fig. 13.3E–G). In the latter two species, the vestibule is nearly spherical (Fig. 13.3F–G) (fig. 6 in [31]). Viewed dorsally, the vestibule bulges dorsolaterally. The lateral semicircular canal arcs smoothly, subparallel to the rounded vestibule. Ventral to the lateral semicircular canal, a large fenestra ovalis opens in the lateral wall of the vestibule and cochlea. In *A. scytale*, *C. maculatus*, and *U. ceylanica*, the fenestra ovalis is open from the ventral margin of the vestibule to the ventral margin of the cochlea. The stapedial footplate rests on the subcircular fenestra ovalis, and the diameter of the stapedial footplate approaches or exceeds the length of the stapedial shaft (Fig. 13.3E–G).

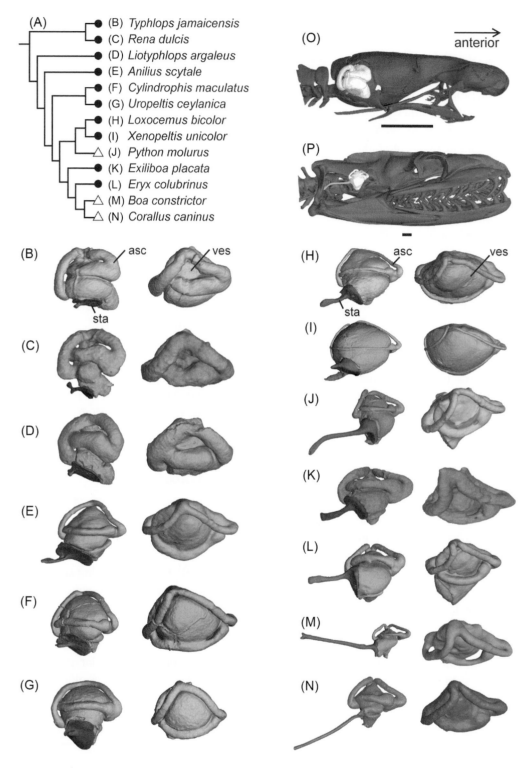

Figure 13.3 The bony labyrinth and stapes of non-caenophidian snakes. (A) Phylogenetic relationships (after [14]) of the species listed in this figure. Black dots denote fossorial species;

13.3.3 Alethinophidia: Pythonidae, Loxocemidae, and Xenopeltidae

Pythonids are terrestrial generalists that utilize a variety of habitats [32]. Their closest extant relatives are two monotypic families, Loxocemidae and Xenopeltidae, both of which are fossorial specialists (Fig. 13.2). These three lineages comprise Pythonoidea.

The habitat generalist *Python molurus* (Fig. 13.3J) has a bony labyrinth slightly different from that of *Ptyas mucosa*, a terrestrial generalist caenophidian (Fig. 13.1B–D). In *P. molurus*, the bony wall of the lateral semicircular canal conjoins with the vestibule, resembling a 'burrowing' ear. However, compared with its burrowing relatives (*Loxocemus bicolor* and *Xenopeltis unicolor*), *P. molurus* lacks a dorsally expanded vestibule and has stouter semicircular canals (Fig. 13.3H–K). With limited sampling of Pythonidae and its closely related clades, a definitive conclusion cannot be drawn as to whether the variations in the cross-sectional profiles are related to habitat and/or phylogeny.

The stapes of *L. bicolor* and *X. unicolor* are distinct from that of *P. molurus* (and other habitat generalists) in that they are shorter and stouter (Fig. 13.1B–D). Little variation is present in the shape of the stapes footplate, and all three species have a footplate that covers a large area in the lateral wall of the bony labyrinth (Fig. 13.3H–J).

13.3.4 Alethinophidia: Boidae, Erycidae, and Ungaliophiidae

Booidea, comprising Boidae and several boa-like clades, is sister to Pythonoidea [14, 27]. Like pythonids, boids are typically medium to large sized snakes that are generalists in terrestrial habitats [32]. Many boids are good swimmers, although exclusively aquatic species are unknown. Erycid and ungaliophiid booids are fossorial to semifossorial snakes that can actively burrow in loose sand and soil [16].

The lateral wall of the vestibule merges with the lateral semicircular canal in the terrestrial *Boa constrictor*, the arboreal *Corallus caninus*, and the two fossorial species *Eryx colubrinus* and *Exiliboa placata* (Fig. 13.3K–N). The same morphology is present in the semiaquatic *Eunectes murinus* (fig. S1 in [4]). Variation in the bony labyrinth is most evident in the region ventral to the lateral semicircular canal. In *E. colubrinus* and *E. placata*, expansion of the vestibule shows a spherical lateral view and a protruding triangular dorsal view (Fig. 13.3K–L). In contrast, lateral expansion of the vestibule is slight in *B. constrictor* (Fig. 13.3M) and inconspicuous in *C. caninus* (Fig. 13.3N). In the middle ear, the stapedial footplate is massive in *E. colubrinus* but diminutive in *C. caninus*. The latter species also has an elongated distal process of the stapes.

Figure 13.3 (*cont.*) white triangles denote terrestrial generalists. The labels (B–N) at the tips of branches in the phylogeny correspond to the labels of the virtual models below. (B–N) Virtual models of the bony labyrinth and stapes of the species in (A). For each species, the bony labyrinth and stapes is shown in lateral (left column) and dorsal (right) view. Not to scale. Anatomical abbreviations: asc, anterior semicircular canal; sta, stapes; ves, vestibule. (O) Skull of *Typhlops jamaicensis* in lateral view, with the bony labyrinth and stapes highlighted. Scale bar = 10 mm. (P) Skull of *Python molurus* in lateral view, with the bony labyrinth and stapes highlighted. Scale bar = 10 mm.

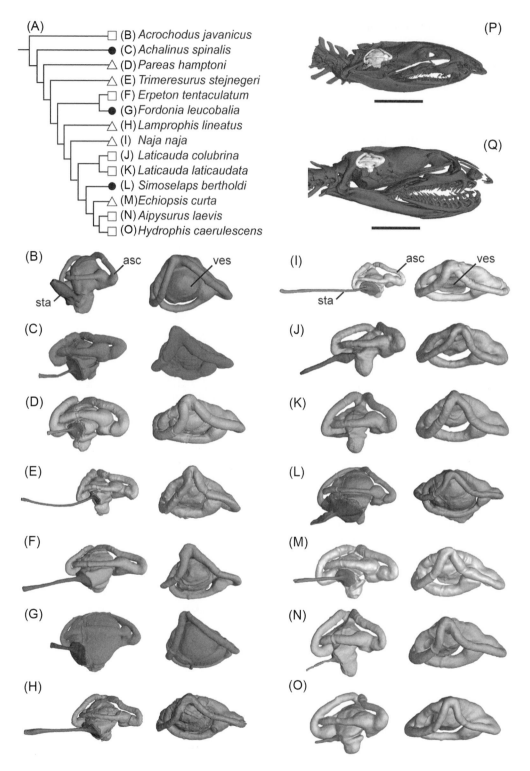

Figure 13.4 The bony labyrinth and stapes of caenophidian snakes. (A) Phylogenetic relationships (after [14, 15]) of the species listed in this figure. Black dots denote fossorial species; white triangles

13.3.5 Caenophidia: Acrochordidae and Xenodermidae

The most shallowly nested extant caenophidians are Acrochordidae and Xenodermidae, two (together paraphyletic) clades with low extant diversity and distinctive ecologies [14, 27]. The three extant species of Acrochordidae are largely piscivorous and fully aquatic in freshwater and/or estuarine/coastal habitats [16, 33]. Xenodermidae includes poorly known, terrestrial and possibly semifossorial species feeding on invertebrates such as earthworms [34].

Acrochordus javanicus inhabits freshwater swamps and preys on fish in the bottom of the water column [35]. This species has been found to aggregate in burrows [16]. Its bony labyrinth shows a combination of characteristics typical of generalist and fossorial taxa. The vestibule of *A. javanicus* bulges dorsally (Fig. 13.4B), resembling the shape in aniliids, uropeltids, and xenopeltids (Fig. 13.3F–I). However, the semicircular canals wrap loosely around the vestibule, resembling those of terrestrial generalists (Fig. 13.1: the colubrid *Ptyas mucosa*). The stapes of *A. javanicus* is stout with a thickened shaft and a small footplate. The stapedial shaft is connected to the quadrate via movable cartilaginous articulations, which resembles the condition in fossorial alethinophidians including *Xenopeltis*, *Loxocemus*, and *Uropeltis* [36]. Although *A. javanicus* is aquatic, the enlarged vestibule and stapes resemble those of burrowing species, perhaps reflecting influences from its bottom-dwelling habit and sedentary lifestyle [16, 33].

In Xenodermidae, *Achalinus spinalis* is nocturnal and fossorial [37], feeding on earthworms [34]. It may burrow into loose soil, but few ecological data are available. The bony labyrinth of *A. spinalis* does resemble that of other burrowers in having a short anterior semicircular canal that does not bend ventrally at the junction with the anterior ampulla (Fig. 13.4C). In addition, the stapes of *A. spinalis* has a large footplate with a short distal shaft. The CT model of the stapes shows a clear fissure in its shaft, suggesting the distal part of the shaft might have had a calcified cartilage dorsal extension from the osseous shaft. Such calcification has been recorded in xenopeltids [36], but more data are needed to assess its presence in *A. spinalis*, and its functional significance, if any.

13.3.6 Caenophidia: Pareidae and Viperidae

Pareas hamptoni and *Trimeresurus stejnegeri* represent separate radiations of arboreal snakes in Pareidae and Viperidae, respectively [16]. Both species are about 1 m in total length and occur in subtropical forests in southeast Asia [16, 38, 39]. The bony labyrinths of *P. hamptoni* (Fig. 13.4D) and *T. stejnegeri* (Fig. 13.4E) share a relatively small vestibule and an elongate and ventrally arching anterior semicircular canal. Although spherical, the vestibule of *P. hamptoni* is much smaller than that of burrowing snakes, relative to the

---

Figure 13.4 (*cont.*) denote terrestrial generalists; white squares denote aquatic species. The labels (B–O) at the tips of branches in the phylogeny correspond to the labels of the virtual models below. (B–O) Virtual models of the bony labyrinth and stapes of the species in (A). For each species, the bony labyrinth and stapes is shown in lateral (left column) and dorsal (right) view. Not to scale. Anatomical abbreviations: asc, anterior semicircular canal; sta, stapes; ves, vestibule. (P) Skull of *Simoselaps bertholdi* in lateral view, with the bony labyrinth and stapes highlighted. Scale bar = 5 mm. (Q) Skull of *Hydrophis caerulescens* in lateral view, with the bony labyrinth and stapes highlighted. Scale bar = 5 mm.

volume of the whole bony labyrinth (compare Fig. 13.4D with Fig. 13.3H: *Loxocemus bicolor* and Fig. 13.4L: *Simoselaps bertholdi*). The stapedial footplate is small in *P. hamptoni* and *T. stejnegeri*, and the stapedial shaft is long in *T. stejnegeri* (Fig. 13.4D–E).

13.3.7 Caenophidia: Elapidae, Homalopsidae, and Lamprophiidae

Elapidae includes extant, fully aquatic sea snakes (including *Aipysurus* and *Hydrophis*) and the semiaquatic sea kraits (*Laticauda*) [27]. The closest extant relatives of sea snakes are terrestrial elapids such as *Echiopsis* (Fig. 13.4A). Therefore, the sea kraits and sea snakes likely represent at least two independent origins of aquatic, marine habits within hydrophiine elapids [20].

The vestibule is generally largest in fossorial elapids and smallest in aquatic elapids. *Simoselaps bertholdi* (Fig. 13.4L), a terrestrial fossorial elapid, has a large vestibule resembling that of other caenophidian burrowers (Fig. 13.4C: *Achalinus spinalis*). In comparison, the aquatic elapids, *Laticauda*, *Aipysurus*, and *Hydrophis* (Fig. 13.4J–K, N–O), have a smaller vestibule relative to the size of the bony labyrinth. The fenestra ovalis and stapedial footplate are also small in aquatic species. *Laticauda* (Fig. 13.4J–K) and *Hydrophis* (Fig. 13.4O) have a particularly tiny stapedial footplate, making the stapes more rod-like than those of terrestrial and fossorial species.

Some fossorial elapids have a small vestibule and a large fenestra ovalis. The fossorial *Brachyurophis australis* inhabits sandy environments, and its large fenestra ovalis extends to the ventral margin of the cochlea (fig. 5 in [6]), resembling that of alethinophidian sand burrowers (Fig. 13.3L: *Eryx colubrinus*).

Homalopsidae comprises aquatic to semiaquatic species collectively called mud snakes [16]. Many species spend much of their time close to the bottom of water bodies, and some species burrow in mud flats. The fully aquatic *Erpeton tentaculatum* (Fig. 13.4F) has a proportionally larger vestibule than that of sea snakes, and it has a stapes similar to those of terrestrial (generalist) caenophidians such as *Lamprophis lineatus* (Fig. 13.4H). *Fordonia leucobalia* (Fig. 13.4G) burrows into mud to prey on crabs, and it has a bony labyrinth resembling that of the burrowing terrestrial elapid *Simoselaps bertholdi* rather than the closely related, more nektonic *E. tentaculatum*. Compared with *E. tentaculatum*, *F. leucobalia* also has a larger vestibule and stouter stapes, with a short distal shaft.

13.3.8 Caenophidia: Colubroidea

Colubroidea consists of nearly 2,000 extant species, the majority of which are terrestrial generalists in contrast to exclusively fossorial or aquatic specialists. Several colubroid species are commonly called water snakes, but none of them is fully aquatic, and they lack the dorsally placed nostrils and laterally compressed tail in sea snakes [16]. The hog-nose dipsadid (*Heterodon platirhinos*) is known as a burrower, but it partly utilizes burrows made by other animals [40] and is less of an active, dedicated burrower than, for example, the non-caenophidian *Uropeltis ceylanica*.

Although extreme aquatic or fossorial adaptations are rare among colubroideans, some have evolved a unique locomotor style in arboreal species: controlled aerial descent. The arboreal southeast Asian *Chrysopelea ornata* glides from trees [41]. Here the bony labyrinth of *C. ornata* is compared with that of arboreal non-gliding snakes, including *Boiga*

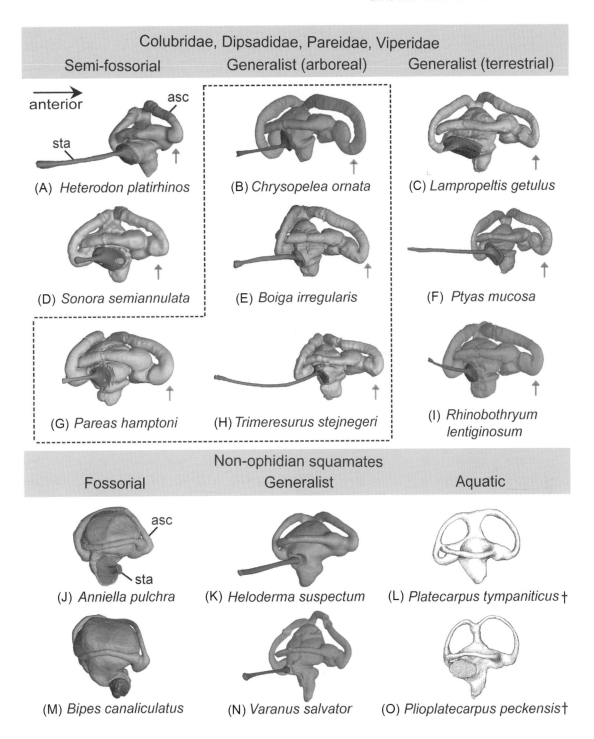

Figure 13.5 Morphological comparisons of squamate bony labyrinth and stapes. (A–J) Caenophidian snakes. Arrows point to the anteroventral margin of the anterior semicircular canal. In arboreal caenophidians, the ventral bending of the anterior semicircular canals is prominent. (K–P) Lizards, including amphisbaenians. The cross sign denotes extinct taxa. Anatomical abbreviations: asc, anterior semicircular canal; f. oval., foramen ovalis; sta, stapes. Ear models not to scale.

irregularis (Colubridae), *Pareas hamptoni* (Pareidae), and *Trimeresurus stejnegeri* (Viperidae). These arboreal species have a large anterior ampulla and a characteristic ventral bend of the anterior semicircular canal (Fig. 13.5B, F, H, I). *Chrysopelea ornata* (Fig. 13.5B) has a larger posterior ampulla than the arboreal non-gliding species.

In the middle ear, the stapes retains a similar shape among arboreal and terrestrial colubroids (Fig. 13.5: *Ptyas mucosa*, and *Rhinobothryum lentiginosum*). An exception is *Lampropeltis getulus* (Fig. 13.5C) that has a shorter stapedial shaft than other terrestrial generalists sampled in this study.

Possibly due to their semifossorial habits, 'burrowing' colubroids lack an enlarged vestibule (Fig. 13.5A, D). In the dipsadid *Heterodon platirhinos* and colubrid *Sonora semiannulata*, the semicircular canals curve loosely around the vestibule, differing from the compact shape in non-colubroid burrowing species described above. However, the two colubroid species do resemble other burrowing snakes in having a short anterior semicircular canal with a simple curve in the anteroventral margin, which is clearly distinguishable from the elaborate anterior semicircular canals in arboreal and ground-dwelling colubroids (Fig. 13.5A–J). The stapes of *S. semiannulata* is as stout as that of other burrowing snakes, yet the stapes of *H. platirhinos* resembles that of terrestrial generalists, with a long shaft and small footplate. More behavioural and anatomical data are required for a greater sampling of colubroids to identify and interpret patterns of morphological variations in their inner ear.

13.4 Ears of Non-Ophidian Squamates ('Lizards')

In snakes, the vestibule is often relatively much larger in fossorial species, smaller in pelagic aquatic species, and intermediate in generalists. A similar pattern exists in lizards. The vestibule is large in volume and expanded dorsally in dedicated burrowers, including *Typhlosaurus lineatus* (Scincomorpha), *Aprasia pulchella* (Gekkota), *Dibamus novaeguineae* (Dibamidae) (fig. S1 in [4]), *Anniella pulchra* (Anguimorpha: Fig. 13.5K), and *Bipes canaliculatus* (Amphisbaenia). Lizard semicircular canals have widely varying shapes, but they are generally more compact in burrowing lizards and more loosely arranged around the vestibule in non-burrowing lizards (Fig. 13.5). Six species of lizards (including amphisbaenians) were sampled in this study (Table 13.1), and information on an additional species was obtained from the literature [7].

Compared with the burrowing *Anniella pulchra* and *Bipes canaliculatus* (Fig. 13.5K, N), surface-dwelling anguimorph lizards have a relatively smaller vestibule and longer semicircular canals (Fig. 13.5L, O). The bony labyrinths of *Heloderma* and *Varanus* are more expanded than those of snakes along the vertical axis, so the dorsal margin of the common crus is located further dorsally from the horizontal plane of the lateral semicircular canal (Fig. 13.5). The bony labyrinth of the semiarboreal mangrove monitor, *Varanus indicus*, resembles that of arboreal snakes in having a small vestibule and an anteroventral bend in the anterior semicircular canal.

Although many extant lizards are good swimmers, none is as aquatic as the extinct, marine Mosasauria (sensu [9]). Two species of *Platecarpus* (Fig. 13.5M, P) have relatively very small vestibules [4, 7], similar to those of sea snakes (Fig. 13.4J, K, N, O). The similarity

Table 13.1. List of sampled specimens.

Family	Species	Specimen	Data source
Bipedidae	*Bipes canaliculatus*	AMNH R-113487	AMNH[a]
Helodermatidae	*Heloderma suspectum*	AMNH R-161167	AMNH
Anniellidae	*Anniella pulchra*	AMNH R-12851	AMNH
Varanidae	*Varanus indicus*	AMNH R-58389	AMNH
	Varanus salvator	AMNH R-94538	AMNH
Typhlopidae	*Typhlops jamaicensis*	USNM 12378	UTCT[b]
Anomalepididae	*Liotyphlops argaleus*	MCZ R-66383	UTCT[c]
Leptotyphlopidae	*Rena dulcis*	TNHC 60638	UTCT[b]
Aniliidae	*Anilius scytale*	USNM 204078	AMNH
Cylindrophiidae	*Cylindrophis maculatus*	AMNH R-126605	AMNH
Uropeltidae	*Uropeltis ceylanica*	AMNH R-43344	AMNH
Loxocemidae	*Loxocemus bicolor*	FMNH 104800	AMNH
Xenopeltidae	*Xenopeltis unicolor*	AMNH R-161693	AMNH
Pythonidae	*Python molurus*	TNHC, to be accessioned	UTCT[b]
Boidae	*Boa constrictor*	FMNH 31182	UTCT[c]
	Corallus caninus	AMNH R-55910	AMNH
Erycidae	*Eryx colubrinus*	FMNH 63117	UTCT[b]
Ungaliophiidae	*Exiliboa placata*	FMNH 207669	UTCT[c]
Acrochordidae	*Acrochordus javanicus*	AMNH R-92269	AMNH
Xenodermidae	*Achalinus spinalis*	AMNH R-34620	AMNH
Pareidae	*Pareas hamptoni*	AMNH R-153711	AMNH
Viperidae	*Trimeresurus stejnegeri*	AMNH R-21057	AMNH
Homalopsidae	*Erpeton tentaculatum*	AMNH R-153799	AMNH
	Fordonia leucobalia	CAS 211970	UTCT[c]
Lamprophiidae	*Lamprophis lineatus*	AMNH R-50646	AMNH
Elapidae	*Naja naja*	FMNH 22468	UTCT[b]
	Laticauda colubrina	AMNH R-28996	AMNH
	Laticauda laticaudata	AMNH R-161778	AMNH

Table 13.1. (*cont.*)

Family	Species	Specimen	Data source
	Simoselaps bertholdi	AMNH R-115427	AMNH
	Echiopsis curta	AMNH R-115397	AMNH
	Aipysurus laevis	AMNH R-5087	AMNH
	Hydrophis caerulescens	AMNH R-86181	AMNH
Colubridae	*Chrysopelea ornata*	AMNH R-161965	AMNH
	Boiga irregularis	AMNH R-69292	AMNH
	Ptyas mucosa	AMNH R-33243	AMNH
	Lampropeltis getulus	AMNH R-95965	AMNH
	Rhinobothryum lentiginosum	AMNH R-52395	AMNH
	Sonora semiannulata	AMNH R-170522	AMNH
Dipsadidae	*Heterodon platirhinos*	FMNH 194529	UTCT[b]

Notes: (a) All AMNH specimens were scanned on site at the MIF lab; specimens sourced from UTCT were scanned at the University of Texas High-Resolution X-ray Computed Tomography Facility and downloaded from either: (b) www.digimorph.org or (c) www.morphosource.org. Institutional abbreviations: AMNH, Herpetology collection, American Museum of Natural History, New York, USA; CAS, Herpetology collection, California Academy of Sciences, San Francisco, USA. FMNH, Amphibian and Reptile Collection, Field Museum of Natural History, Chicago, USA; MACN, Museo Argentino de Ciencias Naturales Bernardino Rivadavia, Buenos Aires, Argentina; MCZ, Museum of Comparative Zoology at Harvard University, Massachussets, USA; QMF, Palaeontological collections, Queensland Museum, Brisbane, Australia; SAMA, South Australian Museum, Adelaide, Australia; TNHC, Texas Natural History Collections, University of Texas at Austin, Texas, USA; USNM, United States National Museum, Smithsonian Institution, Washington, DC, USA.

between the vestibule of the Cretaceous, pelagic [43] *Platecarpus* and modern sea snakes suggests that swimming locomotion strongly affects the shape of the bony labyrinth. Additional sampling in Mosasauria will help understand whether the extremely small vestibule occurs in all mosasaurs, and when it evolved.

13.5 Indicators of Ecology: Quantitative Analyses of the Squamate Bony Labyrinth

Using qualitative characters, it may not be possible to disentangle the impact of ecology and phylogeny on the shape of the squamate bony labyrinth. Using geometric morphometric approaches to generate quantitative data and phylogenetic independent contrast methods, adaptive traits in the bony labyrinth can be identified when phylogenetic signal is controlled for. It has been found that shape convergence occurred in the vestibules of burrowing and swimming squamates [4], and that the shape of the semicircular canals is influenced by phylogeny as well as ecology [6, 22]. The length of the semicircular canals in the fully marine mosasaurs is proportionally as long as terrestrial lizards when compared with skull length [5, 7, 44].

13.5.1 Landmark-Based Multivariate Methods

In geometric morphometric studies, landmarks represent anatomically meaningful points at corresponding positions across a sample. Sets of landmarks and semilandmarks can be used to quantify the geometry of a skull or a single bone [45]. One challenge in landmarking the bony labyrinth is the smooth surface of the vestibule that is deficient in type 1 homologous landmarks, such as intersect points of bone sutures [46, 47]. Other methods of quantifying the shape of the semicircular canals may not apply to the vestibule. For instance, each osseous semicircular canal has a hollow core, and the 'skeletonization' method finds the central path of a semicircular canal by connecting multiple cross-sectional central points along the curve [48–51]. However, the vestibule is approximately spherical in multiple species, the local central points for these vestibules would all be a single point in the center. One solution is to place a small number of type 2 landmarks and a large number of semilandmarks around the lateral surface of the vestibule [4]. A type 2 landmark is a shape-maximum (farthest or nearest) point that is also at the intersect of two morphological structures [46] – for example, the intersection of the lateral semicircular canal and lateral ampulla. Semilandmarks are equidistant points distributed between type 2 landmarks [45, 52].

In order to quantitatively capture the bulging shape in burrowing specialists and the flattened shape in agile marine swimmers (Fig. 13.6A–B), Yi and Norell [4] placed semilandmarks along two curves surrounding the vestibule and lateral semicircular canal. Using principal component analysis (PCA), a group of burrowing squamates were found to occupy a particular region in the morphospace of the vestibule and lateral semicircular canals (fig. 3 in [4]). This morphospace region is populated by obligate burrowers from across Squamata, including members of Elapidae (*Simoselaps*), Xenopeltidae (*Xenopeltis*), Uropeltidae (*Uropeltis*), Cylindrophiidae (*Cylindrophis*), Anguimorpha (*Anniella*), Amphisbaenia (*Bipes*), and Dibamidae (*Dibamus*). A sample of 45 snakes and lizards were subsequently divided into three habitat groups: burrowing, generalist, and aquatic. Using multivarite analysis of variance (MANOVA) of shape coordinates of the vestibule, significant differences were found among the three groups [4]. Additionally, phylogenetic signal was found to be non-significant in the shape data. However, equal branch lengths were used in the phylogenetic test, and this study could be improved by assigning absolute or relative time to the phylogeny [14, 15, 53].

Another approach to landmarking the squamate bony labyrinth has been to focus primarily on the shape of the semicircular canals. Georgi [8] described the bony labyrinth of varanid lizards and mosasaurs, suggesting that semicircular canal shape is indicative of habitat preference. In analysing controlled-aerial descent in iguanian lizards, Boistel et al. [50] found strong phylogenetic signal in general morphometric data of the bony labyrinth, but the shape of the posterior ampulla was affected by locomotor specializations.

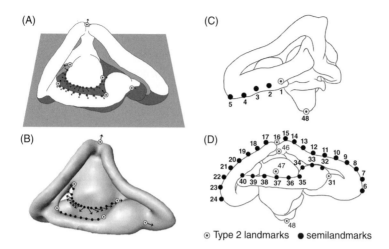

Figure 13.6 Strategies of landmarking the bony labyrinth. (A–B) Landmarks and semilandmarks placed on the vestibule and lateral semicircular canal. Redrawn from Yi and Norell ([4 fig. S2]); (C–D) Landmarks and semilandmarks placed along the semicircular canals; there is one landmark on the lateral surface of the vestibule and one on the ventral tip of the cochlea. Redrawn from Palci et al. ([6 fig. 1]). Ear models not to scale.

Placing dense semilandmarks on the semicircular canals, Palci et al. [6] identified significant phylogenetic signal in the shape of the bony labyrinth of snakes (Fig. 13.6C–D), and a phylogenetic PCA found terrestrial obligate burrowers and the semifossorial mud snakes grouping together in morphospace ([6 fig. 3]). In this same study, canonical variate analysis of bony labyrinth shape data identified three groups comprising fossorial, semi-fossorial, and marine species, but groupings of terrestrial, arboreal, and semiaquatic species were not supported. Arboreal species appear to have generally more elaborate semicircular canals than terrestrial ground-dwelling species, but this distinction was not significant [6].

13.5.2 Semicircular Canal Size and Shape

Mammals exhibit size reduction in the inner ear when adapting to a fully aquatic life – cetaceans have shorter semicircular canals than those of terrestrial mammals [54]. Squamate semicircular canals follow a different trend in transitions from land to water [5, 7]. In the fully marine mosasaurs and sea snakes, semicircular canal length follows a linear relationship with the skull length, independent of phylogeny [7]. In proportion to the skull, the semicircular canals are not significantly shorter in large mosasaurs than in small terrestrial lizards. In studies of the mammalian inner ear, semicircular canal length was regressed with body weight [54, 55]. In studies of the squamate inner ear, the logarithm of the semicircular canal length was regressed with the logarithm of skull size [5, 7], which was shown to be an equivalent method to that used in mammals [5]. In addition to semicircular canal length, Cuthbertson et al. [7] also compared degree of circularity. The anterior and posterior semicircular canals are strongly arched in the mosasaur *Plioplatecarpus*, consistent with the understanding that locomotor styles were unique in mosasaurs among lizards. Comparisons of angles between semicircular canals suggests that mosasaurs were most sensitive to movements in the pitch axis [7].

13.6 Estimating the Ecology of Extinct Snakes

Among extant squamates, it is clear that shape convergence of the bony labyrinth occurred multiple times in habitat specialists. For example, a relatively very large vestibule occurs in fossorial snakes and lizards (Fig. 13.5). Quantitative analysis of the bony labyrinth is a relatively new approach to estimating ecology in snakes for which behavioral and ecological data are lacking and where morphological data might be incomplete, as is the case in most fossil snakes. Most early snake fossils consist of isolated vertebrae and/or incomplete skull material, but the braincases of a few taxa are available for CT reconstructions of the bony labyrinth and stapes (Fig. 13.7).

The Cretaceous *Dinilysia patagonica* [56–58] is a stem snake [9, 59, 60] or possibly nested within the crown group close to the origin of extant alethinophidians [2, 61]. Since its discovery the habit of *D. patagonica* has been under frequent debate. It was suggested to be semiaquatic, resembling modern *Eunectes*, based on a slightly dorsal position of the orbits [58], but this was later called into question [62], because dorsally placed eyes are also present in the fossorial *Xenopeltis*. Additionally, geological data revealed an arid inter-dune environment in which *D. patagonica* was fossilized [62], and its vertebral morphology has been considered to resemble that of extant semifossorial snakes [63].

The otic capsule occupies a large volume in the braincase of *D. patagonica* [4, 59]. The vestibule expands in directions dorsal and ventral to the lateral semicircular canal (Fig. 13.7A), which is as spherical as that of fossorial and semifossorial extant snakes, including *Uropeltis*, *Xenopeltis*, *Simoselaps*, and *Fordonia* (Figs. 13.3 and 13.4). The semicircular canals wrap tightly around the vestibule, which is another signature of extant fossorial specialists (Figs. 13.3–13.5). The anterior semicircular canal is simply shaped at its anteroventral margin, lacking the ventral bend observed in terrestrial generalists. In the middle ear of *D. patagonica*, a large stapedial footplate is present [59, 64], in contrast with the tiny stapedial footplate of extant sea snakes. Thus, there is little resemblance between *D. patagonica* and agile aquatic swimmers in the shape of the vestibule and semicircular canals. The only extant aquatic snakes with a similar vestibule to *D. patagonica* are semifossorial and semiaquatic mud snakes such as *F. leucobalia* (Fig. 13.4G) and *Myron richardsonii* ([6 fig. 5e]).

In the Cenozoic madtsoiids *Yurlunggur* and *Wonambi* [65, 66], the bony labyrinth reconstructed from CT data shows an enlarged vestibule with a large fenestra ovalis [22]. *Yurlunggur* in particular has a spherical vestibule resembling that of extant mud snakes ([6 fig. 5e]) and sand-burrowing boids (Fig. 13.3L), suggesting semifossoriality.

13.7 Summary

Despite sharing an elongate limbless body, extant snakes utilize a wide diversity of habitats. Such diversity is reflected in the morphological variation of their bony

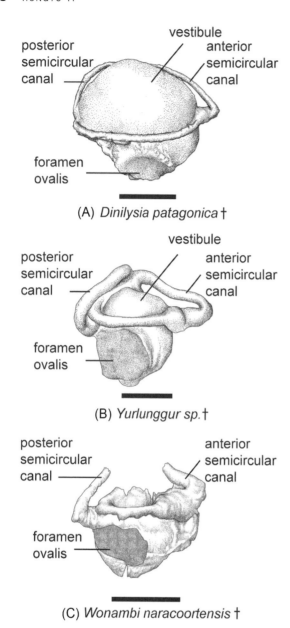

(A) *Dinilysia patagonica* †

(B) *Yurlunggur sp.* †

(C) *Wonambi naracoortensis* †

Figure 13.7 The bony labyrinth of extinct snakes. (A) *Dinilysia patagonica* (MACN-RN 1014), redrawn from Yi and Norell ([4 fig.2]). (b) *Yurlunggur* sp. (QMF 45111–45391), redrawn from Palci et al. ([22 fig. 1a]). (c) *Wonambi naracoortensis* (SAM P30178A), redrawn from Palci et al. ([22 fig. 1b]). The cross sign denotes extinct species. Scale bars = 5 mm.

labyrinths, the site of balance and hearing organs. A large spherical vestibule is observed in dedicated burrowers that mostly live and forage underground. Vestibular expansion towards its dorsal margin and along the horizontal axis occurs repeatedly in the snake phylogeny, notably in Aniliidae, Xenopeltidae, Boidae, and

Elapidae. Not all burrowing snakes have a spherical vestibule, but a spherical vestibule occurs only in dedicated burrowers. At the other end of the morphological spectrum, a small vestibule – reduced in size towards its dorsal margin – is present in marine snakes and in lizards that are active swimmers. Between these two morphological extremes are the ears of habitat generalists in which variation is broad and continuous rather than sharp and categorical.

Virtual models reconstructed from CT data facilitate quantitative estimates of the ecology of extinct snakes, regarding habitat preferences and locomotor styles. Multiple quantitative analyses have shown that different regions of the bony labyrinth, including the vestibule, the semicircular canals, and the cochlea, vary independently. This indicates that different parts of the inner ear are subject to different selection pressures through evolution.

Dinilysia patagonica is the extinct taxon closest to the root of the extant snake tree for which adequate cranial material is available to examine the inner and middle ear, and morphology is suggestive of semifossorial to fossorial habits [4]. Improvements to testing correlation of ear morphology and ecology can be made in the refinement of quantitative approaches to capturing and analysing shape variation, as well as better, less categorical classifications of ecology. Using inner and middle ear morphology to accurately infer the ecology of the ancestral snake will depend also upon robust, well-resolved phylogenies for extinct and extant taxa, as well as much denser taxonomic and ontogenetic sampling, including new fossil discoveries.

Acknowledgements

Thanks to Hussam Zaher and David Gower for the opportunity to participate in this book and the associated meeting. R. Guo provided sketch drawings of ear models in the figures. Susan Evans and David Gower provided constructively critical reviews. H. Yi is funded by the National Natural Science Foundation of China (Grant No. 41702020 and 41688103), and the Chinese Academy of Sciences. The Herpetology Department at the American Museum of Natural History provided specimen access to the majority of extant snake specimens sampled in this study. For non-AMNH specimens of extant snakes, access was provided to CT data by California Academy of Sciences, Field Museum of Natural History, United States National Museum, Smithsonian Institution, Museum of Comparative Zoology–Harvard University, and University of Texas at Austin. Six files of CT data were downloaded from www.digimorph.org; four from www.MorphoSource.org, Duke University. Specimens deposited at the AMNH were scanned on site at the Microscopy and Imaging Facilities, funded by NSF EAR0959384. oVert TCN provided access to the CT data of *Fordonia* with data collection funded by NSF DBI-1701714 and DBI-1701787. oUTCT scan provided access to the CT data of 10 species of snakes originally appearing in ref. [9], with data collection funded by NSF EF-0334961 and data upload to MorphoSource funded by DBI-1902242.

References

1. M. S. Y. Lee, G. L. Bell, and M. W. Caldwell, The origin of snake feeding. *Nature*, 400 (1999), 655–659.

2. T. W. Reeder, T. M. Townsend, D. G. Mulcahy, et al., Integrated analyses resolve conflicts over squamate reptile phylogeny and reveal unexpected placements for fossil taxa. *PLoS ONE*, 10 (2015), e0118199.

3. A. Y. Hsiang, D. J. Field, T. H. Webster, et al., The origin of snakes: revealing the ecology, behavior, and evolutionary history of early snakes using genomics, phenomics, and the fossil record. *BMC Evolutionary Biology*, 15 (2015), 87.

4. H. Yi and M. A. Norell, The burrowing origin of modern snakes. *Science Advances*, 1 (2015), e1500743.

5. H. Yi and M. Norell, The bony labyrinth of *Platecarpus* (Squamata: Mosasauria) and aquatic adaptations in squamate reptiles. *Palaeoworld*, 28 (2019), 550–561.

6. A. Palci, M. N. Hutchinson, M. W. Caldwell, and M. S. Y. Lee, The morphology of the inner ear of squamate reptiles and its bearing on the origin of snakes. *Royal Society Open Science*, 4 (2017), 170685.

7. R. S. Cuthbertson, H. C. Maddin, R. B. Holmes, and J. S. Anderson, The braincase and endosseous labyrinth of *Plioplatecarpus peckensis* (Mosasauridae, Plioplatecarpinae), with functional implications for locomotor behavior. *Anatomical Record*, 298 (2015), 1597–1611.

8. J. A. Georgi, Semicircular canal morphology as evidence of locomotor environment in amniotes. Unpublished PhD thesis (Stony Brook, NY: The Graduate School, Stony Brook University, 2008).

9. J. A. Gauthier, M. Kearney, J. A. Maisano, O. Rieppel, and A. D. B. Behlke, Assembling the squamate tree of life: perspectives from the phenotype and the fossil record. *Bulletin of the Peabody Museum of Natural History*, 53 (2012), 3–308.

10. A. S. Romer, *Osteology of the Reptiles* (Chicago: University of Chicago Press, 1956).

11. S. E. Evans, The lepidosaurian ear: variations on a theme. In J. A. Clack, R. R. Fay and A. N. Popper, eds., *Evolution of the Vertebrate Ear – Evidence from the Fossil Record. Springer Handbook of Auditory Research*, 59 (New York: Springer International Publishing, 2016). pp. 245–84.

12. C. B. Christensen, J. Christensen-Dalsgaard, C. Brandt, and P. T Madsen, Hearing with an atympanic ear: good vibration and poor sound-pressure detection in the royal python, *Python regius. Journal of Experimental Biology*, 215 (2012), 331–342.

13. J. Müller, C. Bickelmann, and G. Sobral, The evolution and fossil history of sensory perception in amniote vertebrates. *Annual Review of Earth and Planetary Sciences*, 46 (2018), 495–519.

14. F. T. Burbrink, F. G. Grazziotin, R. A. Pyron, et al., Interrogating genomic-scale data for Squamata (lizards, snakes, and amphisbaenians) shows no support for key traditional morphological relationships. *Systematic Biology*, 69 (2020), 502–520.

15. H. Zaher, R. W. Murphy, J. C. Arredondo, et al., Large-scale molecular phylogeny, morphology, divergence-time estimation, and the fossil record of advanced caenophidian snakes (Squamata: Serpentes). *PLoS ONE*, 14 (2019), e0216148.

16. H. Greene, *Snakes: The Evolution of Mystery in Nature* (Oakland: University of California Press, 2000).

17. M. R. Miller, The cochlear duct of lizards and snakes. *American Zoologist*, 6 (1966), 421–429.

18. B. F. Simões, F. L. Sampaio, C. Jared, et al., Visual system evolution and the nature of the ancestral snake. *Journal of Evolutionary Biology*, 28 (2015), 1309–1320.

19. T. Konishi, M. W. Caldwell, T. Nishimura, K. Sakurai, and K. Tanoue, A new halisaurine mosasaur (Squamata: Halisaurinae) from Japan: the first record in the western Pacific realm and the first documented insights into binocular vision in

mosasaurs. *Journal of Systematic Palaeontology*, 14 (2016), 809–839.

20. B. F. Simões, D. J. Gower, A. R. Rasmussen, et al., Spectral diversification and trans-Species allelic polymorphism during the land-to-sea transition in snakes. *Current Biology*, 30 (2020), 2608–2615.

21. E. G. Wever, *The Reptile Ear: Its Structure and Function* (Princeton: Princeton University Press, 1978).

22. A. Palci, M. N. Hutchinson, M. W. Caldwell, J. D. Scanlon, and M. S. Y. Lee, Palaeoecological inferences for the fossil Australian snakes *Yurlunggur* and *Wonambi* (Serpentes, Madtsoiidae). *Royal Society Open Science*, 5 (2018), 172012.

23. D. Cundall, V. Wallach, and D. A. Rossman, The systematic relationships of the snake genus *Anomochilus*. *Zoological Journal of the Linnean Society*, 109 (1993), 275–299.

24. J. L. Conrad, Phylogeny and systematics of Squamata (Reptilia) based on morphology. *Bulletin of the American Museum of Natural History*, 310 (2008), 1–182.

25. J. J. Wiens, C. A. Kuczynski, S. A. Smith, et al., Branch lengths, support, and congruence: testing the phylogenomicapproach with 20 nuclear loci in snakes. *Systematic Biology*, 57 (2008), 420–431.

26. R. A. Pyron, F. T. Burbrink, and J. J. Wiens, A phylogeny and revised classification of Squamata, including 4161 species of lizards and snakes. *BMC Evolutionary Biology*, 13 (2013), 93.

27. A. Figueroa, A. D. McKelvy, L. L. Grismer, C. D. Bell, and S. P. Lailvaux, A species-levelphylogeny of extant snakes with description of a new colubrid subfamily and genus. *PLoS ONE*, 11 (2016), e0161070.

28. A. Miralles, L. Marin, D. Markus, et al., Molecular evidence for the paraphyly of Scolecophidia and its evolutionary implications. *Journal of Evolutionary Biology*, 31 (2018), 1782–1793.

29. G. Haas, Anatomical observations on the head of *Liotyphlops albirostris* (Typhlopidae, Ophidia). *Acta Zoologica*, 45 (1964), 1–62.

30. O. Rieppel and J. A. Maisano, The skull of the rare Malaysian snake *Anomochilus leonardi* Smith, based on high-resolution X-ray computed tomography. *Zoological Journal of the Linnean Society*, 149 (200), 671–685.

31. J. C. Olori, *Digital endocasts of the cranial cavity and osseous labyrinth of the burrowing snake Uropeltis woodmasoni* (Alethinophidia: Uropeltidae). *Copeia*, 2010 (2010), 14–26.

32. M. O'Shea, *Boas and Pythons of the World* (Princeton: Princeton University Press, 2007).

33. K. L. Sanders, Mumpuni, A. Hamidy, J. J. Head, and D. J. Gower, Phylogeny and divergence times of filesnakes (*Acrochordus*): inferences from morphology, fossils and three molecular loci. *Molecular Phylogenetics and Evolution*, 56 (2010), 857–867.

34. Y. Yamasaki and Y. Mori, Seasonal activity pattern of a nocturnal fossorial snake, *Achalinus spinalis* (Serpentes: Xenodermidae). *Current Herpetology*, 36 (2017), 28–36.

35. R. Shine, P. Harlow, J. S. Keogh, and Boeadi, Biology and commercial utilization of acrochordid snakes, with special reference to Karung (*Acrochordus javanicus*). *Journal of Herpetology*, 29 (1995), 352–360.

36. D. Cundall and F. Irish, The snake skull. In C. Gans, A. S. Gaunt, and K. Adler, eds., *Biology of the Reptilia*, Volume 20, *Morphology H* (Ithaca: Society for the Study of Amphibians and Reptiles, 2008). pp. 349–692.

37. L. Peng, D. Yang, S. Duan, and S. Huang, Mitochondrial genome of the Common burrowing snake *Achalinus spinalis* (Reptilia: Xenodermatidae). *Mitochondrial DNA, Part B Resources*, 2 (2017), 571–572.

38. P. Guo and E.- M. Zhao, Comparison of skull morphology in nine Asian pit vipers (Serpentes: Crotalinae). *Herpetological Journal*, 16 (2006), 305–313.

39. C. W. You, N. A. Poyarkov, and S. M. Lin, Diversity of the snail-eating snakes *Pareas*

(Serpentes, Pareatidae) from Taiwan. *Zoologica Scripta*, 44 (2015), 349–361.

40. M. Plummer, Observations on hibernacula and overwintering ecology of Eastern hog-nosed snakes (*Heterodon platirhinos*). *Herpetological Review*, 33 (2002), 89–90.

41. M. C. Michener and J. D. Lazell, Distribution and relative abundance of the hognose snake, *Heterodon platirhinos*, in eastern New England. *Journal of Herpetology*, 23 (1989), 35–40.

42. J. J. Socha, Gliding flight in *Chrysopelea*: turning a snake into a wing. *Integrative and Comparative Biology*, 51 (2011), 969–982.

43. J. L. Bell and M. J. Polcyn, *Dallasaurus turneri*, a new primitive mosasauroid from the Middle Turonian of Texas and comments on the phylogeny of Mosasauridae (Squamata). *Netherlands Journal of Geosciences*, 84 (2005), 177–194.

44. H. Yi, D. Sampath, S. Schoenfeld, and M. A. Norell, Reconstruction of inner-ear shape and size in mosasaurs (Reptilia: Squamata) reveals complex aquatic adaptation strategies in secondary aquatic reptiles. *Journal of Vertebrate Paleontology*, 32 Supplement (2012), 198A.

45. D. C. Adams and E. Otárola-Castillo, Geomorph: an R package for the collection and analysis of geometric morphometric shape data. *Methods in Ecology and Evolution*, 4 (2013), 393–399.

46. F. L. Bookstein, *Morphometric Tool for Landmark Data: Geometry and Biology* (Cambridge: Cambridge University Press, 1991).

47. J. T. Richtsmeier, C. H. Paik, P. C. Elfert, T. M. Cole, and H. R. Dahlman, Precision, repeatability, and validation of the localization of cranial landmarks using computed tomography scans. *The Cleft Palate-Craniofacial Journal*, 32 (1995), 217–227.

48. P. Gunz, M. Ramsier, M. Kuhrig, J.-J. Hublin, and F. Spoor, The mammalian bony labyrinth reconsidered, introducing a comprehensive geometric morphometric approach. *Journal of Anatomy*, 220 (2012), 529–543.

49. X. Ni, J. J. Flynn, and A. R. Wyss, Imaging the inner ear in fossil mammals: high-resolution CT scanning and 3-D virtual reconstructions. *Palaeontologica Electronica*, 15 (2012), 1–10.

50. R. Boistel, A. Herrel, R. Lebrun, et al., Shake rattle and roll: the bony labyrinth and aerial descent in squamates. *Integrative and Comparative Biology*, 51 (2011), 957–968.

51. J. A. Georgi, J. S. Sipla, and C. A. Forster, Turning semicircular canal function on its head: dinosaurs and a novel vestibular analysis. *PLoS ONE*, 8 (2013), 1–11.

52. P. Gunz, P. Mitteroecker, and F. L. Bookstein, Semilandmarks in three dimensions. In D. E. Slice, ed., *Modern Morphometrics in Physical Anthropology* (Boston: Springer, 2005), pp. 73–98.

53. Y. Zheng and J. J. Wiens, Combining phylogenomic and supermatrix approaches, and a time-calibrated phylogeny for squamate reptiles (lizards and snakes) based on 52 genes and 4162 species. *Molecular Phylogenetics and Evolution*, 94 (2016), 537–547.

54. F. Spoor, S. Bajpai, S. T. Hussain, and K. Kumar, J. G. M. Thewissen, Vestibular evidence for the evolution of aquatic behaviour in early cetaceans. *Nature*, 417 (2002), 163–165.

55. E. G. Ekdale, Comparative anatomy of the bony labyrinth (inner ear) of placental mammals. *PLoS ONE*, 8 (2013), e66624.

56. A. S. Woodward, On some extinct reptiles from Patagonia, of the genera *Miolania*, *Dinilysia*, and *Genyodectes*. *Proceedings of the Zoological Society of London*, 70 (1901), 169–184.

57. R. Estes, T. H. Frazzetta, and E. E. Williams, Studies on the fossil snake *Dinilysia patagonica* Woodward. I. Cranial morphology. *Bulletin Museum of Comparative Zoology of Harvard University*, 119 (1970), 25–74.

58. T. H. Frazzetta, Studies on the fossil snake *Dinilysia patagonica* Woodward. Part 2. Jaw machinery in the earliest snakes. *Forma et Functio*, 3 (1970), 205–221.

59. H. Zaher and A. Scanferla, The skull of the Upper Cretaceous snake *Dinilysia patagonica* Smith-Woodward, 1901, and its phylogenetic position revisited. *Zoological Journal of the Linnean Society*, 164 (2012), 194–238.

60. T. R. Simões, M. W. Caldwell, M. Tałanda, et al., The origin of squamates revealed by a Middle Triassic lizard from the Italian Alps. *Nature*, 557 (2018), 706–709.

61. H. Zaher, S. Apesteguía, and A. Scanferla, The anatomy of the Upper Cretaceous snake *Najash rionegrina* Apesteguía & Zaher, 2006, and the evolution of limblessness in snakes. *Zoological Journal of the Linnean Society*, 156 (2009), 801–826.

62. M. W. Caldwell and A. M. Albino, Palaeoenvironment and palaeoecology of three Cretaceous snakes: *Pachyophis, Pachyrhachis,* and *Dinilysia. Acta Palaeontologica Polonica,* 46 (2001), 203–218.

63. M. K. Hecht, The vertebral morphology of the Cretaceous snake, *Dinilysia patagonica* Woodward. *Neues Jarhbuch fur Geologie und Palaontologie, Monatshefte,* 1982 (1982), 523–532.

64. M. W. Caldwell and A. Albino, Exceptionally preserved skeletons of the Cretaceous snake *Dinilysia patagonica* Woodward, 1901. *Journal of Vertebrate Paleontology,* 22 (2002), 861–866.

65. J. D. Scanlon, A new large madtsoiid snake from the Miocene of the Northern Territory. *The Beagle: Records of the Museums and Art Galleries of the Northern Territories,* 9 (1992), 49–60.

66. J. D. Scanlon and M. S. Y. Lee, The Pleistocene serpent *Wonambi* and the early evolution of snakes. *Nature* 403 (2000), 416–420.

14

A Glimpse into the Evolution
of the Ophidian Brain

AGUSTÍN SCANFERLA

14.1 Introduction

The brain is the principal organ of the central nervous system of vertebrates. It integrates internal (physiological) and external (environmental) sensory information in order to maintain homeostasis, as well as to coordinate actions to interact with the environment. Besides visual and auditory sensory systems, extant snakes display a panoply of sensory capabilities including an enhanced vomeronasal system that detects pheromones and other chemicals [1], pit organs that sense infrared radiation [2] and tactile organs for mechanoreception [3].

Comparative studies concerning brain anatomy can provide great insight into the evolution of sensory systems, mainly because the relative size and shape of the different segments that conform the brain of reptiles are deeply influenced by sensory organs [4]. Despite the diversity of sensory capabilities, comparative studies of brain gross anatomy and its relation to sensory organs are scarce for snakes, especially the gross anatomy of the different segments of the brain. The seminal work of Senn [5] deserves special mention, because its detailed description of the gross anatomy and cytoarchitecture of dissected brains is the only thorough survey matching these two different morphological levels of the snake brain. Senn was the first to document the contrasting morphology of the brain between the 'basal' burrowing 'Scolecophidia' (hereafter 'blind snakes') and more typically surface-dwelling Alethinophidia (booids, pythonids and relatives, caenophidians). Subsequently, some works dealing tangentially with gross anatomical traits of the snake brain were published, but only as part of studies based mostly on cytoarchitecture or encephalization (e.g., [6, 7]).

The advent and increasing availability of CT scanning has provided a new ability to examine endocranial cavities, especially for fossils. Digital endocasts are a valuable,

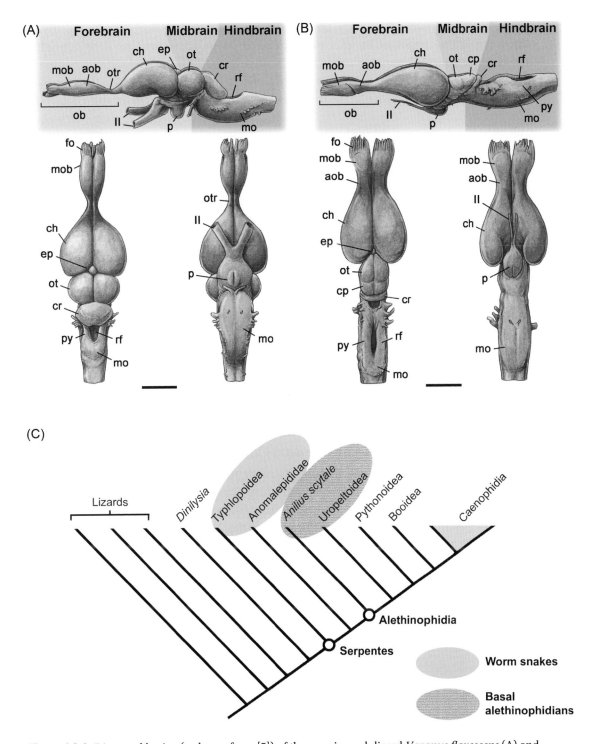

Figure 14.1 Dissected brains (redrawn from [5]) of the anguimorph lizard *Varanus flavescens* (A) and the alethinophidian snake *Xenopeltis unicolor* (B) showing the main gross anatomical features discussed in this chapter. (C) Simplified snake cladogram depicting major groups of snakes

non-destructive approach to study gross anatomy of the brain. Olori [8] was the first to depict the digital endocast of an extant snake, the non-caenophidian alethinophidian *Uropeltis woodmasoni*, and more recently Allemand et al. [9] published the first comparative survey of brain endocasts. To date, only one endocast of a fossil snake has been documented, a natural endocast of the posterior brain and inner ear of the Cretaceous stem snake *Dinilysia patagonica* [10]. That study detected phylogenetic signal as well as ecological trends that shaped the brain during the course of snake evolution. The particular anatomy of the snake braincase, which is almost completely ossified, has supported claims that snake endocasts reflect the overall shape of the brain and the relative sizes of its parts [4, 8, 9, 11]. However, there are no explicit comparative surveys that test whether endocast shape matches actual gross anatomy of the brain.

Despite the efforts mentioned above, the gross anatomy of the snake brain remains largely unexplored and thus most of the evolutionary events behind the anatomical diversity of extant snake brains remain unclear. This chapter has four main aims: (1) to evaluate how reliably endocasts reflect brain gross anatomy in snakes, (2) to assess the most conspicuous features of the gross anatomy of snake brains, including the Cretaceous fossil stem snake *Dinilysia patagonica*, (3) to interpret some gross anatomical traits in light of evolutionary changes in sensory systems, and (4) to discuss these results in the context of the origin and evolution of snakes.

14.2 Gross Anatomy of the Snake Brain

14.2.1 Brain-Head Relations and Endocast Fidelity

The characteristic persistence of a platytrabic condition during the ontogeny of snakes results in a tubular anterior region of the braincase where the olfactory bulb region of the brain is housed. The interorbital space thus defined houses a broad olfactory bulb region in contrast to the thin olfactory peduncle present in most lizards (Fig. 14.1). However, the interorbital space of snakes with a large eye size (ES > 20, where ES, eye size, is eye diameter relative to skull length expressed as a percentage: see Supplementary Appendix 14.S1 and S2) is notably reduced and consequently the olfactory bulbs undergo a remarkable reduction in diameter and diverge anteriorly (Fig. 14.2A–D). Also, the anterior region of the cerebral hemispheres of snakes with large ES is laterally emarginated by the eyes (Fig. 14.2D) and the forebrain therefore acquires an amphora-like shape.

Figure 14.1 (*cont.*) discussed in the chapter, including 'Scolecophidia' (blind snakes), Uropeltoidea (Cylindrophiidae, Uropeltidae and Anomochilidae), Pythonoidea (*Loxocemus bicolor*, *Xenopeltis* spp. and Pythonidae), Booidea (Boidae, Candoiidae, Charinaidae, Erycinae, and Sanziniidae) and Caenophidia. Scale bars equal to 5 mm. II, optic tract; aob, accessory olfactory bulb + olfactory tracts; ch, cerebral hemisphere; cp, colliculus posteriores; cr, cerebellum; ep, epiphysis; fo, fila olfactoria; mo, medulla oblongata; mob, main olfactory bulb; ob, olfactory bulb; ot, optic tectum; otr, olfactory tract; p, pituitary; py, medullary pyramid; rf, rhomboid fossa.

The otic capsules influence the posterior region of the brain in snakes. In blind snakes, the inner ears are located laterally and are widely separated from the midline (Fig. 14.2A), whereas in alethinophidians they are applied dorsolaterally to the hindbrain and close to the midline (Fig. 14.2B). The large otic region of the skull of snakes with a cranial longitudinal shortening (blind snakes, uropeltoids, and some cryptozoic caenophidians) invades and thus embays the posterolateral aspect of the cerebral hemisphere in its posterolateral regions (Fig. 14.2E). Conversely, the otic region of snakes with elongated skulls is located posterior to the forebrain and thus the rear region of the cerebral hemisphere has a spherical shape, even in species with a relatively large otic region (Fig. 14.2F). Notably, the medial parietal pillar of the parietal bone encloses this cerebral region in most alethinophidians, whereas in blind snakes and uropeltids this bony structure is absent and therefore the cerebral hemispheres are in direct contact with the bony otic region (Fig. 14.2E, F).

There are some brain regions where the snake endocast fails to represent the brain gross anatomy. Compared to that of alethinophidians, the anteroventral area of the olfactory bulb region in blind snakes is not completely enclosed by bone. The subolfactory process of the frontal bone fails to surround the olfactory bulb region anteroventrally, and both the medial and lateral frontal flanges are absent. Conversely, these structures, which together form the medial frontal pillar, divide the anterior region of the olfactory bulbs in alethinophidians and thus provides bony boundaries that facilitate endocast segmentation from CT data. The medial parietal pillars are another bony feature with a direct influence on endocast fidelity in the posterior region of the cerebral hemispheres. These bony walls are in close contact with the cerebral hemispheres posteriorly, thus casting the shape of this brain segment (Fig. 14.2F). Conversely, the endocast of some snakes without medial parietal pillars such as blind snakes and uropeltids exhibits a triangular shape in this region ([8]; Fig. 14.2E), which does not conform to the actual spherical shape of the cerebral hemispheres (compare Fig. 14.2E with fig. 4 in [5]). The interaction between the brain and other cephalic structures represents a further source of 'artifacts' in endocasts. The dorsal region of the hindbrain is fully encapsulated by the supraoccipital and otooccipital bones, but there remains a conspicuous space between the skull roof and the region encompassing the posterodorsal region of the midbrain, the cerebellum and the rhomboid fossa. In life it is filled by the choroid plexus (Fig. 14.2G), which covers dorsally the rhomboid fossa and the cerebellum in squamates [12]. Endocasts thus record the space filled by the choroid plexus, masking the posteriormost region of the midbrain roof, the cerebellum and the rhomboid fossa.

14.2.2 Gross Brain Anatomy of Extant Snakes

As in other vertebrates, the various segments of the snake brain are arranged linearly along the longitudinal axis of the skull. Blind snakes and uropeltids exhibit a remarkable shortening along the longitudinal axis in comparison with most alethinophidian snakes [5, 8]. In blind snakes and uropeltids, this shortening can be appreciated through the considerable reduction of space and overlapping between brain segments (Fig. 14.3A). As will be detailed below, this aspect greatly influences the shape of the different brain segments in these fossorial species.

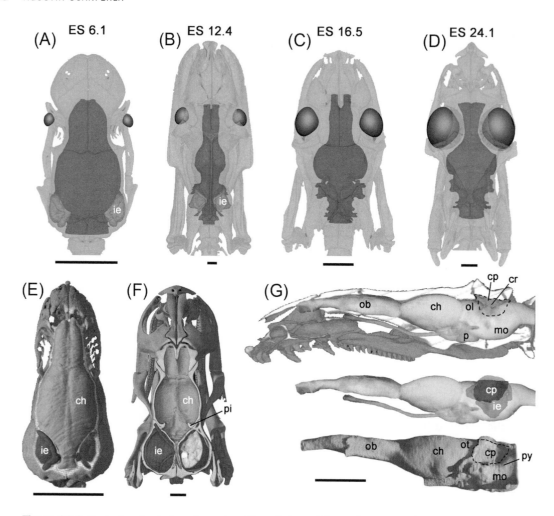

Figure 14.2 Brain-head relations in snakes. Dorsal view of three-dimensional reconstructions based on microCT data of the skull, brain endocast and eyeballs of *Typhlops jamaicensis* (A), *Python molurus* (B), *Tropidophis haetianus* (C), and *Coluber constrictor* (D) figured at the same snout-condyle length. Dorsal view of the horizontal cutaway of the skull of the uropeltoid *Teretrurus sanguineus* (E) and *Xenopeltis unicolor* (F) showing the brain and inner ear endocasts. (G) Lateral views of (from top to bottom) the brain of the 'basal' alethinophidian *Anilius scytale* within the braincase (depicted as a sagittal cutaway), the brain with the inner ear and choroid plexus endocasts in natural position, and the brain endocast. ch, cerebral hemisphere; cp, choroid plexus; ES, relative size of the eyes (diameter as percentage of skull length); ie, inner ear endocast; mo, medulla oblongata; ol, olfactory lobe; p, pituitary gland; pi, medial parietal pillar; py, medullary pyramid. Dashed line delimits the area of influence on the endocast of the choroid plexus. Scale bars equal 2.5 mm.

Forebrain

In general terms, the forebrain of snakes can be divided into the olfactory bulb, cerebral hemispheres, and the thalamus, which includes the pituitary gland as the most conspicuous gross feature. The olfactory bulb is essentially a tubular structure attached to the

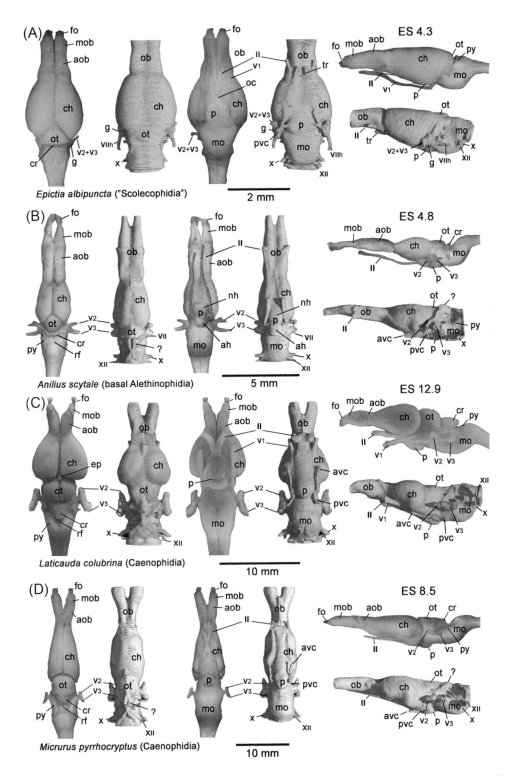

Figure 14.3 Dorsal (left), ventral (centre) and left lateral (right) views of the brain and the endocast of the blind snake *Epictia albipuncta* (A), the 'basal' alethinophidian *Anilius scytale* (B), the marine elapid

retrobulbar region of the cerebral hemispheres. It is composed of the main olfactory bulb anteriorly, and the accessory olfactory bulb plus both olfactory tracts (Fig. 14.1). The olfactory bulb of blind snakes is shorter than the length of the cerebral hemispheres. Conversely, most alethinophidians have an olfactory bulb longer than the cerebral hemispheres. As noted above, eye size strongly influences the shape of the olfactory bulb among snakes. Those species with large ES exhibit marked lateral compression and consequently a reduction in diameter of the olfactory bulb (Fig. 14.3A, C). Notably, cryptozoic caenophidians (often with relatively smaller eye size) also show a marked shortening and widening of the olfactory bulb in much the same fashion as in major lineages of fossorial non-caenophidian alethinophidian snakes (blind snakes, uropeltids) and in contrast to surface-dwelling relatives (Fig. 14.4B, D). There are several filaments at the anterior tip of the main olfactory bulb that form the fila olfactoria (Fig. 14.3A–D), which is composed of unmyelinated axons that enter the olfactory bulb region from the nasal sac and the vomeronasal organ. The fila olfactoria is not represented in endocasts due to the lack of a cribriform plate or other analogous bony structure in snakes. Thus, snake endocasts fail to delimit the anterior tip of the olfactory bulb. The limit between the main olfactory bulb and accessory olfactory bulb regions can be only assessed in dissected brains through a vertical or oblique furrow (Fig. 14.3A–D). Left and right olfactory bulbs cannot be differentiated in snake endocasts due to the poor expression of this furrow. Notably, uropeltids appear to be the only major lineage of snakes with a distinctive configuration of the olfactory bulb, due to the existence of a long and thin peduncle bridging both olfactory bulbs [S5]. Regrettably, this peduncle and the main olfactory bulb cannot be observed in uropeltid endocasts, which only cast the posterior olfactory bulb region ([8]; Fig. 14.2F). The main olfactory bulb is invariably smaller than the accessory olfactory bulb in snakes. This latter structure does not form a conspicuous surface feature, thus its precise size and shape in dissected brains can be determined only through histological studies, which have demonstrated an important morphological diversity among snake groups [6, 13, 14].

The cerebral hemispheres of snakes bulge laterally in their posterior region, but the degree of expansion is variable among clades. Whereas in most alethinophidians examined it bulges strongly (Figs. 14.3C and 14.4A, C), blind snakes and fossorial or cryptozoic alethinophidians lack this posterior widening (Figs. 14.3A, B, D and 14.4B, D). In blind snakes, the posterodorsal region of the cerebral hemispheres (preoptic area) overlaps the anterior aspect of the optic tectum (Fig. 14.3A). Associated with this condition, the preoptic sulcus that separates the cerebral hemispheres from the optic tectum is barely recognizable

Figure 14.3 (*cont.*) *Laticauda colubrina* (C) and the cryptozoic elapid *Micrurus pyrrhocryptus* (D). II, optic tract; V1, ophthalmic branch of the trigeminal nerve; V2, maxillary branch of the trigeminal nerve; V3, mandibular branch of the trigeminal nerve; V2+V3, maxillary and mandibular branches of the trigeminal nerve; VIIh, hyoid ramus of the facial nerve; ah, adenohypophysis, aob, accessory olfactory bulb + olfactory tracts; avc, anterior opening of the Vidian canal; ch, cerebral hemisphere; cr, cerebellum; ep, epiphysis; fo, fila olfactoria; g, ganglionic chamber; mo, medulla oblongata; mob, main olfactory bulb; nh, neurohypophysis; ob, olfactory bulb region; ot, optic tectum; pvc, posterior opening of the Vidian canal; p, pituitary; py, medullary pyramid; rf, rhomboid fossa; tr, trabecula; ?, bilobed structure.

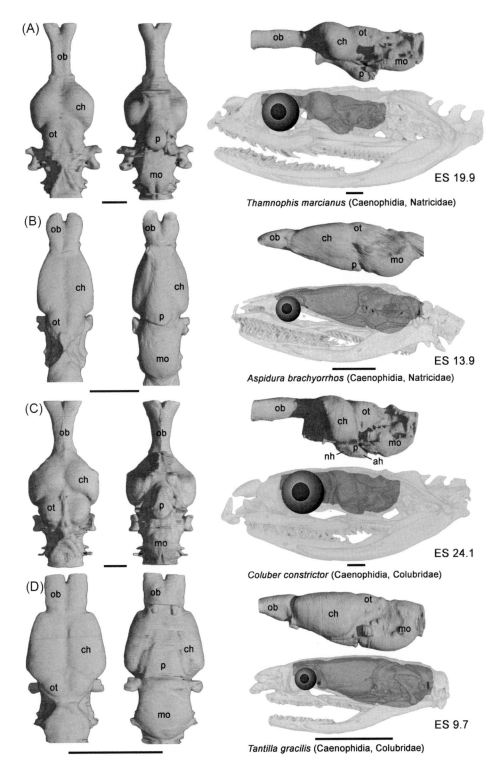

Figure 14.4 Comparison of the brain endocast of surface-dwelling and cryptozoic caenophidians. The natricids *Thamnophis marcianus* (A) and *Aspidura brachyorrhos* (B); colubrids *Coluber constrictor* (C) and *Tantilla gracilis* (D). Skulls are figured at the same snout-condyle length. Scale bars equal 2.5 mm. ah, adenohypophysis; ch, cerebral hemisphere; mo, medulla oblongata; nh, neurohypophysis; ob, olfactory bulb region; ot, optic tectum; p, pituitary.

in blind snake endocasts. Conversely, in most alethinophidians the preoptic sulcus is well developed, and thus the posterodorsal region of the cerebral hemispheres is clearly rounded (Fig. 14.3C, D). The lateral part of this sulcus is occupied by the medial parietal pillar, which separates the cerebral hemisphere from the bony otic region and the optic tectum (Fig. 14.2F). Thus, cerebral hemispheres of alethinophidians are the brain segments best represented in snake endocasts.

The epiphysis is located at the intersection of the preoptic and sagittal sulci. It is clearly visible on some alethinophidian brains (Fig. 14.3C), although it is not represented in endocasts because of its small size and low profile between the cerebral hemispheres and optic tectum.

A variety of morphologies of the pituitary gland can be recognized among surface-dwelling and fossorial snakes [9, 15, 16]. The pituitary of fossorial or cryptozoic blind snakes and non-caenophidian alethinophidians is symmetrical and distinctly flattened, with the neurohypophysis being lateral or posterodorsal to the adenohypophysis (Fig. 14.3A, B). This gland in *Anilius scytale* exhibits a bilobed shape in ventral view (Fig. 14.3B), with the neurohypophysis lying posteriorly to the adenohypophysis. This spatial relationship is not easily identified in the corresponding endocast because the neurohypophysis lobe is barely discernible. Surface-dwelling alethinophidians bear a rounded, well-developed pituitary that settles in a deeply concave sella turcica located in the parabasisphenoid bone (Fig. 14.3C), often posteriorly projected in species with a large gland. Some species exhibit an asymmetry between the size of the adenohypophysis and neurohypophysis (Fig. 14.4A, C), with the neurohypophysis lying dorsolateral to the adenohypophysis. Notably, surveyed species of fossorial and cryptozoic caenophidians show a flattened pituitary (Figs. 14.3D and 14.4B, D), thus mirroring those of blind snakes and 'basal' alethinophidians.

Midbrain

The two prominent features of the midbrain roof in snakes are the optic tectum and to a lesser degree the colliculus posteriores (collicular region), which is located posteriorly. The optic tectum shows a considerable diversity in shape and size among extant snakes. It is usually divided into two spherical structures, the optic lobes, by a more-or-less defined sagittal sulcus. Blind snakes have the smallest eyes among extant snakes; their optic lobes are notably small and lack the typical spherical shape that characterizes alethinophidians (Fig. 14.3A). The anterior region of the optic tectum is embayed anteriorly between the posteromedial aspect of the cerebral hemispheres, thus it acquires a triangular shape in dorsal view. Endocasts of blind snakes fail to track the boundaries of the optic lobes, making it difficult to distinguish this brain segment from the cerebral hemispheres [9]. The fossorial non-caenophidian alethinophidian *Anilius scytale*, which has small eyes (ES 4.8) similar to blind snakes, has a small optic tectum, lacking any signs of division between optic lobes and collicular region (Fig. 14.3B). In contrast, the optic lobes of most surface-dwelling alethinophidians are rounded, well-defined structures behind the cerebral hemispheres. The size of the optic lobes is variable in alethinophidians, ranging from small bulges in fossorial and cryptozoic forms with small eyes (Figs. 14.3D and 14.4B, D) to large spherical structures reaching the dorsal edge of the cerebral hemispheres in surface-dwelling species with larger eyes (ES > 13) (Figs. 14.3C and 14.4A, C). Endocasts reflect these two tendencies, although the sphericity of the optic lobes is less defined laterally. Remarkably, the endocasts of some large-bodied terrestrial alethinophidians such as *Boa imperator* or *Python molurus*, which also have relatively small eyes (ES ~ 12), have barely recognizable optic lobes.

The collicular region of snakes is posterior to the optic tectum. It departs from that of lizards in having two different structures, the torus semicircularis and paratorus [17], which display different degrees of development across the snake tree. Although the collicular region can accurately be identified through histological sections, a pair of bulges immediately behind the optic lobes are present in some species with well-developed tori. In species with a large collicular region, the midbrain roof becomes anteroposteriorly elongated (Figs. 14.1 and 14.3B), whereas it is mostly rounded in other snakes (Fig. 14.3C). In particular, some non-caenophidian alethinophidians, such as the uropeltoid *Cylindrophis* and pythonoid *Xenopeltis,* show a notable development of the collicular region (Fig. 14.1; [5, 17]). The distinction between the optic tectum and the collicular region is not detectable in endocasts, largely due to its small size and the influence of the choroid plexus on endocast shape.

Hindbrain

The hindbrain constitutes the transition between the brain itself and the spinal cord. Cerebellum, pons, and medulla oblongata are the main gross features of the hindbrain, besides the several cranial nerves with an origin in this part of brain.

The cerebellum of snakes is relatively small in comparison with legged lizards (Fig. 14.1). It is located in the large fold created by the cephalic and cervical flexures. Nieuwenhuys [18] described the cerebellum of snakes as a flat lamella projected posteriorly, but this is simplistic and applicable only to some species. Indeed, snakes exhibit at least two basic cerebellar shapes. Blind snakes have the relatively smallest cerebellum among snakes, characterized as a small, thin triangular lamina positioned deep in the fold delimited by the posterior region of the midbrain roof and the medulla oblongata (Fig. 14.3A). The cerebellum of non-caenophidian alethinophidians, although larger than that of blind snakes, is also lamellar shaped and small (Fig. 14.3B). In contrast, other alethinophidians have an anteroposterior thickening of the cerebellum (Fig. 14.3C), thus acquiring a trapezoidal shape. The cerebellum is projected posterodorsally, covering the dorsal region of the rhomboid fossa (Fig. 14.3C). As noted previously, the cerebellum cannot be discerned in snake endocasts.

The hindbrain ventral to the cerebellum is marked by a bulge, the pons. This region, most notable for being the origin of the trigeminal nerve (V), is homogeneous across snakes. The medulla oblongata is located immediately posterior to the pons without any clear signs of this transition. Its shape is strongly influenced by the longitudinal shortening observed in blind snakes and uropeltids, in which the entire hindbrain is particularly compacted (Fig. 14.3A). There are two walls delimiting the rhomboid fossa, which corresponds with the medullary pyramids (Figs. 14.3 and 14.4). The rhomboid fossa is notably small in blind snakes and other fossorial and cryptozoic snakes due to the longitudinal shortening of the brain. Endocasts track most of the shape of the medulla oblongata, with the exception of the rhomboid fossa that is covered by the cast of the choroid plexus.

14.2.3 The Endocast of *Dinilysia patagonica*

Despite the incompleteness of available skull specimens, the segmentation of CT data for their neurocranial cavities allows a satisfactory reconstruction of the endocast of *Dinilysia* (Fig. 14.5A, B). This endocast reconstruction covers all regions of the brain due to the highly ossified condition of the braincase of *Dinilysia* [19, 20].

The endocast of *Dinilysia* resembles that of *Anilius scytale* (Fig. 14.3B), characterized by an elongated shape with no evidence of the longitudinal shortening and overlap between brain segments observed in blind snakes, uropeltids, and cryptozoic caenophidians. The elongate, slender olfactory bulb is distinctly bent downwards similar to those in lizards, thus contrasting with the straight shape observed in extant snakes. The bilobed shape that characterizes this region in extant snakes can only be observed in the anterior third of the *Dinilysia* endocast, through a wide and shallow longitudinal groove observed on the dorsal surface. The optic nerve exits through an opening located in the posterior third of the olfactory bulb, contrasting with the forward position observed in basal alethinophidians. There is no evidence of bulges or grooves that would allow an accurate identification of the different olfactory bulb regions and tracts.

The cerebral hemispheres of *Dinilysia* are relatively small in comparison with extant alethinophidians. The lateral bulge of the posterior region of the cerebral hemispheres is weakly expressed in comparison with most alethinophidians (Fig. 14.5A–C), instead resembling the shape observed in blind snakes and cryptozoic caenophidians. Notably, the posterodorsal part of the cerebral hemispheres is markedly embayed (Fig. 14.5C), a feature not observed in other squamates. The pituitary of *Dinilysia* is well developed and symmetric, housed in a deep sella turcica [20].

The midbrain roof of *Dinilysia* has a smooth, tubular shape and hence neither the optic tectum nor the collicular region can be differentiated. The strong correlation between the orbit and the eye size in snakes (r = 0.985, see electronic supplementary materials) allows an estimation of eye size for *Dinilysia* through a least-squares regression model (Fig. 14.5D). This eye size estimation (ES 12) falls into the range shared by some terrestrial large snakes such as *Python molurus* (ES 12) whose endocast also show a barely differentiated optic tectum due to the moderate size of the optic lobes.

A remarkable feature of the endocast of *Dinilysia*, first reported by Triviño et al. [10], is the large and markedly bilobed structure located just behind the midbrain roof (Fig. 14.5C) and present in all three endocasts examined here. This feature is at the anteroposterior level of the pituitary, in the region corresponding with the choroid plexus, although in a more dorsal position and angled posterodorsally. A deep and wide groove divides this structure into two symmetric and anteroposteriorly elongated protuberances. Notably, a similar but smaller structure is present in some extant snakes such as the 'basal' alethinophidian *Anilius scytale* (Fig. 14.5C) and the homalopsid caenophidian *Erpeton tentaculatum* (fig. 6 in [9]). The more dorsal and angled position of this structure in *Dinilysia* than that of extant

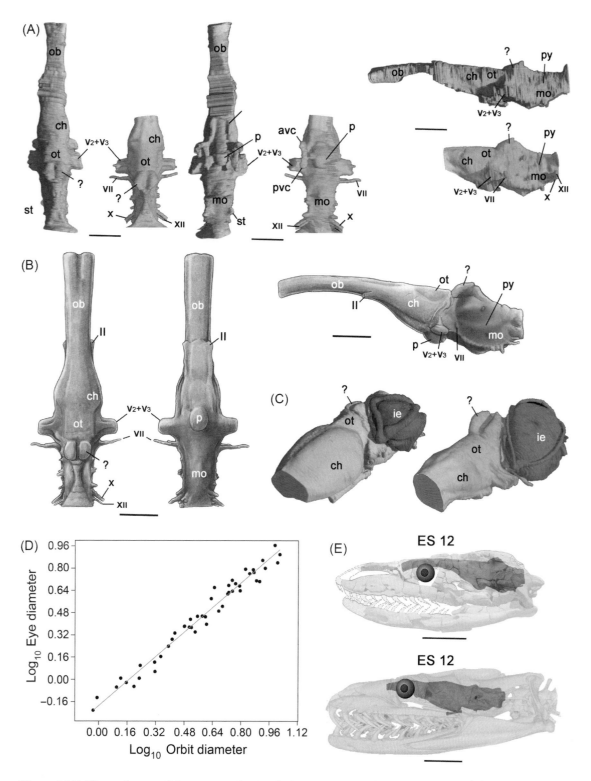

Figure 14.5 The endocast of the stem-snake *Dinilysia patagonica*. Obtained endocasts (A) and reconstruction (B) of the endocast based on specimens MACN RN 1013 and MACN RN 1014 in

snakes seems to be a by-product of the distinct, sizeable inner ear of this extinct taxon (Fig. 14.5C).

The hindbrain of *Dinilysia* is similar to that of extant alethinophidians in having a U-shaped configuration, but it is more pronounced in *Dinilysia* due to the extraordinary size of its otic region. As in other snakes with massive otic regions, such as *Xenopeltis unicolor*, the ventral region of the hindbrain, corresponding with the pons and medulla oblongata, is located below the horizontal plane delimited by the pituitary (Fig. 14.5A, B).

14.3 Discussion

Snakes represent a special case among reptiles because the braincase almost entirely encloses the whole brain, which in turn practically fills the neurocranial cavity. These features make snake endocasts a realistic representation of brain size and shape. Nevertheless, the cerebellum is an exception among the brain segments in snake endocasts because it is masked by the cast of the choroid plexus and probably also other non-neural structures located near to the hindbrain roof. Regardless, endocasts offer a valuable, non-destructive approach to study brain gross anatomy in snakes, although complementary physical dissections and/or contrast-enhanced computed tomography of specimens with soft- and bony tissue are desirable to obtain an accurate and precise picture of the surface of this complex organ.

Previous studies of different groups of vertebrates reported common patterns in brain anatomy that cluster species with similar lifestyles that do not necessarily share close phylogenetic affinities (cf. [21]). In squamates, one of these 'cerebrotypes' was recently established for limbless lizards and some groups of snakes [22]. Based on the results of the present study, I recognize at least two main cerebrotypes in snakes. The 'underground' cerebrotype is characterized by tubular cerebral hemispheres, flattening of the pituitary, and the overlapping of different brain segments. This cerebrotype is characteristic of blind snakes, but also of all the examined cryptozoic species of caenophidians, which also share a short olfactory bulb. Remarkably, 'basal' alethinophidians (with the exception of *Anomochilus*) exhibit long olfactory bulb, thus departing from the pattern present in other basal forms such as blind snakes. On the other hand, the 'surface' cerebrotype is characterized by well-spaced brain segments, a long olfactory bulb, posteriorly bulged cerebral

Figure 14.5 (*cont.*) dorsal (left), ventral (centre) and left lateral (right) view. (C) Brain and inner ear endocasts of *Anilius scytale* (left) and *D. patagonica* (right) in anterolateral view (not to scale). (D) Orbit diameter (log-transformed, x-axis) is plotted against eye diameter (y-axis) of 85 extant snake species (see Supplementary Appendix 14.S1, S2), with a line representing the linear regression model. (E) Lateral view of the skull of *D. patagonica* (upper) and *Python molurus* (lower) figured at the same snout-condyle length and rendered semitransparent to reveal the brain endocast and the eyeball (the latter inferred in *D. patagonica*). Scale bars equal 10 mm. II, optic tract; V2+V3, maxillary and mandibular branches of the trigeminal nerve; VII, facial nerve; avc, anterior opening of the Vidian canal; ch, cerebral hemisphere; ie, inner ear; mo, medulla oblongata; ob, olfactory bulb region; ot, optic tectum; p, pituitary; pvc, posterior opening of the Vidian canal; py, medullary pyramid; ?, bilobed structure.

hemispheres and a bulky pituitary. This cerebrotype is found in terrestrial, aquatic, and arboreal species across the snake tree.

14.3.1 Burrowing and Brain Anatomy

Functional constraints associated with miniaturisation and burrowing in tetrapods induce profound evolutionary modifications in head anatomy [23, 24]. According to previous studies [5, 8] it is likely that these major evolutionary driving forces shaped the underground cerebrotype features. The longitudinal shortening and overlap between brain segments seems to be an efficient way to accommodate the brain into the typically small (possibly miniaturized) skulls of fossorial and cryptozoic snakes. Similarly, the 'spatial packing' hypothesis explains the efficient packaging of relatively larger brains into a limited braincase space in mammals [25].

Among the impressive current diversity of caenophidian snakes, there is a plethora of cryptozoic species in most of recognized major lineages. For example, there are several cryptozoic species across the elapid tree, which consistently exhibit the underground cerebrotype (Fig. 14.6A). Interestingly, the species examined here indicate that this trend is also present in other caenophidian clades. The re-acquisition of the underground cerebrotype in phylogenetically distant groups of caenophidians (and probably other snake clades) suggests that brain gross anatomy is remarkably labile in snakes, a perhaps unexpected condition for such a complex organ.

The transition from surface to underground cerebrotypes and vice versa implies an important change in relative volume of several brain structures. Acceptance of the notion that the amount of neural tissue in a brain structure is proportional to the amount of processing [26] implies there are substantial differences in the ability to process sensory input between snakes with underground and surface cerebrotypes. The implications of this assertion are considered in the next section.

14.3.2 Gross Anatomical Features of the Brain and Their Correspondence with Histology of Some Sensory Processing Structures

Different regions of the brain undergo considerable evolutionary enlargement or reduction in relative size in lineages with surface or underground cerebrotypes. As might be expected, these changes are correlated with variations in volume tissue, expressed as reduction/enlargement of nuclei or degree of layer thickness [5, 6, 13, 26–28].

Forebrain

The principal brain structures related to the chemosensory vomeronasal system (VNS) are the accessory olfactory bulb, dorsal ventricular ridge and nucleus sphericus. The latter is well developed in squamates, occupying up to one-third of the total telencephalic volume in some species [29]. Snakes have the largest nucleus sphericus among squamates [1], which is almost located in the posterior region of the cerebral hemispheres. These three main structures are notably dedifferentiated and small in fossorial and cryptozoic snakes [6, 13, 26], and in limbless burrowing lizards [30]. In particular, the absence of bulging of the cerebral hemispheres in burrowing snakes is plausibly related to the reduction of these

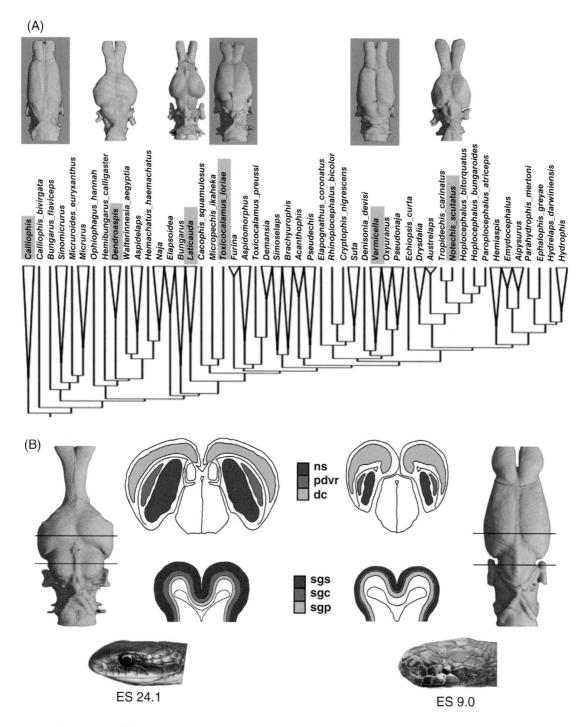

Figure 14.6 (A) Simplified cladogram of elapid snakes (modified from [52]) showing a dorsal view of endocasts of some cryptozoic and surface-dwelling species with their corresponding underground cerebrotype (grey-framed) and surface cerebrotype brains. (B) Dorsal view of the endocast of *Coluber constrictor* (left) and *Achalinus spinalis* (right), with transverse sections depicting the degree of development of major areas in the cerebral hemispheres and optic tectum of a surface cerebrotype and an underground cerebrotype, respectively. dc, dorsal cortex; ns, nucleus sphericus; pdvr, posterior dorsal ventricular ridge; sgc, stratum album + griseum centrale; sgp, stratum album + griseum periventriculare; sgs, stratum album + griseum superficiale.

brain structures, especially the nucleus sphericus (Fig. 14.6B). In reviewing the literature, regrettably, no compelling data were found to clarify whether smaller volume of neural tissue responsible for the processing of vomeronasal information is correlated with lower VNS use or performance. Curiously, the scanty available data on the anatomy of the VNS [31] and on behaviour [32] indicates that chemoreception plays an important role in the biology of fossorial snakes such as blind snakes.

A relatively small pituitary is noteworthy not only in fossorial and cryptozoic snakes but also in burrowing limbless lizards [9, 15, 16]. Although a relatively small pituitary seems to correspond with relatively low tissue volume, the diversity of arrangements of the different subparts that comprise the pituitary raises the question of whether flattening and rearrangement also contributes to its overall smaller size in forms.

Midbrain

Fossorial and cryptozoic snakes have notably thin and dedifferentiated optic tectum layers [5, 6, 26–28, 33]. In gross anatomical terms, this corresponds to notably small and simplified optic lobes (Fig. 14.6B). Senn [5] was the first to infer a causal relationship between a reduction in eye size and reduction of the optic tectum in fossorial and cryptozoic species. Subsequently, several authors [6, 9, 33, 27] concluded that snakes with well-developed eyes have a larger optic tectum than burrowing, eye-reduced forms. Notably, some surface dwellers such as large booids and pythonoids bear relatively small eyes (ES 11–12), but their optic tectum is similar to that of snakes with larger eyes. Although its major input is from the eyes, the optic tectum is a multisensory center that receives information from other sense organs. For example, the optic tectum receives input and integrates information from the infrared pit organs of pythonoids and booids [2], as well as mechanosensory information originated in tactile skin corpuscles or tentacles like in the homalopsid *Erpeton tentaculatum* [34]. Thus, although the anatomy of the optic tectum is clearly related to eye size, an accurate anatomical assessment of this brain segment should be evaluated considering the input of other sensory organs in addition to the eyes.

The collicular region is a very interesting part of the midbrain roof of snakes. Whereas in lizards it is formed only by the torus semicircularis, snakes bear an additional structure, the paratorus [5, 17]. The relative size of these structures varies with phylogeny; in blind snakes they are equal but in most alethinophidians the paratorus is larger [17]. The collicular region, often termed the midbrain auditory roof, is the structure that receives ascending auditory fibers in vertebrates and its variability in both size and complexity reflects the degree of development of the auditory system [35]. Contrary to the idea that their earless condition makes snakes poor hearers, a sound transmitting mechanism mediated by the skull bones suggests a capable system [36]. Furthermore, the presence of multiple auditory pathways in snakes suggests that their auditory system may be quite sophisticated [37]. To my knowledge, the auditory nature of the paratorus has never been thoroughly explored, but I hypothesize its evolution is linked to the peculiar auditory system of snakes. The development of the paratorus in alethinophidians possibly represents an improvement of auditory processing capabilities during the reinvasion of surface macrohabitats from a more fossorial ancestor.

Hindbrain

The cerebellum of limb-reduced and limbless squamates tends to be relatively small in comparison with that of fully limbed species [18, 22, 37], suggesting that limblessness is causally related to its reduction in complexity and size. However, this relationship might not be straightforward in most alethinophidian snakes because the cerebellum of surface-dwelling forms is equivalent in complexity and relative size to that of limbed lizards [22]. The cerebellum receives input from several sensory systems relayed through other regions of the brain and the spinal cord, being responsible for balance and the precise coordination of muscular activity [35]. It is thus reasonable to hypothesize that the increase in relative size and complexity of the cerebellum of surface-dwelling alethinophidians is causally related to the considerable resources that this brain region devotes to the integration of highly specialized sensory systems, the coordination of limbless locomotion, and complex behaviours such as constriction, caudal luring, or rattling.

14.3.3 Brain Gross Anatomy of *Dinilysia*, Its Palaeoecology, and Snake Origins

The singularity of the skull anatomy of *Dinilysia patagonica* has been extensively highlighted (e.g., [20, 38, 39]. Unsurprisingly, the digital endocasts show that the brain of *Dinilysia* also had an unusual gross anatomy. It is characterized by a very long and curved olfactory bulb, relatively small and posterodorsally embayed cerebral hemispheres, a well-developed and symmetrical pituitary, and an undifferentiated midbrain roof preceding a large bilobed structure. Notably, a specimen that preserves a partial natural endocast belonging to the older Patagonian Cretaceous snake *Najash rionegrina* displays the same bilobed structure located just behind the midbrain roof (fig. 1f of [40]). Furthermore, small cerebral hemispheres and the presence of a smooth, undifferentiated midbrain roof can be observed in another natural endocast of *Najash* (figs. 4–6 of [41]), thus suggesting that the unusual brain gross anatomy described for *Dinilysia* was also present in other stem snakes. The anatomical analysis performed here contradicts the interpretation [10] that the bilobed structure corresponds with the cerebellum. The cerebellum of all snake species examined here and previously [5, 18, 22] exhibits a lamellar or trapezoidal shape. Furthermore, comparisons made here clearly demonstrate that the cerebellum is not represented in snake endocasts. Therefore, this feature actually represents another brain region or a vascular structure, probably associated with the choroid plexus. Judging from the shape and position of this structure in extant snakes, this latter inference is considered more likely.

Dinilysia has been the focus of palaeoecological inferences in order to assess the habitat in which snakes first evolved. Recently, Yi and Norell ([42]; see also Chapter 13) analysed the anatomy of the inner ear of *Dinilysia*, highlighting its large size and bulbous shape, similar to the inner ear of fossorial and cryptozoic extant snakes. This resemblance underpinned the inference that *Dinilysia* was a burrowing form and that snakes likely evolved from a burrowing ancestor. This hypothesis was subsequently tested through an extensive analysis of squamate inner ear anatomy [39]. Palci et al. [39] observed that the inner ear morphology of *Dinilysia* is not exclusive to fossorial and cryptozoic extant snakes but it is also present in some surface-dwelling taxa. Furthermore, the inner ear of *Dinilysia* falls outside of typical ecomorphological categories, thus making inner ear morphology less conclusive as an ecological predictor in this case [39]. Curiously, both aforementioned studies overlooked the fact that the inner ear of *Dinilysia* followed an unusual isometric growth trajectory in comparison with extant snakes, in which it follows a typical negative allometric trajectory during postnatal ontogeny [43].

Brain gross anatomy and eye-size estimation for *Dinilysia* reported here provide insight into the paleoecology of this extinct snake. Beyond its uniqueness, the elongated general shape of the brain of *Dinilysia* contrasts with the underground cerebrotype of most fossorial and cryptozoic extant snakes. Furthermore, the estimated eye size suggests that *Dinilysia* bore eyes equal to those of extant large surface-dwelling snakes such as some large pythonoids and booids. In addition, I suggest that body length is a relevant trait that contradicts a generally burrowing lifestyle for *Dinilysia*. Body length of extant fossorial and cryptozoic snakes never surpasses 1.5 m [44], whereas the body size estimations for *Dinilysia* based on skull length (Supplementary Appendix 14.S1, S2) indicates that it reached more than 3 m total length, greatly exceeding previous conjectures [42, 45]. Taken together, the absence of any unequivocal osteological feature related to burrowing, the information provided by brain gross anatomy and inferred eye size, and the estimated large body size suggest that *D. patagonica* was a surface-dwelling form.

The visual system has historically occupied a prominent place in debates about the origin and early evolution of snakes (Gower et al., Chapter 15) Since the influential work of Walls [46] the uniqueness of snake eye anatomy compared with that of lizards (and other reptiles) has been explained through a degeneration of the eye in a burrowing ancestor and a subsequent redevelopment of analogous structures during an evolutionary radiation above ground. Furthermore, many other snake features such as a closed bony braincase and loss of the middle ear were also explained through this evolutionary scenario. However, other studies of the visual system suggested a more complex interplay between burrowing and surface (though possibly nocturnal) stages in the early steps of ophidian evolution [28, 47]. Additionally, data for retinal photopigment genes ([48]; Gower et al., Chapter 13) does not support the interpretation (e.g., [49]) that the ancestral snake was as fossorial as extant blind snakes. In this way, there is a growing body of information suggesting the dawn of snake evolution was not restricted to fossorial animals but may have involved episodes of occupation of diverse macrohabitats coupled with correlated major transformations in skull anatomy and feeding behaviour [44, 50, 51]. The surface-dwelling lifestyle inferred here for *Dinilysia patagonica* and non-fossoriality of at least some other Cretaceous snakes, such as the marine simoliophiids, constitutes additional evidence that at least some of the early history of snakes occurred above ground.

14.4 Concluding Remarks

While compiling this chapter, it became clear to me that any attempt to uncover the diversity and evolution of the ophidian brain is an enormous task, because of the remarkable complexity of this organ and the intricate evolutionary history of this clade. Moreover, the diversity of the sensory paraphernalia that these skilled predators have acquired during their evolution provides further challenges. Although there is little prospect of a comprehensive picture of the evolution of the ophidian brain in the near future, some recent efforts offer new perspectives to achieve that. Broad-scale comparative surveys based on contrast-enhanced CT imagery together with geometric morphometrics are powerful tools. Applying new comparative data and approaches to the relationships between histology, the relative volume of each brain segment, and the condition of corresponding sensory organs across the ecological diversity of snakes will be very informative.

Acknowledgements

This study was conducted in part during the global lockdown imposed to curb COVID-19. My most sincere gratitude to all essential workers (especially health, food, and cleaning services) around the world who daily endangered their lives and allowed me to work. I thank David Gower and Hussam Zaher for inviting me to contribute to this book. I am also grateful to Bhart-Anjan Bhullar, David Gower and an anonymous reviewer for making helpful suggestions. A particular debt of thanks is due to Krister Smith for useful suggestions and corrections to English grammar. Claudia Koch, Jennifer Olori, Filipa Sampaio, Jessie Maisano, and Matt Colbert provided assistance during scanning, and facilitated access to valuable CT data. This gratitude is extended to the staff of Digimorph, Morphosource and Phenome10K digital libraries. Thanks to Gabriel Lio for the skillful drawings of some brains. This work was funded by ANPCyT (PICT 2013-220) and CONICET (PIP 0497).

References

1. K. Schwenk, The evolution of chemoreception in squamate reptiles: a phylogenetic approach. *Brain, Behaviour and Evolution*, 41 (1993), 124–137.

2. R. C. Goris, Infrared organs of snakes: an integral part of vision. *Journal of Herpetology*, 45 (2011), 2–14.

3. J. M. Crowe-Riddell, R. Williams, L. Chapuis, and K. L. Sanders, Ultrastructural evidence of a mechanosensory function of scale organs (sensilla) in sea snakes (Hydrophiinae). *Royal Society Open Science*, 6 (2019), 182022.

4. D. Starck, Cranio-cerebral relations in recent reptiles. In C. Gans, R. G. Northcutt and P. Ulinski, eds., *Biology of the Reptilia* Vol. 9, *Neurology A* (London: Academic Press, 1979), pp. 1–38.

5. D. G. Senn, Über das optische System im Gehirn squamater Reptilien. Eine vergleichend-morphologische Untersuchung, unter besonderer Berucksichtigung einiger Wuhlschlangen. *Acta Anatomica*, 65 (1966), 1–87.

6. H. Masai, K. Takatsuji, and Y. Sato, Comparative morphology of the telencephalon of the Japanese colubrid snakes under consideration of habit. *Journal of Zoological Systematics and Evolutionary Research*, 18 (1980), 310–314.

7. R. Platel, Nouvelles données sur l'encéphalisation des Reptiles Squamates.

Zeitschrift für Zoologische Systematik und Evolutionsforschung, 13 (1975), 161–184.

8. J. C. Olori, Digital endocasts of the cranial cavity and osseous labyrinth of the burrowing snake Uropeltis woodmasoni (Alethinophidia: Uropeltidae). *Copeia*, 2010 (2010), 14–26.

9. R. Allemand, R. Boistel, G. Daghfous, et al., Comparative morphology of snake (Squamata) endocasts: evidence of phylogenetic and ecological signals. *Journal of Anatomy*, 231 (2017), 849–868.

10. L. N. Triviño, A. Albino, M. T. Dozo, and J. D. Williams, First natural endocranial cast of a fossil snake (Cretaceous of Patagonia, Argentina). *The Anatomical Record*, 301 (2018), 9–20.

11. J. A. Hopson, Paleoneurology. In C. Gans, R. G. Northcutt, and P. Ulinski, eds., *Biology of the Reptilia*, Vol. 9, *Neurology A* (New York: Academic Press, 1979), pp. 39–146.

12. H. J. Donkelaar, Reptiles. In R. Nieuwenhuys, H. J. ten Donkelaar and C. Nicholson, eds., *The Central Nervous System of Vertebrates* (Berlin: Springer, 1998), pp. 1313–1524.

13. M. Halpern, The telencephalon of snakes. In S. O. E. Ebbesson, ed., *Comparative Neurology of the Telencephalon* (New York: Plenum Press, 1980), pp. 257–294.

14. W. Rudin, Untersuchungen am olfaktorischen System der Reptilien. III. Differenzierungsformen einiger olfaktorischer Zentren bei Reptilien. *Acta Anatomica*, 89 (1974), 481–515.

15. H. Saint Girons, The pituitary gland. In C. Gans and T. S. Parsons, eds., *Biology of the Reptilia*, Vol. 3, *Morphology C* (London: Academic Press, 1970), pp. 135–199.

16. M. P. Schreibman, The pituitary gland. In P. K. T. Pang and M. P. Schreibman, eds., *Vertebrate Endocrinology: Fundamentals and Biomedical Implications*, Vol. 1 (New York: Academic Press, 1986), pp. 11–55.

17. D. G. Senn, The saurian and ophidian colliculi posteriores of the midbrain. *Acta Anatomica*, 74 (1969), 114–120.

18. R. Nieuwenhuys, Comparative anatomy of the cerebellum. *Progress in Brain Research*, 25 (1967), 1–93.

19. R. Estes, T. H. Frazzetta, and E. E. Williams, Studies on the fossil snake Dinilysia patagonica Woodward. I. Cranial morphology. *Bulletin Museum of Comparative Zoology of Harvard University*, 119 (1970), 25–74.

20. H. Zaher and A. Scanferla, The skull of the Upper Cretaceous snake Dinilysia patagonica Smith-Woodward, 1901, and its phylogenetic position revisited. *Zoological Journal of the Linnean Society*, 164 (2012), 194–238.

21. A. N. Iwaniuk, P. L. Hurd, and D. R. Wylie, Comparative morphology of the avian cerebellum: II. Size of folia. *Brain, Behavior, and Evolution*, 69 (2007), 196–219.

22. S. Macrì, Y. Savriama, I. Khan, and N. Di-Poï, Comparative analysis of squamate brains unveils multi-level variation in cerebellar architecture associated with locomotor specialization. *Nature Communications*, 10 (2019), 5560.

23. J. Hanken and D. B. Wake, Miniaturization of body size: organismal consequences and evolutionary significance. *Annual Review of Ecology, Evolution, and Systematics*, 24 (1993), 501–519.

24. H. C. Maddin, J. C. Olori, and J. S. Anderson, A redescription of Carrolla craddocki (Lepospondyli: Brachystelechidae) based on high-resolution CT, and the impacts of miniaturization and fossoriality on morphology. *Journal of Morphology*, 272 (2011), 722–743.

25. D. E. Lieberman, B. Hallgrímsson, W. Liu, T. E. Parsons, and H. A. Jamniczky, Spatial packing, cranial base angulation, and craniofacial shape variation in the mammalian skull: testing a new model using mice. *Journal of Anatomy*, 212 (2008), 720–735.

26. D. G. Senn and R. N. Northcutt, The forebrain and midbrain of some squamates and their bearing on the origin of snakes. *Journal of Morphology*, 140 (1973), 153–152.

27. O. J. Reperant, D. Miceli, J.-P. Rio, and C. Weidner, The primary optic system in a microphthalmic snake (Calabaria reinhardtii). *Brain Research*, 408, 233–238.

28. O. J. Reperant, J.-P. Rio, R. Ward, et al., Comparative analysis of the primary visual system of reptiles. In C. Gans and P. S. Ulinski, eds., *Sensory Integration. Biology of the Reptilia*, Vol. 17, *Neurology C* (Chicago and London: The University of Chicago Press, 1992) pp. 175–240.

29. R. Platel, Analyse volumétrique comparée des principales subdivisions télencéphaliques chez les reptiles saurines. *Journal für Hirnforschung*, 21 (1980), 271–291.

30. R. G. Northcutt, Forebrain and midbrain organization in lizards and its phylogenetic significance. In N. Greenberg and P. D. Maclean, eds., *Behavior and Neurology of Lizards* (Rockeville: NIMH, 1978), pp. 11–64.

31. A. Gharzi, M. Abbasi, and P. Yusefi, Histological studies on the vomeronasal organ of the worm-like snake, Typhlops vermicularis. *Journal of Biological Sciences*, 13 (2013), 372–378.

32. J. K. Webb and R. Shine, To find an ant: trail-following in Australian blindsnakes (Typhlopidae). *Animal Behaviour*, 43 (1992), 941–948.

33. H. Masai, Structural patterns of the optic tectum in Japanese snakes of the family Colubridae in relation to habit. *Journal für Hirnforschung*, 14 (1973), 367–374.

34. K. C. Catania, The brain and behavior of the tentacled snake. *Annals of New York Academy of Sciences*, 1225 (2011), 83–89.

35. A. B. Butler and W. Hodos, *Comparative Vertebrate Neuroanatomy: Evolution and Adaptation* (Hoboken: John Wiley, 2005).

36. C. B. Christensen, J. Christensen-Dalsgaard, C. Brandt, and P. T. Madsen, Hearing with an atympanic ear: good vibration and poor sound-pressure detection in the royal python, Python regius. *Journal of Experimental Biology*, 215 (2012), 331–342.

37. O. Larsell. The cerebellum of reptiles: lizards and snake. *The Journal of Comparative Neurology*, 41 (1926) 59–94.

38. M. W. Caldwell, *The Origin Of Snakes: Morphology and the Fossil Record* (Boca Raton: Taylor & Francis, 2019).

39. A. Palci, M. N. Hutchinson, M. W. Caldwell, and M. S. Y. Lee, The morphology of the inner ear of squamate reptiles and its bearing on the origin of snakes. *Royal Society Open Science*, 4 (2017), 170685.

40. F. F. Garberoglio, S. Apesteguia, T. Simoes, et al., New skulls and skeletons of the Cretaceous legged snake Najash, and the evolution of the modern snake body plan. *Science Advances*, 5 (2019), eaax5833.

41. F. F. Garberoglio, R. O. Gómez, S. Apesteguia, et al., A new specimen with skull and vertebrae of Najash rionegrina (Lepidosauria: Ophidia) from the early Late Cretaceous of Patagonia. *Journal of Systematic Palaeontology*, 17 (2019), 1533–1550.

42. H. Yi and M. A. Norell, The burrowing origin of modern snakes. *Science Advances*, 1 (2015), e1500743.

43. A. Scanferla and B. A. S. Bhullar, Postnatal development of the skull of Dinilysia patagonica (Squamata-Stem Serpentes). *The Anatomical Record*, 297 (2014), 560–573.

44. A. Scanferla, Postnatal ontogeny and the evolution of macrostomy in snakes. *Royal Society Open Science*, 3 (2016), 160612.

45. M. W. Caldwell and A. M. Albino, Palaeoenvironment and palaeoecology of three Cretaceous snakes: Pachyophis, Pachyrhachis, and Dinilysia. *Acta Palaeontologica Polonica*, 46 (2001), 203–218.

46. G. L. Walls, Ophthalmological implications for the early history of the snakes. *Copeia 1940* (1940), 1–8.

47. G. Underwood, The eye. In C. Gans and T.S. Parsons, eds., *Biology of the Reptilia*, Vol. 2, *Morphology B* (New York: Academic Press, 1970), pp. 1–97.

48. B. F. Simões, F. L. Sampaio, C. Jared, et al., Visual system evolution and the nature of

the ancestral snake. *Journal of Evolutionary Biology*, 28 (2015), 1309–1320.

49. A. Miralles, L. Marin, D. Markus, et al., Molecular evidence for the paraphyly of Scolecophidia and its evolutionary implications. *Journal of Evolutionary Biology*, 31 (2018), 1782–1793.

50. A. Y. Hsiang, D. J. Field, T. H. Webster, et al., The origin of snakes: revealing the ecology, behavior, and evolutionary history of early snakes using genomics, phenomics, and the fossil record. *BMC Evolutionary Biology*, 15 (2015), 87.

51. F. O. Da Silva, A.-C. Fabre, Y. Savriama, et al., The ecological origins of snakes as revealed by skull evolution. *Nature Communications*, 9 (2018), 376.

52. H. Zaher, R. W. Murphy, J. C. Arredondo, et al., Large-scale molecular phylogeny, morphology, divergence-time estimation, and the fossil record of advanced caenophidian snakes (Squamata: Serpentes). *PLoS ONE*, 14 (2019), e0216148.

15

Eyes, Vision, and the Origins and Early Evolution of Snakes

DAVID J. GOWER, EINAT HAUZMAN, BRUNO F. SIMÕES AND RYAN K. SCHOTT

15.1 Introduction

This chapter addresses the origin and early evolution of snakes from the perspective of vision biology. It examines phenotypic and genotypic data for squamate reptiles, and asks what that can tell us about eyes and vision in the ancestral snake and thus, indirectly, about snake origins and early evolution. As advocated in Garth L. Underwood's [1] *A Contribution to the Classification of Snakes*, and beyond, it is informative to draw upon many lines of evidence when addressing questions in evolutionary biology. So, why consider vision in particular? We suggest that this is worthwhile because (1) snake eyes differ notably from those of other squamates (see § 15.4.1), (2) of the role that eyes have played in the history of the subject, being used as evidence in support of fossorial (burrowing), nocturnal and aquatic hypotheses for snake origins (e.g., [2–7]), and (3) vertebrate vision is an intricate yet well-understood system providing a rich source of readily interpretable data. Much is known about the anatomy of vertebrate eyes, and how they function and have evolved. The physiology underlying photon capture, spectral sensitivity, colour discrimination, signal propagation, and the genes and proteins involved, have been largely characterised, making vision one of the best-known biological signal-transduction systems. Additionally, many aspects of eye function and evolution can be identified and quantified, including selective pressures, based on physical first principles and molecular genetics.

Here we provide an overview of the subject. We summarise types of data available, and how they might inform on the visual system of the earliest snakes. We assess the extent to which vision data might enable discrimination among competing hypotheses of ecological aspects of snake origins. We synthesise recent discoveries, identify gaps in knowledge, and assess future prospects.

15.2 The Ancestral Snake and Other Terminology

We use 'ancestral snake' to refer to the most recent common ancestor of all extant snakes, i.e., the most recent common ancestor of crown snakes (Serpentes, as applied here). The same basis applies to our use of 'ancestral alethinophidian', 'ancestral caenophidian' etc. We use 'lizard' to refer informally to any non-snake squamate. The phylogenetic framework for this chapter is illustrated in Figure 15.1. We use the term 'henophidia' to informally refer to the paraphyletic group of all non-caenophidian alethinophidians. We use scolecophidian and Scolecophidia to refer to non-aethinophidian snakes, irrespective of whether or not this is a paraphyletic or monophyletic taxon (see e.g., Ruane & Streicher, Chapter 10). Snake classification otherwise follows Zaher et al. [8], and with the informal term 'colubroid' referring to members of Colubroidea. Names or abbreviations of genes are written in italics, and proteins encoded by genes are in upper case text. See Box 15.1 for a glossary of some vision biology terms used in this chapter.

15.3 Vertebrate Vision

Light reaches the retina after passing through ocular media (see § 15.4.2). Conversion of photons into electrical signals is termed phototransduction, and occurs in the outer segments of the outer-retinal photoreceptor cells. Vertebrate outer-retinal photoreceptors are of two main types, rods and cones, though at least some caenophidian snakes somewhat blur this distinction via photoreceptor evolutionary transmutation. Vertebrates, including snakes [9], also have inner-retinal, intrinsically photosensitive retinal ganglion cells, that express melanopsin and whose function in vision is not as well understood [10, 11]. Hereafter, we use 'photoreceptor' to refer to outer-retinal (rod and cone) photoreceptor cells. Typically, rods mediate low-light (scotopic) vision, and cones bright-light (photopic) and colour vision, with both types contributing to mesopic vision in intermediate light levels, though, again, at least some caenophidian snakes blur this distinction, possibly incorporating cone-like rods in photopic trichromatic colour vision [12–14].

Photons pass through the inner retinal layers before reaching photoreceptor outer segments, where visual pigments lie within stacks of membrane organelles termed discs. Vertebrate visual pigments each comprise a transmembrane opsin protein and a chromophore. The chromophore is derived from vitamin A_1 (retinal) or vitamin A_2 (3,4-dehydroretinal), thus far only the former has been found in snakes. The chromophore sits within a binding pocket formed by the opsin, where it undergoes *cis-trans* photoisomerization when struck by photons, changing the conformation of the opsin to trigger a transduction cascade. Electrical signals generated by phototransduction are propagated from hyperpolarized photoreceptors through to processing neural (bipolar, horizontal, amacrine and ganglion) cells in the retina and onto the brain via the optic nerve. Photoreceptor cells are then returned to a depolarised, dark state via the visual cycle, which includes visual pigments being regenerated with *cis* retinal.

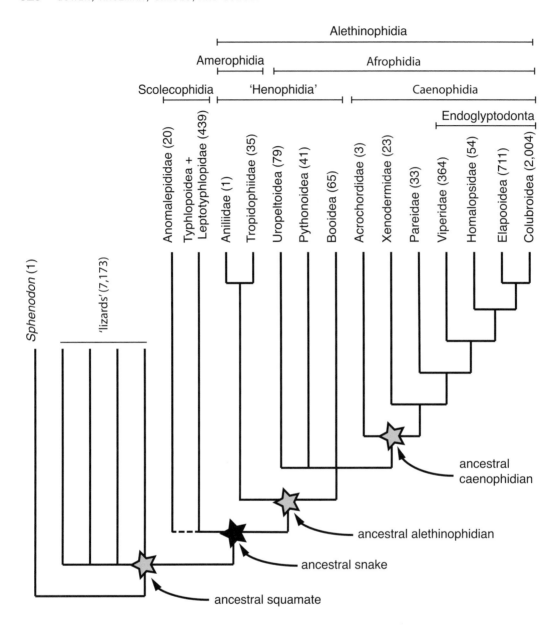

Figure 15.1 Summary of the phylogenetic framework of extant squamates for the chapter. Numbers in parentheses are currently recognized extant species (based on [140]). A few species-poor lineages are not depicted, such as the total of four species of Bolyeriidae and Xenophidiidae, little studied, poorly known, probable close relatives of Pythonoidea. See § 15.2 for notes on taxon names.

Typically, rods are more sensitive and slower to return to a dark state than cones. Rod and cone physiological differences are explained by mostly distinct rod and cone copies of phototransduction and (to a lesser extent) visual-cycle genes such as kinases, ion exchangers, transducins, and arrestins, though differing expression levels and photoreceptor

Box 15.1 Glossary of some vision biology terms used in this chapter. Words that are also titles of entries in the glossary are indicated in bold text

Accommodation: Adjustments by the eye to maintain objects in focus on the retina. In lizards accommodation occurs primarily by deformation of the lens, in snakes by moving the lens relative to the retina.

Acuity: The spatial resolution of image-forming vision. Higher acuity equates to sharper and more detailed images.

Brille: The spectacle – a specialized near-transparent scale overlying the cornea of the eye of most snakes and some lizards.

Chromophore: A molecule linked to an **opsin** protein to form a visual pigment. In most vertebrates, the chromophore is 'retinal' derived from vitamin A_1 that upon capturing a photon undergoes isomerization, triggering a phototransduction cascade. The chromophore is located within a 'binding' pocket within the opsin protein.

Crystallins: The main structural proteins forming the lens of the vertebrate eye.

Eye adnexae: Accessory structures associated with the eye, such as extraocular muscles, eyelids, conjunctiva, lacrimal glands, etc.

Fovea: A cone-rich, high-acuity area of the retina, present in most lizards and absent in most snakes.

Ganglion cells: Retinal ganglion cells are located in the inner retina and their axons form the optic nerve that extends into the brain. They receive visual information from cones and rods via other retinal neurons, such as horizontal, bipolar, and amacrine cells. Some ganglion cells (termed intrinsically photosensitive retinal ganglion cells) are themselves photosensitive and involved in non-visual functions that require light input such as circadian rhythms and the pupillary light reflex.

Melanopsin: An ancient **opsin** present in most extant deuterostomes. Normally responsible for circadian rhythms, and other non-visual responses to light such as the pupillary light reflex, melatonin suppression, skin-colour changes, etc. In vertebrates two paralogs occur, the *Xenopus*-like (Opn4x) and the mammalian-like (Opn4m) melanopsin, which originated from a duplication event early in vertebrate evolution. Snakes have only *opn4x* in their genomes.

Ocular media: Structures of the eye that light passes through before reaching the retina. In vertebrates, these are (from the retina outwards) the vitreous humor, lens, aqueous humor, and cornea. Most snakes also have extraocular media comprising the subspectacular space and **spectacle**. In a few highly fossorial snakes there is no **spectacle** and extraocular media comprise relatively unspecialized head scales and soft tissues between these and the eyes.

Oil droplet: Spherical organelle located in the inner segment of cones (and possibly rarely some rods) of some vertebrates, but not in any snakes. Oil droplets consist of sometimes pigmented lipids. Coloured oil droplets are typically characteristic of the cones of diurnal species, where they act as cut-off filters and increase the potential for colour discrimination. Oil droplets are also absent in other major vertebrate lineages that are thought to have adapted to low-light environments in their evolutionary history, such as crocodilians and mammals.

Opsin: A 7-transmembrane protein member of the G protein-coupled receptor superfamily. Opsins bond with light-sensitive **chromophores** to form visual pigments.

Peak absorbance: The wavelength of light at which absorption by a visual pigment is greatest, typically reported as λ_{max}. See also **spectral tuning**.

Box 15.1 *(continued)*

Photoreceptor: A cell in the retina capable of phototransduction. Although some inner-retinal **ganglion cells** are intrinsically photosensitive, the term 'photoreceptor' is typically applied to rod and cone cells in the outer retina.

Refractive index: The speed that light travels through a particular substance relative to passing through a vacuum. Light passes more slowly through substances with a higher refractive index. The refractive index determines the extent to which light is bent (refracted) as it enters a particular substance obliquely.

Ringwulst: The annular pad – thickened tissue around the equator of the lens of lizards. Absent in snakes, it plays a role in lens-deformation based **accommodation**.

Scleral ossicles: a ring of thin, typically overlapping bones that develop within the sclera (the outermost part of the eyeball). Absent in snakes but present in most lizards, where they likely have a role in **accommodation**. Scleral ossicles are also absent in other major vertebrate lineages that are thought to have adapted to low-light environments in their evolutionary history, such as amphisbaenians, crocodilians, and mammals.

Spectacle: See **Brille**.

Spectral tuning: Variation in the wavelength of **peak absorbance** of a visual pigment, as determined largely by the type of **chromophore** and the amino-acid residues of the **opsin**, especially at influential 'tuning sites'.

Summation: Relationship among **photoreceptor** and bipolar and **ganglion cells** in the retina. High summation occurs when large numbers of **photoreceptors** converge on fewer bipolar cells that transfer signals to **ganglion cells**; this increases sensitivity but reduces **acuity**. Lowest (zero) summation occurs when a single **photoreceptor** contacts a single **ganglion cell** via a single bipolar cell.

Transcriptome: The set of RNA produced by transcription of DNA within a particular cell, organ, organism, or tissue type, such as eyes.

Transmission: The amount of light that passes through a substance or structure, such as components of the **ocular media**.

Transmittance: The amount of light absorbed, scattered, or reflected by a substance or structure.

Two-tiered retina: In some vertebrates, **photoreceptors** change length in relation to changes in ambient light via elongation and contraction of the myoid part of their inner segment. In these taxa, the photoreceptor layer of the retina is periodically notably 'two-tiered' – the outer parts of rods are scleral to those of cones under higher intensity light and vice versa under lower light. Some snakes might have permanently two-tiered retinas because only the latter condition has been observed, and because rods and cones of snakes are not known to undergo such photomechanical (retinomotor) movement.

UVA: Long-wavelength ultraviolet light of wavelengths between 315 and 400 nm.

ultrastructure also play a role. The regeneration of rod visual pigments occurs in the retinal pigment epithelium. For cone visual pigments, Müller glial cells are also involved. Excellent summaries and reviews of vertebrate photoreceptors, phototransduction and visual cycles are available (e.g., [15–18]).

Ancestrally, vertebrates had five visual-opsin genes, *rh1*, *rh2*, *sws1*, *sws2*, and *lws* and these underwent various independent episodes of loss and duplication in multiple lineages (for review see [19]). Of these, four typically occur in cones (*rh2*, *sws1*, *sws2*, *lws*) and one in rods (*rh1*). The ancestral squamate retained the five ancestral vertebrate visual-opsin genes, with *rh2* and *sws2* being lost along the snake phylogenetic stem (e.g., [6, 19–21]).

15.4 Lines of Evidence

15.4.1 The Disparate Morphology of the Snake Visual System

The eyes (and eye adnexae) of extant snakes are highly distinctive compared to other vertebrates, and even to other squamate reptiles [2, 4, 22–26]. Notable differences between the eyes of lizards and snakes are summarised in Table 15.1 and Figure 15.2. There are additional detailed differences. For example, snake double cones differ from those of lizards (Fig. 15.3) in having a peripheral cell paranuclear body and a Müller cell that intervenes between the members of the double cell, and snake double and some single cones also have ellipsoids with refringent bodies, which are absent in lizards [2, 22–24].

The highly distinctive nature of the snake eye and adnexae have been used to infer the ecological origins of snakes. Walls [22] argued for a nocturnal ancestry in contrast to the largely diurnal lizards. Walls [2, 33] (see also [3, 27]) subsequently hypothesized a fossorial ancestry. Underwood [24, 25, 28] modified this, proposing an initial nocturnal phase followed soon after by fossoriality. In contrast, Caprette et al. [4] applied a morphological phylogenetics approach and interpreted their results as indicating that snake visual systems are more similar to those of aquatic rather than fossorial terrestrial vertebrates. A thorough reassessment is beyond the scope of this chapter, but we note that Caprette et al.'s results are not unambiguous. They found snake eyes to be most similar to those of caecilian amphibians, a group that Caprette et al. consider 'primitively aquatic', but which plesio-morphically were likely terrestrial and fossorial as adults, because the few extant caecilian species that have aquatic adults are deeply phylogenetically nested, far from the root of the caecilian tree (e.g., [29, 30]).

Other aspects of the visual system differing between typical snakes and lizards have been interpreted as consistent with hypotheses of a dim-light environment origin of snakes. Superficial neural layers in the optic tectum of the brain of snakes are reduced, with similar conditions in lizards reported only among lineages of typically fossorial, limb-reduced or limbless and small-eyed taxa, though snakes also have unique structural peculiarities [31–34]. Brain anatomy and function is understudied in snakes (Chapter 14).

15.4.2 Ocular-Media Anatomy, Transmittance, and Refractive Index

The ocular media of snakes are (from the retina outwards) the vitreous humour, lens, aqueous humour, cornea, tissue or fluid in the intraconjuctival space between the cornea and the spectacle, and typically the spectacle (= Brille) [2, 26]. The latter is a specialized scale overlying the eye in most snakes, and so it and the space between it and cornea might more accurately be termed extraocular media. All snakes lack moveable eyelids. Some

Table 15.1. Some differences in morphology between the eye (and eye adnexae) of a typical lizard and a typical snake, based primarily on [1, 2, 22, 23, 25, 27, 35, 136–139].

	Typical lizard	**Typical snake**
Eyelids & nictitating membrane	present	absent
Spectacle (Brille)	absent	present
Eyeball	oblate	subspherical
Sclera	chondrified	fibrous
Scleral ossicles	present	absent
Ciliary muscle	present	absent
Conjunctival drainage	two canaliculi	single canaliculus
Retractor bulbi & bursalis muscles	present	absent
Optic nerve	with septa	without septa
Lens	oblate, soft, with Ringwulst	spherical, firm, no Ringwulst
Accommodation mode	lens deformation	lens displacement
Lens pigments	absent	sometimes present
Fovea	present	absent
Retinal photoreceptor morphology	simplex	duplex
Photoreceptor oil droplets	present	absent
Contractile myoids	present	absent
Paraboloids	present	absent
Cone ellipsoids	non-refringent	refringent in some cones
Double cone accessory member	without paranuclear body	with paranuclear body
Retinal circulation	ectodermal conus	mesodermal conus

highly fossorial snakes (e.g., scolecophidians, *Anilius*, some uropeltoids) lack spectacles, their eyes instead lying beneath head scales that may or may not bulge above the surface of the adjacent scale(s) (e.g., [27]). Spectacles have played an important role in inferences about snake origins because in non-snake squamates they are restricted to fossorial taxa or to nocturnal taxa whose eyes are generally close to the substrate [35].

Vertebrate corneas and humours are typically highly transmissive from the infrared to ca. 300 nm (e.g., [36]), but some lenses contain pigments that prevent shorter-wavelength light reaching the retina. As far as we are aware, transmission data for snake corneas have

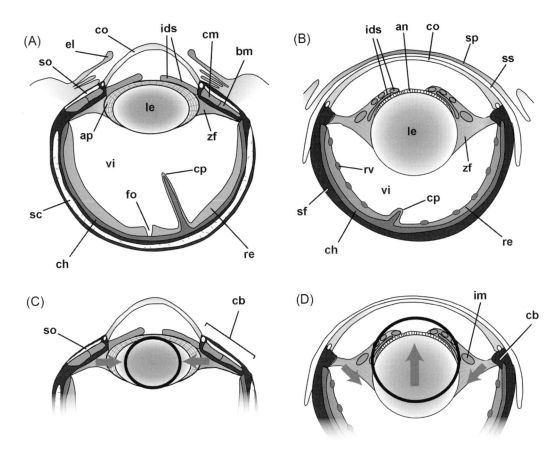

Figure 15.2 Differences between the eyes of typical lizards (A, C) and snakes (B, D), (redrawn from [4]; based on [23]). Lower pair of images show method of accommodation: lizards focus by deforming the lens via contraction of ciliary body muscles (bm, cm); snakes focus by moving their hard lens using changes in vitreous pressure controlled by contraction of iris muscles (im) (though see § 15.4.2). Abbreviations: ap, annular pad (Ringwulst); bm, Brücke's ciliary muscle; cb, ciliary body; ch, choroid; cm, Crompton's ciliary muscle; co, cornea; cp, conus papiliaris; el, eyelid; fo, fovea; ids, iris dilator and sphincter muscles; im, iris muscle; le, lens; sc, scleral cartilage; sf, sclera (fibrous); so, scleral ossicle; re, retina; rv, retinal blood vessels; sp, spectacle (Brille); ss, subspectacular space; vi, vitreous; zf, zonular fibres. See Supplementary Appendix 15.S1 for colour version.

been published only by Hart et al. [37], while spectacle and lens transmission data have been published for tens of species in multiple studies [38–42]. Spectacles and corneas are typically transmissive to wavelengths >350 nm, although there is some taxonomic variation in the former within the 350–400 nm range, at least within Caenophidia (with considerable UV blocking in, for example, *Naja annulifera*: [42]). Variation also occurs during scale-shedding cycles and blood flow through the spectacle [41]. Based on measurements made using optical coherence tomography [43–46], spectacles are thicker in fossorial and aquatic than arboreal and ground-dwelling snakes [45]. Corneas are thickest in diurnal and arboreal snakes [46].

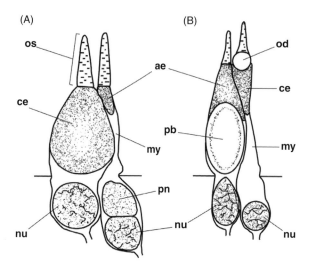

Figure 15.3 Double cone photoreceptor of (A) snake (the dipsadid colubroid *Hypsirhynchus callilaemus*) and (B) lizard (the dactyloid *Anolis lineatopis*) (redrawn from [1]). Abbreviations: ae, ellipsoid of accessory member of double cone; ce, ellipsoid of chief (principal) member of double cone; my, myoid; nu, nucleus; od, oil droplet; os, outer segment; pb, paraboloid; pn, paranuclear body. Unlabelled horizontal straight lines indicate position of outer limiting membrane of the retina.

All nocturnal snakes examined thus far have very transmissive lenses, even to UVA (315–400 nm) (Fig. 15.4N). Some diurnal snakes have similarly transmissive lenses across a broad range of wavelengths, whereas others block some or all light <400 nm (Fig. 15.4N). Evidence for the latter is thus far restricted to caenophidians, mostly highly visual hunters, consistent with the hypothesis that blocking UV enhances acuity (e.g., [36]) as well as perhaps protecting retinal cells from damage. Data are scant but, as far as is known, snakes with UV-blocking lenses have visual pigments with peak sensitivities >400 nm. Lens and spectacle transmission data have not been published for non-caenophidian snakes other than some pythonids and boids, so although the assumption is that the ancestral snake had highly transmissive ocular media, this is open to further testing with data for more extant squamates.

The refractive index of the ocular media of snakes has been measured directly or estimated in only a few caenophidian species [47, 48]. Compared with lizards, the refractive index of snake corneas is much lower and that of snake lenses much higher. Additionally, snake spectacles have greater refractive power than corneas. Snakes have substantially fewer lens crystallin genes than other amniotes, even compared to other lineages that, like snakes, have lost large numbers of ancestral vertebrate visual genes [49]. These losses may be related to differences in lens shape and rigidity, and associated differences in method of accommodation (Table 15.1). However, the implications of these losses, potential functional adaptation in remaining crystallins, and the possible recruitment of new, snake-specific lens crystallins have yet to be studied. Such studies would be worthwhile given that lens-crystallin gene losses and other changes are adaptive in other vertebrates (e.g., [50, 51]). Although generally accepted that the snake lens is rigid and that accommodation occurs by lens displacement along the optical axis (e.g., [2–7]), the mechanism is not fully clear and at least some snakes have a lens that is deformable to some degree – prompting calls for re-evaluation [52].

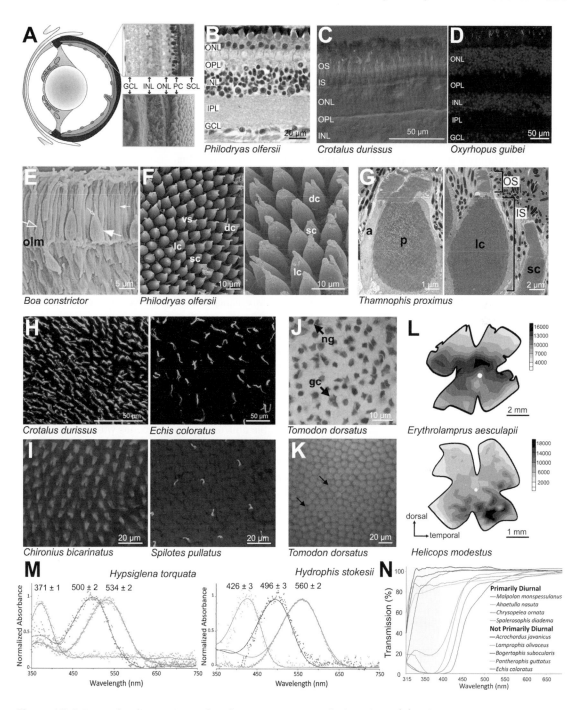

Figure 15.4 Example observations of snake eye anatomy and physiology. (A) Schematic section through snake eye (left) showing position and orientation of retinal histological section (above right) and scanning electron (SE) micrograph (below right) images of natricid colubroid *Thamnophis proximus*. (B) Classical, haematoxylin and eosin-stained histological section of retina (outer or scleral surface at top of image) of a dipsadid colubroid. (C) Immunofluorescence labelling of

15.4.3 Pupils

Snakes have approximately circular pupils when dilated, that close down to varying degrees and shapes, most often slits (mostly vertical, some horizontal) or circles. At least some snakes constrict pupils in response to tactile stimuli [53] and dilate in response to some

---

Figure 15.4 (*cont.*) LWS-opsin (green) and SWS1-opsin (magenta) cone outer segments, superimposed on a differential interference contrast view of a retinal section of a crotaline viperid. (D) Immunofluorescence labelling of LWS-opsin (magenta) and SWS1-opsin (green) cone outer segments in a retinal section of a dipsadid colubroid. (E) SEM image of outer part of the rod-dominated retina of a boid henophidian, labelling outer parts of inner segments of large single cone (large filled arrow), small single cone (large unfilled arrow), and rod (small filled arrow). (F) SE micrographs of retinal photoreceptors of a diurnal dipsadid colubroid, showing different types of cone-like photoreceptors and lack of rod-like photoreceptors; the 'vs' are likely to be transmuted (cone-like) rods (G) Transmission electron micrographs of a double cone (left) and a large and small single cone (the latter probably a transmuted, cone-like rod) (right) of a diurnal natricid colubroid. (H) Confocal image stacks of immunofluorescence labelling of LWS opsin- (green) and SWS1 opsin- (magenta) containing outer segments of cone populations of a crotaline (left) and viperine (right) viperid (rods, not identified by immunolabelling, are much more abundant in the *E. coloratus* retina). (I) Immunofluorescence images of wholemounted retinas of colubrid colubroids with anti-LWS opsin antibody-labelled (red) outer segments of large single and double cones (left image), and anti-SWS1 antibody-labelled (green) outer segments of small single cones (right). (J) Photomicrograph of a Nissl-stained wholemounted retina, focusing on the GCL. (K) Wholemounted retina of a diurnal dipsadid colubroid, focused on inner segments of photoreceptor layer, showing predominance of double and/or large single cones with rare small and very small single cones (examples indicated with red and black arrows, respectively), the latter are likely transmuted (cone-like) rods. (L) Topographic maps showing density distribution of photoreceptors in a terrestrial dipsadid colubroid (above) and semiaquatic dipsadid colubroid (below); white dots indicate position of optic-nerve head; values are photoreceptor cells per mm^2. (M) Absorbance spectra for three visual pigments (shown as violet/blue, black and red curves for SWS1, RH1 and LWS opsin-based pigments, respectively) each for a terrestrial dipsadid colubroid (left) and marine elapid elapoid (right), numbers above curves are peak absorbance (λ_{max}, in nm). (N) Spectral transmission curves for lenses of a selection of snakes showing examples of highly and less-transmissive lenses in not- and primarily diurnal species. Some figure parts reproduced with permission, as follows: A (fig. 2 of [12]), B (fig. 5 of [74]), C (fig. 7a3 of [70]), E (fig. 1B of [69]), F (fig. 6a of [74]), G (fig. S7 of [12]), H (fig. 8. of [70]), M (replotted from data from [39, 66]), N (redrawn from fig. 3 of [38]). *Source*: B and F Copyright © 2014, reprinted with permission from © 2014 S. Karger AG, Basel. C and H by permission of Oxford University Press. E Reprinted with permission from Ellis R. Loew, Jacqueline L. Johnson, Arnold J. Sillman © Wiley, 2001, *Journal of Experimental Zoology* 290(4):359–65. Abbreviations: a, accessory member of double cone; dc, double cone; gc, ganglion cell body; GCL, ganglion cell layer; INL, inner nuclear layer; IS, inner segments of photoreceptors; lc, large single cone; ng, non-ganglion cell; olm, outer limiting membrane; ONL, outer nuclear layer; OPL, outer plexiform layer; OS, outer segments of photoreceptors (rods and cones); p, principal (chief) member of double cone; PC, photoreceptor outer segments; PE, pigment epithelium; sc, small single cone; SCL, sclera; vs, very small single cone. (A black and white version of this figure will appear in some formats. For the colour version, please refer to the plate section).

emotional states [54]. Pupil constriction typically occurs in response to illumination (e.g., [55]) but in at least some semiaquatic snakes it occurs upon entering water [53] despite typically lower light intensity than in air, perhaps to increase depth of field [56]. Pupil dynamics are little studied in reptiles, including snakes [56].

Establishing functional implications and causes of variation in pupil shape across vertebrates is not straightforward and there have been few quantitative, detailed comparative analyses (see review in [56]). Evidence has been presented that foraging mode as well as diel activity explains variation in constricted pupil shape, at least for elapids [57]. Perhaps all highly fossorial (non-caenophidian) snakes have a circular pupil even when constricted, but so do some (but not all) diurnal snakes (e.g., *Spalerosophis diadema*: DJG pers. obs.). Pupils are present in many scolecophidians, but are reportedly absent in some species with very reduced eyes (e.g., [58]). Whatever the extent to which constricted pupil shape predicts ecology, reconstructing pupil shape in the ancestral snake with high confidence is non-trivial because constricted-pupil shape is unknown for many species, especially for small-eyed non-caenophidians. Reports of pupil shape in systematic accounts are often based on preserved specimens or, if in life, often do not report whether pupils were dilated or constricted (e.g., [59, 60]). Many excellent photographs of live snakes, for example in field guides, seemingly do not depict strongly constricted pupils. However, as far as we are aware, all extant non-caenophidian snakes have either circular or vertical slit or elliptical pupils when constricted. Horizontal pupils and more elaborate shapes, such as the 'key-hole'-shaped pupils of *Ahaetulla* spp., are seemingly restricted to Caenophidia and so are likely derived conditions for snakes.

15.4.4 Retinal Anatomy

Major contributions to understanding snake retinal anatomy were undertaken using classical histology and light microscopy [1, 2, 22, 25, 26, 61–65]. These provided evidence demonstrating substantial differences in anatomy between snakes and other vertebrates (see § 15.4.1) and exceptional diversity in the complement of photoreceptor cells within snakes, that contributed to the development of fossorial and/or nocturnal snake-origin hypotheses. According to Walls [2, 23], the ancestral snake was fossorial with a rod-only retina, as in extant scolecophidians, but Underwood [24, 25, 28] viewed the relatively simple duplex (rod and cone) retina seen in 'henophidians' as more likely representative of the ancestral snake condition, implying loss of cones in scolecophidians. Walls, Underwood and others (e.g., [63]) agreed that features such as double cones, transmuted photoreceptors, and a fovea (a specialized, cone-rich, high-acuity retinal region) are derived for snakes, representing retinal elaboration within Caenophidia, and this has thus far survived scrutiny from additional morphological, molecular and phylogenetic studies.

Histology

Classical histology (e.g., Fig. 15.4A, B) data for snakes (and observation of gecko retinas: e.g., [2, 22]) underpinned Walls' Transmutation Theory, that challenged the orthodox view that cones and rods are fixed, divergent photoreceptor types, and proposed instead that, at least to some extent, cones can evolve rod-like form (and perhaps function) and vice versa.

Walls' transition series were based on outdated views of snake phylogeny, but his Transmutation Theory receives compelling support from modern studies of squamate vision (e.g., [12, 13, 20, 66, 67]).

Classical histological studies have not always correctly identified comprehensive photoreceptor complements. For example, comparatively rare and subtly different small cones were overlooked by Walls, Underwood and others in pythonids and boids (see [68, 69]) and in crotaline viperids (see [70]).

Immunohistochemistry

Immunohistochemistry (IHC) involves labelling proteins to visualize them in histological sections (e.g., Fig. 15.4C, D). This technique has been applied to the natricid *Thamnophis* [12, 49], 10 dipsadids and two colubrids [13, 14], one viperine and one crotaline viper [70], and the scolecophidian *Anilios* [21]. Each of these studies immunolabeled particular visual opsins using antibodies or antiserum, combined with general cone- (and sometimes also rod-) agglutinin labels and DAPI 'staining' of nuclei, while three studies [12, 13, 49] focusing on photoreceptor transmutation also labelled rod-specific transducin, a phototransduction protein. These few studies provided several important and novel insights. Transmuted, cone-like rods in colubroids contain RH1 visual pigments and rod-specific transducin [12, 13, 49]. Despite snakes having only two typical cone visual pigments (SWS1, LWS) multiple cone populations can be identified based on varying combinations of morphology and opsin gene expression [14, 70]. Diurnal and nocturnal colubroids differ not only in the relative proportion of rods and cones, but also in the number of layers of nuclei in the outer nuclear layer of the retina [14]. Each member of viper double cones may express SWS1 or LWS within the same retina, a condition unknown elsewhere among vertebrates [70]. At least some scolecophidians retain a functional *lws* expressed in photoreceptor outer segments, even though no morphologically cone-like photoreceptors have been found in scolecophidians [21].

Electron Microscopy

Transmission electron microscopy (TEM) of thin sections (e.g., Fig. 15.4G) enables ultrastructural observations (e.g., [12, 25, 71, 72]), including the relationship between the plasma membranes and cell membrane of photoreceptor outer segments (separate in typical rods, not in typical cones), and the unique nature of snake double cones. These data support Underwood's hypothesis of a duplex retina in the ancestral snake (because of the occurrence of typical rod- and cone-like outer segments in those retinas), and Walls' Transmutation Theory (because of cone-like ultrastructure in photoreceptors that resemble rods at lower magnification). As far as we are aware, TEM has not yet been applied to non-caenophidian snakes.

Scanning electron microscopy (SEM) of critically point dried retinas provides detailed images of 3D structure and arrangement of retinal layers, especially photoreceptors (e.g., Fig. 15.4A, E, F). These techniques have been applied to pythonid, boid and caenophidian snakes [12, 37, 65, 68, 69, 73, 74]. Particular success has been achieved in discovering rare photoreceptor types overlooked by classical histology, and being combined with physiological and/or molecular genetic data to better understand the homology and evolution of photoreceptors and visual pigments. SEM investigations of scolecophidians and amerophidians (and more afrophidian henophidians) would likely help to more accurately and precisely infer the ancestral snake retina.

Retinal Wholemounts

Retinas can be examined as flattened sheets [75, 76]. Such retinal wholemounts (e.g., Fig. 15.4H–L) can yield a variety of (electro)physiological and morphological data, such as limits of acuity based on cone and ganglion-cell spacing [75]. For morphology, wholemounts can provide information on spatial arrangements (topography) of photoreceptors and other cell types (e.g., Fig. 15.4L), and on retinal blind spots, and aspects of retinal circuitry (see below). Regions of higher neuron density ('specializations') occur as two main types. Visual streaks are elongated higher-density regions, usually disposed along the nasal-temporal axis. *Areae centrales* are concentric higher-density regions that can occur in any quadrant of the retina. Variation in topography is often explained by ecology with, for example, nasal–temporal ('horizontal') streaks considered to be adaptations to open environments dominated by a horizon, and *areae centrales* as adaptations to more visually complex environments. However, topography is explained by phylogeny as well as visual ecology (e.g., [77–79]).

Retinal topographic analyses are carried out by quantifying cells of wholemounts labelled with different markers. For example, Nissl staining labels all cell types (nonspecifically) and is used to identify ganglion cells, and immunolabelling can be used to visualize different types of neurons, including photoreceptor types (e.g., [68]; Fig. 15.4H–K). Although widely applied in other vertebrates, very few stereological studies of snake retinas have been published, for only a few species of three caenophidian families: [37, 74, 80, 81].

Three diurnal species, the natricid *Thamnophis sirtalis* and two dipsadid *Philodryas* spp., have relatively low densities of photoreceptors (ca. 10,000 cells mm^{-2}) with low convergence to ganglion cells (ca. 2.5:1 in the *area centralis*) [74, 80]. Ganglion cell populations were described in only four studies [37, 74, 80, 81]. A horizontal streak was documented in the arboreal *P. olfersii*, and a ventral *area centralis* in the ground-dwelling *P. patagoniensis*, with this difference linked potentially causally with habitat use and capacity to locate predators in either the same level (arboreal), or upper part of the visual field (ground-dwelling) [74]. In three marine elapids, a horizontal streak and two *areae centrales* (one nasal and one temporal) were observed, with two species having an additional ventral *area* – this variation explained possibly by differences in habitat use and/or foraging behaviour [37]. In comparisons of densities and distributions of ganglion cells in 12 dipsadids [81], diurnal species had higher densities and estimated upper visual-acuity limits (see § 15.4.7 and § 15.4.11 for other measures of acuity) than nocturnal species (ca. 9,500 cells mm^{-2} and 2.5 cycles degree^{-1} *versus* ca. 7,600 cells mm^{-2} and 1.3 cycles degree^{-1}, respectively). Most diurnal species had a horizontal streak, whereas nocturnal species had an *area centralis* in various regions – a ventral *area* in the semi-arboreal *Dipsas* might benefit viewing aerial predators, while a dorsal *area* in the fossorial *Atractus* might improve spatial resolution of the lower visual field [81].

Retinal Circuitry

Circuitry is the cellular connections and signal pathways within the retina, from the synaptic terminals of the photoreceptors to the optic nerve. Circuitry and signal processing play key roles in overall visual performance. Knowledge of the morphology of snake retinal circuitry comes from histology and retinal wholemounts (and IHC) but data are scant. Beyond sparse data on ganglion-cell densities (see previous section) information on circuitry in snakes is largely restricted to photoreceptor synaptic terminals and to the relative thicknesses and numbers of nuclei in the retinal layers.

Typically, vertebrate rods have small, relatively simple synaptic terminals (spherules) that connect to one or two bipolar or horizontal-cell dendrites, and cones have larger, more complex synaptic terminals (pedicels) connecting to hundreds of dendrites. Underwood was impressed that *Alligator* and geckos have photoreceptors with mostly rod-like outer segments but cone-like pedicels [82, 83], and by his own observations of similar anatomy in *Vipera berus* [24, 25]. This led Underwood [24] to propose 'Pedlar transformation' and 'Walls transformation' as two components of photoreceptor transmutation – namely a synaptic-terminal modification component and an inner- and (especially) outer-segment modification component. In testing hypotheses of transmutation, much greater attention has been paid to overall photoreceptor morphology, especially outer segments, than to synaptic terminals and changes in retinal circuitry.

Substantial variation in the relative thickness and numbers of nuclei in inner and outer layers of caenophidian snake retinas has been reported [2, 14, 22, 25, 63, 84]. Underwood [84] related variation to different relative abundances of oligo- and polysynaptic photoreceptors. However, such inferences are not possible without distinguishing among various bipolar cell types and among bipolar and other retinal neurons [70] – research that has not yet been carried out for snakes. Additionally, even if variation in the ratio between inner and outer nuclear layers (= inner nuclear quotient of [63]) is directly correlated with photoreceptor synaptic–terminal morphology, linking this with visual function and ecology might not be straightforward (see also [22]: 903) despite Underwood's [84] claim that high levels of summation (inferred by a relatively thick outer nuclear layer) are correlated with nocturnality. Despite this, the discovery that diurnal species have relatively much thinner outer nuclear layers with fewer nuclei than nocturnal species in a small sample of colubroids [14] is a striking result that will hopefully prompt further research. Clarifying retinal circuity in the major phylogenetic lineages and in taxa with a range of ecologies and photoreceptor complements would seem worthwhile. It should improve reconstructions of the ancestral snake, and assessments of the extent to which that can be used to predict ancestral ecology.

15.4.5 Microspectophotometry

Microspectrophotometry (MSP) can be used to measure absorbance spectra of both inert and photosensitive pigments. Given that snakes lack photoreceptor oil droplets (e.g., [2, 25, 26]), application of MSP in snakes has been restricted to visual, photosensitive pigments *in situ* in the outer segments of cone and rod photoreceptors. MSP of visual pigments requires fresh retinal tissue dissected in the dark or under dim red light from animals that have been 'dark adapted' for several hours, so that pigments are in their inactivated, 11-*cis*-retinal form (see § 15.3).

The chief features of absorbance spectra of visual pigments (e.g., Fig. 15.4M) that are interpreted are the wavelength of peak absorbance (λ_{max} in nm), and shape of the normalized curve. The latter gives an indication of which chromophore (vitamin A_1 or A_2) is combined with the opsin protein to form the visual pigment. Sources of error include incorrect photoreceptor identification or curve fitting as well as noisy data. Good preparations can allow data to also be generated on cell and segment sizes. Qualitative information can also be obtained on features such as approximate proportions of major cell types, and the presence of double cones, but it can be difficult to obtain precise information on regional variation in photoreceptor densities, and rare cell types can be overlooked.

The first MSP data for snakes were published in 1989 for two viperids, a colubrid and a psammophiid [85]. MSP data have subsequently been published for a handful of snake species: viperids, elapids, dipsadids, colubrids, the natricid *Thamnophis*, *Python*, *Boa*, non-anomalepidid scolecophidians, and *Anilius* [5, 12, 37, 39, 68–70, 73]. These data have been used to identify classes of visual pigments, chromophores and photoreceptors, clarify relationships between opsin amino-acid sequences and pigment λ_{max}, and provide insights into photoreceptor transmutation. They have also been used to support the conclusion that extremely fossorial scolecophidian and aniliid snakes have only rod pigments [5], though this is, at least in part, questioned by genomic data [21].

15.4.6 *In vitro* Regeneration of Visual Pigments

In vitro regeneration of visual pigments allows opsin genes to be heterologously expressed in cultured cells. The opsin protein is harvested from the cells, regenerated with a chromophore (typically 11-*cis*-retinal) and purified, resulting in a functional, light-sensitive visual pigment. The regenerated pigment can then be assayed spectroscopically to determine dark and light state absorbance spectra, analogously to those measurements using *in situ* MSP (see § 15.4.5), which can in turn be used to estimate λ_{max}. Several other functional aspects can also be measured. Visual-pigment activation rates can be measured in the dark by comparing the proportion of inactivated and activated visual pigment over time or by using a transducin-activation assay to compare transducin activation rates in the dark *versus* the light. Properties of light-state activation can be determined using fluorescence spectroscopy to measure the increase in fluorescence resulting from dissociation of the chromophore from the pigment after activation. How visual pigments react to different chemicals can also be measured. For example, hydroxylamine assays can indicate the conformation of the chromophore binding pocket, which typically differs between rod (closed) and cone (open) pigments.

Visual pigments from only a few snakes have been expressed so far. The first were from *Xenopeltis unicolor* [86], for which each of the three visual pigments (LWS, SWS1, and RH1) was expressed and their absorbance spectra measured. They were found to have λ_{max} values of 550, 361, and 497 nm, respectively, slightly lower than the *in situ* MSP values measured for LWS and RH1 in the same study (558–562 and 499 nm, respectively).

Subsequently, expressed *Thamnophis proximus* RH1 was found to have a highly blue-shifted λ_{max} of 481 nm, and to occur in superficially cone-like photoreceptors, providing molecular support for Walls' [2] photoreceptor Transmutation Theory [12]. More recently, all three visual pigments (LWS, SWS1, RH1) from *Pituophis melanoleucus* were expressed and, similar to *T. proximus*, the RH1 was highly blue-shifted (481 nm) and expressed in cone-like rods [13]. Additionally, *P. melanoleucus* RH1 reacted to hydroxylamine, similar to the LWS pigment and unlike the control (bovine) RH1, consistent with a more cone-like function of *P. melanoleucus* RH1 [13].

RH1 has been expressed from four different elapids: the terrestrial *Sinomicrurus japonicus boettgeri*; a laticaudine sea krait, *Laticauda semifasciata*; and two hydrophiine sea snakes, *Emydocephalus ijimae* and *Hydrophis ornatus* [87]. The two hydrophiine RH1s had more-typical vertebrate RH1 λ_{max} values at 494 and 490 nm, respectively, whereas the two other species had blue-shifted spectra similar to those in colubroids, with λ_{max} values of 485 and 482 nm, respectively. The longer hydrophiine RH1 absorbances were inferred to be derived within Elapidae, and are consistent with *in situ* MSP measurements from two other hydrophiines [37, 87]. A functional explanation for this shift is unclear and unlikely related to photoreceptor transmutation because *Hydrophis* spp. also (like the colubroids) appear to have 'transmuted', cone-like rods [37, 66]. The same study [87] also expressed the LWS pigment of *H. ornatus,* which has a λ_{max} of 551 nm similar to that found for *X. unicolor,* and to the MSP data for other hydrophiines. Most recently, *sws1* of the freshwater dipsadid *Helicops modestus* has been expressed [88]. *Helicops* spp. each have two distinct *sws1* forms ([88], see also [38]) which expressed *in vitro* were confirmed UV and violet sensitive. This finding was supported by *in situ* MSP of UV- and violet-sensitive cones, and was linked to a single amino-acid substitution (Phe86Val).

To date, *in vitro* protein expression has not been used to infer properties of the visual system of the ancestral snake but, when combined with codon-based likelihood models for ancestral-sequence reconstruction, offers a powerful tool that could be applied to this purpose once sufficient sequence data for snakes are available to facilitate robust ancestral-sequence reconstruction. Studies of this type have been conducted for other vertebrates (e.g., [89]).

15.4.7 Visual-Pigment Spectral Tuning

Spectral tuning refers to λ_{max} of visual pigments and its relationship to chromophore type (vitamin A_1 or A_2) and opsin amino-acid sequence. Background knowledge on the homology and physiology of vertebrate visual pigments allows predictions of λ_{max} from opsin amino-acid sequences, to varying degrees of precision and confidence (e.g., [90]). Predictions have been largely based on tight correlations between λ_{max} and amino-acid residues at a handful of key 'tuning' sites within opsins, especially in and around the chromophore binding pocket. Predictions of λ_{max} can be tested by *in situ* MSP and/or *in vitro* regeneration (§ 15.4.5– 6).

Where measured and predicted λ_{max} values can be compared for snake visual pigments, predictions have been upheld to within a few nm in most cases [38, 66, 70]. These comparisons of λ_{max} have been undertaken for all three visual pigments found in snakes, based on correlations between amino-acid sequences in other vertebrates for RH1 [90],

SWS1 [91, 92] and LWS ([93], see review [94]). However, snake species yet to be subjected to *in situ* MSP or *in vitro* visual-pigment regeneration have also been found with opsin tuning-site amino acids unknown in other vertebrates – so that confident predictions of λ_{max} for these are not possible (e.g., [38, 70]). Furthermore, a few λ_{max} predictions are substantially different from corresponding MSP data, for example for SWS1 in the viper *Agkistrodon contortrix* (416 nm versus predicted ca. 360 nm: [70]) and for LWS in the pythonid *Python regius* (551 nm versus predicted 560 nm: [68, 86]). There is no universal consensus about the impacts of all tuning-site mutations (e.g., [95]), and mismatches between predicted and measured λ_{max} have prompted proposals of additional, as yet undiscovered mechanisms and/or tuning sites (e.g., [86, 96]).

The application of spectral-tuning predictions from opsin sequences allows reconstruction of ancestral λ_{max} values based on reconstructions of ancestral opsin-gene sequences, and this has been applied in some cases to snakes [38, 70]. In general, tuning-site amino acids are conserved in snakes, making for straightforward ancestral-state reconstructions [38]. The SWS1 pigment is inferred to have been ultraviolet sensitive in the ancestral snake and ancestral alethinophidian [38] – ultraviolet-sensitive cone pigments are often associated with activity in low-light conditions (e.g., [97]). LWS and RH1 are estimated to have had an λ_{max} of ca. 555 nm and 491–496 nm in the ancestral snake and 560 nm and 491–496 nm in the ancestral alethinophidian, respectively [38].

15.4.8 Vision-Gene DNA Sequences, Including Genomes and Transcriptomes

Despite the first sequencing of a vertebrate opsin occurring in the early 1980s [98–100], the first snake visual-opsin sequence data were not published until 2009 [86]. These were *rh1*, *sws1* and *lws* for *Xenopeltis unicolor* and *Python regius*, providing evidence for the loss of *rh2* and *sws2* in the snake stem but without an ancestral reduction of the visual system to a rod-only retina. Subsequently, visual-opsin genes have been sequenced for tens of species of snakes from most major lineages [5, 6, 12, 14, 38, 39, 66, 70, 87, 101, 102].

Snake visual-opsin sequences have not, to the best of our knowledge, been generated from direct protein sequencing. The first snake visual-opsin gene sequences were obtained by PCR amplification and Sanger sequencing using vertebrate opsin-gene primers applied to retinal complementary DNA (cDNA) synthesized from mRNA extracted from macerated eyes [86]. This approach was used to generate most currently available snake visual-opsin gene sequence data [5, 12–14, 38, 66, 70, 87, 102]. We know of only two reports of snake visual-opsin gene sequences generated using PCR and Sanger sequencing from genomic DNA (gDNA: [12, 39]). More recently, visual-opsin sequences have been obtained by Next Generation Sequencing (NGS), which has also provided sequence data for other snake visual-system genes. The first snake-vision data obtained using NGS were generated during whole-genome sequencing for the King cobra *Ophiophagus hannah* [103] and Burmese python *Python bivittatus* [104]. In addition to other snake whole or near-complete genomes (e.g., [49, 105]), visual-system gene sequences for snakes have been generated by NGS through transcriptome sequencing (RNA-Seq) of mRNA from eyes or retinas [6, 21] and from targeted capture (hybrid enrichment) from gDNA [101].

Each of these approaches has different benefits and costs (e.g., [101]). Traditional PCR from eye cDNA followed by Sanger sequencing is efficient, in terms of time and expense, for obtaining sequences for a few genes for tens of samples, and it generates data for functional genes expressed in a given time and place, but it requires freshly preserved eyes and can provide false negatives (inferred gene loss or pseudogenization), at least for small eyes and/ or genes with divergent sequences for which available primers are not suitable. Eye or retinal RNA-Seq avoids the need for primers and produces vastly more data but is substantially more expensive if there is interest in only few genes, and has the same drawbacks as cDNA sequencing. RNA-Seq can also provide information on gene-expression levels but comparison among samples requires replication and controlled sampling, due to high variability in expression. gDNA sequencing removes reliance on fresh tissues and can ameliorate potentially overlooking genes due to low or variable expression but introduces other challenges. Sequencing gDNA using traditional PCR and Sanger sequencing can be challenging and time consuming due to the presence of introns. Whole-genome sequencing using NGS provides more data, but with much greater time and cost investment, and desired sequences may still be missed because of difficulties associated with assembly and the stochastic nature of sequencing. Targeted capture can be used to enrich genomes for sequences of interest, which can reduce costs of genome sequencing when there are a large number of genes and/or species of interest, but this requires considerable up-front time investment and existing resources (e.g., genomes or transcriptomes), to design probes for enrichment (capture).

NGS approaches have determined that snakes retained up to 41 of 44 phototransduction and visual-cycle genes likely present in the ancestral squamate [21, 49, 104]. Only the visual-opsin genes *rh2* and *sws2* and the cone and rod phototransduction gene *grk1* occur in at least some lizards but no snakes ([21] see also [6]). In the only scolecophidians thus far examined using NGS (*Anilios* spp.) 11 visual genes likely present in the ancestral snake are absent, seven involved in cone phototransduction, two in cone and rod phototransduction, and two in the visual cycle [21]. In addition to phototransduction and visual-cycle genes, snakes (at least alethinophidians) have lost at least eight non-visual opsin genes (light-associated genes not directly involved in image-forming vision) and five lens crystallin genes [20, 49].

Although beyond the scope of this chapter, it is worth also considering those genes having other (non-visual) light-associated functions when inferring environments inhabited by the earliest snakes and their closest non-snake relatives. The visual genes and proteins likely lost along the snake stem (Fig. 15.5) are absent also in some other extant vertebrate lineages, many of which are hypothesized to have lost (functional copies of) these genes as adaptations to low-light environments. Examples are dominated by postulated fossorial and/or nocturnal ancestry, such as for crocodylians [106], xenarthran mammals [107], and moles and mole rats [108], but also include cetaceans, especially those that dive to great depths and have echolocation [109]. Snake genomes examined thus far (i.e., excluding scolecophidians) also lack non-vision light-associated genes lost in these other vertebrate lineages [110], providing further evidence that, in their immediate ancestry and/or early history, snakes very likely adapted to low-light environments.

Vision-gene sequence data are useful in many ways. They provide an additional line of evidence for homology assessments (e.g., of visual pigments), allow predictions of

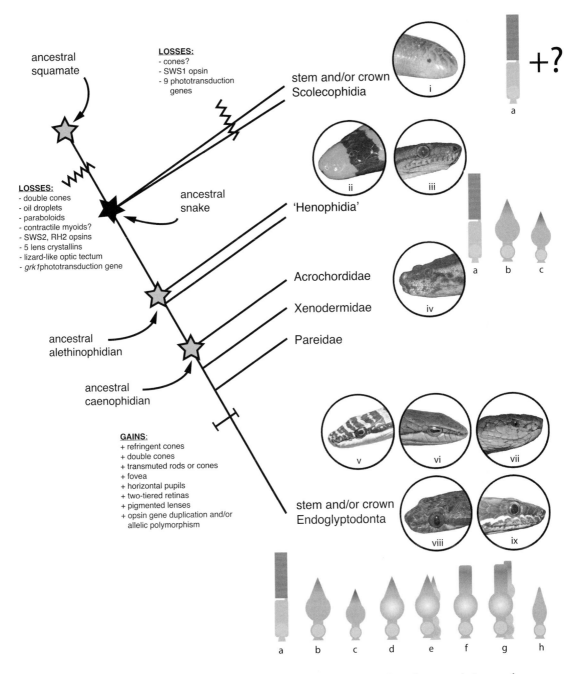

Figure 15.5 Summary of inferred evolutionary events in snake vision plotted onto a phylogeny (see Fig. 15.1). Cartoon photoreceptors show main rod and cone types found in major lineages: a = rod, b = large single cone, c = small single cone, d = large single refringent cone, e = double refringent cone, f = transmuted, rod-like cone, g = transmuted, rod-like double cone, h = transmuted, cone-like rod. Other photoreceptor types in Endoglyptodonta that might be recognized (e.g., those with elongated myoids in two-tiered retinas) are not shown. The precise branch on which changes occurred is not always clear; for example, a spectacle (not listed here) might have been acquired in stem snakes and

physiology (see § 15.4.7), and enable insights into adaptive processes such as gene loss, duplication and expression, and alternative functional pathways that evolved to maintain visual requirements in different environmental, ontogenetic and evolutionary contexts. They also allow quantitative tests of selection (see § 15.4.9). For snake vision, sequence data have provided crucial evidence that the ancestral snake likely had a duplex retina with three visual opsins, retaining most of the genomic machinery for vision that is present in lizards (e.g., [6, 21]), that photoreceptor transmutation has occurred within snakes and results in changes in photoreceptor physiology as well as morphology [12, 13, 66], and that selective-pressure regimes differed in the evolution of visual (especially cone-phototransduction) genes in lineages with transmuted photoreceptors [6].

15.4.9 Selection Tests for Inferring Molecular Genetic Evolution

Selection tests are useful for investigating molecular evolution of protein-coding genes, including those involved in vision. These tests use codon models of sequence evolution to infer selective pressures on genes by estimating nonsynonymous and synonymous substitution rates, d_N and d_S, respectively [111]. A ratio of those rates, denoted ω (= d_N/d_S), of < 1 indicates a conserved protein sequence, under purifying or negative selection. An $\omega \approx 1$ indicates neutral evolution, whereas if changes to the amino-acid sequence are advantageous and being favoured then $\omega > 1$, indicating positive selection. Software packages implementing codon models of molecular evolution offer various models and selection tests [112, 113]. Many selection tests focus on detecting positive selection due to the potentially straightforward link between protein function and particular amino-acid substitutions [111]. Other useful selection tests include those for relaxed selective constraint (e.g., RELAX; [114]), which can indicate changes in the functional importance of genes and, more generally, those that test for shifts in selective pressures among lineages (e.g., clade models; [115]). These models are useful in testing for long-term shifts in selection pressure associated with changes in ecology and function (e.g., [116, 117]).

Most applications of these methods to snake vision evolution have focused on visual-opsin genes because effects of amino-acid substitutions on pigment function are comparatively well understood and can, in some cases, be further tested by *in vitro* expression and MSP [5, 12–14, 38, 39, 66, 70, 87, 88]. These studies have found that selective pressures on snake visual pigments vary considerably among lineages, with many instances of positive selection and shifts in selective pressure associated with different ecological traits, activity patterns, and habitats that suggest functional adaptation to different light environments. More recently, efforts have begun to study changes in selective pressure, and their

---

Figure 15.5 (*cont.*) been lost in scolecophidians and some henophidians, or might have arisen in stem or later alethinophidians. Note also that scolecophidians might be paraphyletic. Representative taxa of major lineages depicted in circles (photographers in parentheses): i = *Indotyphlops* sp. (Mendis Wickramasinghe), ii = *Anilius scytale* and iii = *Corallus hortulanus* (both: Gabriela Bittencourt), iv = *Acrochordus javanicus* (David Gower), v = *Chrysopelea taprobanica*, vi = *Ahaetulla perroteti* and vii = *Trimeresurus macrolepis* (all three: V. Deepak), viii = *Leptodeira annulata* and ix = *Lycognathophis seychellensis* (both: David Gower). See Supplementary Appendix 15.S1 for colour version.

functional implications, in other vision genes and their proteins that function downstream of visual pigments [6, 20, 67], as well as the non-visual opsin, melanopsin [9].

One study [6] applied selection tests to address snake origins by testing Walls' [2] hypothesis that the visual system underwent substantial degeneration during snakes' fossorial ancestry. Profound degeneration predicts relaxed selective constraint in stem snakes, but no evidence for this was found in analyses of opsin and phototransduction genes of alethinophidian snakes and other reptiles [6]. Instead, patterns of selection were more consistent with a dim-light (e.g., nocturnal) origin of snakes that did not involve profound fossoriality [6].

15.4.10 Electrophysiology

Electroretinography measures the sum of electrical responses of retinal cells extracellularly, usually from the cornea. As far as we are aware, there have been only three published reports of electroretinography applied to snakes. Jacobs et al. [118] aimed to detect retinal visual pigments in *Thamnophis* spp., but found only one of the three subsequently discovered by MSP and molecular genetic sequencing (though better results were obtained by increasing stimulus intensity, see [73]: 99). Elliot et al. [119] and Glickman et al. [120] used electroretinography to assess response to laser eye injuries in *Elaphe*, and provided data on acuity (ca. 1 cycle degree^{-1}). Baker et al. [121] considered acuity in *Nerodia* (ca. 5 cycles degree^{-1}) to be 'good for a small eye', approximately the equal of domestic cats, and approximately 20/120 in US human clinical terms. Although it has barely been applied to snakes, there is potential for electroretinography to explore aspects of snake vision, including tests of purported rod monochromats, that could provide insight into the visual biology of the earliest snakes. We know of no studies of intracellular electrophysiological activity of individual retinal cell types for snakes.

15.4.11 Behaviour

Although anatomy, molecular genetics and physiology can be used to make inferences about traits such as colour vision and acuity, verification requires behavioural experimentation. For example, multipigment retinas might be adaptations to increasing sensitivity and/or contrast rather than to polychromacy [122]. There have been few behavioural studies of snake vision, and seemingly none testing for colour vision. In perhaps the only study of its kind published thus far, Rumpff [123] quantified visual acuity in ten species of Indian snakes by optomotor response. Other acuity data for snakes have been derived not from behavioural experiments but from stereology (see § 15.4.4) or electrophysiology (see § 15.4.10). Many snakes feed on relatively large meals infrequently, making them challenging subjects for classical conditioning experiments of colour perception, brightness discrimination, or acuity. Little is known about learning abilities in snakes [124] but a few studies suggest there is potential in this direction. In *Thamnophis sirtalis,* learning of chemosensory stimuli of noxious food was enhanced by offering food with black and yellow versus green forceps [125], though it is unclear whether this is explained by shade and/or colour discrimination. Juvenile *Agkistrodon piscivorus* were successfully trained to locate food indicated by a visible red card [126], but it is unclear whether the stimulus was the card, its colour and/or its brightness.

The importance of vision to snakes, compared to other sensory systems, has largely not been assessed experimentally. Blindfolded *Notechis scutatus* (Elapidae) struggle to capture mobile prey in laboratory conditions, and those naturally blinded by seabirds are able to feed, grow and reproduce successfully without sight only by feeding on abundant and largely immobile seabird chicks [127]. Experimental deprivation of eyes and/or infrared sensing pits in pythonids and boids found that both senses contribute to effective striking at mice, though vision may be the dominant imaging system [128].

15.4.12 Phylogenetics and Ancestral-State Estimation

In the absence of exceptionally-preserved fossil evidence of direct ancestors, ancestral-state estimation relies on having phylogenetic hypotheses and knowledge of trait conditions in terminal taxa. Thus, evolving (hopefully improving) understanding of squamate phylogenetics might require reassessments of previous ancestral estimates. Two prominent examples in the snake-origins literature are fossoriality and macrostomy. Previously, it was simple to infer fossoriality and non-macrostomy for the ancestral snake given the understanding that fossorial, narrow-gaped scolecophidians, aniliids and uropeltoids were together paraphyletic and lying outside a clade that includes all extant macrostomatan snakes (e.g., [129]). However, since Wilcox et al.'s [130] landmark paper, molecular phylogenetics has found increasingly strong evidence that aniliids and the non-fossorial, macrostomatan tropidophiids comprise a clade (Amerophidia; see Fig. 15.1) lying outside all other extant alethinophidians (Afrophidia: uropeltoids and macrostomatan pythonoids, booids and caenophidians; see Fig. 15.1). As a result, evidence for a fossorial, narrow-gape snake ancestor was eroded, and these traits in aniliids and uropeltoids became viewed as apo- rather than plesiomorphies, with these taxa potentially representing 'regressed macrostomatans' (see [131]). Even more recently, molecular phylogenetics has also found evidence for the paraphyly of scolecophidians (see Chapter 10), and this has renewed support for a highly fossorial, narrow-gape ancestral snake with reduced eyes [7].

Shifting views of squamate phylogeny have long had an impact on evolutionary interpretations of the snake eye. Walls [132] initially considered caenophidians with 'all-cone' retinas (now known to include cone-like rods) to be derived from duplex ancestors. Subsequently [2, 22, 23] he adopted the view, seemingly persuaded by herpetologists of the day, that aglyphous colubroids were shallowly phylogenetically nested within Caenophidia, and that the plesiomorphic caenophidian condition was an 'all-cone' retina such as he found in many colubroids – so that all 'rods' observed elsewhere within Caenophidia were instead transmuted cones. Fortunately, Walls nonetheless recognized that the duplex retina of boids and pythonids was more similar to the likely plesiomorphic snake condition, and instances of possible photoreceptor transmutation in caenophidians proved widespread enough that the main component of his Transmutation Theory was not buried and forgotten as understanding of snake phylogeny improved.

Thus far, ancestral states of snake visual-system phenotypic and genotypic traits have been inferred by parsimony, Maximum Likelihood, and Bayesian approaches (e.g., [7, 38, 39, 70]). Estimation of ancestral phenotypes has been carried out using discrete rather than continuous characters. Although not performed using formal analyses, the presence or absence of functional copies of visual-opsin genes along internal phylogenetic branches has been conducted with the assumption that loss of (functional) opsin genes from the genome is irreversible (e.g., [5, 6, 21]).

15.4.13 Palaeontology

Addressing the ecological (rather than only the phylogenetic) origins and early evolution of snakes requires consideration of long-extinct ancestors for which we probably lack (and might never have) direct evidence in the fossil record. Nonetheless, fossils provide the only direct evidence of extinct phenotypes, including combinations of features not necessarily present in extant taxa. Including extinct taxa in phylogenetic analyses often alters, and can improve, estimates of relationships of extant taxa (e.g., [133]). Additionally, the fossil and rock records provide information on palaeoenvironments, and evidence of particular phenotypes in particular locations in time and space (e.g., Head et al., Chapter 4).

The majority of the snake fossil record does not include cranial remains (Chapters 3 and 4). Greater representation of (particularly articulated) crania occurs in marine *Konservat-Lagerstätten* than fossorial or other terrestrial habitat deposits. The majority of fossorial non-caenophidian snake species alive today burrow in moist tropical soils, and so their cranial remains are unlikely to be well fossilized. Extant fossorial snakes are typically relatively small, which likely further reduced the chance of preservation and future discovery of fossils of similar snakes.

In the absence of exceptional preservation including soft tissues and/or DNA, there is a lack of information on the visual systems of extinct snakes. Additionally, orbital bones have been reduced or lost in multiple clades [134], limiting the ability to estimate eye sizes, with some exceptions (Chapter 14). Conversely, snake braincases and vomeronasal chambers are generally well ossified, potentially providing osteological correlates of the visual and other sensory systems in fossils (Chapters 13 and 14). Better understanding of the visual ability and ecology of extant snakes and bony correlates of their visual-system soft tissues will aid palaeobiological research, and the formulation and testing of snake-origin hypotheses.

15.5 The Visual System of the Ancestral and Other Early Snakes

Given their phylogenetic result of scolecophidian paraphyly and their ancestral-state estimations, Miralles et al. [7] proposed a highly fossorial, scolecophidian-like ancestral snake with substantially reduced eyes lacking cones, consistent with Walls' [2, 23] hypothesis (see also [135]). In contrast, consideration of functional visual-gene complements, and a lack of evidence for relaxed selective constraint in phototransduction genes in stem snakes, led others to argue that the eyes of the ancestral snake were not as reduced as those known in any extant scolecophidian, instead retaining typical vertebrate cones and cone-phototransduction and visual-cycle genes [5, 6, 21, 49]. Some of the visual gene-based arguments imply or explicitly acknowledge the assumption that re-evolution of lost genes is much more unlikely than morphological re-elaboration of many aspects of the eye and retina (e.g., [21]). We find it compelling that alethinophidian snakes have not only cones and rods and three visual opsin genes homologous with those of other vertebrates, but also have substantial genetic components of cone phototransduction and the visual cycle found in other vertebrates [6, 21, 67, 101]. Re-elaboration of the alethinophidian visual system from a highly reduced ancestral-snake condition, as envisaged by Walls [2, 23], would be expected to be accompanied by substantially modified retinal-protein networks with novel components, of which there is no evidence in extant snakes. Thus, the reduced visual-system genotype as well as phenotype of extant scolecophidians studied thus far (which lack nearly 60 per cent of the cone phototransduction genes present in alethinophidians and other squamates: [21]) are much more plausibly interpreted as derived rather than plesiomorphic conditions for snakes – and these apomorphic reductions occurred at least twice if scolecophidians are paraphyletic. However, Miralles et al.'s [7] hypothesis of a scolecophidian-like ancestral snake can still find support if some other extant scolecophidians are found to plesiomorphically retain more of the components of the visual system found in extant alethinophidians.

A summary of inferred evolutionary changes in some snake vision characters is provided in Figure 15.5. Limitations of this summary include sparse data for terminal taxa and lingering uncertainty about some phylogenetic relationships (Chapter 10). However, there is good evidence that the ancestral snake had a less-developed visual system than extant diurnal lizards, but retained three visual opsins (RH1, SWS1, LWS) and cones as well as rods. The ancestral alethinophidian (and possibly the ancestral snake) likely had a spectacle covering the eye, a highly transmissive lens, and a UV-sensitive visual pigment. The eyes of extant scolecophidians are clearly derived in many ways, and it is not yet fully clear which of their features might be plesiomorphic for snakes. Among extant taxa, the eyes of henophidians and non-endoglyptodont caenophidians are probably most similar to those of the ancestral snake, having a duplex retina with single cones but dominated by rods.

Although the visual system of the ancestral snake was simpler than that of typical extant lizards, those of many endoglyptodont caenophidians are derived in diverse ways. Too few taxa have been sampled thus far to establish precisely where within the caenophidian tree these features were acquired, but Endoglyptodonta includes the only snakes (and for some features the only vertebrates) known to have, for example, refringent cones, rod and cone photoreceptor transmutation, pigmented lenses, permanently two-tiered retinas (see e.g., [70]), horizontal pupils, foveas, double cones with both members expressing *sws1*, SWS1 polymorphism, and possibly visual gene duplication. These features do not provide much in the way of evidence for snake origins, but clarifying the ancestral snake genotype and phenotype is important for understanding the origin of these novelties, and the extent to which the diversity and distinctiveness of caenophidian eyes can be explained by the simplified Bauplan of the ancestral snake eye.

15.6 Future Directions and Prospects

The hypothesis that we believe is currently best supported, that the ancestral snake had a visual system adapted to dim-light but perhaps not to extreme fossoriality of the sort seen in extant scolecophidians, can be tested further by generating additional, and more detailed, data on more extant scolecophidians and henophidians, especially amerophidians and uropeltoids. Some of these additional data can simply come from conducting similar investigations to those that have already been performed for a few species, but newer approaches, such as single-cell transcriptomics, and studies of aspects of snake eyes and vision that have barely been studied thus far, such as eye development (not covered in this article) and retinal circuitry, and visual aspects of behaviour, will likely all yield useful information. Increasing availability of genomic resources, especially for non-endoglyptodonts, will provide a valuable resource for understanding the origin, early evolution, and function of snake eyes and vision. Closer attention could be paid in future to ways that visual-system traits might allow discrimination among inferring highly fossorial *versus* nocturnal and other dim-light ancestral ecologies. Other aspects for future study include the early evolution of vision-related traits in snakes within the context of other senses, especially their greatly expanded olfactory system.

Though less likely to inform the question of snake ancestry, there are many worthwhile research programmes to undertake on eyes and vision in caenophidian snakes. These include testing the hypothesis of the non-homology of snake and lizard double cones, investigating the transmutation of cones to cone-like rods (only the reverse of this has been studied in detail thus far, for a couple of species), and the origin and function of two-tiered retinas. The exceptional disparity and diversity of caenophidian eyes offers the opportunity to learn more about the function and evolvability of vertebrate vision.

Although snake eye and vision research has already contributed much in the way of evidence and ideas to snake-origins research, we believe it still has lots more to offer. The phenotypes and genotypes of only a few extant species have been studied, particularly for those key lineages that will have greatest impact on inferring ancestral-snake traits.

Acknowledgements

We thank Colin McCarthy for sharing an unpublished G.L. Underwood manuscript (*A survey of snake retinas, with systematic implications* – presented as a poster at the 2002 Joint Meeting of Ichthyologists and Herpetologists) that drew attention to Walls' shifting ideas of evolution of the snake eye mentioned in § 15.4.12. Mendis Wickramasinghe, Gabriela Bittencourt and Deepak Veerappan are thanked for providing images of snakes used in Figure 15.5. This chapter was improved by constructive reviews of the submitted manuscript by Ron Douglas and Jeff Streicher, and helpful suggestions from Nihar Bhattacharyya, David Cundall and Jason Head. DJG thanks his coauthors plus Jim Bowmaker, Belinda Chang, Ron Douglas, Nathan Hart, David Hunt, Ellis Loew, Julian Partridge, Leo Peichl and Hans-Joachim Wagner for patiently helping him to learn about vertebrate vision.

References

1. G. Underwood, *A Contribution to the Classification of Snakes* (London: British Museum (Natural History), 1967).
2. G. L. Walls, *The Vertebrate Eye and Its Adaptive Radiation* (Bloomfield Hills, MI: Cranbrook Institute of Science, 1942).
3. A. D'A. Bellairs and G. Underwood, The origin of Snakes. *Biological Reviews*, 26 (1951), 193–237.
4. C. L. Caprette, M. S. Y. Lee, R. Shine, A. Mokany, and J. F. Downhower, The origin of snakes (Serpentes) as seen

through eye anatomy. *Biological Journal of the Linnean Society*, 81 (2004), 469–482.

5. B. F. Simões, F. L. Sampaio, C. Jared, et al., Visual system evolution and the nature of the ancestral snake. *Journal of Evolutionary Biology*, 28 (2015), 1309–1320.

6. R. K. Schott, A. Van Nynatten, D. C. Card, T. A. Castoe, and B. S. W. Change, Shifts in selective pressures on snake phototransduction genes associated with photoreceptor transmutation and dim-light ancestry. *Molecular Biology and Evolution*, 35 (2018), 1376–1389.

7. A. Miralles, J. Marin, D. Markus, et al., Molecular evidence for the paraphyly of Scolecophidia and its evolutionary implications. *Journal of Evolutionary Biology*, 31 (2018), 1782–1793.

8. H. Zaher, F. G. Grazziotin, J. E. Cadle, et al., Molecular phylogeny of advanced snakes (Serpentes, Caenophidia) with an emphasis on South American xenodontines: a revised classification and descriptions of new taxa. *Papéis Avulsos de Zoologia*, 49 (2009), 115–153.

9. E. Hauzman, V. Kalava, D. M. O. Bonci, and D. F. Ventura, Characterization of the melanopsin gene (*Opn4x*) of diurnal and nocturnal snakes. *BMC Evolutionary Biology*, 19 (2019), 174.

10. M. L. Aranda and T. M. Schmidt, Diversity of intrinsically photosensitive retinal ganglion cells: circuits and functions. *Cellular and Molecular Life Sciences*, 78 (2021), 889–907.

11. N. M. Díaz, L. P. Morera, and M. E. Guido, Melanopsin and the non-visual photochemistry in the inner retina of vertebrates. *Photochemistry and Photobiology*, 92 (2016), 29–44.

12. R. K. Schott, J. Müller, C. G. Y. Yang, et al., Evolutionary transformation of rod photoreceptors in the all-cone retina of a diurnal garter snake. *Proceedings of the National Academy of Sciences, USA*, 113 (2016), 356–361.

13. N. Bhattacharyya, B. Darren, R. K. Schott, V. Tropepe, and B. S. W. Chang, Cone-like rhodopsin expressed in the all-cone retina of the colubrid pine snake as a potential adaptation to diurnality. *Journal of Experimental Biology*, 220 (2017), 2418–2425.

14. E. Hauzman, D. M. O. Bonci, E. Y. Suárez-Villota, M. Neitz, and D. F. Ventura, Daily activity patterns influence retinal morphology, signatures of selection, and spectral tuning of opsin genes in colubrid snakes. *BMC Evolutionary Biology*, 17 (2017), 249–263.

15. T. D. Lamb, Evolution of phototransduction, vertebrate photoreceptors and retina. *Progress in Retinal and Eye Research*, 36 (2013), 52–119.

16. K. Palczewski and P. D. Kiser, Shedding new light on the generation of the visual chromophore. *Proceedings of the National Academy of Sciences, USA*, 117 (2020), 19629–19638.

17. E. H. Choi, A. Daruwalla, S. Suh, H. Leinonen, and K. Palczewski, Retinoids in the visual cycle: role of the retinal G protein-coupled receptor. *Journal of Lipid Research*, 62 (2020), 100040.

18. A. Morshedian, J. J. Kaylor, S. Y. Ng, et al., Light-driven regeneration of cone visual pigments through a mechanism involving RGR opsin in Müller glial cells. *Neuron,* 102 (2019), 1172–1183.

19. W. I. L. Davies, S. P. Collin, and D. M. Hunt, Molecular ecology and adaptation of visual photopigments in craniates. *Molecular Ecology*, 21 (2012), 3121–3158.

20. N. J. Gemmell, K. Rutherford, S. Prost, et al., The tuatara genome reveals ancient features of amniote evolution. *Nature*, 584 (2020), 403–409.

21. D. J. Gower, J. F. Fleming, D. Pisani, et al, Eye-transcriptome and genome-wide sequencing for Scolecophidia: implications for inferring the visual system of the ancestral snake. *Genome Biology and Evolution*, 13(2021), evab253.

22. G. L. Walls, The reptilian retina: I. A new concept of visual-cell evolution. *American*

Journal of Ophthalmology, 17 (1934), 892–915.

23. G. L. Walls, Ophthalmological implications for the early history of the Snakes. *Copeia*, 1 (1940), 1–8.
24. G. Underwood, Some suggestions concerning vertebrate visual cells. *Vision Research*, 8 (1968), 483–488.
25. G. Underwood, The eye. In C. Gans, T. S. Parson, eds., *Biology of the Reptilia, Morphology B.* (New York: Academic Press, 1970), pp. 1–97.
26. A. Rochon-Duvigneaud, *Les yeux et la vision des vertébrés* (Paris: Masson, 1943).
27. B. C. Mahendra, Some remarks on the phylogeny of the Ophidia. *Anatomischer Anzeiger*, 86 (1938), 347–356.
28. G. Underwood, On lizards of the family Pygopodidae. A contribution to the morphology and phylogeny of the Squamata. *Journal of Morphology*, 100 (1957), 207–268.
29. M. Wilkinson, D. S. Mauro, E. Sherratt, and D. J. Gower, A nine-family classification of caecilians (Amphibia: Gymnophiona). *Zootaxa*, 64 (2011), 41–64.
30. D. San Mauro, D. J. Gower, H. Müller, et al., Life-history evolution and mitogenomic phylogeny of caecilian amphibians. *Molecular Phylogenetics and Evolution*, 73 (2014), 177–189.
31. D. G. Senn, Uber das optische System im Gehirn squamater Reptilien; eine vergleichendmorphologische Untersuchung, unter besonderer Berucksichtigung einiger Wuhlschlangen. *Acta Anatomica*, 65 (1966), 1–87.
32. W. Stingelin and D. G. Senn, Morphological studies on the brain of Sauropsida. *Annals of the New York Academy of Sciences*, 167 (1969), 156–163.
33. D. G. Senn and R. G. Northcutt, The forebrain and midbrain of some squamates and their bearing on the origin of snakes. *Journal of Morphology*, 140 (1973), 135–151.
34. J. Repérant, J. P. Rio, R. Ward, et al., Comparative analysis of the primary visual system of reptiles. In C. Gans and S. Ulinski, eds., *Biology of the Reptilia*, Vol. 17 (Chicago: University of Chicago Press, 1992), pp. 175–240
35. G. L. Walls, The significance of the reptilian 'spectacle'. *American Journal of Ophthalmology*, 17 (1934), 1045–1047.
36. R. H. Douglas and G. Jeffery, The spectral transmission of ocular media suggests ultraviolet sensitivity is widespread among mammals. *Proceedings of the Royal Society B: Biological Sciences*, 281 (2014), 20132995.
37. N. S. Hart, J. P. Coimbra, S. P. Collin, and G. Westhoff, Photoreceptor types, visual pigments, and topographic specializations in the retinas of hydrophiid sea snakes. *Journal of Comparative Neurology*, 520 (2012), 1246–1261.
38. B. F. Simões, F. L. Sampaio, R. H. Douglas, et al., Visual pigments, ocular filters and the evolution of snake vision. *Molecular Biology and Evolution*, 33 (2016), 2483–2495.
39. B. F. Simões, D. J. Gower, A. R. Rasmussen, et al., Spectral diversification and trans-species allelic polymorphism during the land-to-sea transition in snakes. *Current Biology*, 30 (2020), 2608–2615.
40. R. H. Douglas and N. J. Marshall, A review of vertebrate and invertebrate ocular filters. In S. N. Archer, M. B. A. Djamgoz, E. R. Loew, J. C. Partridge, and S. Vallerga, eds., *Adaptive Mechanisms in the Ecology of Vision* (The Netherlands: Springer, 1999), pp. 95–162.
41. K. Van Doorn and J. G. Sivak, Blood flow dynamics in the snake spectacle. *Journal of Experimental Biology*, 216 (2013), 4190–4195.
42. K. Van Doorn and J. G. Sivak, Spectral transmittance of the spectacle scale of snakes and geckos. *Contributions to Zoology*, 84 (2015), 1–12.
43. M. A. O. Da Silva, S. Heegaard, T. Wang, J. R. Nyengard, and M. F. Bertelsen, The spectacle of the ball python (*Python regius*): a morphological description. *Journal of Morphology*, 275 (2014), 489–496.

44. H. Lauridsen, M. A. O. Da Silva, K. Hansen, et al., Ultrasound imaging of the anterior section of the eye of five different snake species. *BMC Veterinary Research*, 10 (2014), 1-6.

45. M. A. O. Da Silva, S. Heegaard, T. Wang, et al., Morphology of the snake spectacle reflects its evolutionary adaptation and development. *BMC Veterinary Research*, 13 (2017), 1-8.

46. M. A. O. Da Silva, J. T. Gade, C. Damsgaard, et al., Morphology and evolution of the snake cornea. *Journal of Morphology*, 281 (2020), 240-249.

47. J. G. Sivak, The role of the spectacle in the visual optics of the snake eye. *Vision Research*, 17 (1977), 293-298.

48. C. L. Caprette, Conquering the cold shudder: the origin and evolution of snakes eyes. Unpublished PhD thesis, Ohio State University (2005).

49. B. W. Perry, D. C. Card, J. W. McGlothin, et al., Molecular adaptations for sensing and securing prey and insight into amniote genome diversity from the garter snake genome. *Genome Biology and Evolution*, 10 (2018), 2110-2129.

50. G. J. Wistow, *Molecular Biology and Evolution of Crystallins: Gene Recruitment and Multifunctional Proteins in the Eye Lens* (Heidelberg: Springer-Verlag, 1995).

51. B. Röll, Multiple origin of diurnality in geckos: evidence from eye lens crystallins. *Naturwissenschaften*, 88 (2001), 293-296.

52. M. Ott, Visual accommodation in vertebrates: mechanisms, physiological response and stimuli. *Journal of Comparative Physiology A*, 192 (2006), 97-111.

53. C. L. Fontenot, Variation in pupil diameter in North American Gartersnakes (*Thamnophis*) is regulated by immersion in water, not by light intensity. *Vision Research*, 48 (2008), 1663-1669.

54. D. F. Munro, Vertical position of the pupil in the Crotalidae. *Herpetologica*, 5 (1949), 106-108.

55. Y. L. Werner, Extreme adaptability to light, in the round pupil of the snake *Spalerosophis*. *Vision Research*, 10 (1970), 1159-1160.

56. R. H. Douglas, The pupillary light responses of animals: a review of their distribution, dynamics, mechanisms and functions. *Progress in Retinal and Eye Research*, 66 (2018), 17-48.

57. F. Brischoux, L. Pizzatto, and R. Shine, Insights into the adaptive significance of vertical pupil shape in snakes. *Journal of Evolutionary Biology*, 23 (2010), 1878-1885.

58. R. A. Pyron and V. Wallach, Systematics of the blindsnakes (Serpentes: Scolecophidia: Typhlopoidea) based on molecular and morphological evidence. *Zootaxa*, 3829 (2014), 1-81.

59. D. J. Gower, A. Captain, and S. S. Thakur, On the taxonomic status of *Uropeltis bicatenata* (Günther) (Reptilia: Serpentes: Uropeltidae). *Hamadryad*, 33 (2008), 64-82.

60. V. B. Giri, D. J. Gower, A. Das, et al., A new genus and species of natricine snake from northeast India. *Zootaxa*, 4603 (2019), 241-264.

61. G. Underwood, On the visual-cell pattern of a homalopsine snake. *Journal of Anatomy*, 100 (1966), 571-575.

62. G. Underwood, A comprehensive approach to the classification of higher snakes. *Herpetologica*, 23 (1967), 161-168.

63. J. B. Rasmussen, The retina of *Psammodynastes pulverulentus* (Boie, 1827) and *Telescopus fallax* (Fleischmann, 1831) with a discussion of their phylogenetic significance (Colubroidea, Serpentes). *Journal of Zoological Systematics and Evolutionary Research*, 28 (1990), 269-276.

64. J. B. Rasmussen, An intergeneric analysis of some boogie snakes – Bogert's (1940) Group XIII and XIV (Boifinae, Serpentes). *Vidensk Meddelelser fra Dansk Naturhistorisk Foren*, 141 (1979), 97-155.

65. J. B. Rasmussen, A re-evaluation of the systematics of the African rear-fanged snakes of Bogert's Groups XII-XVI, including a discussion of some

evolutionary trends within Caenophidia. In *Proceedings of the International Symposium on African Vertebrates Zoologisches Forschungsinstitut und Museum Alexander Koenig* (Bonn, 1985), pp. 531–548.

66. B. F. Simões, F. L. Sampaio, E. R. Loew, et al., Multiple rod–cone and cone–rod photoreceptor transmutations in snakes: evidence from visual opsin gene expression. *Proceedings of the Royal Society B: Biological Sciences*, 283 (2016), 20152624.

67. R. K. Schott, N. Bhattacharyya, and B. S. W. Chang, Evolutionary signatures of photoreceptor transmutation in geckos reveal potential adaptation and convergence with snakes. *Evolution*, 73 (2019), 1958–1971.

68. A. J. Sillman, J. K. Carver, and E. R. Loew, The photoreceptors and visual pigments in the retina of a boid snake, the ball python (*Python regius*). *Journal of Experimental Biology*, 202 (1999), 1931–1938.

69. A. J. Sillman, J. L. Johnson, and E. R. Loew, Retinal photoreceptors and visual pigments in *Boa constrictor imperator*. *Journal of Experimental Zoology*, 290 (2001), 359–365.

70. D. J. Gower, F. L. Sampaio, L. Peichl, et al., Evolution of the eyes of vipers with and without infrared-sensing pit organs. *Biological Journal of the Linnean Society*, 126 (2019), 796–823.

71. O. Munk and J. B. Rasmussen. Note on the rod-like photoreceptors in the retina of the snake *Telescopus fallax* (Fleischmann, 1831). *Acta Zoologica*, 74 (1993), 9–13.

72. W. H. Miller and A. W. Snyder, The tiered vertebrate retina. *Vision Research*, 17 (1977), 239–255.

73. A. J. Sillman, V. I. Govardovskii, P. Rohlich, J. A. Southard, and E. R. Loew, The photoreceptors and visual pigments of the garter snake (*Thamnophis sirtalis*): a microspectrophotometric, scanning electron microscopic and immunocytochemical study. *Journal of Comparative Physiology A*, 181 (1997), 89–101.

74. E. Hauzman, D. M. O. Bonci, S. R. Grotzner, et al., Comparative study of photoreceptor and retinal ganglion cell topography and spatial resolving power in dipsadidae snakes. *Brain Behavior and Evolution*, 84 (2014), 197–213.

75. J. F. P. Ullmann, B. A. Moore, S. E. Temple, E. Fernández-Juricic, and S. P. Collin, The retinal wholemount technique: a window to understanding the brain and behaviour. *Brain, Behavior and Evolution*, 79 (2012), 26–44.

76. J. Stone, *The Whole Mount Handbook. A Guide to the Preparation and Analysis of Retinal Wholemounts* (Sydney: Maitland Publications, 1981).

77. J. Stone, Parallel Processing in the Visual System (New York: Plenum, 1983).

78. A. Hughes, The topography of vision in mammals of contrasting life style: comparative optics and retinal organisation. In F. Crescitelli, C. A. Dvorak, D. J. Eder, et al., eds., *The Visual System in Vertebrates. Handbook of Sensory Physiology.* Vol. 7 (Heidelberg, Berlin: Springer, 1977), pp. 613–756.

79. L. Schmitz, Evolution of retinal topography in coral reef fishes. *Journal of Morphology*, 280 (2019), S69.

80. R. O. Wong, Morphology and distribution of neurons in the retina of the American garter snake *Thamnophis sirtalis. Journal of Comparative Neurology*, 283 (1989), 587–601.

81. E. Hauzman, D. M. O. Bonci, and D. F. Ventura, Retinal topographic maps: a glimpse into the animals' visual world. In T. Heinbockel, ed., *Sensory Nervous System* (London: IntechOpen, 2018), pp. 101–126.

82. M. Kalberer and C. Pedler, The visual cells of the alligator: an electron microscopic study. *Vision Research*, 3 (1963), 323–329.

83. C. Pedler and R. Tilly, The nature of the gecko visual cell. *Vision Research*, 4 (1964), 499–510.

84. G. Underwood, An overview of venomous snake evolution. In R. S. Thorpe, W. Wüster and A. Malhotra, eds., *Venomous Snake. Ecology, Evolution and Snakebite, The Zoological Society of London* (Oxford: Clarendon Press, 1997), pp. 1–13.

85. V. I. Govardovskii and N. I Chkheidze, Retinal photoreceptors and visual pigments in certain snakes. *Biological Abstracts*, 90 (1989), 1036.

86. W. L. Davies, J. A. Cowing, J. K. Bowmaker, et al., Shedding light on serpent sight: the visual pigments of henophidian snakes. *Journal of Neuroscience*, 29 (2009), 7519–7525.

87. T. Seiko, T. Kishida, M. Toyama, et al., Visual adaptation of opsin genes to the aquatic environment in sea snakes. *BMC Evolutionary Biology*, 20 (2020), 1–13.

88. E. Hauzman, M. E. Pierotti, N. Bhattacharyya, et al., Simultaneous expression of UV and violet SWS1 opsins expands the visual palette in a group of freshwater snakes. *Molecular Biology and Evolution*, 38 (2021), 5225–5240.

89. A. Van Nynatten, G. M. Castiglione, E. A. Gutierrez, N. R. Lovejoy, and B. S. W. Chang, Recreated ancestral opsin associated with marine to freshwater croaker invasion reveals kinetic and spectral adaptation. *Molecular Biology and Evolution*, 38 (2021), 2076–2087.

90. S. Yokoyama, Evolution of dim-light and color vision pigments. *Annual Review of Genomics and Human Genetics*, 9 (2008), 259–282.

91. S. Yokoyama, W. T. Starmer, Y. Takahashi, and T. Tada, Tertiary structure and spectral tuning of UV and violet pigments in vertebrates. *Gene*, 365 (2006), 95–103.

92. D. M. Hunt and L. Peichl, S cones: evolution, retinal distribution, development, and spectral sensitivity. *Vision Neuroscience*, 31 (2014), 115–138.

93. S. Yokoyama, H. Yang, and W. T. Starmer, Molecular basis of spectral tuning in the red- and green-sensitive (M/LWS) pigments in vertebrates. *Genetics*, 179 (2008), 2037–2043.

94. D. M. Hunt and S. P. Collin, The evolution of photoreceptors and visual photopigments in vertebrates. In D. M. Hunt, M. W. Hankins, S. P. Collin, and J. N. Marshall, eds., *Evolution of Visual and Non-visual Pigments* (Boston: Springer, 2014), pp. 163–217.

95. F. E. Hauser, I. Van Hazel and B. S. W. Chang, Spectral tuning in vertebrate short wavelength-sensitive 1 (SWS1) visual pigments: can wavelength sensitivity be inferred from sequence data? *Journal of Experimental Zoology Part B: Molecular and Developmental Evolution*, 322 (2014), 529–539.

96. M. Martin, J. F. Le Galliard, S. Meylan, and E. R. Loew, The importance of ultraviolet and near-infrared sensitivity for visual discrimination in two species of lacertid lizards. *Journal of Experimental Biology*, 218 (2015), 458–465.

97. C. C. Veilleux and M. E. Cummings, Nocturnal light environments and species ecology: implications for nocturnal color vision in forests. *Journal of Experimental Biology*, 215 (2012), 4085–4096.

98. P. A. Hargrave, J. H. McDowell, D. R. Curtis, et al., The structure of bovine rhodopsin. *Biophysics of Structure and Mechanism*, 9 (1983), 235–244.

99. J. Nathans and D. S. Hogness. Isolation, sequence analysis, and intron-exon arrangement of the gene encoding bovine rhodopsin. *Cell*, 34 (1983), 807–814.

100. Y. A. Ovchinnikov, Rhodopsin and bacteriorhodopsin: structure-function relationships. *FEBS Letters*, 148 (1982), 179–191.

101. R. K. Schott, B. Panesar, D. C. Card, et al., Targeted capture of complete coding regions across divergent species. *Genome Biology and Evolution*, 9 (2017), 398–414.

102. C. Katti, M. Stacey-Solis, N. A. Coronel-Rojas, and W. I. L. Davies, Opsin-based photopigments expressed in the retina of a South American pit viper, *Bothrops atrox* (Viperidae). *Visual Neuroscience*, 35 (2018), e027.

103. F. J. Vonk, N. R. Casewell, C. V. Henkel, et al., The king cobra genome reveals dynamic gene evolution and adaptation in the snake venom system. *Proceedings of the*

National Academy of Sciences, USA, 110 (2013), 20651–20656.

104. T. A. Castoe, A. P. J. de Koning, K. T. Hall, et al., The Burmese python genome reveals the molecular basis for extreme adaptation in snakes. *Proceedings of the National Academy of Sciences, USA*, 110 (2013), 20645–20650.

105. W. Yin, Z. Wang, Q-.Y. Li, et al., Evolutionary trajectories of snake genes and genomes revealed by comparative analyses of five-pacer viper. *Nature Communications*, 7 (2016), 1–11.

106. C. A. Emerling, Archelosaurian color vision, parietal eye loss, and the crocodylian nocturnal bottleneck. *Molecular Biology and Evolution*, 34 (2016), 666–676.

107. C. A. Emerling and M. S. Springer, Genomic evidence for rod monochromacy in sloths and armadillos suggests early subterranean history for xenarthra. *Proceedings of the Royal Society B: Biological Sciences*, 282 (2014), 20142192.

108. C. A. Emerling and M. S. Springer, Eyes underground: regression of visual protein networks in subterranean mammals. *Molecular Phylogenetics and Evolution*, 78 (2014), 260–270.

109. M. S. Springer, C. A. Emerling, N. Fugate, et al., Inactivation of cone-specific phototransduction genes in rod monochromatic cetaceans. *Frontiers in Ecology and Evolution*, 4 (2016), 61.

110. C. A. Emerling, Genomic regression of claw keratin, taste receptor and light-associated genes provides insights into biology and evolutionary origins of snakes. *Molecular Phylogenetics and Evolution*, 115 (2017), 40–49.

111. M. Anisimova and C. Kosiol, Investigating protein-coding sequence evolution with probabilistic codon substitution models. *Molecular Biology and Evolution*, 26 (2009), 255–271.

112. Z. Yang, PAML 4: Phylogenetic analysis by maximum likelihood. *Molecular Biology and Evolution*, 24 (2007), 1586–1591.

113. S. L. Kosakovsky Pond, S. D. W. Frost, and S. V. Muse, HyPhy: Hypothesis testing using phylogenies. *Bioinformatics*, 21 (2005), 676–679.

114. J. O. Wertheim, B. Murrell, M. D. Smith, S. L. Kosakovsky Pond, and K. Scheffler, RELAX: Detecting relaxed selection in a phylogenetic framework. *Molecular Biology and Evolution*, 32 (2015), 820–832.

115. J. P. Bielawski and Z. Yang, A maximum likelihood method for detecting functional divergence at individual codon sites, with application to gene family evolution. *Journal of Molecular Evolution*, 59 (2004), 121–132.

116. R. K. Schott, S. P. Refvik, F. E. Hauser, H. López-Fernández, and B. S. W. Chang, Divergent positive selection in rhodopsin from lake and riverine cichlid fishes. *Molecular Biology and Evolution*, 31 (2014), 1149–1165.

117. J. L. Baker, K. A. Dunn, J. Mingrone, et al., Functional divergence of the nuclear receptor NR2C1 as a modulator of pluripotentiality during hominid evolution. *Genetics*, 203 (2016), 905–922.

118. G. H. Jacobs, J. A. Fenwick, M. A. Crognale, and J. F. Deegan, The all-cone retina of the garter snake: spectral mechanisms and photopigment. *Journal of Comparative Physiology A*, 170 (1992), 701–707.

119. W. R. Elliott, R. D. Glickman, H. Rentmeister-Bryant, and H. Zwick, Functional assessment of snake retina using pattern ERG. *Investigative Ophthalmology & Visual Science*, 44 (2003), 2706–2706.

120. R. D. Glickman, W. R. Elliott III, and N. Kumar, Functional and cellular responses to laser injury in the rat snake retina. In *Optical Interactions with Tissue and Cells* XVIII, 643511, International Society for Optics and Photonics (2007).

121. R. A. Baker, T. J. Gawne, M. S. Loop, and S. Pullman, Visual acuity of the midland banded water snake estimated from evoked telencephalic potentials. *Journal of*

Comparative Physiology A, 193 (2007), 865–870.

122. J. N. Lythgoe, *Ecology of Vision* (Oxford: Oxford University Press, 1979).

123. H. Rumpff, Experimental studies on vision in Indian snakes. *Journal of the Bombay Natural History Society*, 76 (1979), 475–480.

124. B. Szabo, D. W. A. Noble, and M. J. Whiting, Learning in non-avian reptiles 40 years on: advances and promising new directions. *Biological Reviews*, 96 (2021), 331–356.

125. T. D. Terrick, R. L. Mumme, and G. M. Burghardt, Aposematic coloration enhances chemosensory recognition of noxious prey in the garter snake *Thamnophis radix*. *Animal Behaviour*, 49 (1995), 857–866.

126. R. Friesen, *Spatial learning of shelter locations and associative learning of a foraging task in the Cottonmouth*, (*Agkistrodon piscivorus*). Unpublished MSc thesis, Missouri State University (2017).

127. F. Aubret, X. Bonnet, D. Pearson, and R. Shine, How can blind tiger snakes (*Notechis scutatus*) forage successfully? *Australian Journal of Zoology*, 53 (2005), 283–288.

128. M. S. Grace and A. Matsushita, Neural correlates of complex behavior: vision and infrared imaging in boas and pythons. In R. Henderson and R. Powell, eds., *Biology of the Boas, Pythons and Related Taxa* (Eagle Mountain: Eagle Mountain Publishing), pp. 271–285.

129. M. S. Y. Lee, A. F. Hugall, R. Lawson, and J. D. Scanlon, Phylogeny of snakes (Serpentes): combining morphological and molecular data in likelihood, Bayesian and parsimony analyses. *Systematics and Biodiversity*, 5 (2007), 371–389.

130. T. P. Wilcox, D. J. Zwickl, T. A. Heath, and D. M. Hillis, Phylogenetic relationships of the dwarf boas and a comparison of Bayesian and bootstrap measures of phylogenetic support. *Molecular Phylogenetics and Evolution*, 25 (2002), 361–371.

131. O. Rieppel, 'Regressed' Macrostomatan Snakes. *Fieldiana Life and Earth Sciences*, 2012 (2012), 99–103.

132. G. L. Walls, Visual purple in snakes. *Science*, 75 (1932), 467–468.

133. N. M. Koch, R. J. Garwood, and L. A. Parry, Fossils improve phylogenetic analyses of morphological characters. *Proceedings of the Royal Society B*, 288 (2021), 20210044.

134. D. Cundall and F. Irish, The snake skull. In C. Gans, A. S. Gaunt and K. Adler, eds., *Biology of the Reptilia*, Vol. 20, *Morphology H* (Ithaca, NY: Society for the Study of Amphibians and Reptiles, 2008), pp. 349–692.

135. J. J. Wiens, C. R. Hutter, D. G. Mulcahy, et al., Resolving the phylogeny of lizards and snakes (Squamata) with extensive sampling of genes and species. *Biology Letters*, 8 (2012), 1043–1046.

136. J. B. Atkins and T. A. Franz-Odendaal, The sclerotic ring of squamates: an evo-devo-eco perspective. *Journal of Anatomy*, 229 (2016), 503–513.

137. V. Franz, Vergleichende anatomie des wirbeltierauges. In L. Bolk, E. Goppert, E. Kallius and W. Lubosch, eds., *Handbuch der vergleichenden anatomie der wirbeltiere* (Urban and Schwarzenberg, Berlin, 1934), pp. 989–1292.

138. A d'A. Bellairs and J. D. Boyd, The lachrymal apparatus in lizards and snakes.-II. The anterior part of the lachrymal duct and its relationship with the palate and with the nasal and vomeronasal organs. *Proceedings of the Zoological Society of London*, 120 (1950), 269–310.

139. A d'A. Bellairs and J. D. Boyd, The lachrymal apparatus in lizards and snakes – I. The brille, the orbital glands, lachrymal canaliculi and origin of the lachrymal duct. *Proceedings of the Zoological Society of London*, 117 (1947), 81–108.

140. P. Uetz, P. Freed, R. Aguilar and J. Hošek, *The Reptile Database*. www.reptile-database.org (accessed March 10, 2021).

PART V

Anatomical and Functional Morphological Perspectives

16

Diversity and Evolution of Squamate Hemipenes

An Overview with Particular Reference to the Origin and Early History of Snakes

Giovanna G. Montingelli, David J. Gower, and Hussam Zaher

16.1 Introduction

The evolution of an intromittent organ capable of transferring sperm into the female reproductive tract is a key component in shifts from external to internal fertilization in many animal lineages [1, 2]. Recent studies indicate a unique origin of the copulatory organ in Amniota, with the ancestral amniote likely having a single midline phallus with an open and endodermally derived sulcus spermaticus [3]. Within Amniota, phallus morphology is especially diverse in Squamata, the diapsid radiation comprising lizards and snakes.

16.1.1 The Squamate Phallus

Squamates have a bilateral pair of copulatory organs termed hemipenes, with each hemipenis being a hollow organ that lies inverted and encapsulated in the base of the tail when not in use [4, 5]. In early stages of development, the external genitalia in mammals, lepidosaurs, turtles, and archosaurs emerge as two genital swellings anteriorly to the cloacal aperture that merge to form a single midline bud [3]. Squamate hemipenes originated from a modification of this ancestral developmental program, in which paired buds mature independently as two phallic structures and the sulcus spermaticus develops from an invagination of ectoderm on each hemipenis instead of the endoderm as in other amniotes [5, 6]. The absence of hemipenes in extant tuatara and lack of relevant fossil data for extinct rhynchocephalians (as far as we are aware) prevents unambiguous inference of the presence or absence of hemipenes in the most-recent common ancestor of extant lepidosaurians [3], so the hemipenis is currently considered a squamate synapomorphy.

Each hemipenis opens to the exterior through an aperture on the lateral margin of the posterior lip of the cloacal opening [4]. Associated with the hemipenis are a series of specialized propulsor and retractor muscles, namely the *m. retractor penis magnus*, *retractor penis parvus*, and *retractor penis basalis* [4, 7]. In species with bilobed hemipenes, the *retractor penis magnus* can be divided for some distance before its insertion [4]. Arnold [7] offered a broader review of squamate hemipenial muscle variation, highlighting its potential for investigating higher-level relationships, and stressing the specialized nature of this musculature in snakes (and amphisbaenians).

During hemipenial eversion, lymph sinuses permeating the sulcate layer, and the blood sinus bordering the central lumen of the organ are filled with lymph and with blood, respectively. Simultaneously, the retractor muscles relax and the propulsors contract [4]. Only a basal portion of the organ is everted prior to insertion into the female cloaca, following which the organ becomes fully everted [8]. The external surface of the hemipenis (internal when inverted) bears the sulcus spermaticus, an open groove in which semen travels [4]. When completely everted, ornaments sometimes present on the external surface help to anchor the organ inside the female cloaca. Completely everted, elaborately ornamented hemipenes typically cannot be withdrawn from the female before being deflated [8]. For a brief consideration of aspects of hemipenial form and function, see Supplementary Appendix 16.S1.

Hemipenial morphology provides important systematic characters, first studied and used for this purpose by Cope [8, 9]. Cope combined observations of hemipenes and lungs to differentiate major groups of snakes, and he explored hemipenial variation within several families of lizards, noting a considerable variety of structures showing some parallelism with snakes. Subsequently, hemipenial characteristics have been used widely for systematics at multiple taxonomic levels (e.g., [4, 10, 11] and references therein).

Cope's observations were of dissections of inverted (retracted) organs. More than 60 years after Cope's studies, standardized techniques for examination and description of everted hemipenes of preserved specimens began to be disseminated, starting with Dowling and Savage's [4] landmark study. Most subsequent studies used everted organs for descriptions and comparisons, though complementary study of inverted and everted conditions has been encouraged [12]. Several techniques to evert, inflate, and stain hemipenes have been proposed ([12–16] and references therein).

Despite a substantial literature, wide gaps in knowledge of squamate hemipenial diversity remain. Although snake and lizard hemipenes have been the focus of much (especially systematic) investigation, and some authors have carried out studies of both groups [17–21], as far as we are aware there have been no considerations of hemipenial diversity and evolution across all squamates (though see, e.g., [22]), particularly to infer states likely to have been present across the lizard–snake transition and at basal divergences within Serpentes. Previous hemipenial studies have typically been conducted at much finer scale, within rather than among major lineages. In this chapter, we provide an overview and summarize hemipenial morphological variation among major lineages of lizards and snakes. Comparisons were based on preparation of hemipenes from museum specimens, and from data compiled from literature. Based on the literature and our observations, we address some key questions, including: (1) are there differences between snake and lizard hemipenes? (2) does hemipenial morphology characterize major groups of lizards and snakes? (3) can we infer the morphology of the ancestral squamate and snake hemipenis? (4) what are the major trends in hemipenial evolution across the lizard–snake transition?

We do not document hemipenial morphology in all squamates but instead provide an overview of what is known for each major lineage. The amount of information available varies substantially depending on the group, and variation in the use of terms further complicates straightforward comparisons and summaries. For some larger extant groups (e.g., Anguiformes, Teiioidea, Endoglyptodonta) there is too much information for us to review comprehensively and we have likely inadvertently overlooked some works. We mapped some major hemipenial characters (Fig. 16.1) onto a recent phylogenetic hypothesis for Squamata [23], aiming to provide an overview of squamate hemipenial morphology. We provide photographs of hemipenes of representatives of major squamate lineages, including gekkotans, scincomorphs, lacertibaenians (Fig. 16.2), teiioids, iguanians, anguiforms (Figs. 16.3 and 16.4), and snakes (Figs. 16.4–16.6). Except where noted (Figs. 16.2–16.6), hemipenes illustrated in this chapter were newly prepared from preserved specimens (see Supplementary Appendix 16.S1) following Zaher and Prudente's [13] methodology. Higher-level classification follows Burbrink et al. [23] and Zaher et al. [24]; the number of currently recognized extant species reported for each group follows Uetz et al. [25]. We refer to all non-snake squamates informally as 'lizards'.

16.2 Hemipenial Terminology

Squamate hemipenial terminology varies substantially. Although snakes benefit from a reasonably stable and unified terminological framework first idealized by Dowling and Savage [4], names used for the structures found in lizards are in a greater state of flux [10, 26–28], and it would be helpful for studies such as this to eventually have greater terminological consistency across squamates.

The vast majority of squamates share a general configuration comprising a hemipenial body and hemipenial lobe(s), with a generally sulcate and asulcate surface and a pair of lateral surfaces, and this can form the basis of a squamate-wide terminology. Because a major morphofunctional subdivision into body and lobular regions can be observed throughout Squamata, we consider it unnecessary here to apply a different terminology for lizards than for snakes (e.g., [10, 27]; see also § 16.2.1). We refer to both outer surfaces lying between the sulcate and asulcate surfaces as lateral (rather than one medial and one lateral) because we prefer a terminology based on the organ rather than its orientation to the body of its owner. The sulcate and asulcate surfaces are as broad or broader than the lateral surfaces. The sulcate surface is that on which the sulcus reaches the lobular region.

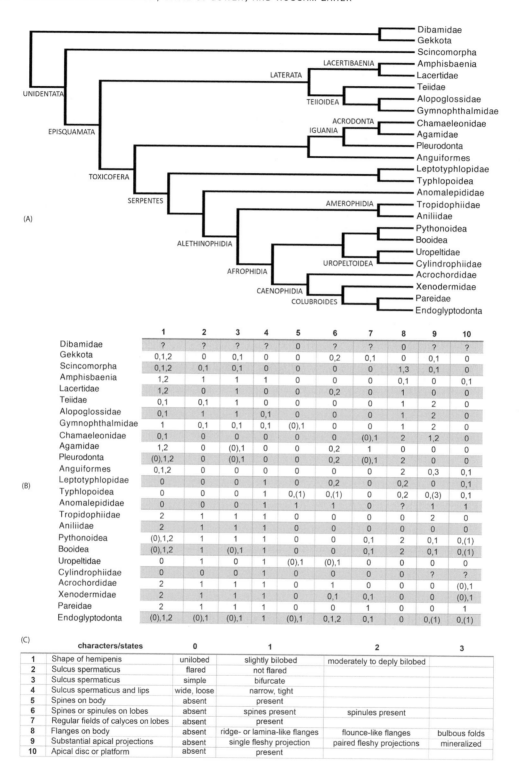

Figure 16.1 Phylogenetic relationships [23] and hemipenial diversity of Squamata. States in parentheses are rare conditions for lineages. See Supplementary Appendix 16.S1 for character optimizations.

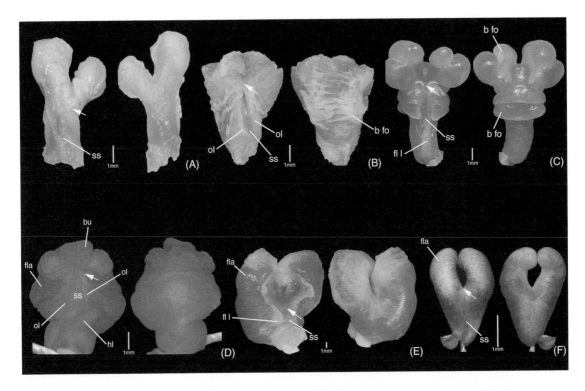

Figure 16.2 Hemipenes of (A) *Delma inornata* (Pygopodidae); (B) *Pygomeles braconnieri* (Scincidae; possibly incompletely everted); (C) *Mabuya* sp. (Scincidae; Reprinted from Sánchez-Martínez et al. with permission © 2019 American Association for Anatomy [36]); (D) *Platysaurus* sp. (Cordylidae); Reprinted from Nunes et al. with permission © 2014 Wiley Periodicals, Inc.; (E) *Gallotia stehlini* (Lacertidae; incompletely inflated); (F) *Amphisbaena brasiliana* (Amphisbaenidae; reproduced with permission [14]). Sulcate (left) and asulcate (right) views. See Supplementary Appendix 16.S1 for colour version, and Box 16.1 for abbreviations.

Box 16.1 Abbreviations used in figures

a ar, apical area; **a aw,** apical awn; **a cp,** apical cup; **a di,** apical disc; **a la,** apical lamellae; **b fo,** bulbous fold; **bu,** bulbous-fold like structure; **ca,** calyces; **fla,** ridge- or lamina-like flange; **fl l,** flapped lip of the sulcus; **flo,** flounce; **hb,** hemibaculum; **hl,** hypertrophied lip of ss; **il,** inner lip of ss; **ol,** outer lip of ss; **pr,** paryphasmata; **ro,** rotula; **se lo,** secondary lobe; **sp,** spine; **ss,** sulcus spermaticus; **wlp,** wing-like process; arrow indicates point at which ss expands or bifurcates.

16.2.1 Hemipenial Lobes

Hemipenial lobes are typically characterized morphofunctionally, being the region distal from the point where the sulcus spermaticus generally divides or expands (see arrows in Figs. 16.1–16.6) and where there is a different ornamentation than on the hemipenial body. Although hemipenial lobes are clearly distinct structures in most snakes, they may not be so well delimited in lizards. Not all indicators of the boundary necessarily occur at exactly the

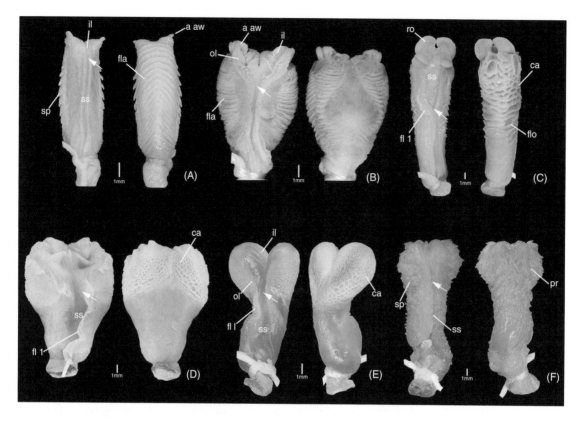

Figure 16.3 Hemipenes of (A) *Heterodactylus imbricatus* (Gymnophthalmidae); (B) *Teius teyou* (Teiidae; courtesy P. Nunes); (C) *Chamaeleo* sp. (Chamaeleonidae); (D) *Uromastyx aegyptia* (Agamidae; incompletely everted); (E) *Agama paragama* (Agamidae); (F) *Anguis fragilis* (Anguidae). Sulcate (left) and asulcate (right) views. See Supplementary Appendix 16.S1 for colour version, and Box 16.1 for abbreviations.

same distance from the base of the hemipenis, but the junction between the hemipenial body and the lobular region is most clear in cases where the hemipenis is bilobed, the lobes are expanded transversely, the sulcus spermaticus is flared or bifurcate (see below), and/or where the lobes are capitate (demarcated from the body by a waist, groove or overhang) and differently ornamented than the body (e.g., Fig. 16.3E; Supplementary Figure 16.S1a). Rarely, the hemipenis is unilobed and subcylindrical (columnar) without substantial transitions in ornamentation along its length, the sulcus is simple, and/or lobes are not capitate – such that the body-lobe boundary is particularly unclear (e.g., Uropeltidae). In most instances a lobular region can be identified even if a fixed, discrete lobe–body boundary cannot be located. The hemipenial body may also possess differentiated ornamentations of its own, unless the hemipenis and its structures are evidently simplified or reduced, in which case the distinction between lobes and body also tends to fade or disappear (e.g., Psammophiidae). The extremities of lobes are sometimes distinctly regionalized morphologically, such as the textured cap-like apices in many amphisbaenids [29, 30], or the flattened to concave, sometimes rimmed apical platforms of varanoids

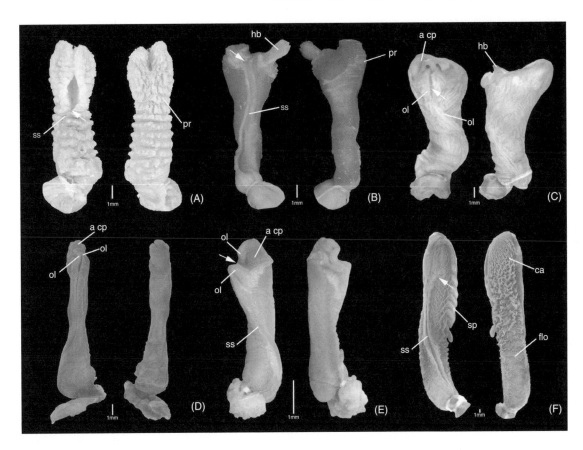

Figure 16.4 Hemipenes of (A) *Pseudopus apodus* (Anguidae); (B) *Varanus prasinus* (Varanidae); (C) *Amerotyphlops minuisquamus* (Typhlopidae); (D) *Amerotyphlops paucisquamus* (Typhlopidae; possibly incompletely inflated); (E) *Amerotyphlops brongersmianus* (Typhlopidae); (F) *Amerotyphlops reticulatus* (Typhlopidae). Sulcate (left) and asulcate (right) views; (C–F) Reprinted from Graboski et al. © 2018 Royal Swedish Academy of Sciences [45]. See Box 16.1 for abbreviations. (A black and white version of this figure will appear in some formats. For the colour version, please refer to the plate section).

(e.g., [31]) or apical discs of some snakes (e.g., [11, 17]). We view apices as distinct, specialized parts of lobular regions that are absent in many squamates. In contrast, lobular regions, whether or not they include distinct apices, are identifiable in almost all squamates, which underlies our preference (see also § 16.2) for a bipartite (body, lobe) rather than tripartite (pedicel, truncus, apex: [27]) subdivision of the hemipenis, though in some taxa (particularly some varanids) the lobular region and a distinct apical region are essentially the same.

The degree of lobular division can be challenging to define. We recognize hemipenes as being unilobed (e.g., Figs. 16.2D, 16.3C, F, 16.4C–F, 16.5A–C, 16.6F), shallowly or slightly bilobed (e.g., Figs. 16.2B, 16.3A, B, D, 16.4B, 16.6A, B, D), or moderately to deeply bilobed (e.g., Figs. 16.2A, C, E, F, 16.3E, 16.4A, 16.5D–F, 16.6C, E).

Some hemipenial morphologies that have been discussed for snakes and given unifying terms have seemingly attracted no or much less attention in lizards. The reverse of this situation might also occur, but our (arguably biased) perception is that terminology applied to lizard hemipenes is generally less unified. A prominent example is the terminology applied to snakes describing lobular capitation. Lobes are semicapitate if their capitula have a basal waist on only one side (when seen in sulcate or asulcate views), and bicapitate if the lobes are clearly waisted on both sides [11]. Capitate lobes in snakes are especially obvious when calyculate [11]. Some lizard hemipenes, even including those without calyces on their lobes, might be considered capitate (e.g., *Callopistes flavopunctatus*: [28: fig. 54]) although not traditionally termed as such in the literature (but see e.g., [32]).

16.2.2 Sulcus Spermaticus

The sulcus spermaticus is a conspicuous homologous structure occurring throughout squamates (though unconfirmed in Dibamidae). It functions to transmit semen and comprises a continuous channel bordered by paired 'lips'. The sulcus runs from the base of the hemipenial body distally to the lobular region, terminating at or towards the tip(s) of the organ. It can be wide or narrow, and shallow or deep and this can vary along its length. Sulcal lips can be very closely or more distantly spaced, vary from low, inconspicuous ridges to more prominent and sometimes frilled flaps that may at least partially cover the sulcus, and be symmetrical or highly asymmetrical (Fig. 16.2E). Whether the hemipenis is uni- or bilobed, the sulcus can be forked (bifurcate) or unforked (simple), and the unforked condition can be with largely parallel sulcal lips throughout, or flared (expanded) with distally diverging lips (Fig. 16.3C, D). In the bifurcate condition, the two distal sulcal branches each have a pair of lips, an inner and outer lip, with the inner lips typically less well developed (e.g., [33]; Fig. 16.3B, E). Bifurcation or flaring occurs at or close to the crotch between the bases of the lobes in bilobed hemipenes. Flared (expanded) sulci are superficially similar to bifurcate sulci (and can be difficult to tell apart in some cases), but as the outer lips diverge they are not paired by opposing inner lips to form a pair of grooves, and the open area bordered by the divergent (outer) lips forms a more or less conspicuous naked (unornamented) region. This region has been termed a distal cup or apical platform in varanids [10, 18], apical lobe in teiids [28], and sulcal pad in pygopodids [34]. Such regions often co-occur with lobular apical ornamentations in the form of folds, awns, and (rarely) mineralized projections (hemibacula) (see § 16.4.3).

We emphasize that we restrict the term 'bifurcate sulcus' to instances where a forked sulcus retains paired (inner and outer) lips along most of its length—some previous articles also refer to some flared or expanded sulci (with divergent outer lips but lacking inner lips) as forked or bifurcate (e.g., [35]). However, each branch of a bifurcate sulcus may also expand (flare) at its distal end, as occurs in many lizards (e.g., *Mabuya altamazonica*: fig. 1d of [36] and a few snakes [37]). Bifurcate sulci on bilobed hemipenes can be centripetal, bifurcating in the crotch with each branch extending along the medial (inner) surfaces of the lobes, or centrolineal or centrifugal, bifurcating slightly or clearly before (proximal to) the crotch and running along the sulcate or lateral surfaces of the lobes, respectively [38]. We apply the same terminology to the outer lips of flared (expanded, but not bifurcate) sulci on bilobed hemipenes. On a unilobed hemipenis, a simple unflared sulcus or a bifurcate sulcus can also be described as centrolineal, if extending distally approximately along the midline on the sulcate surface, or centrifugal if shifting onto either lateral surface as it extends distally. A simple, expanded sulcus on a unilobed hemipenis can also be described using these terms based on the path taken by the flared outer lips.

16.2.3 Ornamentation

Hemipenial ornamentation refers to structural, typically raised features on the external surface of the everted organ (other than lobes and sulcus lips). The long list of terms employed to describe hemipenial ornamentation, paucity of histological investigations of tissue composition, and rarity of thorough homology assessments complicates comparisons across Squamata, especially at greater phylogenetic distances. Hemipenes are mostly ornamented with a variety of soft-tissue features of the lobes and (mostly distal part of) the body. Some of these soft-tissue ornaments may themselves bear smaller ornamentations such as fleshy papillae or mineralized spinules or spicules. Others are somewhat stiffened, in parts, by regions of denser, unmineralized connective tissue. Regions lacking ornamentation are termed naked.

In snakes, approximately transverse, fleshy ridge-, fold- or flap-like structures are universally termed flounces, but in lizards, superficially similar ornamentations on the lobes and body have been given many terms, including plicae [39], frills [18], paryphasmata [10], laminae [28], lamellae [35], creases [40], ridges [30], flounces and welts [9], with preference for terms seemingly tied partly to how prominent and/or substantial these structures are. In both snakes and lizards, flounces (and similar structures) may be more or less papillate and/or be reticulate, with linked longitudinal and transverse flanges bounding concavities that are termed calyces (e.g., Figs. 16.3C–E and 16.6B, C). The distinctive, well-developed transverse ornaments of many anguiforms are corrugated or ribbed flounce-like structures with longitudinal ridges and are typically termed paryphasmata (e.g., [10]), the longitudinal ridges sometimes giving them a superficially calyculate appearance. Where investigated histologically, paryphasmata (Figs. 16.3F and 16.4A, B) are composed of heterogeneously distributed areas of looser and denser connective tissue [41]. See § 16.4.5 for further discussion of terminology and possibly homology of transverse ornamental structures.

Mineralized ornamentation occurs in a few major lineages. Paired mineralized structures adorning the lobular extremities of hemipenes in varanids and some gekkonids have been given various terms, including cornua [26], horns [18], os penes [42], hemipenial bones [43], and hemibacula [10] (Fig. 16.4B). Some of these have been described as bony but without supporting histological data [44]. Where investigated histologically, as far as we are aware, hemibacula are cartilaginous [41]. Macroscopic mineralized, pointed spines projecting from the lobes and/or hemipenial body, are reported only within Gymnophthalmidae and Serpentes (e.g., [11, 14, 45]; Figs. 16.3A and 16.4F). Smaller mineralized spines (spinules) are also documented in snakes and gymnophthalmids, in the latter they always project from serial soft-tissue flanges ([11, 14]; and references therein). Mineralized spinules occur also in some other lizard groups, where they are typically even smaller and mainly confined to the lobes (e.g., pygopodids and a few iguanians: [10, 34]). To the best of our knowledge, it has not been suggested that the microscopic spinose structures in Lacertidae that have been termed spines or spiny epithelium or tubercles (e.g., [10, 33]) are mineralized, and they may be keratinized [30].

Beyond the mineralized lobular projections of varanids and some gekkonids, several major lineages have unmineralized lobular ornaments. These include large fleshy papillae [35] and awns [28] (Fig. 16.3A, B). Chamaeleonids bear particularly distinctive fleshy projections adorning the apical region of their hemipenes termed pedunculi, rotulae, and auriculae [27] (Fig. 16.3C).

Some superficially similar ornamentations are likely not homologous across squamates, such as lobular calyces of pygopodids and snakes or apical discs of xenopeltids and dipsadids. However, it seems reasonable to use a unified terminology for these structures in providing readily comparable descriptions, at least until more detailed investigations of diversity and homology have been conducted. Other probably not universally homologous but similar structures that might benefit from general terms that can be applied across Squamata include spinules and papillae, though note that the latter term has been applied to unmineralized but distinctive projections on lobular tips as well as on the hemipenial body and main parts of lobes. In the taxon accounts below (§ 16.3) we generally report the terms for ornamentation features used in the cited literature. See Discussion (§ 16.4) for further consideration of the meaning and utility of aspects of hemipenial terminology.

16.3 Diversity of the Squamate Hemipenis

Space and time constraints limit the scope of this article to summaries of main hemipenial features on the external surface of the everted organs of single or small numbers of adult (or probably adult) specimens. We largely avoid discussion of hemipenial musculature or armature (*sensu* [39]). There have been few studies of intraspecific geographical, individual and/or ontogenetic variation in squamate hemipenial morphology (e.g., [46–49]). There is evidence of seasonal variation in hemipenial size and/or ornamentation in at least some lacertid, gekkotan, and chamaeleonid lizards (e.g., [47, 50–52]), that indicates that further study of intraspecific variation is warranted.

16.3.1 Dibamidae

Most recent DNA studies find the small, fossorial dibamids (*Dibamus* and *Anelytropsis*: ca 25 extant species) [53], as most closely related to gekkotans, together forming the sister group of all other extant squamates, or as lone sister to all other squamates [23, 54]. Information is scarce on dibamid hemipenial morphology [10, 53, 55], restricted to brief description and illustration for *Dibamus greeri*: 'quite smooth conic formations tapering to the apex with a small hollow near the tip' [55]. It appears to have unilobed, naked hemipenes, but it is unclear if the described organs are fully everted. The presence of a sulcus spermaticus was not mentioned by Darevsky [55] and cannot be confirmed from the published figures.

16.3.2 Gekkota

Gekkotans comprise ca 2,000 extant species in seven families, Carphodactylidae (ca 32 species), Diplodactylidae (ca 157 species), Eublepharidae (ca 43 species), Gekkonidae (ca 1,400 species), Phyllodactylidae (ca 151 species), Pygopodidae (ca 46 species), and Sphaerodactylidae (ca 228 species). Hemipenial morphology is better known for the last four of these families. With exceptions, gekkotans generally have a slightly to deeply bilobed hemipenis, with a flared (simple) or bifurcate sulcus spermaticus (with the outer lips diverging more or less close to the crotch) and distinct ornamentation mainly confined to the lobes [10, 19, 34, 56]. Limbed gekkotans tend to have a unilobed or bilobed hemipenis with a long hemipenial body and short rounded, 'club-shaped' lobes (e.g., *Gekko*: [57]). The lobes can be naked but are typically ornamented with small, densely packed calyces restricted to the lobular region [56, 58–60] but which can extend proximally onto the hemipenial body [52, 57, 61], though the latter is mainly naked. Well-developed hemibacula were described in the sphaerodactylid *Aristelliger* and gekkonid *Uroplatus lineatus* [60]. Gekkonids tend to have a broadly flared sulcus on moderately bilobed hemipenes [57, 61].

Members of the near-limbless Pygopodidae generally have simplified hemipenes compared to other gekkotans [10], interpreted as resulting from an 'ancestral miniaturization event' [34]. However, the characteristic club-shaped condition with a voluminous lobular region of other gekkotans occurs also in many pygopodids [34]. Most pygopodids have asymmetrically bilobed hemipenes (Fig. 16.2A), unilobed organs occurring only in species of *Aprasia* and *Delma*. A slightly inflated lobular crotch between the main lobes is reported for *Pygopus nigriceps* (the 'tri-lobed' condition of [34]). Hemipenes can be naked (Fig. 16.2A) or bearing spinules and calyces, that are mainly distributed on the lobes and distal part of the body. Calyces can be diminutive or large and deep (e.g., *Pygopus lepidonotus*: Supplementary Figure 16.S1a). The sulcus spermaticus is described as bifurcate for most pygopodids [10, 34], though most species seem to have a flared (non-bifurcate) sulcus with a sometimes very wide naked area ('sulcal pad' of [34]) bordered by the sulcal outer lips (e.g., Supplementary Figure 16.S1a). When the sulcus is bifurcate, the branches are broad (expanded) and shallow (e.g., Fig. 16.2A).

16.3.3 Scincomorpha

Scincomorpha is an ecomorphologically diverse and speciose lineage comprising four families: Scincidae (ca 1,700 species), Cordylidae (ca 70 species), Gerrhosauridae (ca 40 species), and Xantusiidae (ca 35 species). Little hemipenial information is available for Scincomorpha, and most of it is for Scincidae [10, 19, 36, 62, 63], where it is far from representative of this highly diverse lineage.

Hemipenes of the few scincids described – for the sphenomorphine, eugongyline, scincine, mabuyine, and lygosomine radiations [36, 62–65] (Fig. 16.2C) – are typically bilobed, but can be unilobed as in the lygosomine *Riopa guentheri* [64] and the Australian '*Egernia* group' characterized by a 'columnar hemipenis' [65]. Bilobed scincid hemipenes appear to possess a sulcus that bifurcates centripetally on a pair of lobes that are either short and bulbous or long and slender, and unilobed hemipenes possess a simple sulcus spermaticus that expands within the lobe to form a naked area delimited by the outer lips [36, 65]. Although scincid hemipenes are basically smooth, they exhibit different types of conspicuous, substantially inflated (bulbous) protuberances that are mainly transversely oriented (e.g., [36, 66]) on the asulcate and lateral surfaces of the organ, but can also be longitudinally oriented (e.g., [9]) on the lateral surfaces of the body as in *Pinoyscincus* [62] – these have been termed 'folds' (e.g., [36]). The bulbous nature of these structures is observable only in well-everted and expanded hemipenes [36]. Although seemingly thin walled and soft, some of these bulbous folds bear streaks of what might be denser connective tissue (e.g., 'ridges' labeled in fig. 1 of [36]). Longitudinal folds on the asulcate surface of the lobes and distal half of the hemipenial body occur in representatives of the limbless Australian and Malagasy scincids *Tropidoscincus* and *Pygomeles* [67]. The hemipenis of *Pygomeles* is here illustrated for the first time, for a fully everted but incompletely expanded example (Fig. 16.2B): it is slightly bilobed with a proximally naked body and an expanded lobular region. On the sulcate surface, the sulcus is broad and delimited by lips that tend to broaden at the base of the lobular region where the sulcus flares (without inner lips) to form a naked area surrounded by expanded lobular tissues. A series of incompletely expanded transverse folds is visible on the asulcate surface, on the lobes and distal part of the body.

Little information is available for cordylids and gerrhosaurids [9, 10, 68, 69]. Based on the literature, cordylids have bulbous, seemingly unilobed (to possibly slightly bilobed) hemipenes [10]. The simple sulcus is oriented obliquely at its proximal end where one lip is hypertrophied, with the distal end of this hypertrophied section terminating lateral to the sulcus as a projection on the sulcate surface of the body (Fig. 16.2D). The sulcus is wide and shallow and fades out some distance from the distal end of the organ. The distal end is heavily creased and seemingly thin walled. The organ is devoid of spines or spinules. The body and lobes may bear approximately transverse irregular folds or ridges on especially the asulcate and lateral surfaces. Böhme [10] was unable to confirm Cope's [9] description of calyces in any cordylid. A hemipenis of *Platysaurus* sp., is here illustrated for the first time (Fig. 16.2D), its morphology largely consistent with Böhme's [10] description. The simple sulcus runs longitudinally on the main part of the body but is notably deflected obliquely at its proximal end. In the longitudinal part, the sulcus is very broad and shallow with low ridge-like lips, but in the more oblique section the more proximal lip is greatly hypertrophied into a fleshy flap with a free, somewhat lobular distal end. The distal end of the organ bears three bulbous and naked lobe-like structures conferring the appearance of a somewhat bilobed organ with an inflated bulbous fold in the crotch. The distal end of the sulcus fades out at the base of the midline bulb. Proximal to the trilobate distal end of the organ, a series of irregular, loose and soft, approximately transverse fold-like flanges run circumferentially around the hemipenis (except across the sulcus), being less prominent on the asulcate surface. A stiff plate of connective tissue was described underlying the floor of the sulcus of cordylids [10] that is not apparent in this specimen.

Gerrhosaurids have very similar hemipenes to cordylids [10, 68]. Descriptions are available for *Gerrhosaurus*, *Zonosaurus*, and *Tracheloptychus* but, as far as we know, illustrations only for *G. flavigularis* and *Z. quadrilineatus* [10, 21]. The organ is less bulbous than those described for cordylids, slightly bilobed in *G. flavigularis*. The asulcate surface has a distinctive pocket overhung by the base of an apical area, which might be a synapomorphy of the family. Together, cordylids and gerrhosaurids appear to be diagnosed by the synapomorphic condition of a highly distinctive broad, shallow sulcus spermaticus that fades out distally and which has a proximally oblique, hypertrophied lip with a free distal projection.

It is difficult to reconcile Cope's [9] description of the hemipenis of an unnamed xantusiid with Böhme's [10] description of *Xantusia henshawi*. Hemipenial morphology in this family is even less well known than in cordylids and gerrhosaurids.

16.3.4 Teiioidea

This large lineage of New World lizards is characterized by a diverse, but highly characteristic, hemipenial morphology, from weakly ornamented to very elaborate organs [10, 14, 35, 70]. Among its major clades, gymnophthalmids (ca 270 species) are the only teiioids bearing prominent mineralized spines on the hemipenial body and lobes ([14]; Fig. 16.3A). Spines in gymnophthalmids are embedded as very regularly and densely disposed series of single rows in the distal edges of serial (and also densely and regularly spaced), approximately transverse soft-tissue structures on the body and lobes described by terms such as plicae, lamellae, and flounces [14, 35]. Even in cases where spines appear at first sight to be isolated (e.g., *Echinosaura*: [14]; *Psilops*: [71]), they are nonetheless regularly arranged and lie within vestigial soft-tissue flanges. Gymnophthalmids have unilobed to slightly bilobed hemipenes, often emphasized by a pair of fleshy apical awns or large papillae on their apical extremities or lobes (when differentiated) that may carry the distal extremities of the branches of the sulcus spermaticus. The outer lips of the sulcus spermaticus typically diverge a short distance from the distal tip of the organ, with a short median prominence often forming the inner lips of the sulcus, which is thus bifurcate rather than simply flared [14, 35] (Fig. 16.3A). The hemipenial crotch of *Arthrosaura montigena* bears the opening ('orificium') of a channel of unknown function and seemingly unknown elsewhere in squamates [72].

Hemipenial morphology within Teiidae (ca 170 species) has been documented for several species [9, 10, 21, 28, 70, 73]. The hemipenis is unilobed to slightly bilobed, with a centrolineal sulcus spermaticus, with the outer lips of the sulcus diverging a short distance from the crotch and with a bifurcate sulcus in most or perhaps all cases (e.g., [28:fig. 54]; Fig. 16.3B). Distally, each lobe typically possesses a more or less distinct pair of eversible, prominent apical awns ([28]; fleshy papillae of [9]) between which lies the distal extremity of the sulcus spermaticus branch (Fig. 16.3B). Awns can be very elaborate (*Alexandresaurus*: [74]) or rarely highly reduced; where reduced there may be fields of small papillae (*Kentropyx*: [28]; Supplementary Figure 16.S1c). The lateral sides and the asulcate surface bear subparallel, densely spaced, approximately transverse, low and rigid flanges (laminae of [28]; imbricate, transverse laminae of [9]). The flanges are generally unornamented macroscopically (bearing papillae in some *Cnemidophorus*: [28]), though microscopic mineralized spinules occur in at least some taxa [10, 28]. Naked areas with longitudinal folds that largely flatten out in the everted condition ('expansion pleats' of [28]) have been described in the middle of the sulcate and asulcate surfaces [28]. Hemipenes in Alopoglossidae (29 species), the sister taxon to Gymnophthalmidae, are uni- to slightly bilobed and resemble those of Teiidae in lacking macroscopic mineralized spines [14, 75, 76]. The sulcus is bifurcate (contra [75]) with each branch extending between a pair of fleshy apical awns. The inner pair of awns ('lobes' of [75]) are hypertrophied and fused (or nearly so) along the midline (the midline crease between them is misidentified as the distal end of an unbifurcate sulcus by Ribeiro-Júnior et al. [75]). Although hemipenes of all three major teiioid lineages are each distinct, they all share a bifurcate sulcus that extends to between a pair of apical awns.

16.3.5 Lacertibaenia

Lacertibaenia is a mainly molecularly supported clade comprising Lacertidae (ca 350 species) and the fossorial Amphisbaenia (ca 200 species). Much research on lacertid hemipenes has focused on musculature, armature, and microornamentation. Images and detailed descriptions of the external features of fully everted and inflated lacertid hemipenes are surprisingly rare, limiting comparisons within Lacertidae and between lacertids and other squamates. Lacertids generally are reported to have slightly to moderately bilobed hemipenes that may be complexly folded while inverted [39, 77]. Lobes are covered radially by 'plicae' (stiffened flanges) bearing microornamentation in the form of spinulate tubercles of varied shapes forming a spiky or spiny, keratinized epithelium [33, 47, 51, 78]. The tips of the lobes in *Gallotia* are also covered by a series of large fleshy papillae [33] (the hemipenis in Fig. 16.2E is only partially everted distally, and the papillae remain inverted inside the organ). Plicae may extend in coverage to the distal region of the hemipenial body ([77]; Fig. 16.2E), though it is difficult to precisely define their extent based on the available descriptions from dissected inverted organs [39, 51]. The rest of the hemipenial body is naked [51]. The sulcus is usually described as bifurcate, being generally bordered by well-developed, flapped lips [33: fig. 5]. Inner and outer lip prominence and form varies substantially, with inner lips being sometimes poorly defined and barely visible (e.g., Fig. 16.2E). Flappy lips on the lobular region are characteristic of lacertids, though in *Takydromus* the branches of the sulcus are absent or highly reduced with small lips [51]. Reduction of the size of the lobes and their ornamentation outside breeding seasons has been reported for some lacertids [39, 50, 51]. Some groups, such as *Zootoca* (Lacertini) and Eremiadini, bear an internal, cartilaginous 'armature' – a plate-like structure embedded in the dorsal surface of the *retractor magnus* muscle, that also becomes filled, providing additional support for the everted organ [50, 51].

The vast majority of published information on amphisbaenian hemipenes is for Amphisbaenidae (183 species). Hemipenial morphology of amphisbaenids is generally relatively simple compared to other lacertibaenians and other squamates [14, 29, 30]. The hemipenis is deeply bilobed, with the lobes sometimes disposed perpendicular to the animal's body. The sulcus is typically bifurcate, extending to (or close to) the tips of the lobes. In that respect it is superficially snake like, though the lips are generally much less well defined. The sulcus is centripetal in amphisbaenids (Fig. 16.2F). The hemipenis can be free of ornaments (e.g., *Amphisbaena carli*: [79]), or bear regularly spaced, low transverse ridges (with a 'pleated, accordion-like appearance' [80]) on at least the lateral surfaces of the body and on the lobes, as in species of *Amphisbaena, Chirindia, Leposternon,* and *Monopeltis* [9, 10, 30]. Mineralized ornamentation is not reported for any amphisbaenian. In most amphisbaenids the tips of the lobes bear a discrete apical complex of tightly packed (apparently keratinized: [30]) lamellae or plates, as in some species of *Amphisbaena, Leposternon,* and *Monopeltis* [30, 80] (Fig. 16.2F). The hemipenis in Trogonophiidae (six species), at least in *Trogonophis* [29], is deeply bilobed with densely packed, low lamellar ridges ornamenting the asulcate and lateral surfaces of the lobes – these are extremely similar to the keratinized apical disc lamellae of amphisbaenids other than not being associated with the branches of the sulcus. In contrast, the very scant information available for Bipedidae (three species) and Blanidae (seven species) suggests a hemipenis that might not be bilobed (at least not deeply), and that might lack transverse body or lobe ornamentation, and have a simple, flared (rather than bifurcate) sulcus (*Bipes* and *Blanus*: [29]). We are unaware of descriptions or images of the hemipenes of Cadeidae (two species) or of Rhineuridae (one species). Thus, plesiomorphic states of hemipenial characters for Amphisbaenia are unclear.

16.3.6 Toxicofera

Toxicofera unites snakes with anguiforms and iguanians, a clade with strong support in recent molecular phylogenetic analyses (see Chapter 10). Here we summarize hemipenial morphology within the three major toxicoferan lineages, Iguania (ca 1,970 species), Anguiformes (ca 240 species) and Serpentes (ca 3,870 species).

Iguania

Iguania comprises two major clades, Acrodonta and Pleurodonta. Acrodonta comprises Chamaeleonidae (217 species) and Agamidae (534), while Pleurodonta (ca 1,000 species) comprises a diverse group of 11 New World, Fijian, and Malagasy iguanian families (e.g., [23]). Chamaeleonid hemipenes have been relatively well studied [10, 27, 81–86]. They are unilobed or slightly bilobed and conical, though more strongly expanded and slightly clavate in at least some brookesiines. The body and base of the lobes in at least most chamaeleonines is ornamented by calyces (or combinations of calyces and flounce-like ornamentations) varying in size and shape [27] (Fig. 16.3C). The lobes (except for the tips) and body of the hemipenes of brookesiines are typically naked [27, 82, 83]. The chamaeleonid sulcus spermaticus is simple and mostly greatly expanded, with outer lips sometimes asymmetrically developed and/or flap like [10]. The expanded outer lips delimit a differentiated apical region that is naked except for various unique, generally bilateral projections (Fig. 16.3C) termed rotulae, papillary fields, crests, horns, pedunculi, auriculae, and papillae in elongate pairs or tufts [10, 27, 81]. Agamids generally have stout, clavate, slightly to moderately bilobed hemipenes (Fig. 16.3D, E; Supplementary Figure 16.S1b). The lobes bear calyculate ornamentation, though this may grade into papillae and/or 'flounces' towards the bases of lobes [87, 88]. Some *Calotes* bear a median fleshy projection at the asulcate base of the lobular region [88]. The hemipenial body, often short, typically lacks differentiated ornamentation. At least some species of *Draco* have notably spinose ridges [10]. The sulcus is often bifurcate [10, 87–89] but some appear to be simple, flaring centrolineally onto the lobes with a large, well-defined naked (or finely calyculate [10]) region formed between the outer sulcal lips (Fig. 16.3D). In many species, the sulcal lips are strongly asymmetrical, with one a prominent flap that may overlie the sulcus (e.g., *Sitana gokakensis*: [87]; Fig. 16.3D, E). Some species lack discernable, raised lips (e.g., *Lyriocephalus scutatus, Ceratophora stoddarti*: [10]). The calyculate distal end of the organ has been described as having four segments or bulges [10, 88] because the two lobes are often subdivided along their lengths by the outer margins of the substantially flared, centripetal, or centrolineal sulcus spermaticus.

Information on pleurodont hemipenes is very uneven for this speciose clade, with relatively little data (and very few images) available for some large groups. Pleurodont hemipenes are mostly slightly to moderately bilobed, with deeply bilobed organs (probably apomorphic [10]) restricted to a few families that neither subtend the pleurodont tree basally nor which are closely related (Tropiduridae, Polychrotidae, Leiosauridae [10], some Dactyloidae [90]). Unilobed hemipenes appear to be rare within Pleurodonta (e.g., *Anolis gaigei*: [91]). The sulcus spermaticus can be simple and centrolineally flared (e.g., *Iguana*: [10]; *Oplurus*: [92]), though is typically bifurcate and centripetal in members of the families having moderately to deeply bilobed hemipenes (e.g., the leiosaurids *Diplolaemus* and *Pristidactylus*: [10]; tropidurids: [48, 93]). The sulcus bifurcates at the base of the lobular region before flaring substantially on each lobe in many dactyloids [90, 91]. Sulcal lips vary in prominence and can be strongly asymmetric (e.g., Iguanidae: [10]), those in *Brachylophus* form a tube-like sulcus [40]. The sulcus typically reaches the tips of the lobes or close to them [10, 48, 90, 94]. Pleurodont hemipenes are rarely unornamented (e.g., *Anolis gaigei*: [91]), most commonly the lobes and (mostly the distal part of the) body bear calyces (perhaps absent only in Polychrotidae: [10]), sometimes grading proximally into transverse lamellae or flounces [9, 10, 40, 48, 94]. Calyces are sometimes restricted to tips of lobes (e.g., *Crotaphytus*: [9]) or occur only on the asulcate surface and/or more proximally (e.g., [94]). In *Polychrus marmoratus*, the lobes bear spinose papillae [10]. Lobes of some iguanids (e.g., *Iguana iguana*) have stiffened ridges (welts of [9]; Stützsäume of [10]). Mineralized ornaments have not been reported. Lobes in some taxa bear fleshy apical projections that seem to be expanded from the distal tips of the sulcal lips (e.g., [94]), reminiscent of gymnophthalmid hemipenes. Species in multiple families have a median projection in the crotch [10]. A slightly to moderately bilobed hemipenis with a flared sulcus (whether simple or bifurcate) and largely calyculate lobes is similar to that of agamids, and thus some or all of these features might be plesiomorphic for Iguania.

Anguiformes

The basal divergence within Anguiformes is between Neoanguimorpha (Helodermatidae, Xenosauridae, Diploglossidae, Anniellidae and Anguidae: 5, 12, 51, 6, and 80 extant species, respectively) and Paleoanguimorpha (Shinisauridae, Lanthanotidae and Varanidae: 1, 1, and 83 species) [23]. A major feature of anguiform hemipenes is the presence of paryphasmata [10] (Figs. 16.3F and 16.4A, B) – transverse, flounce-like ornaments on the body and lobes that are distinctly ribbed with longitudinal ridges of denser connective tissue [41]. Many anguiforms have an apical projection at the extremity of each lobe, but only in

Varanidae are these mineralized to form hemibacula [31, 81] (Fig. 16.4B). Most anguiform hemipenes are slightly to deeply bilobed (with the slightly bilobed forms accentuated by substantial and often rigid apical projections), with a simple, flared sulcus spermaticus [10, 18, 31]. Helodermatids depart from this in having unilobed organs (and undivided retractor muscles) lacking apical projections [18, 31]. Additionally, helodermatids have a wide sulcus, and paryphasmata that are asymmetrically disposed on both the proximal and more distal parts of the body [31]. Xenosaurids lack paryphasmata, but have denticulate sulcal lips and a denticulate apical process [10, 31]. Within Diploglossidae, *Diploglossus* spp. and *Ophiodes* spp. have a moderately to deeply bilobed hemipenis, with the lobes (and distal body in *Diploglossus*) ornamented by paryphasmata ('pleated flounces' of [14, 95]). The longitudinal ridges on the flounces of the paryphasmata of diploglossids are often prominent, with paryphasmata superficially appearing calyculate [14, 95]. At least *D. nigropunctatus* possesses a 'solid awn' [95] projecting from each apex – though it is unknown if these are mineralized (= hemibacula). The centrolineal sulcus spermaticus of diploglossids is not bifurcate, it expands below the lobular division where each outer lip runs along the respective lobe, the inner surfaces of which are naked [95]. Anguids have generally similar hemipenes, though the free edges of the paryphasmata are spinose in some taxa (*Anguis* and some *Ophisaurus*: [81]; Figs. 16.3F and 16.4A), but without evidence of mineralization [14] and perhaps likely to be extensions of the denser ridges of connective tissue sequentially arranged in paryphasmata. Hemipenes of the limbless Anniellidae have been described only very briefly, as undivided organs ornamented with transverse and wrinkled folds or flounces [9: p. 466], lacking mineralized structures [14].

Within Paleoanguimorpha, *Shinisaurus* has papillate paryphasmata and fleshy apical horns. *Lanthanotus* (Supplementary Figure 16.S1d) and *Varanus* (Fig. 16.4B) have very slightly bilobed (possibly unilobed in some *Varanus*) hemipenes with often elongate bodies, a simple sulcus, and naked tips of the lobes. However, there are several differences between species of the two genera. Apical projections are mineralized only in *Varanus* – earlier suggestions of mineralized elements in *Lanthanotus* [10, 18] having been subsequently questioned or rejected [31, 96]. Additionally, *Varanus* hemipenes have more or less asymmetrical lobes, ornamentation, and path of the sulcus, whereas *Lanthanotus* hemipenes are symmetrical. The naked tips of the lobes in many *Varanus* form a more or less well-defined apical platform, the border of which is continuous with the flared sulcal lips. Among *Varanus* there is variation in features such as the form of the hemibacular elements (shaft and cusps, degree of asymmetry, and number of mineralized elements), the form of the apical platform, prominence and symmetry of the sulcal lips, and the disposition and form of paryphasmata (e.g., [10, 18, 31, 96]). *Varanus indicus* and *V. prasinus* and close relatives have particularly strongly asymmetrical hemipenes, including a sulcus that extends to the tip of only that lobe bearing paryphasmata and a smaller hemibaculum, instead of flaring at the base of the lobular region and/or terminating at an apical disc [31, 97].

Serpentes

Relationships among major lineages of extant snakes have long been debated on the basis of morphological and molecular data (see Chapter 10). The traditional basal division into

Scolecophidia and Alethinophidia has come under substantial pressure, with most recent molecular analyses supporting scolecophidian paraphyly. Although alethinophidian monophyly is not controversial, Macrostomata ('wide-gaped' alethinophidians; see Chapters 9 and 19) is considered para- or polyphyletic, with the non-macrostomatan Aniliidae and Uropeltoidea being the sister groups of Tropidophiidae and either Pythonoidea or all other afrophidians, respectively [23, 98, 99].

'Scolecophidia' – Here we treat this putatively paraphyletic group of mostly small and highly fossorial snakes as three main lineages: Typhlopoidea, Leptotyphlopidae, and Anomalepididae. In recent molecular phylogenies, the former two are sister taxa, together sister of all other extant snakes, with Anomalepididae the sister of Alethinophidia. Until recently, only a limited number of scolecophidian hemipenial descriptions were available [17, 100–106], more becoming available in the last decade [45, 107–114] though information remains scattered. Many published descriptions are of incompletely everted and/or inflated organs (e.g., [17, 102, 110]), an expected outcome due to small absolute size, acknowledged by most authors. Although the morphology of partially everted and inflated organs should be interpreted carefully, the available information is generally informative ([11, 13]; this chapter). In general, scolecophidian hemipenes are unilobed with a simple sulcus spermaticus. The sulcus typically appears to flare distally before reaching the lobular tip (in those taxa without filiform hemipenes) which is a lizard-like condition not seen in the vast majority of other extant snakes (Figs. 16.4C–F and 16.5A, B) except for some derived and obviously convergent cases (e.g., *Phyllorhynchus*: [37]). The hemipenial body tends to be naked or ornamented with large and well-spaced flounces. Calyces, papillae and spines occur in a few taxa.

For Leptotyphlopidae (ca 140 species) hemipenes have been described mainly for species of the Neotropical Epictinae (e.g., [102, 106, 108–111, 113], and a few species of the African and Arabian *Leptotyphlops* and *Myriopholis* [17, 107]. Leptotyphlopid hemipenes are unilobed with a simple sulcus (Fig. 16.5B). The organ can be somewhat tubular throughout or have a bulbous basal region from which extends a long and thin distal needle-like portion with no observable lobular region or apical specializations [106, 108, 111], or finger-like with a sulcus flaring distally on a slightly expanded tip sometimes ornamented by a fleshy papilla ('awn' of [17]). When the distal region is not needle- or finger-like, the sulcus can flare to drain on fleshy walls interconnected by a series of septa well below the tip (*Trilepida macrolepis*; Fig. 16.5B) or flare at an apical disc or a central mound ornamented with papillae, without reaching the distal tip of the organ. The body is naked (e.g., *Epictia munoai*, *Rena segrega*: [109]) or ornamented with fleshy transverse flanges, papillae, and/or sometimes spinose calyces (*Trilepida brasiliensis*, *Epictia* spp.: [109, 111]). It is unclear whether microscopic spines or spinules described for some leptotyphlopids (e.g., [103, 111] are mineralized. The apical disc or central mound can be surrounded by feebly developed fleshy flounces, papillae or calyces.

Typhlopoidea comprises Gerrhopilidae (21 species), Xenotyphlopidae (one species) and Typhlopidae (ca 420 species). As far as we are aware, there is no published information on hemipenial morphology in the former two families. Typhlopid hemipenes are known mainly from species of Typhlopinae [45, 115–117], Afrotyphlopinae [17, 115] and Asiatyphlopinae [100, 104, 105, 112, 115, 118]. Typhlopid hemipenes are of two main types

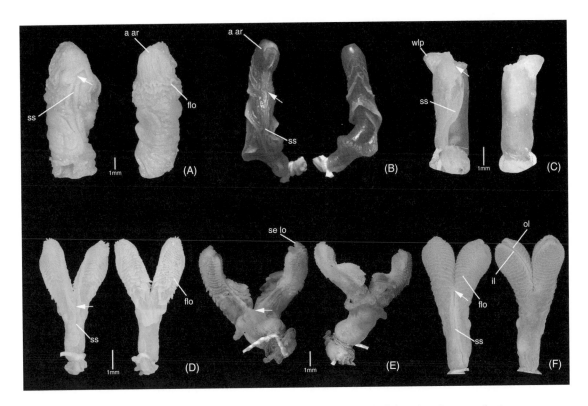

Figure 16.5 Hemipenes of (A) *Afrotyphlops lineolatus* (Typhlopidae); (B) *Trilepida macrolepis* (Leptotyphlopidae: Reprinted from Martins et al. with permission from © 2019 Blackwell Verlag GmbH [109]; (C) *Cylindrophis melanotus* (Cylindrophiidae; possibly incompletely everted); (D) *Trachyboa boulengeri* (Tropidophiidae); (E) *Tropidophis paucisquamis* (Tropidophiidae: Reprinted from Curcio et al. with permission from © 2012 by The Herpetologists' League, Inc. from fig. 12 of [120]); (F) *Anilius scytale* (Aniliidae; courtesy A. Prudente). Sulcate (left) and asulcate (right) views. See Supplementary Appendix 16.S1 for colour version, and Box 16.1 for abbreviations.

(e.g., [100, 115]), 'eversible', as in other squamates, and 'protrusible', characteristic of Oceanian asiatyphlopines. Protrusible hemipenes comprise a permanently everted, stiffened awn that is sheathed helically within an eversible, fleshy hemipenial body [104]. The rigid awn is formed by stiffened connective tissue capable of bending at its extremity when retracted [104, 105]. Protrusible hemipenes are generally very long and slender (<1.0 mm diameter), with awns that taper gradually to pointed or slightly expanded tips [112]. The awn is shallowly excavated externally by the sulcus spermaticus. Protrusible hemipenes are invariably present in species of *Ramphotyphlops*, *Acutotyphlops*, and *Anilios*, and variably present in *Indotyphlops*, possibly representing a synapomorphy of that larger group (being secondarily lost within *Indotyphlops*).

Afrotyphlopines and typhlopines retain the typical eversible organ of other squamates [115]. The hemipenes of the afrotyphlopines *Afrotyphlops* and *Rhinotyphlops* are known from only a few partially everted and inflated organs that are likely unilobed with a simple sulcus, a mostly naked body or with two large fleshy basal papillae, and low, loosely spaced calycal flanges on the lobes [17]. A flap-like expanded part of the sulcal lip may occur in *R. lalandei* [17], but confirmation requires examination of fully everted and inflated organs. The fully everted and partially expanded organ of *Afrotyphlops lineolatus* available to us (Fig. 16.5A) is unilobed with a simple sulcus terminating proximal to the tip, flaring into a domed naked apical area that is longitudinally wrinkled and bordered proximally by three or four flounces; the hemipenial body is naked. Typhlopine hemipenes occur in four main shapes: (1) distally expanded (conical or trumpet shaped), (2) uniformly cylindrical, (3) attenuate and tapering distally, (4) oblique [45, 117]. Oblique hemipenes are reported only in Antillean species [117]. Typhlopines have a similar overall hemipenial morphology, with a unilobed hemipenis and simple sulcus that ends proximal to the apex (*Amerotyphlops reticulatus*; Figs. 16.4F) or drains into a terminal depression – the apical disc or distal cap (Fig. 16.4C–E). The distal end of the sulcus of *A. reticulatus* is unusual, curving on the lobular region and slightly flaring into a region of densely packed irregular flanges and calyces. The typhlopine hemipenial body is usually naked or bears irregular and weakly defined flounces. Within typhlopines, only *A. minuisquamus* (Fig. 16.4C) and *A. reticulatus* (Fig. 16.4F) are known to exhibit relatively large, mineralized spines, the former bearing a pair of spines projecting distally from the apical disc and the latter a row of large curved spines arranged longitudinally on the distal half of the organ [45, 119]. The spines of *A. reticulatus* are similar to the hemipenial body spines of alethinophidians, but the pair of mineralized apical spines of *A. minuisquamus* resemble the hemibacula projecting from the apical platform of anguiforms – compare, for example, Fig. 16.4C with *Varanus niloticus* (fig. 3 of [18]) and *V. mitchelli* (fig. 46 of [31]). *Amerotyphlops brongersmianus* and *A. paucisquamus* have a small apical cup, the figured example of the latter species (Fig. 16.4D) is probably not fully inflated distally.

Knowledge of hemipenial morphology in anomalepidids (20 extant species) is restricted to the description of an everted (and likely fully inflated) organ of *Liotyphlops albirostris* [101]. The hemipenis is unilobed, bulbous, and 'capitate', bearing irregular spines on the asulcate side and on the tip. The sulcate surface is mainly unornamented, except for a small region of spinules at the base of the 'capitulum'. The sulcus spermaticus is simple and distally flared. The expanded, subcircular intrasulcular area was possibly damaged but based on the only available figure [101] it resembles the naked, apical disc of some other scolecophidians and some lizards. Branch [17: p. 295] added that spines are 'papillose and appear non-ossified (pers. obs.) although they may have subsequently decalcified'.

Tropidophiidae – Hemipenes in Tropidophiidae (35 species) are deeply bilobed with a centripetal (*Tropidophis*) to slightly centrolineal (*Trachyboa*) bifurcate sulcus spermaticus dividing at or just below the lobular crotch and running on the internal surface of the lobes to reach their tip (Fig. 16.5D, E: the sulcus is mistakenly described as centrifugal by [120]). *Tropidophis* and *Trachyboa* have a largely naked hemipenial body, and lobes ornamented by subtransverse, more or less papillate, flounces. Unlike *Trachyboa*, the hemipenis of *Tropidophis* spp. is described as 'quadrifurcate' with the main ('primary') lobes each bearing a pair of apical projections ('secondary lobes'), and the retractor muscle also being quadrifurcate [120, 121]. The sulcus tip drains at the base of the 'secondary lobes'. In

Tropidophis paucisquamis (Fig. 16.5E), the inner surface of each lobe also bears a substantial, 'wing-like fold' [120] ornamented with papillae.

Aniliidae – The hemipenis of the sole extant aniliid, *Anilius scytale* [21, 122] (Fig. 16.5F) is similar to those of tropidophiids. The organ is deeply bilobed, with lobes about the same size as the body, with a centripetally bifurcate sulcus extending to the tip of each lobe (Fig. 16.5F). The hemipenial body is largely devoid of ornaments. Differently from its extant sister, Tropidophiidae, *Anilius* has flounces that become irregularly calyculate at the lobe tips (particularly on the asulcate side), and lacks notable apical projections, and chevron-shaped and somewhat papillate flounces.

Uropeltoidea – Uropeltoidea comprises Cylindrophiidae (15 species), Anomochilidae (three species) and Uropeltidae (ca 60 species). Most of the scarce information available is for partially everted hemipenes of uropeltids, which have a unilobed, stout (*Melanophidium*) or slender, subcylindrical to slightly tapering (*Rhinophis* spp., *Uropeltis grandis*) hemipenis, with a simple sulcus spermaticus that has closely juxtaposed lips and extends to the tip without flaring or bifurcating [123–126]. *Rhinophis melanoleucus* has an additional basal, medial lobe-like process with irregular longitudinal folds but no evidence of a sulcus [125]. A spineless hemipenis with convoluted folds has been reported for *M. punctatum*, and a spinose hemipenis occurs in *Rhinophis* and *Uropeltis* species (at least *R. dorsimaculatus*, *R. lineatus*, *U. grandis*), with spines varying in size and distribution [123–126]. An illustration of perhaps the only fully everted hemipenis thus far documented for uropeltoids shows small spines throughout its length (*U. grandis*: [21: fig.106.1]); spines are also widely dispersed in *R. melanoleucus* [125]; *R. lineatus* has curved and closely-spaced spines on the distal one-third of the organ [127]; *R. dorsimaculatus* has fine spines along most of the organ's length [124].

Among cylindrophiids, *Cylindrophis engkariensis* has been described as having a unilobed, naked, coniform hemipenis [128]. *Cylindrophis aruensis* has a short hemipenis without ornamentation and with prominent sulcal lips [122]. Our preparation of the hemipenes of two *C. melanotus* (Fig. 16.5C) largely agrees with these descriptions, the organ is unilobed, columnar, smooth, and lacks spines, and the simple sulcus is bordered by fleshy, well-developed lips that become increasingly prominent distally. However, the distal end of the organ bears a single wing-like projection. The sulcus appears to flare on the terminus of the organ, though very few cylindrophiid hemipenes have been described thus far and further work is clearly needed to establish hemipenial features for this family. We know of no published information on the hemipenis of *Anomochilus,* the only extant anomochilid genus.

Bolyeriidae – For the monotypic, possibly recently extinct, Mauritius endemic *Bolyeria*, the only available information regarding the hemipenis is given by Underwood [129], who described it as 'very deeply divided'. This is suggestive of a bifurcate sulcus spermaticus, but nothing further can be concluded. We are unaware of hemipenis data for the other extant bolyeriid, *Casarea dussumieri*.

Pythonoidea – Pythonoidea comprises Xenopeltidae (two species), Loxocemidae (one species) and Pythonidae (38 species) (e.g., [23, 99, 130]). Their hemipenes are slightly to deeply bilobed with a centrolineal, bifurcate sulcus spermaticus, ornamented by large fleshy calyces and flounces but lacking spines or spinules (Fig. 16.6A, B). The hemipenes

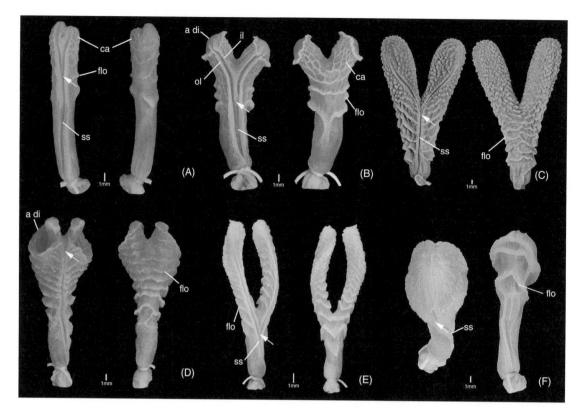

Figure 16.6 Hemipenes of (A) *Morelia spilota* (Pythonidae); (B) *Xenopeltis unicolor* (Xenopeltidae); (C) *Epicrates cenchria* (Boidae; courtesy P. Passos); (D) *Charina bottae* (Charinaidae); (E) *Candoia bibroni* (Candoiidae); (F) *Calabaria reinhardtii* (Calabariidae). Sulcate (left) and asulcate (right) views, except left of (F), an apical view of the sulcate surface. See Supplementary Appendix 16.S1 for colour version, and Box 16.1 for abbreviations.

of *Loxocemus* and *Xenopeltis* are similar in shape and ornamentation. Both have a deeply bilobed organ, with a centrolineal sulcus bifurcating below the crotch and extending distally to a broad apical disc delimited by a broad fleshy fringe; the body is naked proximally but otherwise ornamented by a few large flounces; the lobes tend to bear large and shallow calyces [17, 122, 131] (Fig. 16.6B).

Pythonids have unilobed (*Morelia viridis*: Supplementary Figure 16.S1e), slightly bilobed (e.g., *Python anchietae*), or deeply bilobed hemipenes (e.g., *Python sebae*) [17, 132], with a centrolineally (rarely centrifugally, e.g., *P. curtus*: [122]) bifurcate sulcus, and ornamentation in the form of calyces and/or often substantial flounces. The lobes can bear papillate terminal awns (e.g., *P. curtus, Liasis olivaceus*: [17, 131, 133]) carrying the end of the branches of the sulcus, and with a subterminal region of calyces grading into flounces [122]. Different pythonid groups have different patterns of flounces and/or calyces [122], and the body may also bear isolated papillae [123].

Booidea – Booidea comprises Calabariidae (one species), Sanziniidae (four species), Charinaidae (four species), Ungaliophiidae (three species), Erycidae (13 species),

Candoiidae (five species), and Boidae (65 species). Hemipenes within this group are unilobed [17, 21, 131, 134–136] (Fig. 16.6F) to feebly, moderately or deeply bilobed [21, 137–139] (Fig. 16.6C–E; Supplementary Figure 16.S1f). Unilobed organs occur convergently in *Eryx* and *Calabaria* [17, 136] (Fig. 16.6F). The sulcus spermaticus generally bifurcates to reach the lobular tips and is centrolineal or slightly centrifugal. In *Calabaria reinhardtii* and *Eryx jaculus* the sulcus is simple and unflared [131, 136, 140]. In some candoids and in sanziniids, the branches of the bifurcate sulcus do not extend to the lobe tips [137]. Spines or other mineralized structures are absent and the hemipenial body is typically ornamented with large flounces often extending onto the lobes. The lobes can be ornamented with calyces and papillae, and the tip(s) may bear apical discs or awns [129, 137, 138]. Lobes are mainly calyculate in *Boa*, and are papillate in *Corallus, Epicrates,* and *Eunectes* [137–139] (Fig. 16.6C). Apical discs are domed in erycids [131], and concave in charinaids with shallow lateral (*Lichanura*) or deep cup-like structures (*Charina*) [131, 136, 137] (Fig. 16.6D). In *C. bottae* (Fig. 16.6D), the sulcus is simple, its lips flaring distally in a naked area between the laterodistal cups. Confirmation is needed of whether the sulcus is bifurcate or simply flared in particular species. The hemipenes of ungaliophiids have been briefly described (*Ungaliophis*: [141]) and illustrated (*Exiliboa*: [131]), being long, bilobed and with a bifurcate sulcus. Ungaliophiids typically retain large transverse flounces around the distal half of the hemipenial body and proximal region of the lobes, with the more distal parts of the lobes bearing calyces [131, 141]. A small distal awn at the tip of the lobes in *Exiliboa* [131] is not confirmed in *Ungaliophis* due to the partially everted condition of the only hemipenis described [141].

Caenophidia – Spinules on the lobes is considered a synapomorphy of Caenophidia, and spinules and spines extending proximally onto the hemipenial body as a synapomorphy of Colubroides [11, 24]. Hemipenes of non-endoglyptodont caenophidians (ca 66 extant species: Acrochordidae, Xenodermidae, and Pareidae [including Xylophiinae: [142]]) retain the plesiomorphic hemipenial characteristics of very long, cylindrical lobes, and naked bodies (e.g., [143]). The presence of well-defined calyces arranged in rows on the lobes is considered a synapomorphy of Colubroides [24, 143]. The most pronounced hemipenial variation in caenophidians occurs in the diverse Endoglyptodonta (ca 3,140 species: Viperidae, Homalopsidae, Elapoidea, Colubroidea: [143]), a well-supported clade characterized by hemipenial bodies bearing mineralized, pointed to hooked spines, and shortened lobes (with some exceptions, e.g., *Pseudaspis cana*) [143]. Body spines vary considerably in shape and number and can be reduced or absent in some groups (e.g., Psammophiidae; [143]). In viperids, spines often occur on the lobes. In homalopsids and some pseudoxyrhophiids, bulbous lobes are uniquely covered with densely disposed spinules [143].

Hemipenes in Colubroides are typically deeply bilobed. Unilobed hemipenes occur only in lineages deeply phylogenetically nested within Endoglyptodonta. The sulcus spermaticus is invariably bifurcate in many endoglyptodont lineages, including Viperidae, Homalopsidae, Pseudaspididae, Lamprophiidae, and Pseudoxenodontidae. Within Elapoidea, the sulcus of Prosymnidae and Psammophiidae is likely apomorphically simple, associated with a secondarily unilobed hemipenis. Colubroidea includes taxa with a bifurcate or simple sulcus spermaticus – the bifurcate form characterizing Dipsadidae (with rare exceptions, such as *Amnesteophis melanauchen*; see [144], for discussion), Pseudoxenodontidae, Calamariidae, and Grayiidae, and the simple sulcus characterizing Sibynophiidae and Colubridae. Both a simple and centripetally bifurcate sulcus occur in Natricidae [11, 37, 145, 146], though instances of a simple sulcus in natricids are considered non-homologous to those of Colubridae or Sibynophiidae, instead resulting from the loss of the right branch of the bifurcate sulcus spermaticus ([11, 146]; see [37] and references therein for a detailed review of natricid hemipenial morphology). Colubridae is diagnosed by a simple sulcus derived from the loss of the left branch of a bifurcate sulcus [11, 37, 145, 146]. Sibynophiids bear a simple sharply curved, U-shaped convolution of the sulcus [147].

16.4 Discussion

Despite the paucity of hemipenial information for some major squamate lineages (e.g., Dibamidae, Scincomorpha, Cordyloidea, Anomalepididae) some inferences about major patterns in squamate hemipenial evolution can be made. Some external features vary substantially within as well as among major squamate clades, and/or vary in ways that make homology determination or identification of discrete states particularly challenging. Such features include absolute size, the form and distribution of many types of ornamentation, the orientation of the sulcus spermaticus on the body, the relative lengths of lobes, and the distance between and prominence of sulcal lips. This is sometimes compounded by deficiencies in terminology that can be readily applied across Squamata. Several features can be difficult to assess from published descriptions and 2D images and illustrations, especially if comparisons include dissected (inverted) and (possibly incompletely) everted and inflated hemipenes. Lack of histological data also limits thorough homology assessments for several features, especially ornamentations. Finally, determining patterns of variation across Squamata would benefit in future from more thorough considerations of phylogenetic uncertainty and more explicit ancestral-state estimations within as well as among major lineages. Such work, beyond the scope of this chapter, would help test some of our assumptions about what constitutes relatively 'minor' and 'major' variation within and among major squamate lineages, and about character-state polarities across the tree.

Because our knowledge of hemipenial variation in major squamate groups is very patchy, hemipenial features that seemingly vary substantially within as well as among major lineages are largely excluded from further discussion here. Instead, we focus on major evolutionary patterns for states of characters that can be traced with more confidence, with particular attention paid to features inferred for the ancestral snake and early crown snakes. States for these characters are plotted for terminal taxa in Figure 16.1.

16.4.1 Hemipenial Shape

Slightly to moderately bilobed hemipenes (and bifurcate retractor muscles) seem to be the dominant condition in squamates, occurring in most gekkotans, scincomorphs, lateratans, iguanians, anguiforms, and snakes for which we have information, suggesting that this morphology arose early within squamates and was possibly present in their most recent common ancestor. However, the lobular condition is not clear in the basally diverging Dibamidae, for which only the partially everted organ of *Dibamus greeri* is known so far (see § 16.3.1), and the plesiomorphic condition in some major lineages is not well established. Instances of unilobed hemipenes occur within all major lineages of lizards and snakes, seemingly as independently derived states (though interpreted as plesiomorphic for some major lineages: e.g., [81]). Where they occur, deeply bilobed hemipenes also appear to have evolved independently in several main lineages, within scincomorphs (e.g., sphenomorphines), iguanians (tropidurids), amphisbaenians (amphisbaenids), and snakes (alethinophidians). If scolecophidians are paraphyletic then the unilobed condition was either acquired independently in both major lineages (Anomalepididae and Leptotyphlopidae + Typhlopoidea) or the bilobed condition is a secondarily derived synapomorphy of Alethinophidia. With bilobed hemipenes being the likely ancestral condition of Alethinophidia, the unilobed hemipenes of uropeltoids (and some caenophidians) are interpreted as apomorphic.

16.4.2 Sulcus Spermaticus

A distally bifurcate or flared (expanded) sulcus spermaticus occurs in most major squamate lineages (e.g., Figs. 16.2C, E and 16.3C, D). In the flared condition, the sulcus spermaticus widens distally at or towards the apex of a typically unilobed or slightly bilobed organ. This occurs in most gekkotans, scincomorphs, lateratans, and iguanians (except those with deeply bilobed hemipenes), and perhaps all anguiforms and scolecophidians. Teiioids, amphisbaenians, and alethinophidians are exceptions, in which the sulcus spermaticus tends to retain parallel lips throughout its length (with rare exceptions), likely representing a synapomorphy of Teiioidea and evolving independently in the other two groups. Amphisbaenians and alethinophidians (except e.g., *Cylindrophis*) are similar in having narrow sulci that extend to the extremities of the lobes without opening or broadening into a naked region. In amphisbaenian, teiioid, and snake bilobed hemipenes, the sulcus may bifurcate in the crotch (centripetally) to form two distinct branches that reach (or almost reach) each lobular extremity. In alethinophidians, the sulcus may also divide on the sulcate surface of the hemipenial body to reach the distal extremity of the lobe or lobes centrifugally or centrolineally.

The section of the sulcus spermaticus on the hemipenial body tends to form a broader gutter with often flap-like lips in limbed squamates (e.g., Figs. 16.2E and 16.3D, E) while in limb-reduced, serpentiform groups that section of the sulcus is often more tightly enclosed by less flap-like outer lips (e.g., Figs. 16.2F and 16.3F). Formulating and plotting discrete character states describing variation in the sulcal lips, such as the spacing between the outer lips (on at least the hemipenial body), the extent to which they protrude, and their thickness, is difficult. However, closely spaced, well-defined lips (and a relatively narrow, clearly demarcated, regular sulcus) throughout the hemipenial body might be a synapomorphy of snakes, with lizard sulci typically having more widely spaced and/or less-well defined lips (with a few, moderate exceptions: e.g., Fig. 16.3A). This condition seems to be present (convergently) also in amphisbaenians.

Within snakes, a bilobed hemipenis with a narrow sulcus spermaticus that bifurcates centripetally with branches ending at the tips of lobes is considered the likely ancestral condition of Alethinophidia (e.g., Fig. 16.5F). Scolecophidians (and possibly *Cylindrophis*) have a lizard-like sulcus spermaticus that flares near the tip of the hemipenis (e.g., *Liotyphlops albirostris, Amerotyphlops reticulatus, Afrotyphlops lineolatus*: § 16.3.6; Figs. 16.4C–F and 16.5A, B), sometimes into an apical cup that resembles the condition present in varanids (e.g., *Amerotyphlops* spp.; Fig. 16.4C, E). Although proximally the sulcal lips are closely spaced and never flap like (as also occurs in other snakes, unlike the condition in many lizards), none of the described scolecophidian sulci maintain closely juxtaposed lips reaching the distal extremity of the lobular region, unlike the condition in (almost all) alethinophidian snakes (exceptions being within booids and colubroideans and possibly *Cylindrophis*).

Among alethinophidians, a centripetal sulcus spermaticus occurs in the basally diverging clade Amerophidia while the remaining non-caenophidian alethinophidians have a sulcus (where bifurcate) that is more centrifugal. Tropidophiids and *Anilius* share a similar, deeply bifurcated hemipenis, a sulcus that runs throughout a mainly naked hemipenial body and bifurcates centripetally, and lobes covered with flounces. This combination of characters is not known in other non-caenophidian clades, with densely disposed flounces on the lobes possibly representing a hemipenial synapomorphy of Amerophidia (appearing convergently within Caenophidia). In the remaining alethinophidians, centripetally bifurcating sulci appear in only a few endoglyptodont groups (e.g., Homalopsidae, Elapidae, Natricidae, Atractaspididae, and a single dipsadid).

Among snakes a simple sulcus spermaticus occurs in scolecophidians, uropeltoids, calabariids, cyclocorids, psammophiids, sibynophiids, colubrids, and within natricids and dipsadids [8, 11]. Except for scolecophidians that likely retained a plesiomorphic lizard condition, a simple sulcus in snakes probably evolved multiple times independently from a bifurcate condition.

16.4.3 Lobular Apical Ornamentations Intimately Associated with the Distal Extremity of the Sulcus Spermaticus

In lizards, the extremity of sulcal branches is sometimes closely associated with substantial distal lobular projections. Alethinophidians lack such projections at the extremity of the sulci, except for rare, convergently derived instances (e.g., *Tropidophis, Python, Exiliboa, Xenochrophis, Aspidura*). As far as we know, substantial apical lobular projections are absent also in pleurodonts, lacertids, most gekkotans and scincomorphs. Specialized lobular projections (often associated with the end of the sulcus) occur in some anguiforms, some gekkotans and some scolecophidians in the form of mineralized hemibacula, in chamaeleonids in the form of unique unmineralized ornaments [31], and in teiioids in the form of fleshy awns [28] and papillae [35]. These features are likely synapomorphies of their respective groups, except for scolecophidians, where hemibacula are currently known in only a single species. Interestingly, *Amerotyphlops minuisquamus* (Fig. 16.4C) has a hemipenial morphology that is very similar to some varanids (compare with *Varanus mitchelli*; [31]), including a pair of mineralized, hemibacula-like projections from within the space of the apical cup flanking the flared sulcal lips. Amphisbaenids present serially lamelate lobular tips forming a discrete patch around each sulcal branch extremity that are unique to the group (e.g., [30]). Although this lamellar region of amphisbaenids lacks substantial apical projections, these apical ornamentations are closely associated with the distal extremity of the sulcus. More investigation of scolecophidian hemipenes is required, but our current interpretation is that the most recent common ancestors of extant snakes, of extant alethinophidians and of extant caenophidians lacked notable lobular apical ornamentation associated with the distal end(s) of the sulcus spermaticus.

16.4.4 Well-Developed Spines on the Hemipenial Body

In general, lizard hemipenial bodies are not ornamented with mineralized spines, except for gymnophthalmids in which they occur in at least 21 genera (see [14] and references therein). Mineralized spines are also present in snakes, presumably convergently – the hemipenial body spines of gymnophthalmids differ from those of snakes in being more regularly arranged and typically associated with (embedded within) approximately transverse, laminar bands of soft tissue (see § 16.3.4). Among non-caenophidian snakes, substantial spiny structures on the hemipenial body are known in at least one anomalepidid (in the only hemipenis documented for the family – unclear whether these structures are mineralized), a single typhlopid, and perhaps in most uropeltids, being absent in Amerophidia, Cylindrophiidae, Pythonoidea and Booidea. Hemipenial spines on at least the lobes are a putative synapomorphy for Caenophidia, being secondarily lost independently in Pareidae, Psammophiidae and Calamariidae [11]. Spines extend onto the hemipenal body in viperids, homalopsids, colubroids and elapoids, being a synapomorphy of Endoglyptodonta (excepting rare cases of secondary loss; see § 16.3.6 Caenophidia). Given the absence of mineralized body spines in toxiceferan lizards and almost all scolecophidians, we consider it likely that the spiny body considered typical of snakes is a synapomorphy of Caenophidia that is present convergently in some scolecophidians and uropeltids.

16.4.5 Soft-Tissue Ornamental Ridges and Lamellae on the Hemipenial Body and Lobes

This section addresses soft-tissue ornamentations excluding the bulbous folds of scincomorphs and papillae (other than instances of papillate flanges). Many squamate hemipenes are ornamented with raised, predominantly transversely oriented soft-tissue structures organized serially on the body and lobes. These have been described in lizards using a plethora of terms, but in snakes have been described primarily with the single term 'flounce' despite their great diversity in that group. Although exceptions can often be found, and variation is somewhat continuous, we recognize two main morphotypes of predominantly transversely oriented soft-tissue ornaments in squamates: relatively thin and limp (soft) versus more stout and stiffened. In addition, we perceive major axes of variation in these structures in terms of how closely and regularly they are arranged, how prominent they are, and whether or not they bear mineralized spines. If we provisionally assume that these structures are homologous across Squamata to some extent (though see e.g., [10] for a different view), then what term(s) is (are) best to apply? We consider crease to be best used to describe a rather superficial, perhaps somewhat irregular groove or ridge. The terms fold and plica might be appropriate but we lack sufficient histological data to understand the extent to which these features are genuinely folded. Thus, we prefer terms such as flounce, lamella, lamina, or welt to more neutrally describe a projecting ridge or flap that may or may not be formed of folded tissue. Flounce and especially frill might be best restricted to describe a flange with a longer free than attached edge. While recognizing the stability of the use of flounce as a general term in snake hemipenial literature, we here use 'flange' as a general term for all similar structures across Squamata.

In snakes, flounces can be short (low) or long (prominent) and more or less frilly, but are typically limp (soft), and often disposed in somewhat irregularly and/or relatively widely spaced series. The same general form (relatively prominent and often not tightly packed flanges) occurs in many non-agamid toxicoferan lizards (Chamaeleonidae, Pleurodonta, and Anguiformes). Among toxicoferan lizards paryphasmata (flounces stiffened with longitudinal ridges) are a synapomorphy of Anguiformes. In Laterata, flanges are generally disposed in highly regular and densely spaced series, with each flange typically short (not prominent), forming a stiff ridge or lamina (often with a transverse, distal band of denser tissue) that may be fringed with embedded mineralized spines (Gymnophthalmidae) or microscopic spinules (Teiidae) in highly regular series. Data are scant for Scincomorpha and interpretation of homology is complicated by the disparate and obviously derived bulbous folds of scincids, but there is evidence of densely spaced series of flanges (somewhat like those of lateratans) in at least cordylids. Non-unidentatan squamates (dibamids and gekkotans) lack flanges on the hemipenial body. Thus, ornamental flanges on the hemipenial body are likely a synapomorphy of Unidentata. Within snakes, booids and pythonoids share a synapomorphy of large, fleshy and smooth flounces on the distal half of the hemipenial body and proximal region of the lobes – a condition also seen in some scolecophidians and higher caenophidians, but absent in most snakes. Regular fields of calyces, with their bounding flanges sometimes bearing projecting and seemingly mineralized spinules, occur on mostly the lobes in representatives of Serpentes, Gekkota, and Iguania. Although being a synapomorphy of each of these major lineages (or subsets thereof), homology of these structures among these major lineages is doubtful (though calyces have been considered homologous across lizards and convergently acquired in snakes: [10]).

16.4.6 The Hemipenis of the Ancestral Snake

As summarized above, lizard and snake hemipenes are similar in many ways, but with notable differences in gross morphology and apparent evolutionary trends. Lizards exhibit several features that are absent in snakes, including: flap-like and sometimes notably asymmetric sulcal lips, sulcus wide on body, densely spaced series of stout or stiffened flanges, flanges fringed with embedded mineralized spines, and inflated, bulbous folds. However, summarizing general conditions for lizards and for snakes (or other major squamate groups) is challenging because there are almost always exceptions, and because plesio- and apomorphic conditions are not yet clearly established for many lineages.

Based on the preceding sections of this Discussion and the ten traits summarized in Figure 16.1, we propose that the most-recent common ancestor of alethinophidians had a moderately to deeply bilobed hemipenis with lobular flounces, lacking spines, and a sulcus spermaticus bifurcating centripetally and reaching the distal tip of the lobes, with closely spaced and symmetrical lips. Most or all of these traits might be synapomorphies of Alethinophidia. Deeper in the squamate tree, the ancestral toxicoferan likely had slightly bilobed hemipenes lacking spines and with a simple, distally flared sulcus. The ancestral toxicoferan possibly also had flounce-like flanges on the distal part of the body, that were lost in agamids and reduced in some pleurodonts – this interpretation would receive further support if anguiforms were the extant sister group of snakes (relationships among Iguania, Anguiformes and Serpentes are not yet compellingly resolved: e.g., [23]). Thus, notable changes in hemipenial morphology occurred between the basal divergence within Toxicofera and the basal divergence within Alethinophidia. Clarifying precisely where in the tree these changes occurred, and inferring with confidence the form of the hemipenis in the most recent common ancestor of extant snakes is more challenging. However, all described scolecophidians resemble the inferred ancestral toxicoferan in having a simple, flared sulcus, and thus this is likely the plesiomorphic condition for snakes. Given scole-cophidian paraphyly, the most parsimonious interpretation is also that the ancestral snake had a unilobed hemipenis (and a weakly differentiated lobular region) lacking spines or regular fields of calyces and with a simple sulcus spermaticus that flared distally into a naked apical region – though this is only minimally more parsimonious that the ancestral snake having a somewhat bilobed hemipenis and two main extant scolecophidian lineages acquiring their unilobed hemipenes independently.

Patchy morphological data make it particularly challenging to interpret the evolution of three major hemipenial features in the early history of snakes. Given the phylogenetic framework considered here [23], it is most parsimonious to interpret: (1) hemipenial body spines of caenophidians, some uropeltids, anomalepidids, and *Amerotyphlops reticulatus*, (2) hemibacula of anguiforms and *A. minuisquamus*, and (3) distally flared sulci of toxicoferan lizards, scolecophidians, and *Cylindrophis melanotus* as non-homologous, but they remain curious and striking instances of morphological similarity relatively close to the base of the snake tree. Given the patchy taxonomic sampling thus far, there remains great scope to further investigate these features. Phylogenetic uncertainty also plays a role. The (im)probabilities of these three homologies change if, for example, Anguiformes is the extant sister to snakes, if scolecophidians are monophyletic, or if uropeltoids are the extant sister to all other extant afrophidians – though we note that some of these contentious hypotheses about long-debated regions of the squamate tree are not strongly supported by the most recent analyses of large molecular data sets (Chapter 10).

16.4.7 Future Directions

There remains much to learn about the anatomy, function, diversity, and evolution of squamate hemipenes. Anatomical descriptions and/or illustrations are lacking or superficial for many species. Special attention should be given to particularly poorly sampled major lineages, especially gekkotans, dibamids, scincoids, cordyloids, pleurodonts, scolecophidians, bolyeriids, and uropeltoids. As a rule, descriptions should state the extent to which hemipenes are fully everted and fully inflated. It is challenging to formulate discrete character states for some hemipenial features, but it would be a step forwards to compile databases and map character states onto detailed phylogenies to more rigorously estimate ancestral states – something beyond the scope of this study.

The focus of this article has been the gross anatomy of the external surfaces of the hemipenis, but studies of tissue anatomy and of structures such as retractor and propulsor muscles will likely help to resolve outstanding issues about the homology (and effective terminology) and function of hemipenial traits. In pursuing this, modern imaging techniques such as DiceCT can complement more traditional approaches such as histology. In terms of testing hypotheses that explain the taxonomic patterns of traits, improved and/or more densely sampled phylogenies will be needed in addition to unambiguously formulated characters and states – not only for hemipenial features, but also for a broader suite of ecomorphological traits encompassing female reproductive tract anatomy (see e.g., Chapter 17), tail morphology, and reproductive behaviour and ecology.

Acknowledgements

See Supplementary Appendix 16.S1. In addition, we thank V. Deepak, Pedro M. S. Nunes and Ana L. C. Prudente for constructive reviews of the submitted manuscript.

References

1. M. L. Gredler, C. E. Larkins, F. Leal , et al., Evolution of external genitalia: insights from reptilian development. *Sexual Development*, 8 (2014), 3113–3126.

2. M. L. Gredler, Developmental and Evolutionary origins of the amniote phallus. *Integrative and Comparative Biology*, 56 (2016), 694–704.

3. T. J. Sanger, M. L. Gredler, and M. J. Cohn, Resurrecting embryos of the tuatara, *Sphenodon punctatus*, to resolve vertebrate phallus evolution. *Biology Letters*, 11 (2015), 20150694.

4. H. G. Dowling and D. E. Savage, A guide to the snake hemipenis: a survey of basic structure and systematic characteristics. *Zoologica*, 45 (1960), 17–28.

5. F. Leal and M. J. Cohn, Development of hemipenes in the Ball Python snake *Python regius*. *Sexual Development*, 9 (2014), 6–20.

6. A. Raynaud and C. Pieau. Embryonic development of the genital system. In C. Gans, F. Billett, eds., *Biology of the Reptilia* (New York: John Wiley and Sons, 1985), pp. 149–300.

7. E. N. Arnold. Variation in the cloacal and hemipenial muscles of lizards and its bearing on their relationships. In M. W. J. Ferguson, ed., *The Structure, Development and Evolution of Reptiles (Symposium of the Zoological Society of London 52)* (London: Academic Press, 1984), pp. 47–85.

8. E. D. Cope, The classification of the Ophidia. *Transactions of the American Philosophical Society*, 18 (1895), 186–219.

9. E. D. Cope, On the hemipenes of the Sauria. *Proceedings of the Academy of Natural Sciences of Philadelphia*, 48 (1896), 461–467.

10. W. Böhme, Zur Genitalmorphologie der Sauria: funktionelle un stammesgeschichtliche Aspekte. *Bonner Zoologische Monographien*, 27 (1988), 1–176.

11. H. Zaher, Hemipenial morphology of the South American xenodontine snakes, with a proposal for a monophyletic Xenodontinae and a reappraisal of colubroid hemipenes. *Bulletin of the American Museum of Natural History*, (1999), 1–168.

12. C. W. Myers and J. E. Cadle, On the snake hemipenis, with notes on *Psomophis* and techniques of eversion: a response to Dowling. *Herpetological Review*, 34 (2003), 295–302.

13. H. Zaher and A. L. C. Prudente, Hemipenes of *Siphlophis* (Serpentes, Xenodontinae) and techniques of hemipenial preparation in snakes: a response to Dowling. *Herpetological Review*, 34 (2003), 302–307.

14. P. M. Nunes, F. F. Curcio, J. G. Roscito, and M. T. Rodrigues, Are hemipenial spines related to limb reduction? A spiny discussion focused on gymnophthalmid lizards (Squamata: Gymnophthalmidae). *Anatomical Record*, 297 (2014), 482–495.

15. P. R. Manzani and A. S. Abe, Sobre dois métodos de preparo do hemipênis de serpentes. *Memórias do instituto Butantan*, 50 (1988), 15–20.

16. O. S. Pesantes, A method for preparing the hemipenis of preserved snakes. *Journal of Herpetology*, 28 (1994), 93–95.

17. W. R. Branch, Hemipenial morphology of african snakes: a taxonomic review. Part 1. Scolecophidia and Boidae. *Journal of Herpetology*, 20 (1986), 185–299.

18. W. R. Branch, Hemipeneal morphology of platynotan lizards. *Journal of Herpetology*, 16 (1982), 16–38.

19. F. Zhang, Studies on morphological characters of hemipenes of the Chinese lizards [in Chinese]. *Acta Herpetologica Sinica*, 5 (1986), 254–259.

20. F. Zhang, S. Q. Hu, and E. M. Zhao, Comparative studies and phylogenetic discussions on hemipenial morphology of the Chinese Colubrinae (Colubridae). *Acta Herpetologica Sinica*, 3 (1984), 23–44.

21. H. G. Dowling and W. E. Duellman, *Systematic Herpetology: A Synopsis of Families and Higher Categories* (New York: HISS Publications, 1978).

22. C. McCann, The hemipenis in reptiles. *Journal of the Bombay Natural History Society*, 46 (1946), 347–368.

23. F. T. Burbrink, F. G. Grazziotin, R. A. Pyron, et al., Interrogating genomic-scale data for Squamata (lizards, snakes, and amphisbaenians) shows no support for key traditional morphological relationships. *Systematic Biology*, 69 (2020), 502–520.

24. H. Zaher, F. G. Grazziotin, J. E. Cadle, et al., Molecular phylogeny of advanced snakes (Serpentes, Caenophidia) with an emphasis on South American xenodontines: a revised classification and descriptions of new taxa. *Papéis Avulsos de Zoologia (São Paulo)*, 49 (2009), 115–153.

25. P. Uetz, P. Freed, R. Aguilar, and J. Hošek. *The Reptile Database*. www.reptile-database.org2021 (accessed 1 May 2021)

26. J. M. Savage, On terminology for the description of the hemipenes of squamate reptiles. *Herpetological Journal*, 7 (1997), 23–25.

27. C. Klaver and W. Böhme, Phylogeny and classification of the Chamaeleonidae (Sauria) with special reference to hemipenis morphology. *Bonner Zoologische Monographien*, 22 (1986), 1–64.

28. M. B. Harvey, G. N. Ugueto, and R. L. Gutberlet, Review of teiid morphology with a revised taxonomy and phylogeny of the Teiidae (Lepidosauria: Squamata). *Zootaxa*, 3459 (2012), 1–156.

29. W. Böhme, Zur systematischen Stellung der Amphisbanen (Reptilia: Squamata), mit besonderer Berücksichtigung der Morphologie des Hemipenis. *Journal of Zoological Systematics and Evolutionary Research*, 27 (1989), 330–337.

30. H. I. Rosenberg, M. J. Cavey, and C. Gans, Morphology of the hemipenes of some

Amphisbaenia (Reptilia: Squamata). *Canadian Journal of Zoology*, 69 (1991), 359–368.

31. T. Ziegler and W. Böhme, Genitalstrukturenund Paarungsbiologie bei squamaten Reptilien, speziell den Platynota, mit Bemerkungen zur Systematik. *Mertensiella*, 8 (1997), 1–210.

32. V. Parra, P. M. Nunes, and O. Torres-Carvajal, Systematics of *Pholidobolus* lizards (Squamata, Gymnophthalmidae) from southern Ecuador, with descriptions of four new species. *ZooKeys*, 954 (2020), 109–156.

33. E. N. Arnold, Relationships of the Palaearctic lizards assigned to the genera *Lacerta*, *Algyroides* and *Psammodromus* (Reptilia: Lacertidae). *Bulletin of the British Museum (Natural History), Zoology*, 25 (1973), 289–366.

34. I. G. Brennan and A. M. Bauer, Notes on hemipenial morphology and its phylogenetic implications in the Pygopodidae Boulenger, 1884. *Bonn Zoological Bulletin*, 66 (2017), 15–28.

35. C. W. Myers, G. Rivas Fuenmayor, and R. C. Jadin, New species of lizards from Auyantepui and La Escalera in the Venezuelan Guayana, with notes on 'microteiid' hemipenes (Squamata: Gymnophthalmidae). *American Museum Novitates*, 3660 (2009), 1–31.

36. P. M. Sánchez-Martínez, M. P. Ramírez-Pinilla, E. Meneses-Pelayo, and P. M. Nunes, Hemipenial morphology of nine South American species of *Mabuya* (Scincidae: Lygosominae) with comments on the morphology of the family. *Anatomical Record*, 11 (2020), 2917–2930.

37. J. E. Cadle, Hemipenial morphology in the North American snake genus *Phyllorhynchus* (Serpentes: Colubridae), with a review of and comparisons with natricid hemipenes. *Zootaxa*, 3092 (2011), 1–25.

38. C. W. Myers and J. A. Campbell, A new genus and species of colubrid snake from the Sierra Madre del Sur of Guerrero, Mexico. *American Museum Novitates*, 2708 (1981), 1–20.

39. E. N. Arnold, The hemipenis of lacertid lizards (Reptilia: Lacertidae): structure, variation and systematic implications. *Journal of Natural History*, 20 (1986), 1221–1257.

40. D. F. Avery and W. W. Tanner, Evolution of the iguanine lizards (Sauria, Iguanidae) as determined by osteological and myological characters. *Science Bulletin, Brigham Young University*, 12 (1971), 1–79.

41. A. Y. Al-ma'Ruf, R. P. Sari, I. Mostofa, et al., Morphology and histology of paryphasmata and hemibaculum of *Varanus salvator* based on sexual maturity. Open *Veterinary Journal*, 11 (2021), 330–336.

42. M. A. Smith, The Fauna of British India, including Ceylon and Burma. Reptilia and Amphibia. Vol. II. *Sauria*. (London: Taylor and Francis, 1935).

43. G. M. Shea and G. L. Reddacliff, Ossifications in the hemipenes of varanids. *Journal of Herpetology*, 20 (1986), 566–568.

44. Y. L. Werner, Are hemipenial 'ossifications' of Gekkonidae and Varanidae ossified? *Israel Journal of Zoology*, 35 (1988), 99–100.

45. R. Graboski, J. C. Arredondo, F. G. Grazziotin, et al., Molecular phylogeny and hemipenial diversity of South American species of *Amerotyphlops* (Typhlopidae, Scolecophidia). *Zoologica Scripta*, 48 (2019), 139–156.

46. R. C. Jadin and R. B. King, Ontogenetic effects on snake hemipenial morphology. *Journal of Herpetology*, 46 (2012), 393–395.

47. W. Böhme, über das Stachelepithel am Hemipenis lacertider Eidechsen und seine systematische Bedeutung. *Journal of Zoological Systematics and Evolutionary Research*, 9 (1971), 187–223.

48. A. K. S. De-Lima, I. P. Paschoaletto, L. O. Pinho, et al., Are hemipenial traits under sexual selection in *Tropidurus* lizards? Hemipenial development, male and female genital morphology, allometry and coevolution in *Tropidurus torquatus*

(Squamata: Tropiduridae). *PLoS ONE*, 14 (2019), e0219053.

49. H. Zaher and A. L. C. Prudente, Intraspecific variation of the hemipenis in *Siphlophis* and *Tripanurgos*. *Journal of Herpetology*, 33 (1999), 698–702.

50. E. N. Arnold, Why copulatory organs provide so many useful taxonomic characters: the origin and maintenance of hemipenial differences in lacertid lizards (Reptilia: Lacertidae). *Biological Journal of the Linnean Society*, 29 (1986), 263–281.

51. E. N. Arnold, O. Arribas, and S. Carranza, Systematics of the palaearctic and oriental lizard tribe Lacertini (Squamata: Lacertidae: Lacertinae), with descriptions of eight new genera. *Zootaxa*, 1430 (2007), 1–86.

52. F. Glaw, J. Kosuch, F.-W. Henkel, et al., Genetic and morphological variation of the leaf-tailed gecko *Uroplatus fimbriatus* from Madagascar, with description of a new giant species. *Salamandra*, 42 (2006), 129–144.

53. A. E. Greer, The relationships of the lizard genera *Anelytropsis* and *Dibamus*. *Journal of Herpetology*, 19 (1985), 116–156.

54. J. W. Streicher and J. J. Wiens, Phylogenomic analyses of more than 4000 loci resolve the origin of snakes among lizard families. *Biology Letters*, 13 (2017), 20170393.

55. I. S. Darevsky, Two new species of the worm-like lizard *Dibamus* (Sauria: Dibamidae) with remarks on the distribution and ecology of *Dibamus* in Vietnam. *Asiatic Herpetological Research*, 4 (1992), 1–12.

56. M. Das and J. Purkayastha, Insight into the hemipnenial morphology of five species of *Hemidactylus* Oken, 1817 (Reptilia: Gekkonidae) of Guwahati, Assam, India. *Hamadryad*, 36 (2012), 32–37.

57. Z.-T. Lyu, C.-Y. Lin, J.-L. Ren, et al., Review of the *Gekko* (*Japonigekko*) *subpalmatus* complex (Squamata, Sauria, Gekkonidae), with description of a new species from China. *Zootaxa*, 4951 (2021), 236–258.

58. H. G. Dowling and F. W. Gibson, The hemipenis of the Onion-Tail gecko *Thecadactylus rapicaudus* (Houttuyn). *Herpetological Review*, 3 (1971), 110.

59. J. Purkayastha, M. Das, A. M. Bauer, et al., Notes on the *Hemidactylus bowringii* complex (Reptilia: Gekkonidae) in India, and a change to the national herpetofaunal list. *Hamadryad*, 35 (2010), 20–27.

60. H. Rösler and W. Böhme, Peculiarities of the hemipenes of the gekkonid lizard genera *Aristelliger* Cope, 1861 and *Uroplatus* Duméril, 1806. Proceedings of the 13th Congress of the Societas Europaea Herpetologica (Bonn: SEH, 2006).

61. H. Rösler, A. Bauer, M. P. Heinicke, et al., Phylogeny, taxonomy, and zoogrography of the genus *Gekko* Laurenti, 1768 with the revalidation of *G. reevesii* Gray, 1831 (Sauria: Gekkonidae). *Zootaxa*, 2989 (2011), 1–50.

62. C. W. Linkem, A. C. Diesmos, and R. M. Brown, Molecular systematics of the Philippine forest skinks (Squamata: Scincidae: *Sphenomorphus*): testing morphological hypotheses of interspecific relationships. *Zoological Journal of the Linnean Society*, 163 (2011), 1217–1243.

63. V. S. Vergilov, B. Zlatkov, and N. D. Tzankov, Hemipenial differentiation in the closely related congeners *Ablepharus kitaibelii* (Bibron & Bory de Saint-Vincent, 1833) and *Ablepharus budaki* Göçmen, Kumlutas & Tosunoglu, 1996. *Herpetozoa*, 30 (2017), 39–48.

64. A. K. Bhilala, K. Ashaharraza, M. Ingle, et al., Records of Günther's gracile skink, *Riopa guentheri* (Peters, 1879) (Reptilia: Scincidae: Lygosominae) from Central India. *Records of the Zoological Survey of India*, 121 (2021), 47–53.

65. A. E. Greer, A phylogenetic subdivision of Australian skinks. *Records of the Australian Museum*, 32 (1979), 339–371.

66. G. K. Noble and H. T. Bradley, The mating behavior of lizards; its bearing on the theory of sexual selection. *Annals of the New York Academy of Sciences*, 35 (1933), 25–100.

67. A. E. Greer, *The Biology and Evolution of Scincid lizards*. www.academia.edu/35305801/The_Biology_and_Evolution_of_Scincid_Lizards.doc: Academia (2007).

68. C. A. Domergue, Observations sur les hémipênis des ophidiens et sauriens de Madagascar (1). *Bulletin de l'Académie Malgache*, [1963] (1963), 21–33.

69. M. Lang, Generic relationships within Cordyliformes (Reptilia : Squamata). *Bulletin de l'Institut Royal des Sciences Naturelles de Belgique*, 61 (1991), 121–188.

70. W. Presch, Descriptions of the hemipenial morphology in eight species of microteiid lizards (Family Teiidae, Subfamily Gymnophthalmidae). *Herpetologica*, 34 (1978), 108–112.

71. M. T. Rodrigues, R. Recoder, M. Teixeira Jr., et al., A morphological and molecular study of Psilops, a replacement name for the Brazilian microteiid lizard genus *Psilophthalmus* Rodrigues 1991 (Squamata, Gymnophthalmidae), with the description of two new species. *Zootaxa*, 4286 (2017), 451–482.

72. C. W. Myers and M. A. Donnelly, The summit herpetofauna of Auyantepui, Venezuela: report from the Robert G. Goelet American Museum–Terramar expedition. *Bulletin of the American Museum of Natural History*, 308 (2008), 1–147.

73. M. B. da Silva, G. R. de Lima-Filho, Á. A. Cronemberger, et al., Description of the hemipenial morphology of *Tupinambis quadrilineatus* Manzani and Abe, 1997 (Squamata, Teiidae) and new records from Piauí, Brazil. *ZooKeys*, 361 (2013), 61–72.

74. P. M. Nunes. Morfologia hemipeniana dos lagartos microteídeos e suas implicações nas relações filogenéticas da família Gymnophthalmidae *(Teiioidea: Squamata)*. Vols. I and II. (São Paulo, Universidade de São Paulo, 2011).

75. M. A. Ribeiro-Júnior, P. M. Sánchez-Martínez, L. J. C. de Lima Moraes, et al., Uncovering hidden species diversity of alopoglossid lizards in Amazonia, with the description of three new species of *Alopoglossus* (Squamata: Gymnophthalmoidea). *Journal of Zoological Systematics and Evolutionary Research*, 59 (2021), 1322–1356.

76. C. Hernández Morales, M. J. Sturaro, P. M. Nunes, et al., A species-level total evidence phylogeny of the microteiid lizard family Alopoglossidae (Squamata: Gymnophthalmoidea). *Cladistics*, 36 (2020), 301–321.

77. O. J. Arribas, Hemipenial morphology and microornamentation in *Iberolacerta* Arribas, 1997 (Squamata: Lacertidae). *Butlletí de la Societat Catalana d'herpetologia*, 24 (2017), 12–23.

78. K. Klemmer, Untersuchungen zur Osteologie und Taxonomie der europäischen Mauereidechsen. *Abhandlungen der Senckenberg Gesellschaft für Naturforschung*, 496 (1957), 1–56.

79. P. H. Pinna, A. F. Mendonça, A. Bocchiglieri, and D. S. Fernandes, A new two-pored Amphisbaena Linnaeus from the endangered Brazilian Cerrado biome (Squamata: Amphisbaenidae). *Zootaxa*, 2569 (2010), 44–54.

80. R. Thomas and S. B. Hedges, Two new species of *Amphisbaena* (Reptilia: Squamata: Amphisbaenidae) from the Tiburon Peninsula of Haiti. *Caribbean Journal of Science*, 42 (2006), 208–219.

81. W. Böhme and T. Ziegler, A review of iguanian and anguimorph lizard genitalia (Squamata: Chamaeleonidae; Varanoidea, Shinisauridae, Xenosauridae, Anguidae) and their phylogenetic significance: comparisons with molecular data sets. *Journal of Zoological Systematics and Evolutionary Research*, 47 (2009), 189–202.

82. F. Glaw, J. Köhler, O. Hawlitschek, et al., Extreme miniaturization of a new amniote vertebrate and insights into the evolution of genital size in chameleons. *Scientific Reports*, 11 (2021), 2522.

83. F. Glaw, M. Vences, T. Ziegler, et al., Species distinctness and biogeography of

the dwarf chameleons *Brookesia minima*, *B. peyrierasi* and *B. tuberculata* (Reptilia: Chamaeleonidae): evidence from hemipenial and external morphology. *Journal of Zoology, London*, 247 (1999), 225–238.

84. D. F. Hughes, C. Kusamba, M. Behangana, and E. Greenbaum, Integrative taxonomy of the Central African forest chameleon, *Kinyongia adolfifriderici* (Sauria: Chamaeleonidae), reveals underestimated species diversity in the Albertine Rift. *Zoological Journal of the Linnean Society*, 181 (2017), 400–438.

85. C. J. Raxworthy and R. A. Nussbaum, Systematics, speciation and biogeography of the dwarf chameleons (*Brookesia*; Reptilia, Squamata, Chamaeleontidae) of northern Madagascar. *Journal of Zoology, London*, 235 (1995), 525–558.

86. H. I. Rosenberg, A. M. Bauer, and A. P. Russell, External morphology of the developing hemipenes of the dwarf chameleon, *Bradypodion pumilum* (Reptilia: Chamaeleonidae). *Canadian Journal of Zoology*, 67 (1989), 884–890.

87. V. Deepak, A. Khandekar, R. Chaitanya, and P. Karanth, Descriptions of two new endemic and cryptic species of *Sitana* Cuvier, 1829 from peninsular India. *Zootaxa*, 4434 (2018), 327–365.

88. K. Maduwage and A. Silva, Hemipeneal morphology of Sri Lankan dragon lizards (Sauria: Agamidae). *Ceylon Journal of Science (Biological Sciences)*, 41 (2012), 111–123.

89. V. Deepak, F. Tillack, N. B. Kar, et al., A new species of *Sitana* (Squamata: Agamidae) from the Deccan Peninsula Biogeographic Zone of India. *Zootaxa*, 4948 (2021), 261–274.

90. A. B. D'Angiolella, J. Klaczko, M. T. Rodrigues, and L. J. Avila-Pires, Hemipenial morphology and diversity in South American anoles (Squamata: Dactyloidae). *Canadian Journal of Zoology*, 94 (2016), 251–256.

91. G. Köhler, A. Batista, M. Vesely, et al., Evidence for the recognition of two species of *Anolis* formerly referred to as *A. tropidogaster* (Squamata: Dactyloidae). *Zootaxa*, 3348 (2012), 1–23.

92. C.-P. Blanc, Reptiles Sauriens Iguanidae. *Faune de Madagascar*, 45 (1977), 1–195.

93. H. G. Dowling, T. C. Majupuria, and F. W. Gibson, Hemipenial morphology of the tree lizard, *Plica plica* (Linnaeus). *Herpetological Review*, 3 (1971), 91–92.

94. M. Quipildor, A. S. Quinteros, and F. Lobo, Structure, variation, and systematic implications of the hemipenes of liolaemid lizards (Reptilia: Liolaemidae). *Canadian Journal of Zoology*, 96 (2018), 987–995.

95. R. Thomas and S. B. Hedges, New anguid lizard (*Diploglossus*) from Cuba. *Copeia*, 1998 (1998), 97–103.

96. W. Card and A. G. Kluge, Hemipeneal skeleton and varanid lizard systematics. *Journal of Herpetology*, 29 (1995), 275–280.

97. V. Weijola, S. C. Donnellan, and C. Lindqvist, A new blue-tailed Monitor lizard (Reptilia, Squamata, *Varanus*) of the *Varanus indicus* group from Mussau Island, Papua New Guinea. *ZooKeys*, 568 (2016), 129–154.

98. S. M. Harrington and T. W. Reeder, Phylogenetic inference and divergence dating of snakes using molecules, morphology and fossils: new insights into convergent evolution of feeding morphology and limb reduction. *Biological Journal of the Linnean Society*, 121 (2017), 379–394.

99. H. Zaher and K. T. Smith, Pythons in the Eocene of Europe reveal a much older divergence of the group in sympatry with boas. *Biology Letters*, 16 (2020), 20200735.

100. S. B. McDowell, A catalogue of the snakes of New Guinea and the Solomons, with special reference to those in the Bernice P. Bishop Museum, Part I. Scolecophidia. *Journal of Herpetology*, 8 (1974), 1–57.

101. C. W. Myers and L. Trueb, The hemipenis of an anomalepidid snake. *Herpetologica*, 23 (1967), 235–238.

102. M. Fabrezi, A. Marcus, and G. Scrocchi, Contribución al conocimiento de los Leptotyphlopidae de Argentina. I. *Leptotyphlops weyrauchi* y *Leptotyphlops albipuncta*. *Cuadernos de Herpetología*, 1 (1985), 1–20.

103. B. R. Orejas-Miranda, Descripción del hemipenis de *Leptotyphlops munoai* Orejas-Miranda, 1961. *Comunicaciones Zoologicas del Museo de Historia Natural de Montevideo*, 97 (1962), 1–9.

104. J. Robb, The internal anatomy of *Typhlops* Schneider (Reptilia). *Australian Journal of Zoology*, 8 (1960), 181-216.

105. J. Robb, The generic status of the Australasian typhlopids (Reptilia: Squamata). *Annals and Magazine of Natural History*, 13 (1966), 106–108.

106. J. A. Peters and B. R. Orejas-Miranda, Notes on the hemipenis of several taxa in the family Leptotyphlopidae. *Herpetologica*, 26 (1970), 320–324.

107. D. G. Broadley and V. Wallach, A revision of the genus *Leptotyphlops* in northeastern Africa and southwestern Arabia (Serpentes: Leptotyphlopidae). *Zootaxa*, 1408 (2007), 1–78.

108. A. C. Ferreira, J. Klaczko, and A. Martins, Hemipenial morphology of *Epictia vellardi* (Laurent, 1984) (Leptotyphlopidae, Serpentes) with the proposition and discussion of two general hemipenial patterns within the genus *Epictia*. *Zoomorphology*, 140 (2020), 143–150.

109. A. Martins, C. Koch, R. Pinto, et al., From the inside out: Discovery of a new genus of threadsnakes based on anatomical and molecular data, with discussion of the leptotyphlopid hemipenial morphology. *Journal of Zoological Systematics and Evolutionary Research*, 57 (2019), 840–863.

110. R. Pinto and F. F. Curcio, On the generic identity of *Siagonodon brasiliensis*, with the description of a new leptotyphlopid from central Brazil (Serpentes: Leptotyphlopidae). *Copeia*, 2011 (2011), 53–63.

111. V. Wallach, Morphological review and taxonomic status of the *Epictia phenops* species group of Mesoamerica, with description of six new species and discussion of South American *Epictia albifrons*, *E. goudotii*, and *E. tenella* (Serpentes: Leptotyphlopidae: Epictinae). *Mesoamerican Herpetology*, 3 (2016), 216–374.

112. A. H. Wynn, R. P. Reynolds, D. W. Buden, et al., The unexpected discovery of blind snakes (Serpentes: Typhlopidae) in Micronesia: two new species of *Ramphotyphlops* from the Caroline Islands. *Zootaxa*, 3172 (2012), 39–54.

113. P. Passos, U. Caramaschi, and R. R. Pinto, Redescription of *Leptotyphlops koppesi* Amaral, 1954, and description of a new species of the *Leptotyphlops dulcis* group from Central Brazil (Serpentes: Leptotyphlopidae). *Amphibia–Reptilia*, 27 (2006), 347–357.

114. P. Passos, U. Caramaschi, and R. R. Pinto, Rediscovery and redescription of *Leptotyphlops salgueiroi* Amaral, 1954 (Squamata, Serpentes, Leptotyphlopidae). *Boletim do Museu Nacional*, 520 (2005), 1–10.

115. R. A. Pyron and V. Wallach, Systematics of the blindsnakes (Serpentes: Scolecophidia: Typhlopoidea) based on molecular and morphological evidence. *Zootaxa*, 3829 (2014), 1–81.

116. R. Thomas, The relationships of Antillean *Typhlops* (Serpentes: Typhlopidae) and the description of three new Hispaniolan species. *Biogeography of the West Indies*, 1989 (1989), 409–432.

117. R. Thomas. Systematics of the Antillean Blind Snakes of the Genus *Typhlops* (Serpentes: Typhlopidae). (LSU Historical Dissertations and Theses 1976).

118. A. H. Wynn and A. E. Leviton, Two new species of blind snake, genus *Typhlops* (Reptilia: Typhlopidae), from the Philippine archipelago. *Proceedings of the Biological Society of Washington*, 106 (1993), 34–45.

119. J. Dixon and F. S. Hendricks, The wormsnakes (family Typhlopidae) of the Neotropics, exclusive of the Antilles. *Zoologische Verhandelungen*, 173 (1979), 1–39.

120. F. F. Curcio, P. M. Nunes, A. J. S. Argolo, et al., Taxonomy of the South American dwarf boas of the genus *Tropidophis* Bibron, 1840, with the description of two new species from the Atlantic forest (Serpentes: Tropidophiidae). *Herpetological Monographs*, 26 (2012), 80–121.

121. F. W. Gibson, The 'quadrifurcate' hemipenis of *Tropidophis. Herpetological Review*, 2 (1970), 29–30.

122. S. B. McDowell, A catalogue of the snakes of New Guinea and the Solomons, with special reference to those in the Bernice P. Bishop Museum. Part II. Anilioidea and Pythoninae. *Journal of Herpetology*, 9 (1975), 1–79.

123. M. A. Smith, The fauna of British India, Ceylon and Burma, including the whole of the Indo-Chinese sub-region. *Reptilia and Amphibia. Vol III. – Serpentes.* (London: Taylor and Francis, 1943).

124. D. J. Gower and J. L. M. Wickramasinghe, Recharacterization of *Rhinophis dorsimaculatus* Deraniyagala, 1941 (Serpentes: Uropeltidae), including description of new material. *Zootaxa*, 4158 (2016), 203–212.

125. V. P. Cyriac, S. Narayanan, F. L. Sampaio, et al., A new species of *Rhinophis* Hemprich, 1820 (Serpentes: Uropeltidae) from the Wayanad region of peninsular India. *Zootaxa*, 4778 (2020), 329–342.

126. R. A. Pyron, S. R. Ganesh, A. Sayyed, et al., A catalogue and systematic overview of the shield-tailed snakes (Serpentes: Uropeltidae). *Zoosystema*, 38 (2016), 453–506.

127. D. J. Gower and K. Maduwage, Two new species of *Rhinophis* Hemprich (Serpentes: Uropeltidae) from Sri Lanka. *Zootaxa*, 2881 (2011), 51–68.

128. R. Stuebing, A new species of *Cylindrophis* (Serpentes: Cylindrophiidae) from Sarawak, Western Borneo. *Raffles Bulletin of Zoology*, 42 (1994), 967–73.

129. G. Underwood, A Contribution to the Classification of Snakes (London: *British Museum of Natural History*, 1967).

130. N. Vidal, A.-S. Delmas, and S. B. Hedges. The higher-level relationships of alethinophidian snakes inferred from seven nuclear and mitochondrial genes. In R. W. Henderson, R. Powell, eds., *Biology of the Boas and Pythons* (Eagle Mountain, Utah: Eagle Mountain Publishing, 2007), pp. 27–33.

131. H. G. Dowling. The Nearctic snake fauna. In H. G. Dowling, ed., 1974 *Yearbook of Herpetology* (New York: HISS Publications, 1974), pp. 191–202.

132. W. Böhme and U. Sieling, Zum Zusammenhang zwischen Genital Struktur, Paarungsverhalten und Fortpflanzungserfolg bei squamaten Reptilien: erste Ergehnisse. *Herpetofauna*, 15 (1993), 15–23.

133. A. G. Kluge, *Aspidites* and the phylogeny of pythonine snakes. *Records of the Australian Museum*, 19 (1993), 1–77.

134. C. A. Domergue, Observations sur les pénis des ophidiens (deuxième partie). *Bulletin de la Société des Sciences Naturelles et Physiques du Maroc*, 42 (1962), 87–105.

135. A. R. Hoge, A new genus and species of Boinae from Brazil. *Xenoboa cropanii*, gen. nov., sp. nov. *Memorias do Instituto Butantan*, 25 (1953), 27–34.

136. A. G. Kluge, *Calabaria* and the phylogeny of erycine snakes. *Zoological Journal of the Linnean Society*, 107 (1993), 293–351.

137. W. R. Branch, Hemipenes of the Madagascan boas *Acrantophis* and *Sanzinia*, with a review of hemipenial morphology in the Boinae. *Journal of Herpetology*, 15 (1981), 91–99.

138. S. B. McDowell, A catalogue of the snakes of New Guinea and the Solomons, with special reference to those in the Bernice P. Bishop Museum. Part III. Boinae and Acrochordoidea (Reptilia, Serpentes). *Journal of Herpetology*, 13 (1979), 1–92.

139. P. Passos and R. Fernandes, Revision of the *Epicrates cenchria* complex (Serpentes: Boidae). *Herpetological Monographs*, 22 (2008), 1–30.

140. K. Andonov, N. Natchev, Y. V. Kornilev, and N. Tzankov, Does sexual selection influence ornamentation of hemipenes in Old World snakes? *Anatomical Record*, 300 (2017), 1680–1694.

141. C. M. Bogert, The variations and affinities of the dwarf boas of the genus *Ungaliophis*. *American Museum Novitates*, 2340 (1968), 1–26.

142. V. Deepak, S. Ruane, and D. J. Gower, A new subfamily of colubroid fossorial snakes from the Western Ghats of peninsular India. *Journal of Natural History*, 52 (2019), 2919–2934.

143. H. Zaher, R. W. Murphy, J. C. Arredondo, et al., Large-scale molecular phylogeny, morphology, divergence-time estimation, and the fossil record of advanced caenophidian snakes (Squamata: Serpentes). *PLoS ONE*, 14 (2019), e0216148.

144. C. W. Myers, A new genus and new tribe for *Enicognathus melanauchen* Jan, 1863, a neglected South American snake (Colubridae: Xenodontinae), with taxonomic notes on some Dipsadinae. *American Museum Novitates*, 3715 (2011), 1–33.

145. D. A. Rossman and W. G. Eberle, Partition of the genus *Natrix*, with preliminary observations on evolutionary trends in natricine snakes. *Herpetologica*, 33 (1977), 34–43.

146. S. B. McDowell, [Review of] Systematic division and evolution of the colubrid snake genus *Natrix*, with comments on the subfamily Natricinae, by Edmond V. Malnate. *Copeia*, 1961 (1961), 502–506.

147. H. Zaher, F. G. Grazziotin, R. Graboski, et al., Phylogenetic relationships of the genus *Sibynophis* (Serpentes: Colubroidea). *Papéis Avulsos de Zoologia (São Paulo)*, 52 (2012), 141–149.

The Evolution of Sperm-Storage Location in Squamata with Particular Reference to Snakes

Henrique B. Braz and Selma M. Almeida-Santos

17.1 Introduction

Female sperm storage is characterized by a temporary maintenance of viable sperm between insemination and oocyte fertilization [1]. Sperm storage provides various advantages for female squamate reptiles, such as decreased copulation frequency and costs, fertilization assurance in cases of not finding a mate, production of multiple clutches, enhancement of offspring quality through sperm competition, and synchronization of sperm release and fertilization with optimal periods [1–5]. Therefore, female sperm storage has clear implications for the evolution of squamates by dramatically influencing the evolution of life histories, mating systems, post-copulatory sexual selection (e.g., cryptic female choice and sperm competition), and sexual conflict [1]. Importantly, sperm must remain viable in the female reproductive tract during the storage period, until ovulation and fertilization [1]. Accordingly, sperm-storage sites and structures should be under strong selection.

The anatomy and evolution of sperm storage in female squamates have been periodically reviewed [6–14]. The occurrence of female sperm storage is conserved across squamate phylogeny, but the sites of sperm storage in the oviduct vary considerably among lineages, with multiple evolutionary transitions between character states [11]. In the last decade, numerous studies addressing the anatomy of female sperm storage have been conducted in various squamate lineages, particularly in snakes [15–39]. This increased number of studies, combined with improved understanding of squamate phylogeny [40–42], creates a great opportunity to reassess the diversity and evolution of sperm storage in squamates.

Our major aim here is to provide insights into the evolution of female sperm storage in squamates, with particular consideration to the origin and early evolution of snakes. For this, we summarize the literature on the anatomy of female sperm storage and reconstruct

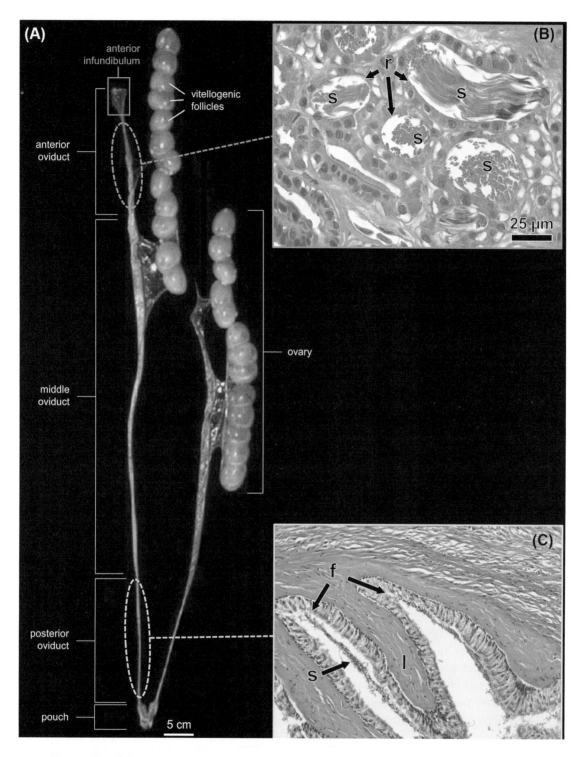

Figure 17.1 (A) Reproductive system of female lancehead snake (the crotaline viperid *Bothrops jararaca*). Three general oviductal regions are recognized in the squamate oviduct: the posterior

the evolution of sperm-storage location. We then propose a scenario for the evolution of female sperm storage in snakes. Higher-level snake classification follows Zaher et al. [41] and Burbrink et al. [42].

17.2 Sites and Structures for Sperm Storage

Oviductal morphology and sperm storage have been reported for at least 111 squamate species (70 lizards and 41 snakes) from various lineages (Supplementary Table 17.S1). Oviductal morphology is rather conserved among squamates (reviewed in [6, 7, 9, 10, 13]). Histologically, the oviduct is formed by three tissue layers: (1) an inner mucosa, formed by a luminal epithelium with ciliated and/or nonciliated cells and an underlying lamina propria, which may contain glands; (2) a middle muscularis, consisting of inner circular and/or outer longitudinal muscle layers; and (3) an outer serosa (reviewed in [6, 7, 9, 10, 13]). These layers vary widely in thickness and function throughout the oviduct. Three general regions are recognized grossly. The posterior oviduct is a thick-walled portion that contains a region referred to as the nonglandular uterus or uterovaginal junction, with a deeply folded mucosa. The middle oviduct, which contains the glandular uterus (where eggs reside), is the longest portion and is often slightly folded. The anterior oviduct can be divided into two regions: the posterior infundibulum (a folded and opaque region) and the anterior infundibulum (a less folded, translucent, and funnel-shaped region; Fig. 17.1).

Sperm storage occurs in the posterior oviduct, anterior oviduct, or both [8–10] (Fig. 17.1). In the posterior oviduct (hereafter: posterior sperm storage), sperm are stored in deep furrows formed by longitudinal projections of the lamina propria, in crypts in the furrows, or in tubules formed by invaginations of the luminal epithelium [8–10] (Fig. 17.1). In the anterior oviduct (hereafter: anterior sperm storage), sperm are typically stored in tubules formed by invaginations of the luminal epithelium [8, 9, 43, 44] (Fig. 17.1). The tubules of both the posterior and anterior oviduct have been given various terms such as sperm-storage tubules, sperm-storage receptacles, seminal receptacles, or '*fausses glandes de la trompe*' [9]. It is unclear whether these terms reflect any morphological or functional difference. For consistency with previous reviews [8, 11], we use the term sperm-storage tubules (SSTs), even in the absence of spermatozoa.

In lizards, posterior sperm storage occurs in tubules in iguanians [8, 43, 45] and in crypts in other groups [46–49]. In snakes, however, posterior sperm storage occurs mostly in crypts

---

Figure 17.1 (*cont.*) oviduct, the middle oviduct, and the anterior oviduct. Sperm storage in squamates occurs in the posterior oviduct, anterior oviduct, or both. (B) Histological section (haematoxylin-eosin) of the anterior oviduct (posterior infundibulum) of the dipsadid snake *Apostolepis gaboi* showing sperm stored in infundibular glands (sperm receptacles). Note the parallel alignment of the spermatozoa in the sperm receptacles. (C) Histological section (haematoxylin-eosin) of the posterior oviduct of *B. jararaca* showing sperm stored in the furrows. f: furrows, l: lamina propria, r: sperm receptacle, s: spermatozoa. Photomicrographs: Karina N. Kasperoviczus. See Supplementary Figure 17.S1 for colour version.

or furrows [22, 36, 38, 50, 51] (see also Supplementary Table 17.S1). Exceptions include the natricid *Liodytes pygaea* [52, 53] and the elapids *Micrurus corallinus* and *M. frontalis* [17] in which posterior sperm storage occurs in tubules. These tubules are undifferentiated from the epithelium in *L. pygaea* [52, 53] but differentiated in *Micrurus* spp. [17]. In the typhlopid scolecophidian *Amerotyphlops brongersmianus* and the leptotyphlopid scolecophidian *Trilepida koppesi*, no crypts or SSTs have been observed in the posterior oviduct [25, 32]. The extent to which the absence of crypts and SSTs in the posterior oviduct is widespread in scolecophidians remains to be determined via investigation of more species. Therefore, most snake species examined so far either have unspecialized (i.e., undifferentiated from the epithelium) or lack sperm-storage structures in the posterior oviduct. However, the lack of specialized structures does not mean that the posterior oviduct is unfit for sperm storage, because sperm may reside in this region for prolonged periods in various snakes [50, 54].

Some snakes have evolved morphological adaptations to hold sperm in the posterior oviduct. In many viperids, the posterior oviduct is convoluted and contracted in newly mated females. This convolution was initially described as a region of prolonged sperm storage in the crotaline *Crotalus viridis viridis* [55], but later interpreted as a mechanism preventing additional inseminations in the viperine *Vipera berus* [56]. A subsequent study of the crotaline *Crotalus durissus* corroborated that the posterior oviduct becomes convoluted soon after autumnal mating [54]. This convolution has subsequently been termed 'uterine muscular twisting' and reported in various crotalines [15, 19, 23, 27, 30, 33, 35, 57–59]. A recent anatomical study has shown that, in *C. durissus*, this convolution is formed by a coiling (rather than a twisting) of the inner layers of the posterior oviduct [24]. Because oviductal morphology is very conserved in snakes [6, 9], we expect that the coiling pattern described in *C. durissus* likely occurs also in other viperids previously reported to exhibit uterine muscular twisting. Thus, we apply the term 'uterine muscular coiling' (UMC) rather than 'uterine muscular twisting'. In *C. durissus*, UMC persists until spring, when the oviduct relaxes and releases sperm to migrate up the oviduct and fertilize the oocytes [54]. Thus, UMC functions as a mechanism of long-term sperm storage by keeping sperm alive and viable in the posterior oviduct for at least five months. The formation and maintenance of UMC in *C. durissus* may be related to an oestradiol/progesterone balance with a possible participation of arginine vasotocin [60, 61]. However, nothing is known about the hormonal control of UMC in other crotalines. The question remains as to the effectiveness of UMC in preventing spermatozoa from additional inseminations from entering the posterior oviduct and/or keeping them confined to this region [62, 63]. In *V. berus*, UMC apparently decreases but does not prevent sperm from other copulations to penetrate the coiled region [62, 64]. In some lanceheads (*Bothrops*), sperm have been observed in the infundibular SSTs during the mating season, suggesting that some spermatozoa are able to escape the trap created by UMC and ascend the infundibulum long before ovulation [19, 23, 33] or remain stored there for successive breeding seasons. Even if UMC does not completely block sperm, it is likely relevant in controlling the movement of sperm in and out of the posterior oviduct.

Unlike the posterior oviduct, SSTs in the anterior oviduct (infundibulum) are often differentiated (as uniformly ciliated or uniformly secretory) from the surrounding epithelium in most squamates [9, 10, 17, 19, 36, 37]. Infundibular SSTs vary morphologically across species, appearing as branched tubular, simple to branched alveolar, or branched to compound alveolar glands, and their distal portions function as sperm receptacles [9, 43, 65]. In lizards, infundibular SSTs occur in most gekkotans and the few lacertids and anguids examined but are absent in the iguanians, amphisbaenians, and scincids studied thus far (Supplementary Table 17.S1). In contrast, infundibular SSTs are almost ubiquitous in snakes (Supplementary Table 17.S1). The only known exception is the colubrid *Boiga irregularis* [66]. A recent study has also reported the absence of infundibular SSTs in the homalopsid *Cerberus rynchops* collected in January [67]. Because infundibular SSTs are formed seasonally in some snake species [25, 68, 69], the lack of these structures in *C. rynchops* requires confirmation.

17.3 Sperm Transit in the Oviduct

Sperm aggregations in the posterior oviduct are suggested to reflect the large amount of ejaculate released into the contiguous pouch in recent matings and/or the bottleneck created by the luminal narrowing between the pouch and oviducts [9]. Indeed, abundant sperm may be seen in the lumen of the posterior oviduct within 24 hours after mating [50, 56, 68]. Sperm initially fill the lumen and crypts or furrows of the posterior oviduct but are later seen only in the crypts or furrows and tubules, where they reside for varied periods in different species [28, 36, 43].

In various squamates, males deposit a gelatinous copulatory plug in the female cloaca [70-75]. Although copulatory plugs are commonly interpreted as a paternity assurance mechanism [76, 77], they may also play a role in sperm transfer and transit in the oviduct. In the natricid *Thamnophis sirtalis*, copulatory plugs reduce sperm leakage from the female cloaca and function as a spermatophore that release sperm gradually [78]. Whether copulatory plugs exhibit similar functions in other squamates has yet to be demonstrated.

Sperm later migrate up to the anterior oviduct, apparently attracted by chemotactic stimulation and/or ciliary beating of cells lining the infundibulum and SSTs epithelium [9, 65, 79]. Additional factors that may contribute to attracting sperm to the infundibulum and SSTs include intrinsic sperm motility, mobilization by carrier matrices, contractions of the oviduct smooth muscle, or some combination thereof [6, 50, 65, 80, 81]. In species exhibiting only anterior sperm storage, sperm migrate promptly to the posterior oviduct (e.g., the dipsadid *Erythrolamprus miliaris* [37]). As sperm reach the anterior oviduct (infundibulum), they usually fill the most posterior receptacles first so that the most anterior receptacles may never be filled, as observed in the natricid *Thamnophis* [65, 79]. It is unclear how widespread this pattern of SST filling is in squamates. Devine [76] suggested that if a female mated multiple times, sperm from a second (or a third) mating might fill the most anterior SSTs, thus taking precedence to fertilize the eggs. From the female perspective, this strategy may be advantageous by promoting sperm competition and cryptic choice. A female would have the potential to ensure that a subsequent preferred male was more likely to sire her offspring, thus benefiting from multiple matings.

In some squamates, sperm reach the posterior infundibulum but have never been observed within the SSTs. Examples include the lacertid *Podarcis muralis* [70] and the viperid *Lachesis muta* [30]. In other squamates, despite the existence of infundibular SSTs, sperm have been found only in the posterior oviduct, as in some leptotyphlopid and typhlopid scolecophidians [82], the elapids *Micrurus* spp. [17], the viperid *Crotalus durissus* (S. M. Almeida-Santos, pers. obs.), and the dipsadids *Atractus pantostictus* [34] and *Tomodon dorsatus* [21]. The non-detection of sperm in the infundibulum or infundibular SSTs may obviously reflect a failure to find sperm in these species. However, except for the scolecophidians, the above species have been reasonably well investigated. Thus, the observations reported above suggest that if sperm do enter the infundibular SSTs in these species, they apparently remain there for a very brief period. Infundibular SSTs would then function as a site for short-term sperm storage in these species. Some authors have suggested that infundibular SSTs function to attract sperm and then provide mechanical protection during ovulation [9, 65].

Sperm can reside in the oviduct for up to two months from mating to fertilization (i.e., short-term sperm storage) or for longer periods (i.e., long-term sperm storage) [2, 50]. Eckstut et al. [11] suggested that sperm-storage duration is correlated with sperm-storage location based on the observation of sperm storage lasting up to four months in three taxa exhibiting posterior sperm storage and more than six months in three taxa exhibiting anterior sperm storage. However, it should be noted that long-term (> 2 months) sperm storage has also been observed in the posterior oviduct of some squamates [50, 54, 83], and short-term sperm storage has also been observed or suggested in the anterior oviduct of others (e.g., [18, 25, 34]). Although the hypothesis of association between sperm-storage location and sperm-storage duration is plausible, many more species need to be investigated before this can be adequately tested.

Little is known of the underlying mechanisms that sustain sperm in the oviduct for prolonged periods, but it certainly involves biochemical, physiological, and anatomical specializations of the oviduct and the sperm [7, 9, 65]. The synthesis of neutral carbohydrate secretions in the infundibular SSTs increases during mating season or in the presence of sperm in some species, suggesting that these components might play some role in maintaining sperm [36, 37, 50]. Similarly, the luminal epithelium of the UMC region in viperids secretes substances that may help to maintain sperm viability during the storage period [19, 84]. Sperm sustenance may also be influenced by secretions deposited in the seminal fluid by males [85], because secretory material of putatively male origin has been observed in the posterior oviduct (e.g., [86]). However, male contribution is likely to aid only in short-term sperm storage. For example, in the viperid *Crotalus durissus*, the composition of the seminal fluid changes after it reaches the posterior oviduct [84], suggesting no lasting contribution from male-produced components.

Sperm often appear as unordered masses in the lumen of the posterior and anterior oviduct but as packed, parallel bundles with their heads aligned and oriented toward the epithelial cells of the crypts and SSTs [18, 43, 50, 65]. This ordered arrangement is suggested to reduce sperm metabolism and thus conserve energy [65]. In various snake species, long-term sperm storage occurs during winter [50, 54, 57], and lower body temperatures and metabolic rates might also facilitate prolonged sperm storage [87].

17.4 Historical Overview of the Evolution of Sperm-Storage Location in Squamata

The evolution of sperm-storage location and structures in squamates was previously addressed by some authors. Sever and Hamlett [8] reviewed the anatomy of sperm-storage structures and phylogenetically estimated ancestral sperm-storage location. They inferred that the ancestral squamate had infundibular SSTs and stored sperm in the anterior oviduct, with posterior sperm storage evolving once, in the ancestor of Iguania. Their ancestral-state estimations also indicated that the ancestral snake had infundibular sperm storage structures and stored sperm in the anterior oviduct.

Eckstut et al. [11] inferred the evolution of sperm-storage traits of approximately 30 squamate species (27 lizards and 3 snakes) on three competing phylogenies. Regardless of the phylogeny used, most characters investigated showed strong phylogenetic conservatism, including occurrence of sperm storage, presence and structure of SSTs, embedding of sperm within the sperm-storage structures, and sperm storage during gravidity or pregnancy. However, the location of sperm-storage structures, location of secretory cells in sperm-storage structures, and sperm-storage duration varied greatly [11], and different scenarios for the evolution of sperm-storage location in squamates were recovered. Assuming the posterior oviduct as the ancestral location of sperm storage in reptiles, this condition was also inferred in the ancestral squamate. However, anterior sperm storage was inferred to have evolved once and posterior sperm storage to have re-evolved once on the morphological phylogeny, whereas anterior sperm storage evolved independently three times on the molecular phylogeny [11]. Moreover, one evolutionary origin of posterior sperm storage (while retaining anterior storage) and one of anterior storage (while retaining posterior storage) were also inferred on the morphological and molecular phylogenies. Assuming instead anterior sperm storage as the ancestral sperm-storage location in reptiles, this state was also inferred in the ancestral squamate, with multiple independent origins of posterior sperm storage. Regardless of the phylogeny and outgroup, the ancestral site of sperm storage in snakes was inferred to be the anterior oviduct, as proposed by Sever and Hamlett [8].

In their review of female sperm storage, Siegel et al. [9] considered that both the Eckstut et al. and Sever and Hamlett studies had limitations because they considered only the site where sperm-storage structures occur and not whether such sites had prolonged sperm aggregations. They hypothesized that posterior sperm storage was the ancestral condition in Lepidosauria and that sperm-storage structures likely evolved independently in the major squamate lineages. They also noticed that all snakes that had been examined by that time (except the colubrid *Boiga irregularis*) had infundibular SSTs, but that they also retained the plesiomorphic condition of posterior sperm storage.

The evolution of UMC and sperm-storage location in viperids has also been discussed repeatedly in the literature. Due to the occurrence of UMC in various crotalines and one viperine, this mechanism has been suggested as an ancestral strategy in viperids, used to retain sperm in the posterior oviduct from autumn mating to spring ovulation [30, 57, 60].

However, other authors have suggested instead that the anterior oviduct is the ancestral site of sperm storage in viperids [63].

17.5 A Revised Perspective on the Evolution of Sperm-Storage Location in Squamata

To provide fresh insight into the evolution of female sperm storage in squamates, we estimated ancestral states of infundibular SSTs and sperm-storage location. We also mapped potential origins of SSTs in the posterior oviduct. We used mostly species for which the two sperm-storage regions were examined for most of the annual cycle. We did not consider the mere presence of sperm in the lumen of the female reproductive tract as evidence of sperm storage. Strong evidence that sperm are being stored (and are not only in transit) includes one or more of the following conditions: (1) sperm aggregations occurring in specialized structures; (2) consistent finding of sperm in pre-existing sites (if modified structures are lacking); and (3) an organized sperm arrangement, suggesting interaction with the epithelial cells of the female reproductive tract [1]. We formulated two states for the character 'infundibular SSTs' (present or absent) and four states for the character 'sperm storage location' (posterior oviduct, anterior oviduct, both regions, and none). Thus, for posterior sperm storage, we did not distinguish between anterior and posterior nonglandular uterus, although this distinction may be worth examining in future studies. We reconstructed character states using maximum parsimony with 'unordered' states in Mesquite, version 3.61 [88]. We used the phylogeny of Burbrink et al. [42] as a backbone topology and positioned Phyllodactylidae (absent in their tree) as the sister group of Gekkonidae, as recovered in various studies (e.g., [40, 89]). Relationships among genera and species followed Zaher et al. [41] for Colubroidea, Alencar et al. [90] for Viperidae, and Pyron et al. [40] for Gekkonidae, Scincidae, and Lacertidae. As outgroup, we used the tuatara, *Sphenodon punctatus*, the only living representative of Rhynchocephalia, the only extant non-squamatan member of Lepidosauria. In previous reviews, *S. punctatus* was judged to exhibit long-term sperm storage because of the long dissociation between mating in autumn and egg-laying in spring [81, 91]. No specialized sperm-storage structures have been found in the posterior and anterior oviduct of female tuatara [92]. In view of this, it has been hypothesized that long-term sperm storage would occur in the lumen of the posterior oviduct [81, 91]. However, female *S. punctatus* actually ovulate 1–2 months after mating, and embryos enter preovipositional arrest during winter until oviposition [93]. Thus, there is no evidence of long-term sperm storage, although sperm still may reside in the female reproductive tract for a short period (1–2 months) until ovulation [93]. We then reconstructed the evolution of sperm storage location considering two possibilities for *Sphenodon*: (1) no sperm storage and (2) posterior sperm storage. Assuming that *S. punctatus* lacks sperm storage, parsimony reconstruction required 21 steps and retrieved ambiguous ancestral states for most major clades (Lepidosauria, Squamata, Unidentata, Episquamata, Toxicofera, Anguidae + Iguania, Serpentes, and Endoglyptodonta;

Supplementary Figure 17.S2). The assumption that *S. punctatus* has posterior sperm storage retrieved a slightly more parsimonious solution (20 steps) and retrieved unambiguous states for most major lepidosaurian clades. Accordingly, we describe the results of the reconstruction following Saint Girons' [81, 91] suggestion of posterior sperm storage in *S. punctatus* pending confirmation as to whether true sperm storage occurs and where.

As previously noted [11], sperm-storage location varies considerably across squamate phylogeny (Fig. 17.2B). Parsimoniously, the ancestor of both Lepidosauria and Squamata lacked infundibular SSTs and stored sperm in the posterior oviduct (Fig. 17.2A, B), an inference that disagrees with Sever and Hamlett [8], partially agrees with Eckstut et al. [11], and fully agrees with Siegel et al. [9]. Various major clades of Squamata (Unidentata, Episquamata, Scincomorpha, and Laterata) are inferred as retaining the plesiomorphic absence of infundibular SSTs and presence of posterior sperm storage (Fig. 17.2A, B). Absence of infundibular SSs and presence of posterior sperm storage are also inferred as plesiomorphic for Iguania, though it is ambiguous whether the lack of infundibular SSTs in this clade is a retention of a squamatan plesiomorphy or loss of these structures that might have evolved in the ancestor of Toxicofera and been present in the ancestor of Iguania + Anguidae (Fig. 17.2A, B).

Infundibular SSTs appear to have evolved independently multiple times (at least 3–4 origins) among the sampled squamates (Fig. 17.2A). SSTs in the posterior oviduct apparently appeared only in iguanians, *Micrurus*, and *Liodytes pygaea* (Fig. 17.2A). In most sperm-storing lizards, storage occurs either in the anterior or posterior oviduct. In the ancestor of Gekkota, the origin of infundibular SSTs was accompanied by a shift from posterior to anterior sperm storage (Fig. 17.2A, B). Some gekkotan lineages (the diplodactylid *Woodworthia maculata* and the gekkonid *Heteronotia binoei*) apparently re-evolved posterior sperm storage while retaining anterior sperm storage. It is unclear whether the same occurred in the gekkonid *Christinus marmoratus* (which has at least posterior sperm storage) because it is unknown whether this species has infundibular SSTs (Fig. 17.2A, B). A second origin of infundibular SSTs occurred in the ancestor of Lacertidae, but only *Acanthodactylus hardyi* has been shown thus far to store sperm in these structures (Fig. 17.2A, B). Most Laterata and the scincids *Eutropis carinata*, *Trachylepis atlantica*, and *Mabuya* spp., which lack infundibular SSTs, lost posterior sperm storage (Fig. 17.2A, B). An additional evolutionary origin of infundibular SSTs is inferred to have occurred either in the ancestor of Toxicofera (and retained in Anguidae), or there were two separate origins in Anguidae and in Serpentes (Fig. 17.2A).

Based on our analyses of available data, the ancestral snake unequivocally had infundibular SSTs, as previously proposed [9, 82]. However, the reconstruction of sperm-storage location recovered three equally parsimonious solutions for the ancestral snake. Scenario 1 suggests that the ancestral snake stored sperm only in the posterior oviduct (Fig. 17.2B). This scenario seems unlikely given that the ancestral snake unequivocally possessed infundibular SSTs (Fig. 17.2A), and presumably anterior sperm storage. Scenario 2 suggests that anterior sperm storage (with retention of posterior storage) occurred in the ancestral snake, posterior storage was lost in typhlopoid + leptotyphlopid scolecophidians, and anterior and posterior storage was retained in Endoglyptodonta (Fig. 17.2C). Under this

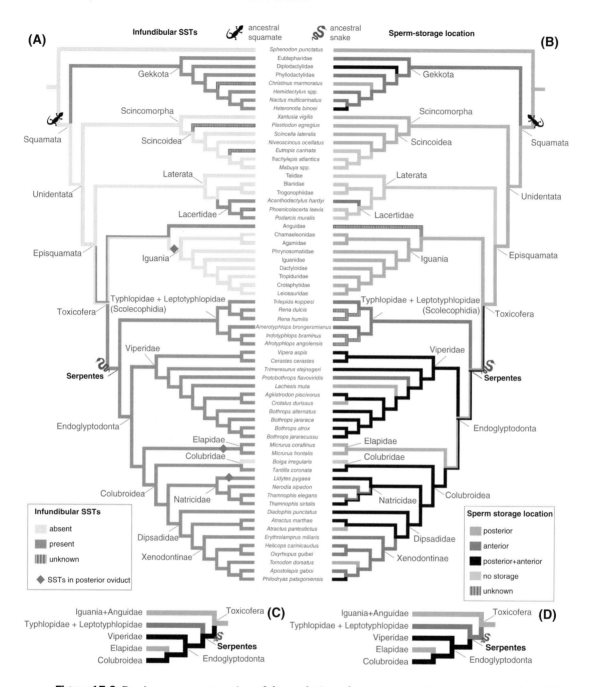

Figure 17.2 Parsimony reconstruction of the evolution of sperm storage in squamate reptiles with particular focus on snakes. (A) Ancestral-state reconstruction of the presence or absence of infundibular glands (sperm-storage tubules, SSTs). (B) Ancestral-state reconstruction of sperm-storage location. Terminal taxon names and branching order are identical in both trees. (C–D) Alternative solutions for the evolution of sperm-storage location in the ancestral snake considering (C) posterior + anterior sperm storage and (D) anterior sperm storage. (A black and white version of this figure will appear in some formats. For the colour version, please refer to the plate section).

scenario, the apparent loss of posterior oviduct crypts in typhlopoid + leptotyphlopid scolecophidians (as discussed in § 17.2) accompanied the transition from posterior to anterior storage. Scenario 3 suggests that a shift from posterior to anterior sperm storage occurred in the ancestral snake, this condition was retained in typhlopoid + leptotyphlopid scolecophidians, and posterior storage (while retaining anterior storage) re-evolved in the ancestor of Endoglyptodonta (Fig. 17.2D). Under this scenario, the loss of posterior oviduct crypts in typhlopoid + leptotyphlopid scolecophidians occurred subsequently to the transition from posterior to anterior sperm storage. A subsequent loss of infundibular SSTs occurred unequivocally in the colubrid *Boiga irregularis*. Data for anomalepidid scolecophidians might help to discriminate among the three scenarios outlined here.

Posterior + anterior sperm storage was unambiguously inferred as the ancestral state of both Viperidae and Colubroidea (and for the ancestral state of the constituent families of the latter clade, i.e., Natricidae, Dipsadidae, and Colubridae), and this condition was conserved in most of the viperids and some of the colubroids examined so far (Fig. 17.2B). Loss of posterior sperm storage (and retention of anterior sperm storage) is unequivocally inferred to have originated at least twice in Endoglyptodonta; once in the viperid *Protobothrops flavoviridis* and once in the ancestral xenodontine dipsadid (Fig. 17.2A). Because of the uncertainty that infundibular SSTs are used to store sperm in some species (as discussed in § 17.3), we refrain from interpreting the occurrence of sperm storage exclusively in the posterior oviduct in some endoglyptodonts (Fig. 17.2B) as a result of either the loss of anterior storage or a shift to posterior storage.

Unlike previous suggestions that the ancestral viperid stored sperm in the posterior [57] or anterior oviduct [63], our parsimony reconstruction infers that the ancestral viperid stored sperm in both oviductal regions (Fig. 17.2B). The assumption that UMC is a plesiomorphic mechanism for sperm storage in viperids [57] requires confirmation. Viperidae contains three subfamilies, with Viperinae being sister to Azemiopinae + Crotalinae [90]. Thus far, UMC has been described only in some crotalines (*Agkistrodon*, *Bothrops*, *Crotalus*, and *Lachesis*) and one viperine (*Vipera berus*), but most genera have yet to be examined (Fig. 17.3).

17.6 Sperm Storage and the Early Evolution of Snakes

The most parsimonious evolutionary scenario suggests that the ancestral squamate lacked SSTs in the anterior oviduct (infundibulum) and thus stored sperm in the posterior oviduct only (in crypts or furrows). Moreover, parsimony analysis also suggests that infundibular SSTs evolved 3–4 times within Squamata, and these origins are associated with either a shift from posterior to anterior sperm storage or the maintenance of the two sperm storage sites. An interesting pattern emerging from our reconstructions is that sperm-storage location differs between lizards and snakes. While most lizards store sperm either in the posterior or anterior oviduct, most snakes store (or potentially store) sperm in both oviductal regions (Fig. 17.2). What can our findings tell us about the origin and early evolution of snakes?

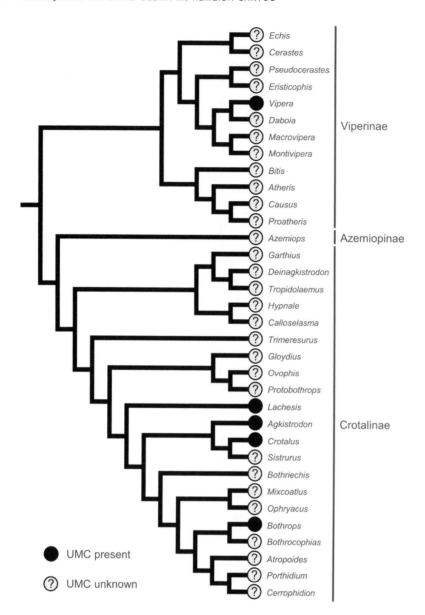

Figure 17.3 Genus-level cladogram adapted from Alencar et al. [90] showing the phylogenetic distribution of uterine muscular coiling (UMC) in the family Viperidae.

Based on parsimonious estimation of ancestral states for current data, the ancestral snake unequivocally had infundibular SSTs, but it may have stored sperm only in the anterior oviduct or in the anterior + posterior oviduct, because both hypotheses are equally parsimonious. Recent molecular phylogenies have found Scolecophidia to be paraphyletic, with anomalepidids recovered as sister to alethinophidians, and leptotyphlopids + typhlopoids as sister to all other snakes ([42, 94, 95], see also Ruane and Streicher, Chapter 10). Scolecophidians share many morphological and ecological traits associated with profound fossoriality, such that their paraphyly suggests that many traits of extant scolecophidians were already present in the ancestral snake ([95]; but see [96] and Gower et al., Chapter 15). In keeping with this interpretation, the scenario proposing that the ancestral snake stored sperm only in infundibular SSTs of the anterior oviduct (Fig. 17.2D), which is the plesiomorphic condition for typhlopoid + leptotyphlopoid scolecophidians, seems more plausible, as opposed to anterior + posterior storage (Fig. 17.2C). We emphasize that sperm storage has thus far been studied among scolecophidians only for typhlopids and lepto-typhlopids, but not anomalepidids.

Profound fossoriality imposes constraints on morphology, physiology, and life history [97, 98]. Fossoriality may also constrain activity and, indeed, various scolecophidians are highly seasonal, synchronizing their (surface) activity and food intake with the warmest (spring-summer) seasons [32, 99–102]. Accordingly, reproductive phenology also seems to be constrained. At least one anomalepidid (*Liotyphlops beui*) and virtually all typhlopids and leptotyphlopids studied thus far exhibit highly seasonal reproductive cycles, with mating and vitellogenesis concentrated in spring, and egg-laying occurring up to mid-summer [25, 32, 99, 100, 103, 104]. A major advantage of long-term female sperm storage is that it allows mating and fertilization to be temporally dissociated [2–4]. The narrow activity window and the association between the mating season and vitellogenesis may have made long-term sperm storage (i.e., sperm retention in the posterior oviduct) unnecessary in scolecophidians, requiring that sperm migrate up to the infundibulum immediately after mating and SSTs offer, at a minimum, physical protection at ovulation. If the ancestral snake was fossorial, re-invasion of surface habitats and the subsequent occupation of different substrates by early alethinophidians possibly created opportunities for the timing of mating to evolve separately from the timing of vitellogenesis. Indeed, mating and fertilization are often largely asynchronous in Endoglyptodonta, and females of this group often exhibit long-term sperm storage, including some of the longest recorded sperm storage in squamates [54, 57, 105, 106]. This scenario may have favoured the re-evolution of posterior sperm storage (while retaining anterior storage) and the evolution of reproductive specializations, such as UMC and copulatory plugs.

17.7 Future Directions

Our major aim here was to provide insights into the evolution of sperm-storage location in female squamates, with particular attention to the origin and early evolution of snakes. Testing the scenario outlined above requires investigating female sperm storage in add-itional squamate lineages. Despite the recent increase in anatomical studies on squamate sperm storage (particularly in snakes; Supplementary Table 17.S1), the number of species investigated relative to squamate diversity remains low. Many diverse and speciose squa-mate families are still poorly studied (e.g., Gekkonidae, Scincidae, Lacertidae, Elapidae, Colubridae, Dipsadidae), and many others are completely unstudied (e.g., Dibamidae, Teiidae, Amphisbaenidae, Varanidae, Liolaemidae, Boidae, Pythonidae). Based on the temporal dissociation between mating and ovulation, female sperm storage (short- or long-term) probably occurs in some of these families, such as scincids [107], boids [108], dipsadids [109], colubrids [110], and elapids [111, 112]. We strongly recommend that future studies on the anatomy of female sperm storage address these poorly or unstudied major lineages. In snakes, particularly, we emphasize the need to study groups that will allow for a substantially expanded phylogenetic sampling to improve ancestral state reconstructions for sperm-storage location for the earliest divergences among snakes. Priority lineages include more scolecophidians (especially the main lineage not yet sampled, Anomalepididae – key for establishing states in the ancestral snake, especially if 'Scolecophidia' is paraphyletic), Aniliidae, Tropidophiidae, Uropeltoidea, Booidea, Pythonidae, Loxocemidae, Xenopeltidae, Acrochordidae, Xenodermidae, Pareidae, and Homalopsidae. Similarly, more viper species need to be studied to adequately test the hypothesis that UMC is plesiomorphic in Viperidae (Fig. 17.3). Of particular priority are more viperines, *Azemiops,* less deeply phylogenetically nested crotalines, and species that are reported to lack long-term sperm storage (e.g., *Protobothrops flavoviridis*: [113]). Anatomical studies of the oviduct of other snake families are also needed to verify whether UMC is indeed restricted to viperids.

Identifying sperm-storage sites in squamates is relatively straightforward using standard microscopic techniques (e.g., histology and electron microscopy), and future studies should be designed to assess the entire oviduct throughout the annual cycle. These studies may shed light on the diversity of sperm-storage sites and sperm-storage duration. As knowledge of sperm storage accumulates, future studies should also investigate ancestral-state estimations using other methods (e.g., Likelihood and Bayesian approaches). Ideally, studies of reproductive ecology and phenology of poorly studied lineages should also be conducted so that a wider range of reproductive traits can be recorded, which allow richer ancestral-state reconstructions and more thorough testing of causal hypotheses. Lastly, many physiological and functional questions on female sperm storage deserve further investigation, such as the underlying mechanisms that sustain sperm in the oviduct and the function of UMC and copulatory plugs.

Acknowledgements

We are thankful to David Gower and Hussam Zaher for inviting us to contribute this chapter and for their kindness and patience during the process. We also thank all the members of the Squamata Reproduction Research Group for all the productive discussions on sperm storage over the years. For this chapter, we are particularly grateful to Erick Bassi, Karina Kasperoviczus, Rebeca Khouri, and Eletra de Souza for discussions on female sperm storage in their study systems. We also thank Christopher Friesen, Martha Ramirez-Pinilla, and David Gower for their reviews that substantially improved the manuscript, Livia Santos, Leonardo de Oliveira, Daniel Blackburn, and Lorenzo Alibardi for sharing difficult-to-access literature, and Karina Kasperoviczus for kindly providing us photomicrographs of sperm storage.

References

1. T. J. Orr and P. L. R. Brennan, Sperm storage: distinguishing selective processes and evaluating criteria. *Trends in Ecology and Evolution*, 30 (2015), 261–272.

2. G. W. Schuett, Is long-term sperm storage an important component of the reproductive biology of temperate pitvipers? In J. A. Campbell and E. D. Brodie, eds., *Biology of the Pitvipers* (Tyler: Selva Publishing, 1992), pp. 169–184.

3. R. D. Aldridge and D. Duvall, Evolution of the mating season in the pitvipers of North America. *Herpetolological Monographs*, 16 (2002), 1–25.

4. T. R. Birkhead and A. P. Møller, Sexual selection and the temporal separation of reproductive events: sperm storage data from reptiles, birds and mammals. *Biological Journal of the Linnean Society*, 50 (1993), 295–311.

5. C. R. Friesen, A. F. Kahrl, and M. Olsson, Sperm competition in squamate reptiles. *Philosophical Transactions of the Royal Society, B Biological Sciences*, 375 (2020), 20200079.

6. D. G. Blackburn, Structure, function, and evolution of the oviducts of squamate reptiles, with special reference to viviparity and placentation. *Journal of Experimental Zoology*, 282 (1998), 560–617.

7. J. E. Girling, The reptilian oviduct: a review of structure and function and directions for future research. *Journal of Experimental Zoology*, 293 (2002), 141–170.

8. D. M. Sever and W. C. Hamlett, Female sperm storage in reptiles. *Journal of Experimental Zoology*, 292 (2002), 187–199.

9. D. S. Siegel, A. Miralles, R. E. Chabarria, and R. D. Aldridge, Female reproductive anatomy: cloaca, oviduct, and sperm storage. In R. D. Aldridge and D. M. Sever, eds., *Reproductive Biology and Phylogeny of Snakes* (Enfield: Science Publishers, 2011), pp. 347–409.

10. D. S. Siegel, A. Miralles, J. L. Rheubert, and D. M. Sever, Female reproductive anatomy: cloaca, oviduct and sperm storage. In J. L. Rheubert, D. S. Siegel, and S. E. Thauth, eds., *Reproductive Biology and Phylogeny of Lizards and Tuatara* (Boca Raton: CRC Press, 2014), pp. 144–195.

11. M. E. Eckstut, D. M. Sever, M. E. White, and B. I. Crother, Phylogenetic analysis of sperm storage in female squamates. In L. T. Dahnof, ed., *Animal Reproduction: New Research Developments* (New York: Nova Science Publishers, 2009), pp. 185–218.

12. B. Howarth, Sperm storage: as a function of the female reporductive tract. In A. D. Johnson and C. W. Foley, eds., *The Oviduct and Its Functions* (New York: Academic Press, 1974), pp. 237–270.

13. H. Fox, Urinogenital System. In C. Gans and T. S. Parsons, eds., *Biology of the Reptilia*, Vol. 6. *Morphology* (London and New York: Academic Press, 1977), pp. 81–96.

14. D. H. Gist and J. M. Jones, Storage of sperm in the reptilian oviduct. *Scanning Microscopy*, 1 (1987), 1839–1849.

15. V. A. Barros, L. R. Sueiro, and S. M. Almeida-Santos, Reproductive biology of the neotropical rattlesnake *Crotalus durissus* from northeastern Brazil: a test of phylogenetic conservatism of reproductive patterns. *Herpetological Journal*, 22 (2012), 97–104.

16. V. A. Barros, C. A. Rojas, and S. M. Almeida-Santos, Reproductive biology of *Bothrops erythromelas* from the Brazilian Caatinga. *Advances in Zoology*, 2014 (2014), 1–11.

17. E. A. Bassi, R. Z. Coeti, and S. M. Almeida-Santos, Reproductive cycle and sperm storage of female coral snakes, *Micrurus corallinus* and *Micrurus frontalis*. *Amphibia-Reptilia*, 41 (2019), 1–15.

18. H. B. Braz, K. N. Kasperoviczus, and T. B. Guedes, Reproductive biology of the

fossorial snake *Apostolepis gaboi* (Elapomorphini): a threatened and poorly known species from the Caatinga region. *South American Journal of Herpetology*, 14 (2019), 37–47.

19. K. M. P. Silva,V. A. Barros, C. A. Rojas, and S. M. Almeida-Santos, Infundibular sperm storage and uterine muscular twisting in the Amazonian lancehead, *Bothrops atrox*. *Anatomical Record*, 303 (2020), 3145–3154.

20. K. M. P. Silva, K. B. Silva, L. R. Sueiro, M. E. E. S. Oliveira, and S. M. Almeida-Santos, Reproductive biology of *Bothrops atrox* (Serpentes, Viperidae, Crotalinae) from the Brazilian Amazon. *Herpetologica*, 74 (2019), 198–207.

21. L. Loebens, S. M. Almeida-Santos, and S. Z. Cechin, Reproductive biology of the sword snake *Tomodon dorsatus* (Serpentes: Dipsadidae) in South Brazil: comparisons within the tribe Tachymenini. *Amphibia-Reptilia*, 41 (2020), 1–15.

22. L. E. Gualdrón-Durán, M. F. Calvo-Castellanos, and M. P. Ramírez-Pinilla, Annual reproductive activity and morphology of the reproductive system of an Andean population of *Atractus* (Serpentes, Colubridae). *South American Journal of Herpetology*, 14 (2019), 58–70.

23. K. M. P. Silva, H. B. Braz, K. N. Kasperoviczus, O. A. V. Marques, and S. M. Almeida-Santos, Reproduction in the pitviper *Bothrops jararacussu*: large females increase their reproductive output while small males increase their potential to mate. *Zoology*, 142 (2020), 125816.

24. D. F. Muniz-da-Silva, J. Passos, D. S. Siegel, and S. M. Santos-Almeida, Caudal oviduct coiling in a viperid snake, *Crotalus durissus*. *Acta Zoologica*, 101 (2020), 69–77.

25. R. S. Khouri, S. M. Almeida-Santos, and D. S. Fernandes, Anatomy of the reproductive system of a population of *Amerotyphlops brongersmianus* from southeastern Brazil (Serpentes: Scolecophidia). *Anatomical Record*, 303 (2020), 2485–2496.

26. M. Villagrán-Santa, E. Mendoza-Cruz, G. Granados-González, J. L. Rheubert, and O. Hernández-Gallegos, Sperm storage in the viviparous lizard *Sceloporus bicanthalis* (Squamata: Phrynosomatidae), a species with continuous spermatogenesis. *Zoomorphology*, 136 (2017), 85–93.

27. V. A. Barros, C. A. Rojas, and S. M. Almeida-Santos, Is rainfall seasonality important for reproductive strategies in viviparous Neotropical pitvipers? A case study with *Bothrops leucurus* from the Brazilian Atlantic Forest. *Herpetological Journal*, 24 (2014), 69–77.

28. G. C. Melo, L. B. Nascimento, and C. A. B. Galdino, Lizard reproductive biology beyond the gonads: an investigation of sperm storage structures and renal sexual segments. *Zoology*, 135 (2019), 125690.

29. S. N. Migliore, Biologia reprodutiva de *Enyalius perditus* (Jackson, 1978) e *Enyalius iheringii* Boulenger, 1885 (Squamata: Leiosauridae). Unpublished Masters dissertation, Universidade de São Paulo, 2016.

30. E. Souza and S. M. Almeida-Santos, Reproduction in the bushmaster (*Lachesis muta*): Uterine muscular coiling and female sperm storage. *Acta Zoologica*, 2020.

31. S. N. Migliore, Estratégias reprodutivas de lagartos Mabuyinae do Brasil. Unpublished PhD thesis, Universidade de São Paulo, 2021.

32. R. S. Khouri, B. F. Fiorillo, H. B. Braz, et al., Reproductive ecology of the Amaral's Blind Snake *Trilepida koppesi* in an area of Cerrado in south-eastern Brazil. *Herpetological Journal*, 32 (2022), 70–79.

33. Kasperoviczus KN. Evolução das estratégias reprodutivas de *Bothrops jararaca* (Serpentes: Viperidae). Unpubished PhD thesis, Universidade de São Paulo, 2013.

34. F. C. de Resende and L. B. Nascimento, The female reproductive cycle of the Neotropical snake *Atractus pantostictus* (Fernandes and Puorto, 1993) from south-eastern Brazil. *Anatomia Histologia Embryologia*, 44 (2015), 225–235.

35. F. M. Amaral, Estratégias reprodutivas da serpente *Bothrops alternatus*: influência de fatores ambientais. Unpublished Masters dissertation, Universidade de São Paulo, 2015.

36. C. A. Rojas, V. A. Barros, and S. M. Almeida-Santos, Sperm storage and morphofunctional bases of the female reproductive tract of the snake *Philodryas patagoniensis* from southeastern Brazil. *Zoomorphology*, 134 (2015), 577–586.

37. C. A. Rojas, V. A. Barros, and S. M. Almeida-Santos, A histological and ultrastructural investigation of the female reproductive system of the water snake (*Erythrolamprus miliaris*): Oviductal cycle and sperm storage. *Acta Zoologica*, 100 (2019), 69–80.

38. S. M. Almeida-Santos, V. A. Barros, C. A. Rojas, L. R. Sueiro, and R. H. C. Nomura, Reproductive biology of the Brazilian lancehead, *Bothrops moojeni* (Serpentes, Viperidae), from the state of São Paulo, southeastern Brazil. *South American Journal of Herpetology*, 12 (2017), 174–181.

39. L. Loebens, C. A. Rojas, S. M. Almeida-Santos, and S. Z. Cechin, Reproductive biology of *Philodryas patagoniensis* (Snakes: Dipsadidae) in south Brazil: Female reproductive cycle. *Acta Zoologica*, 99 (2018), 105–114.

40. R. A. Pyron, F. T. Burbrink, and J. J. Wiens, A phylogeny and revised classification of Squamata, including 4161 species of lizards and snakes. *BMC Evolutionary Biology*, 13 (2013), 93.

41. H. Zaher, R. W. Murphy, J. C. Arredondo, et al., Large-scale molecular phylogeny, morphology, divergence-time estimation, and the fossil record of advanced caenophidian snakes (Squamata: Serpentes). *PLoS ONE*, 14 (2019), e0216148.

42. F. T. Burbrink, F. G. Grazziotin, R. A. Pyron, et al., Interrogating genomic-scale data for Squamata (lizards, snakes, and amphisbaenians) shows no support for key traditional morphological relationships. *Systematic Biology*, 69 (2020), 502–520.

43. O. Cuellar, Oviducal anatomy and sperm storage structures in lizards. *Journal of Morphology*, 119 (1966), 7–19.

44. W. Fox, Special tubules for sperm storage in female lizards. *Nature*, 198 (1963), 500–501.

45. T. R. Shantakumari, H. B. D. Sarkar, and T. Shivanandappa, Histology and histochemistry of the oviductal sperm storage pockets of the agamid lizard *Calotes versicolor*. *Journal of Morphology*, 203 (1990), 97–106.

46. A. P. Amey and J. M. Whittier, The annual reproductive cycle and sperm storage in the bearded dragon, *Pogona barbata*. *Australian Journal of Zoology*, 48 (2000), 411.

47. M. C. A. Uribe, S. R. Velasco, L. J. Guillette, and E. F. Estrada, Oviduct histology of the lizard, *Ctenosaura pectinata*. *Copeia*, 1988 (1988), 1035–1042.

48. L. J. Guillette and R. E. Jones. Ovarian, oviductal, and placental morphology of the reproductively bimodal lizard, *Sceloporus aeneus*. *Journal of Morphology*, 184 (1985), 85–98.

49. Z. Yaron, Effects of ovariectomy and steroid replacement on the genital tract of the viviparous lizard, *Xantusia vigilis*. *Journal of Morphology*, 136 (1972), 313–325.

50. A. P. Halpert, W. R. Garstka, and D. Crews, Sperm transport and storage and its relation to the annual sexual cycle of the female red-sided garter snake, *Thamnophis sirtalis parietalis*. *Journal of Morphology*, 174 (1982), 149–159.

51. R. D. Aldridge, Oviductal anatomy and seasonal sperm storage in the southeastern crowned snake (*Tantilla coronata*). *Copeia*, 1992 (1992), 1103–1106.

52. D. M. Sever and T. J. Ryan, Ultrastructure of the reproductive system of the black swamp snake (*Seminatrix pygaea*): Part I. Evidence for oviducal sperm storage. *Journal of Morphology*, 241 (1999), 1–18.

53. D. M. Sever, T. J. Ryan, T. Morris, D. Patton, and S. Swafford, Ultrastructure of the

reproductive system of the black swamp snake (*Seminatrix pygaea*). II. Annual oviducal cycle. *Journal of Morphology*, 245 (2000), 146–160.

54. S. M. Almeida-Santos and M. G. Salomão, Long-term sperm storage in the female Neotropical rattlesnake *Crotalus durissus terrificus* (Viperidae: Crotalinae). *Japanese Journal of Herpetology*, 17 (1997), 46–52.

55. M. Ludwig and H. Rahn, Sperm storage and copulatory adjustment in the prairie rattlesnake. *Copeia*, 1943 (1943), 15–18.

56. G. Nilson and C. Andrén, Function of renal sex secretion and male hierarchy in the adder, *Vipera berus*, during reproduction. *Hormones and Behavior*, 16 (1982), 404–413.

57. S. M. Almeida-Santos and M. G. Salomão, Reproduction in Neotropical pitvipers, with emphasis on species of the genus *Bothrops*. In G. W. Schuett, M. Höggren, M. E. Douglas, et al., eds., *Biology of the Vipers* (Carmel: Eagle Mountain Publishing, 2002), pp. 445–462.

58. S. F. Nunes, I. L. Kaefer, P. T. Leite, and S. Z. Cechin, Reproductive and feeding biology of the pitviper *Rhinocerophis alternatus* from subtropical Brazil. *Herpetological Journal*, 20 (2010), 31–39.

59. K. N. Kasperoviczus, Biologia reprodutiva da jararaca ilhoa, *Bothrops insularis* (Serpentes: Viperidae), da Ilha da Queimada Grande, São Paulo. Unpublished Masters dissertation, Universidade de São Paulo, 2009.

60. N. Yamanouye, P. F. Silveira, F. M. F. Abdalla, et al., Reproductive cycle of the Neotropical *Crotalus durissus terrificus*: II. Establishment and maintenance of the uterine muscular twisting, a strategy for long-term sperm storage. *General and Comparative Endocrinology*, 139 (2004), 151–157.

61. S. M. Almeida-Santos, F. M. F Abdalla, P. F. Silveira, et al., Reproductive cycle of the Neotropical *Crotalus durissus terrificus*:

I. Seasonal levels and interplay between steroid hormones and vasotocinase. *General and Comparative Endocrinology*, 139 (2004), 143–150.

62. B. Stille, T. Madsen, and M. Niklasson, Multiple paternity in the adder, *Vipera berus*. *Oikos*, 47 (1986), 173–175.

63. D. S. Siegel and D. M. Sever, Utero-muscular twisting and sperm storage in viperids. *Herpetological Conservation and Biology*, 1 (2006), 87–92.

64. C. Andrén and G. Nilson, The copulatory plug of the adder, *Vipera berus*: Does it keep sperm in or out? *Oikos*, 49 (1987), 230.

65. W. Fox, Seminal receptacles of snakes. *Anatomical Record*, 124 (1956), 519–539.

66. K. H. Bull, R. T. Mason, and J. Whittier, Seasonal testicular development and sperm storage in tropical and subtropical populations of the brown tree snake (*Boiga irregularis*). *Australian Journal of Zoology*, 45 (1997), 479.

67. L. Thongboon, S. Senarat, J. Kettratad, et al., Morphology and histology of female reproductive tract of the dog-faced water snake *Cerberus rynchops* (Schneider, 1799). *Maejo International Journal of Science and Technology*, 14 (2020), 11–26.

68. H. Saint Girons, Le cycle sexuel chez *Vipera aspis* (L.) dans l'ouest de la France. *Bulletin Biologique de la France et de la Belgique*, 91 (1957), 284–350.

69. H. Saint Girons, Le cycle reproducteur de la Vipère à cornes *Cerastes cerastes* (L.) dans la nature et en captivité. *Bulletin de la Société Zoologique de France*, 87 (1962), 41–51.

70. H. Saint Girons and R. Duguy, Le cycle sexuel de *Lacerta muralis* L. en plaine et en montagne. *Bulletin du Muséum National d'Histoire Naturelle*, 42 (1970), 690–695.

71. M. Devine, Copulatory plugs in snakes: enforced chastity. *Science*, 187 (1975), 844–845.

72. M. C. Uribe, G. González-Porter, B. D. Palmer, and L. J. Guilette, Cyclic histological changes of the oviductal-cloacal junction in the viviparous snake

Toluca lineata. Journal of Morphology, 237 (1998), 91–100.

73. V. A. Barros, C. A. Rojas, and S. M. Almeida-Santos, Mating plugs and male sperm storage in *Bothrops cotiara* (Serpentes, Viperidae). *Herpetological Journal*, 27 (2017), 63–67.

74. H. A. J in den Bosch, First record of mating plugs in lizards. *Amphibia-Reptilia*, 15 (1994), 89–93.

75. K. N. Kasperoviczus and S. M. Almeida-Santos, Copulatory plugs in Neotropical viperid snakes. 7th World Congress of Herpetology, Vancouver, abstracts (2012), 11–12.

76. M. C. Devine, Potential for sperm competition in reptiles: behavioral and physiological consequences. In R. L. Smith, ed., *Sperm Competition and the Evolution of Animal Mating Systems* (Orlando: Academic Press, 1984), pp. 509–521.

77. R. Shine, M. Olsson, and R. Mason, Chastity belts in gartersnakes: the functional significance of mating plugs. *Biological Journal of the Linnean Society*, 70 (2000), 377–390.

78. C. R. Friesen, R. Shine, R. W. Krohmer, and R. T. Mason, Not just a chastity belt: the functional significance of mating plugs in garter snakes, revisited. *Biological Journal of the Linnean Society*, 109 (2013), 893–907.

79. L. H. Hoffman and W. A. Wimsatt, Histochemical and electron microscopic observations on the sperm receptacles in the garter snake oviduct. *American Journal of Anatomy*, 134 (1972), 71–95.

80. L. Jacobi, Ovoviviparie bei einheimischen Eidechsen. *Zeitschrift für wissenschaftliche Zoologie*, 148 (1936), 401–464.

81. H. Saint Girons, Sperm survival and transport in the female genital tract of reptiles. In E. S. E. Hafez and C. G. Thibault, eds., *The Biology of Spermatozoa* (Basel: Karger, 1975), pp. 105–113.

82. W. Fox and H. C. Dessauer, The single right oviduct and other urogenital structures of female *Typhlops* and *Leptotyphlops*. *Copeia*, 1962 (1962), 590–597.

83. S. R. Srinivas, S. N. Hegde, H. B. D. Sarkar, and T. Shivanandappa, Sperm storage in the oviduct of the tropical rock lizard, *Psammophilus dorsalis. Journal of Morphology*, 224 (1995), 293–301.

84. C. E. Marinho, S. M. Almeida-Santos, S. C. Yamasaki, and P. F. Silveira, Peptidase activities in the semen from the ductus deferens and uterus of the neotropical rattlesnake *Crotalus durissus terrificus. Journal of Comparative Physiology B*, 179 (2009), 635–642.

85. R. D. Aldridge, B. C. Jellen, D. S. Siegel, and S. S. Wisniewski, The sexual segment of the kidney. In R. D. Aldridge and D. M. Sever, eds., *Reproductive Biology and Phylogeny of Snakes* (Enfield: Science Publishers, 2011), pp. 477–509.

86. D. S. Siegel and D. M. Sever, Sperm aggregations in female *Agkistrodon piscivorus* (Reptilia: Squamata): A histological and ultrastructural investigation. *Journal of Morphology* 269 (2008), 189–206.

87. H. Rahn, Sperm viability in the uterus of the garter snake, *Thamnophis. Copeia*, 1940 (1940), 109–115.

88. W. P. Maddison and D. R. Maddison, Mesquite: A modular system for evolutionary analysis. Version 3.61, http://mesquiteproject.org (2019).

89. Y. Zheng and J. J. Wiens, Combining phylogenomic and supermatrix approaches, and a time-calibrated phylogeny for squamate reptiles (lizards and snakes) based on 52 genes and 4162 species. *Molecular Phylogenetics and Evolution*, 94 (2016), 537–547.

90. L. R. V. Alencar, T. B. Quental, F. G. Grazziotin, et al., Diversification in vipers: Phylogenetic relationships, time of divergence and shifts in speciation rates. *Molecular Phylogenetics and Evolution*, 105 (2016), 50–62.

91. H. Saint Girons, Deplacements et survie des spermatozoides chez les reptiles. *Colloques de l'Institut National de la Santé*

et de la Recherche Médicale, 26 (1973), 259–282.

92. M. Gabe and H. Saint Girons, *Contribution a l'histologie de Sphenodon punctatus Gray* (Paris: Éditions du Centre National de la Recherche Scientifique, 1964).

93. A. Cree, J. F. Cockrem, and L. J. Guillette, Reproductive cycles of male and female tuatara (*Sphenodon punctatus*) on Stephens Island, New Zealand. *Journal of Zoology* 226 (1992), 199–217.

94. J. J. Wiens, C. R. Hutter, D. G. Mulcahy, et al., Resolving the phylogeny of lizards and snakes (Squamata) with extensive sampling of genes and species. *Biology Letters*, 8 (2012), 1043–1046.

95. A. Miralles, J. Marin, D. Markus, et al., Molecular evidence for the paraphyly of Scolecophidia and its evolutionary implications. *Journal of Evolutionary Biology*, 31 (2018), 1782–1793.

96. B. F. Simões, F. L. Sampaio, C. Jared., et al., Visual system evolution and the nature of the ancestral snake. *Journal of Evolutionary Biology*, 28 (2015), 1309–1320.

97. P. C. Withers, Physiological correlates of limblessness and fossoriality in scincid lizards. *Copeia*, 1981 (1981), 197–204.

98. C. Gans, Studies on amphisbaenids (Amphisbaenia, Reptilia). 1. A taxonomic revision of the Trogonophinae, and a functional interpretation of the amphisbaenid adaptive pattern. *Bulletin of the American Museum of Natural History*, 119 (1960), 129–204.

99. J. K. Webb, R. Shine, W. R. Branch, and P. S. Harlow, Life-history strategies in basal snakes: reproduction and dietary habits of the African thread snake *Leptotyphlops scutifrons* (Serpentes: Leptotyphlopidae). *Journal of Zoology*, 250 (2000), 321–327.

100. J. K. Webb, W. R. Branch, and R. Shine, Dietary habits and reproductive biology of typhlopid snakes from southern Africa. *Journal of Herpetology* 35 (2001), 558–567.

101. R. J. Sawaya, O. A. V. Marques, and M. Martins, Composição e história natural das serpentes de Cerrado de Itirapina, São Paulo, sudeste do Brasil. *Biota Neotropica*, 8 (2008), 127–149.

102. L. Parpinelli and O. A. V. Marques, Seasonal and daily activity in the pale-headed blindsnake *Liotyphlops beui* (Serpentes: Anomalepidae) in southeastern Brazil. *South American Journal of Herpetology*, 3 (2008), 207–212.

103. R. Shine and J. K. Webb, Natural history of Australian typhlopid snakes. *Journal of Herpetology*, 24 (1990), 357–363.

104. L. Parpinelli and O. A. V. Marques, Reproductive biology and food habits of the blindsnake *Liotyphlops beui* (Scolecophidia: Anomalepididae). *South American Journal of Herpetology*, 10 (2015), 205–210.

105. H. Saint Girons, Reproductive cycles of male snakes and their relationships with climate and female reproductive cycles. *Herpetologica*, 38 (1982), 5–16.

106. R. D. Aldridge, S. R. Goldberg, S. S. Wisniewski, A. P. Bufalino, and C. B. Dillman, The reproductive cycle and estrus in the colubrid snakes of temperate North America. *Contemporary Herpetology*, 2009 (2009), 1–31.

107. M. Smyth and M. J. Smith, Obligatory sperm storage in the skink *Hemiergis peronii*. *Science*, 161 (1968), 575–576.

108. M. Bertona and M. Chiaraviglio, Reproductive biology, mating aggregations, and sexual dimorphism of the Argentine boa constrictor (*Boa constrictor occidentalis*). *Journal of Herpetology*, 37 (2003), 510–516.

109. S. Z. Cechin and J. L. Oliveira, *Sibynomorphus ventrimaculatus* (Southern Snail-eater). Mating. *Herpetological Review*, 34 (2003), 73.

110. O. A. V. Marques, D. F. Muniz-Da-Silva, F. E. Barbo, et al., Ecology of the colubrid snake *Spilotes pullatus* from the Atlantic Forest of southeastern Brazil. *Herpetologica*, 70 (2014), 407–416.

111. R. Shine, Comparative ecology of three Australian snake species of the genus

Cacophis (Serpentes: Elapidae). *Copeia*, 1980 (1980), 831–838.

112. R. Shine, G. V. Haagner, W. R. Branch, P. S. Harlow, and J. K. Webb, Natural history of the African shieldnose snake *Aspidelaps scutatus* (Serpentes, Elapidae).

Journal of Herpetology, 30 (1996), 361–366.

113. F. Yokoyama and H. Yoshida, The reproductive cycle of the female habu, *Trimeresurus flavoviridis. Journal of Herpetology*, 28 (1994), 54–59.

An Overview of the Morphology of Oral Glands in Snakes

Leonardo de Oliveira and Hussam Zaher

18.1 Introduction

The oral cavity of snakes receives secretions from eight distinct cephalic glands traditionally referred to as oral glands: the labial (supra- and infralabial), temporomandibular, rictal, sublingual, premaxillary, accessory, and dental glands (the latter also known as venom and Duvernoy's glands). Oral glands and their venom production have undergone substantial modification in snakes, most notably in association with the diversification of feeding strategies [1, 2]. Their variable morphology provides a rich source of data for systematics and is believed to be responsible for an increase in prey-handling efficiency, a potential key innovation underpinning the adaptive radiation of the group [3–5]. Recently, several important contributions have focused on the oral glands of snakes, and more specifically on the origin and diversification of venom delivery systems, which include the venom glands, differentiated maxillary teeth and associated adductor musculature [e.g., 6, 7].

Among the glands that release their secretions into the oral cavity, the most studied are the venom, sublingual, and supra- and infralabial glands [8]. As in other reptiles, the oral glands of snakes are usually classified by their histochemical properties and the ultrastructure of their secretory granules. Kochva [3] used their histochemical reactions to recognize three types of secretory cells in oral glands: serous cells with protein-rich and Periodic Acid-Schiff (PAS) negative secretions, serous cells with PAS-positive secretions, and mucous cells with little or no protein and acidic mucosubstances. Underwood [5] differentiated oral glands into mucous and serous categories – or equivalent – and indicated that mucous cells secrete only mucus while venom glands always contain some type of serous cells. Here, we follow Underwood [5] and classify the different cell types of oral glands as mucous or serous.

In this chapter, we review historical and recent developments, synthesize data, and present a perspective on the potential and limitations of oral-gland anatomy for snake

Table 18.1 List of specimens used in this study.

Specimens belong to the following collections: IBSP: Coleção Herpetológica 'Alphonse Richard Hoge', Instituto Butantan; MNHNCU: Museo Nacional de Historia Natural, Havana, Cuba; MZUSP: Museu de Zoologia da Universidade de São Paulo, São Paulo, Brazil; NHM, UK: The Natural History Museum, London, UK.

Family	Species	Adult/ Embryo	Technique	Collection numbers
Colubridae	*Pantherophis guttatus*	Embryo	Histology	MZUSP 22481
Dibamidae	*Dibamus novaeguineae*	Adult	Tomography	NHM, UK 1998.621
Diploglossidae	*Ophiodes striatus*	Adult	Histology	MZUSP 97009
Dipsadidae	*Helicops modestus*	Adult	Histology	IBSP 76610
	Imantodes cenchoa	Adult	Histology, dissection	MZUSP 15944
	Leptodeira annulata	Adult	Histology	MZUSP 16958; 16948
	Sibynomorphus mikanii	Embryo	Histology	MZUSP 18716; 18717
	Sibynomorphus neuwiedi	Embryo	Histology	MZUSP 22458
Elapidae	*Hydrophis cyanocinctus*	Adult	Tomography	NHM, UK 1956.1.13.16
	Micrurus corallinus	Adult	Tomography, dissection	MZUSP 11060
Gekkonidae	*Hemidactylus mabouia*	Adult	Histology	MZUSP 49310
Gymnophthalmidae	*Colobosaura modesta*	Adult	Histology	MZUSP 39165
Leptotyphlopidae	*Leptotyphlops fuliginosus*	Adult	Tomography	MZUSP 11513
Scincidae	*Varzea bistriata*	Adult	Histology	MZUSP 89405
Tropidophiidae	*Tropidophis melanurus*	Adult	Tomography	MNHNCU 5107
Typhlopidae	*Amerotyphlops brongersmianus*	Adult	Tomography	MZUSP 14674

systematics, in addition to trying to assess interpretations of the ancestral snake. We also conducted some new histological sections of the entire heads of adults and embryos of snakes (Table 18.1) in order to describe and illustrate oral glands and associated structures. For general structure, we stained paraffin histological sections with hematoxylin-eosin

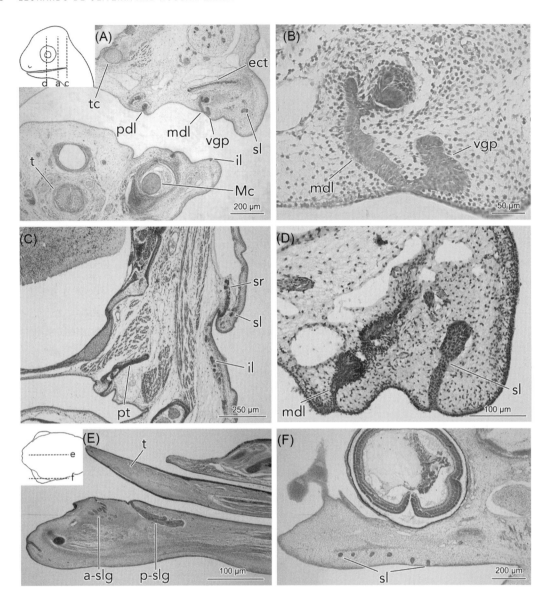

Figure 18.1 Embryonic development of oral glands in snakes. (A) Transverse section of head of *Pantherophis guttatus* (embryo, 22 days after oviposition) at the post-ocular level showing dental and glandular laminae. (B) Detail of the previous figure showing the joint origin of the venom gland primordium (vgp) and maxillary dental lamina (mdl). (C) Transverse section of head of *Sibynomorphus mikanii* (embryo, 41 days) at the level of the corner of the mouth, showing the invagination corresponding to the upper rictal gland (sr) in a more medial region than the supralabial gland (sl). (D) Transverse section of head of *Sibynomorphus mikanii* (embryo, 32 days) at the level of the eyes showing the invagination of the supralabial gland (sl) spatially separate from the maxillary dental lamina (mdl). (E) Sagittal section of head of *Sibynomorphus mikanii* (embryo, 32 days) showing the invagination of the anterior and posterior parts of the sublingual gland (respectively, a-slg and p-slg). Histochemical reactions of PAS and AA. (F) Sagittal section of head of embryo of

(H&E). Sections were also stained using combined alcian blue (pH 2.5) and PAS [9, 10]. PAS and alcian blue were used for identification of neutral and acid mucosubstances, respectively. Moreover, we examined the heads of some species (Table 18.1) using high-resolution tomography (X-ray micro CT scan). For this, entire heads were stained with a 1 per cent iodine-ethanol solution for about one week to increase tissue X-ray contrast [11]. The studied specimens were scanned using a Phoenix v|tome|x m tomography (General Electric Company) at the Microtomography Laboratory of the Museum of Zoology of the University of São Paulo (Brazil) and an X-Tek HMXST 225 X-ray tomography (Nikon Metrology) at the Natural History Museum, London (UK).

Terminology of the oral glands follows Taub [8], Kochva [3], Underwood [12], and terminology of the adductor muscles follows Zaher [13]. Although the venom and Duvernoy's glands are homologues [1, 3, 5], they exhibit notable morphological differences and their nomenclature has been intensely debated (see [14, 15] for details). In a traditional view, venom glands occur only in front-fanged snakes (some atractaspidids, all viperids and elapids), while Duvernoy's glands occur in a range of Endoglyptodonta (*sensu* [16, 17]) with or without rear fangs – or simply in non-front-fanged snakes [18–21]. However, considering the homology of these glands and regardless of their morphofunctional differences, some authors have proposed to synonymize the terms 'Duvernoy's glands' and 'venom glands' [15, 22]. Here, we follow this proposal and use the term 'venom gland' to refer to all oral glands that develop from a single invagination on the caudal portion of the dental lamina of the maxillary bone. Phylogenetic relationships and classification follows the recent proposals by Zaher et al. [17] for Caenophidia and Burbrink et al. [23] for Squamata.

18.2 Labial Glands

Labial glands have been found in the upper and lower jaws (supra- and infralabial glands, respectively) of all snakes studied so far [3, 24]. These glands develop from various serial and non-compound epithelial thickenings, followed by the invagination of the epithelial laminae into the mesenchyme (Fig. 18.1, [25, 26]). The pattern of embryonic development and location of the duct opening distinguish the labial from the dental glands. While labial glands have multiple ducts (polystomatic) opening along the margins of the lips, the ducts of the dental glands open at the bases of the teeth (Fig. 18.1, [5, 27]). The epithelium of labial glands is usually composed of mucous cells, although mixed and serous cells have also been described in many species [3, 18]. The cell types of supra- and infralabial glands

---

Figure 18.1 (*cont.*) *Sibynomorphus neuwiedi* showing the serial and non-compound invaginations of the supralabial gland (sl). Sections a, b, c, d, and f were stained with H&E. Abbreviations: ect, ectopterygoid; il, invagination of the infralabial gland; Mc, Meckel's cartilage; pdl, pterygoid dental lamina; pt, pterygoid; t, tongue; tc, trabecula. The panel in the upper left corner denotes the position of the sections. Figures (C) and (D) modified from [26]. (A black and white version of this figure will appear in some formats. For the colour version, please refer to the plate section).

are often similar. Labial glands usually include several rows of small glands with short ducts arranged along the lips and separated from each other by connective tissue septa [3]. However, some variation can be observed in their organization, as detailed below.

18.2.1 Supralabial Glands

Supralabial glands are located along the lateral margin of the maxilla, underneath the labial scales [8]. In the paraphyletic assemblage 'scolecophidians', two groups of supralabial glands have been described in the rostral region of two typhlopids (*Afrotyphlops* and *Indotyphlops*) and in *Rena dulcis* (Leptotyphlopidae), and a third gland located at the level of the eye only in typhlopids [28]. In *Afrotyphlops punctatus* and *Indotyphlops braminus*, mucous and seromucous cells are grouped into different lobules [28]. An additional 'accessory supralabial' gland with a single duct has been described in *Liotyphlops albirostris* (Anomalepididae) [29]. Saint-Girons [30] proposed that Typhlopidae could be differentiated from other 'scolecophidians' by the existence of deep folds in the oral epithelium filled with large mucous crypts. We confirm the presence of deep folds in the oral epithelium of the typhlopid *Amerotyphlops brongersmianus* (Fig. 18.2B). We also show that, anteriorly in the mouth, the supralabial gland is composed of two cell types grouped into different lobules, which are seen on the computed tomography slides in two shades of grey (Fig. 18.2A, B).

Underwood [12] described some structural variation in the supra- and infralabial glands of some 'henophidian' snakes. He argued that although most of these glands are largely (if not entirely) mucous, a few species have serous tubules leading to mucous, as in *Casarea dussumieri* and *Tropidophis haetianus*. In *Cylindrophis ruffus*, he also reported serous cells opening into mucous tubules, and described an extensive supralabial gland extending back to the corner of the mouth and separated from the anterior part by a curtain of horizontal smooth-muscle fibres. In Boidae, supra- and infralabial glands are composed of mixed cells containing only a small amount of proteins [30, 31].

In Caenophidia, Taub [18, 32] described substantial variation in the upper jaw glands of non-front-fanged snakes [18, 32]. Taub recognized six glandular patterns, ranging from a single and purely mucous supralabial gland to two separated glands composed of a mucous supralabial gland and a serous tubuloacinous venom gland. In an intermediate category, Taub included xenodermids (*Fimbrios*, *Achalinus*, and *Xenodermus*), which have labial glands formed by alternating lobules of serous and mucous cells along the supralabial ridge. However, subsequent studies found no evidence of alternating mucous and serous cells in the supralabial gland, but identified supralabial serous tubules leading to mucous tubules [12]. In *Acrochordus*, supralabial glands are mucous [12, 15, 28]. In Pareidae, the condition of the supralabial gland is unclear because it presents a group of serous cells in its posterior portion. According to Fry et al. [33], *Pareas carinatus* has a greatly atrophied venom gland, while Taub [18] reported that *P. stanleyi* has venom glands composed of mucous cells intermingled with serous cells. In contrast, Underwood and Kochva [34] found neither venom glands nor quadrato-maxillary ligaments in several species of *Pareas* (including *P. stanleyi*), but they recognized a horizontal gland that appears to open backwards into the long rictal groove. Underwood [12] also described supralabial glands

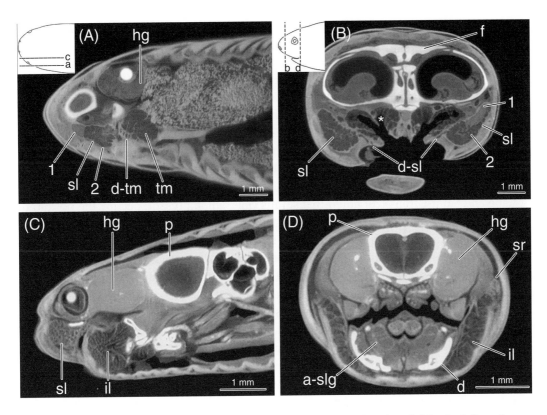

Figure 18.2 High-resolution computed tomography of the head of snakes. (A) Sagittal slice of *Amerotyphlops brongersmianus* showing the position of the supralabial gland (sl) consisting of two types of cells (1 and 2) that may be distinguished by the different shades of grey and the temporomandibular gland (tm) below the eye. Note the presence of ducts (d-tm) in the anteroinferior part of the gland. (B) Transverse slice of *A. brongersmianus* in the anterior part of the head showing the position of the supralabial gland (sl) composed of two different cell types and the presence of ducts (d-sl) opening into the mouth. Note the existence of deep folds (*) in the oral epithelium. (C) Sagittal slice of *Leptotyphlops fuliginosus* showing the position of the supralabial (sl) and infralabial glands (il). (D) Transverse slice of *L. fuliginosus* at the level of the corner of the mouth showing the position of the infralabial gland (il) lateral to the dentary (d) and the presence of an upper rictal gland (sr) with a duct opening into the mouth. Abbreviations: a-slg, anterior sublingual gland; f, frontal; hg, harderian gland; p, parietal. The panel in the upper left corner denotes the position of the sections.

constituted by branched serous tubules passing into branched mucous tubules with columnar epithelium, and suggested that a double-row portion of the supralabial gland might be the Duvernoy's glands (venom glands) reported by Taub [18]; however, he concluded that there is no evidence of a venom gland in *Pareas*.

In Viperidae, the supralabial gland seems to occupy the entire region of the upper jaw. The infra and supralabial glands of *Bothrops jararaca* are reportedly formed by secretory tubules composed of mucous and mucous-serous cells [35]. In *Echis*, an external upper labial gland that opens at the lower half of the last supralabial scale has been described [36, 37]. However, this supralabial gland does not appear to be homologous with typical supralabial glands but rather corresponds to an epidermoid labial gland, due to the duct that opens out of the mouth [37]. Many elapid species have supralabial glands divided in the anterior and posterior parts of the mouth, which is particularly observed in coral snakes, which exhibit ample morphological variation in the supralabial gland (Fig. 18.4C; [38]). In *Micrurus corallinus*, the supralabial gland is essentially composed of mucous cells, but it also has some serous cells [39]. A few elapids have a small supralabial gland restricted to the anteriormost part of the mouth, including the sea krait *Laticauda colubrina*, in which the supralabial gland reaches only the infraorbital region [40]. Reduction in the size of the typical labial glands in *L. colubrina* and other species may be related to their aquatic lifestyle [30].

Gabe and Saint-Girons [28] noted that serous cells, when present, occupy the bottom of the secretory units of the labial glands of most species. This seems to occur in the dipsadid *Helicops modestus*, in which the upper part of the supralabial gland is formed by serous cells, whereas the ventral part of the gland and its ducts are mucous (Fig. 18.4E, F). In non-front-fanged snakes with venom glands, these glands often overlap the posterior third of the supralabial gland, reducing or altering it locally (Fig. 18.4A; [3]). In some dipsadines, the posterior portion of the subralabial glands is substantially enlarged below the orbit (e.g., *Sibynomorphus* and *Geophis*), but no venom glands have been observed [41, 42]. Additionally, a (presumably juxtanarinal) gland lies adjacent to the supralabial gland that borders posteroventrally each external nostril in the dipsadines *Chersodromus*, *Ninia*, and *Nothopsis* [43]. It is unclear whether a similar structure in *Plesiodipsas perijanensis* [44] is novel or part of the supralabial gland.

18.2.2 Infralabial Glands

Infralabial glands are located laterally to the dentary and are usually composed of mucous cells [8]. The infralabial gland consists of a series of lobules that usually open directly by short ducts along the margin of the lip, but which can also open obliquely forwards [12]. In non-caenophidian snakes, the infralabial glands are often more developed than the supra-labial glands [24, 45]. Accessory infralabial glands have been observed in the 'scolecophidian' *Liotyphlops albirostris* [29]. In Typhlopidae and Leptotyphlopidae, the infralabial glands increase in size posteriorly, forming a temporomandibular gland [3, 24]. In *Xerotyphlops vermicularis*, this accessory infralabial gland has not been confirmed [8]. Infra and supralabial glands apparently composed of mucous and serous cells have been reported in *Anomalepis aspinosus* [46]. Recently, a double infralabial accessory gland has been described in *Mitophis leptepileptus*, and single infralabial accessory glands have been reported in some species of *Trilepida*, both genera belonging to the Leptotyphlopidae [47]. However, no information on cell types has been provided. Underwood [12] reported that infralabial glands divided into two parts in *Cylindrophis ruffus* (Cylindrophiidae); the second infralabial part lacked associated muscles, and was interpreted as unlikely to be a mandibular gland. In *Pareas*, a caenophidian genus of snakes, the muscle *levator anguli oris* originates from the external surface of the maxillary and covers the infralabial glands [48, 49]. Moreover, this muscle always inserts directly on the infralabial gland and acts as a compressor muscle [50, 51]. Underwood [12] reported that supra- and infralabial glands of

Pareas carinata and *P. monticola* consist of branched serous tubules passing into branched mucous tubules, with columnar epithelium.

In elapids, infralabial glands have a sac-like structure lined by mucous cells in the anterior portion (Fig. 18.4C; [39]). Although there is little information about this structure, it apparently accommodates the fangs when the mouth closes [3]. The sac-like structure also occurs in the front-fanged atractaspidid *Homoroselaps lacteus* (L. Oliveira, pers. obs.), which seems to reinforce the morphological similarities of the venom delivery system components between the genus *Homoroselaps* and the elapids [34, 60, 82]. In *M. corallinus*, the infralabial glands are composed of mucous cells and a smaller number of serous cells [39].

In 'goo-eater' dipsadines, infralalabial glands are highly developed and predominantly composed of serous cells [42]. Some authors have suggested that these glands participate in the peculiar feeding mechanism of these snakes, which are specialist in preying on soft-bodied invertebrates, particularly slugs, snails, and annelids [8, 52]. In *Sibynomorphus neuwiedi,* the infralabial glands show a higher protein content than the supralabial and venom glands, and the infralabial extract causes paralysis and death in slugs [53]. In the congeneric *S. mikanii*, the infralabial glands also show a higher protein content than the supralabial and venom glands [54]. However, the extract of these three glands causes intense mucus release, body contortions, and immobilization in slugs, and these reactions are more intense in topical than injected applications [54]. In this species, however, the venom gland extract causes a longer immobilization time, whereas secretions from the supra- and infralabial glands cause more intense local effects [54]. Thus, Salomão and Laporta-Ferreira [54] suggested that, in *S. mikanii*, the venom gland secretion acts mainly to immobilize slugs, whereas the supra- and infralabial secretions act in the beginning of the digestive process. More recently, Oliveira et al. [55] demonstrated that the infralabial glands of *S. mikanii* and *Dipsas indica* are predominantly composed of serous cells, and the infralabial glands of *Atractus reticulatus* (a goo-eater specialist preying on earthworms) are formed by mixed acini composed of mucous and serous cells.

In *Dipsas* and *Sibynomorphus*, infralabial glands are highly modified and composed of two parts. The thinnest, most dorsolateral part of the infralabial (il1) is arranged along the labial region and consists of several short ducts that open along the margin of the lower lip. The most hypertrophied part of the infralabial (il2) is located in the ventromedial portion of the mandible and has a single large duct that opens in the epithelium of the mouth floor, level with the intermandibular raphe [42]. In *Sibynomorphus*, il2 develops independently from il1 and derives from a hypertrophied invagination, whereas il1 derives from several small invaginations [26]. In *Geophis*, the infralabial glands are also highly modified, but their condition differs greatly from that of *Dipsas* and *Sibynomorphus* [41]. Although the infralabial glands are also divided into two different portions (il1 and il2) in *Geophis*, these portions are compressed by different muscles and have a large lumen within il2 where secretion is stored [41]. *Enulius* and *Enuliophis* share with *Geophis* the condition of the infralabial glands associated with the adductor muscles [51].

Although the pattern of typical infralabial glands is largely derived in goo-eaters, little is known about the function of these glands in these snakes. In addition to lubrication, their infralabial glands may produce toxins that act specifically on ingested prey (as detailed above) or function as a protein-secreting system that controls mucous secretion and helps to ingest elongated, flexible, and highly viscous prey [41, 42, 55]. Preliminary transcriptomic results indicate that the infralabial glands of *S. mikanii* have a 'non-front-fanged venom gland-like' protein profile dominated by metalloproteinases and C-type lectins, whereas the infralabial glands of *D. indica* have a low expression of venom components, consisting of only 9 per cent of typical toxins [56], suggesting that their secretions might not have a primary envenomation function. The fact that the secretion is carried out and empties in the mouth through a duct that is not associated with a tooth position supports this hypothesis [42]. However, further research is needed on these highly specialized lower jaw systems to understand their structure and function, particularly the high-pressure system of *Geophis*.

18.3 Temporomandibular Glands

Phisalix [57] described the 'glande temporo-mandibulaire' in *Afrotyphlops punctatus* (Typhlopidae) and considered it as different from the labial, rictal, and venom glands. Smith and Bellairs [24] did not recognize this gland in *Argyrophis diardi* and considered it a posterior prolongation of the inferior labial, which extends backwards and upwards in the temporal region. Gabe and Saint-Girons [28] argued that typhlopids are characterized by the presence of a temporomandibular gland that represents the well-developed caudal part of the infralabial mass. They also argued that the temporomandibular gland of Typhlopidae represents only the posterior, undifferentiated part of the infralabial mass and its classification as a distinct gland is unjustified, as previously noticed by Smith and Bellairs [24]. Haas [29] found an accessory supralabial gland in the anomalepidid *Liotyphlops albirostris* but was unsure whether this gland is comparable to the temporomandibular gland described by Phisalix. Kochva [3] argued that the posterior region of the infralabial glands of typhlopids and *Leptotyphlops* increases in size to form the temporomandibular gland. Taub [8] followed Phisalix's [57] proposal and recognized the temporomandibular glands as individualized glands. More recently, Jackson et al. [15] reported a large gland below the eyes of the typhlopid *Anilios guentheri*, which presumably is the Phisalix's temporomandibular gland. In addition, the authors suggested that this gland might be a dental gland and compared it with the accessory supralabial gland of *Liotyphlops albirostris* described by Haas [29]. Our preliminary tomographic images of *Amerotyphlops brongersmianus* indicate a large gland located below the eyes, posteriorly in the infralabial glands. Although this gland has some (approximately four) duct openings at the back of the mouth (Fig. 18.2A), their small size precluded us from identifying their exact location or from identifying the infralabial or rictal glands. However, the duct openings do not appear to be close to the bases of the maxillary teeth, indicating that this gland is not a dental gland. Further investigations are needed to better understand the identity and the morphological diversity of the temporomandibular glands of 'scolecophidians'.

18.4 Rictal Glands

Phisalix and Caius [57, 58] described a 'glande temporale antérieure' (= rictal gland) in *Anilius scytale*, *Eryx johnii*, and *Platyplectrurus*, and listed it as present in other species. They also noted that this gland occurs only in the 'primitive' snake families: Boidae, Aniliidae, and Uropeltidae. Haas [48] described the rictal gland in some species of *Uropeltis* and *Xenopeltis*. Smith and Bellairs [24] found this gland in representatives of 'primitive' families, such as *Uropeltis ceylanica* and *Cylindrophis ruffus*, and some caenophidian snakes of the family Colubridae. Moreover, they [24] presented serial sections showing where the ducts of the rictal glands open into the mouth, and highlighted two important conditions of these glands: they are often difficult to visualize by dissection and appear to vary considerably within genera. Gabe and Saint-Girons [28] described rictal glands in other caenophidians, emphasizing their broad taxonomic distribution. Kochva ([59]; see also Fig. 18.1C) illustrated the primordium of the rictal gland (initially named by Kochva as posterior gland) of *Telescopus fallax* developing mesially to the quadrato-maxillary ligament. McDowell [60] proposed the term rictal gland and provided significant comparative observations, describing these glands in a range of caenophidians. McDowell also proposed homology of the rictal gland of snakes and the rictal fold of lizards (not followed by subsequent studies) and suggested that rictal glands were precursors of venom glands in front-fanged snakes. This was rejected by Wollberg et al. [61]. Underwood and Kochva [5, 34] and Cundall and Rossman [62] reported new observations about rictal glands of caenophidian and non-caenophidian ('primitive') snakes made mainly by dissections. Underwood [5] argued that rictal glands may rise mesially or laterally to the quadrato-maxillary ligament and that they are serous, with some mucous cells in the ducts, that open into the corner of the mouth. He also demonstrated substantial morphological variation in rictal glands that may occur exclusively on the upper jaw (superior rictal gland), lower jaw (inferior rictal gland), in both jaws, or neither. Wollberg et al. [61] provided the first histological evidence of the inferior rictal glands, mainly in atractaspidid snakes. Underwood [12] studied a large number of 'henophidians' preserved in museums and provided detailed descriptions of the dissections and histology of their rictal glands. This was Underwood's last article in life and, although it focused on 'henophidians', it remains possibly the most comprehensive study of rictal glands.

More recently, Jackson et al. [15] reviewed the morphology of the oral glands of snakes and provided some data on the superior rictal glands of elapids as determined by dissections. These authors reported considerable variation, but did not investigate the inferior rictal glands. Oliveira et al. [63] described infralabial glands divided into two distinct regions (anterior and posterior) in the elapids *Micrurus corallinus* and *M. lemniscatus*. The posterior region of the infralabial gland was later interpreted as the inferior rictal gland ([39]; see also Fig. 18.4C), evidencing considerable morphological variation in the lower jaw glands of elapids. Previous descriptions had already shown infralabial glands divided into two regions in *M. mosquitensis* [64]. Moreover, the extract from the posterior region of the infralabial gland (properly inferior rictal gland) is toxic in mice, producing typical symptoms of coral snake envenomation and death when injected intraperitoneally [64]. These results possibly represent the first evidence on the function of the inferior rictal gland of snakes and, added to Phisalix and Caius' [58] report on toxicity in the rictal glands of 'Erycinae' and Uropeltidae (and the fact that the rictal glands of *Cylindrophis ruffus* exhibit the same pattern of gene expression as the venom glands: [65]), constitute the limited information available thus far on the function of the rictal glands in snakes.

Although rictal glands have been reported in 'henophidians' and caenophidians, their presence and form in the paraphyletic 'scolecophidians' is poorly established. Haas [46] described an infralabial gland with ducts concentrated at the lip commissure in *Anomalepis aspinosus* (Anomalepididae), and Underwood [12] suggested that this part appears to have the anatomical relationships of an inferior rictal gland. Martins et al. [47] identified for the first time the presence of rictal glands in the leptotyphlopids *Epictia tenella* and *Trilepida bilineatum*, located posteriorly to the rictal plate and the supralabial gland and recognized by their location, texture, and colour differences in relation to the supralabials. However, no histological or microanatomical evidence was provided, nor a report of where the ducts open into the mouth, an important characteristic to identify rictal glands. However, our preliminary tomographic observations of some 'scolecophidians' reveal the presence of a gland clearly separated from the labial glands that opens through a single duct in the corner of the mouth in *Leptotyphlops fuliginosus* (Fig. 18.2C), suggesting that it is a superior rictal gland. This preliminary observation requires further confirmation, but it may be additional evidence for the presence of upper rictal glands in 'scolecophidians', particularly in leptotyphlopids.

18.5 Sublingual Glands

Sublingual glands occur on the floor of the mouth and are composed of two distinct parts. The anterior part has been found in all snakes studied thus far and consists of paired, pear-shaped, paramedian structures at the front of the lower jaws. Each pair of the anterior part of the sublingual gland contains a series of short ducts that open in the midline of the floor of the mouth, and its posterior portion lies more deeply in the jaw tissues and is almost entirely surrounded by muscles (see Fig. 18.3; [24]). The posterior part of the sublingual gland has been found only in Caenophidia. It consists of a single, elongated structure in the medial region of the floor of the mouth, from the anterior region of the mandible to the posterior region of the glottis [3, 24, 30]. The posterior part of the sublingual gland can also be divided into two regions: rostral and caudal. The rostral portion is composed of large tubules that open anteriorly, and the caudal portion is broader and contains small tubules that open in two parasagittal rows [3]. The posterior part of the sublingual gland has evolved three times independently into salt glands, in three lineages that underwent an evolutionary transition to marine life: in Acrochordidae, and in hydrophiine and laticaudine elapids [66]. In these taxa, sublingual gland secretions contain sodium chloride in higher concentrations than in seawater [3, 67, 68].

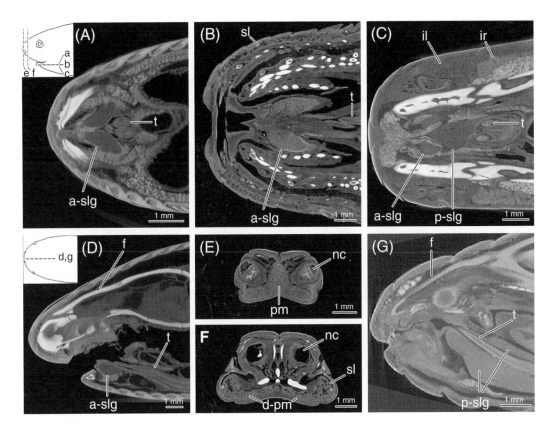

Figure 18.3 High-resolution computed tomography of the heads of various snakes showing the position of the sublingual and premaxillary glands. (A) Frontal slice showing the position of the anterior sublingual gland (a-slg) of *Amerotyphlops brongersmianus*. (B) Frontal slice showing the position of the anterior sublingual gland (a-slg) of *Tropidophis melanurus*. (C) Frontal slice showing the position of the anterior (a-slg) and posterior part of the sublingual gland (p-slg) of *Micrurus corallinus*. (D) Sagittal slice showing the position of the anterior sublingual gland (a-slg) of *Amerotyphlops brongersmianus*. (E) Transverse slice of the rostral region of *Tropidophis melanurus* showing the triangular-shaped premaxillary gland (pm) located between the anterior portion of the nasal cavity (nc), and (F), the opening of its ducts (d-pm) in a more posterior region. (G) Sagittal slice of *Hydrophis cyanocinctus* showing the enlarged posterior sublingual gland (p-slg) completely surrounding the tongue. Abbreviations: f, frontal; il, infralabial gland; ir, inferior rictal gland; sl, supralabial gland; t, tongue. The panel in the upper left corner denotes the position of the sections.

Almost no studies of the development of snake sublingual glands have been conducted. Taub [8] reported that the correct identification of the homologies of these glands remains a problem, because some authors consider them as two separate glands with a paired anterior gland and a single posterior gland. The anterior and posterior parts of the sublingual gland seem to develop simultaneously, but both portions are divided from each other from the beginning of their development. The anterior part of the sublingual gland shows diverse and short invaginations, the posterior part shows a more developed invagination (Fig. 18.1E).

Amerotyphlops brongersmianus and *Tropidophis melanurus* have a pair of long anterior sublingual glands (Fig. 18.3A, B). In *Micrurus corallinus*, the anterior part of the sublingual gland is formed by serous cells arranged in tubules, while the posterior part is composed of serous and mucous cells. As the posterior part of the sublingual extends towards the back of the mouth, the mucous cells become scarce, and only serous cells are seen in these glands ([39]; see Fig. 18.3C). As in other hydrophiine elapids (see [40]), the posterior sublingual gland in *Hydrophis cyanocinctus* is well-developed and surrounds the tongue, forming the 'supralingual gland' (Fig. 18.3G). In addition to several sea snakes, this gland also occurs in some species of *Vipera* and may be interpreted as an extension of the posterior part of the sublingual gland [3].

18.6 Premaxillary Glands

Premaxillary glands are located at the junction of the maxilla and premaxilla and are composed of mucous cells [8]. According to Taub [8], the structure, position, and embryology of the premaxillary glands suggests that they are an anterior expansion of the supralabial glands. Smith and Bellairs [24] agree that this gland may merely be a rostral expansion of the supralabial gland, although in some species it appears during dissection to be separate from the supralabial glands. Smith and Bellairs indicated that the ducts of these glands open inside the labial margin external to the anterior teeth. Burns and Pickwell [40] reported that some elapids (*Laticauda colubrina*, *Hydrophis ornatus*, and *H. platurus*) exhibit a dorsomedial extension of the premaxillary region of the supralabial gland ending beneath the anterior margins of the nasal scales. In the marine homalopsid *Cerberus rhynchops*, a premaxillary salt gland distinct from that of the supralabial gland is documented [69], representing a fourth independent evolutionary origin of salt glands in snakes (see [66]).

In 'scolecophidians', the supralabial gland is often well-developed in the rostral region, and no anterior expansion or any other differentiation between the 'typical' supralabial gland and the premaxillary gland has been observed (Fig. 18.2). In many other snakes, the single triangular-shaped premaxillary gland lies beneath the rostral scale, superficially to the premaxillary bone, but can reach the tip of the nasal bone and cover part of the nasal capsule (see Fig. 18.3E, [15, 31]). In *Tropidophis melanurus*, approximately ten premaxillary gland ducts open in the anteriormost region of the mouth (Fig. 18.3F); and in *Micrurus corallinus*, only six ducts occur in the same position. However, apart from the anterior expansion, no other difference between the supralabial and premaxillary glands is recognizable in our tomographic images.

18.7 Dental Glands

Dental glands differ from labial glands by having ducts that open at the base of the teeth instead of the margins of the lips [5, 27]. Because dental glands are found in some groups of lizards and snakes, these glands are at the centre of the discussion on the origin and evolution of the venom function in reptiles (see also Chapter 12) and are considered an anatomical synapomorphy of Toxicofera [7, 15]. Although the embryonic origin of lizard and snake dental glands appears remarkably similar (see Fig. 18.1; [72]), insufficient comparative developmental information is available to test the homology of this gland across Squamata. In snakes, dental glands occur in the upper jaw (venom glands) of Endoglyptodonta [17]. In lizards, they occur as upper and lower jaw glands in Iguania and as specialized lower jaw glands (also named mandibular glands) in Anguiformes (= Anguimorpha), with a notable exception in *Pseudopus apodus*, which has glands in the lower and upper jaws [70]. In helodermatid and varanid anguiforms, the lower jaw system is formed by two individualized glands: a typical infralabial gland and a specialized dental gland. In anguid anguiforms, there is no separation between the two glandular types, and these glands have been interpreted as typical infralabials by some authors (e.g., [28, 30]), but as dental glands by others (e.g., [70, 71]). These glands are usually mixed, with a serous portion occupying the bottom of the gland, and a mucous portion occupying the part above (see Fig. 18.6; [3, 7, 70, 71]. In the anguid *Ophiodes striatus*, the infralabial gland opens close to but not at the base of the dentary teeth, precluding its interpretation as a typical dental gland (Fig. 18.5E).

In non-toxicoferan lizards, only typical infralabial glands have been observed in the lower jaw, while, with some rare exceptions, supralabial glands are absent (see Fig. 18.5; [3, 7, 28, 30]). In *Dibamus novaeguineae*, we initially observed what appears to be a thin, tubular, unbranched gland in the lower jaw (Fig. 18.5A). This extends along the entire length of the mandible, but openings of the ducts were not visible in the tomographic images. In addition, a second gland unconnected to the infralabial was seen in the most posterior part of the lower jaw, but its identity was not determined. Further investigation is necessary to fully document oral glands in this group of lizards.

Dental glands are the most studied cephalic glands in snakes, particularly in front-fanged species [1, 21, 22, 73]. These glands are located along the caudal portion of the maxilla and are primarily serous, developing from a single invagination on the caudal portion of the dental lamina of the maxillary bone (see Fig. 18.1A, B). The concomitant origin of venom glands and fangs has been demonstrated in various snake lineages, including viperids, elapids, and non-front-fanged species [25, 74–76). Thus, some authors prefer the term dental glands to designate all of them [5]. Recent studies suggest that these glands have evolved once in Caenophidia as a morpho-functional adaptive novelty that facilitated the great radiation of this group, currently comprising approximately 2,500 extant species [6, 19, 76].

The front-fanged system is suggested to have evolved independently at least four times in snakes, two at the family level (Viperidae and Elapidae), and two at the genus level (*Atractaspis* and *Homoroselaps*) within the family Atractaspididae [60, 73]. As suggested earlier (e.g., [77]), recent studies have corroborated that the front-fanged system is derived from a rear-fanged condition [76].

In Viperidae, the venom gland is large and generally triangular-shaped in lateral view, with the longest side of the triangle along the upper lip and a rounded dorsal apex [6]. Viperid venom glands are composed of four distinct portions: the venom gland (the posterior and largest part); the primary duct; the accessory gland (the anterior and

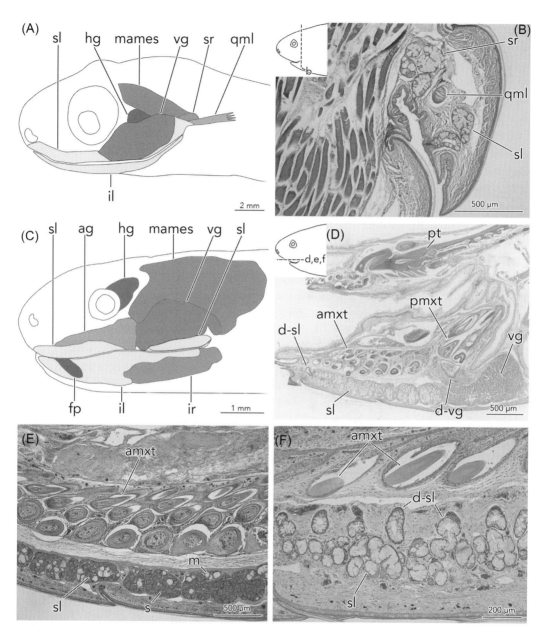

Figure 18.4 (A) Schematic drawing of skinned head of *Imantodes cenchoa* showing the location of the oral glands, the harderian gland (hg), and the *adductor mandibulae externus superficialis* (mames). (B) Transverse section of the head of *I. cenchoa* at the level of the corner of the mouth showing the superior rictal gland (sr) medial to the quadrato-maxillary ligament (qml) and the supralabial gland (sl) lateral to the ligament. (C) Schematic drawing of skinned head of *Micrurus corallinus* showing the location of the oral glands, the harderian gland (hg), and the *adductor mandibularis externus superficialis* (mames). (D) Frontal slice of head of *Leptodeira annulata* showing the position of the supralabial gland (sl) and its ducts (d-sl) and the position of the venom gland (vg) and the opening of its single duct (d-vg) close to the base of the posterior maxillary teeth (pmxt). (E) Frontal

narrowest part); and the secondary duct connecting to the fang sheath [3]. The main venom gland has a complex tubular structure and is divided into several lobules by infoldings of the outer sheath. The lumen of the main venom gland is large and stores a large amount of venom. The accessory gland is globular and is located anteriorly on the venom gland duct [34]. Both the main venom gland and the accessory gland are formed by several cell types [78, 79]. The viperid venom gland is compressed by muscle fibres derived from the *adductor externus medialis* [6, 74], although some fibres of the *adductor externus profundus* and a branch of the *pterygoideus* are also associated with different parts of the gland [3]. Elongated venom glands occur in species of *Causus* [80].

The general pattern of elapid venom glands has been described by numerous authors [40, 81–84]. Generally, these glands are ovoid or pear-shaped in lateral view and occupy practically the entire postocular region, extending from the posterior edge of the eyes to the level of the corner of the mouth (Fig. 18.4C). As in viperids, the venom glands of elapids are composed of two portions, the venom and accessory gland [1]. Unlike viperids, however, the duct that drains the venom gland passes through the accessory gland and opens into the sheath of the fang [80]. The central lumen is relatively small, and the secretion is stored mainly in the secretion granules inside the cells [1]. The elapid venom gland is compressed by the fibres derived from the *adductor externus superficialis*, which in most elapids has a dorsal and a ventral branch [13, 60, 83]. Elongated venom glands evolved at least twice in elapids, occurring in some coral snake species of the genus *Calliophis* (specifically in the *bivirgatus* group) and species of *Toxicocalamus* [15, 85].

The venom glands of *Atractaspis* differ from those of elapids and viperids [1], being cylindrical with secretory tubules arranged around an elongated main lumen [86]. There is no distinctive mucous accessory gland, but mucous cells are found in each secretion tubule near the central lumen [34]. The *Atractaspis* venom gland is compressed by a long arc of muscle fibres that seem to derive from the *adductor externus medialis* [6]. Like other groups of venomous snakes, some species of *Atractaspis* have elongated venom glands extending into the neck [34]. The venom glands of *Homoroselaps* are similar to those of elapids, being composed of two parts: the main venom gland and the accessory gland. In addition, the venom gland of *Homoroselaps* is compressed by the fibres derived from the *adductor externus superficialis* [34, 60, 83]. Although the phylogenetic position of *Homoroselaps* has been intensely debated (see [34, 60, 83]), recent molecular phylogenies have indicated that it is an atractaspidid [17], indicating an independent derivation of its glandular morphology.

Within the diverse and polyphyletic assemblage of non-front-fanged snakes, venom glands show great structural variability, including examples of all transitional forms

Figure 18.4 (*cont.*) slice of head of *Helicops modestus* showing the upper part of the supralabial gland (sl) composed mostly of serous cells (s). (F) Frontal slice of head of *H. modestus* showing the ventral part of the supralabial gland (sl) composed mostly of mucous cells and the opening of the ducts (d-sl). Sections b, d, e, and f were stained with H&E. Abbreviations: ag, accessory gland; amxt, anterior maxillary teeth; fp, fang pocket; il, infralabial gland; ir, inferior rictal gland; m, mucous cells; pt, pterygoid. The panel in the upper left corner denotes the position of the sections. (A black and white version of this figure will appear in some formats. For the colour version, please refer to the plate section).

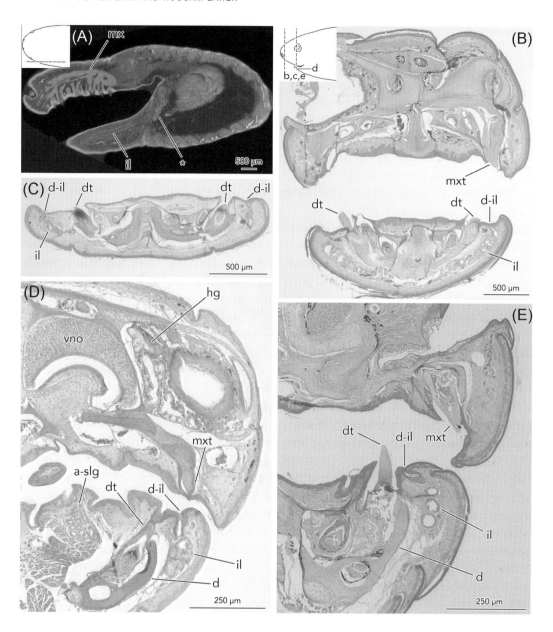

Figure 18.5 Morphology of the oral glands in lizards. (A) High-resolution computed tomography of the head of *Dibamus novaeguineae* (Dibamidae) in sagittal section showing the position of the tubular infralabial gland (il) and the second gland (*) unconnected to the infralabial gland in the posterior portion of the lower jaw. (B) Transverse slice of the head of *Varzea bistriata* (Scincidae) at the nostril level showing the position of the infralabial gland (il) and the opening of its duct (d-il). (C) Transverse slice of the lower jaw of *Hemidactylus mabouia* (Gekkonidae) at the anterior level showing the position of the infralabial gland (il) and the opening of its duct (d-il). (D) Transverse slice of the head of *Colobosaura modesta* (Gymnophthalmidae) at the anterior level of the eyes showing the position of the infralabial gland (il) and the opening of its duct (d-il). (E) Transverse slice of the head of *Ophiodes striatus* (Diploglossidae) at the nostril level showing the position of the infralabial gland (il) and the

between the complete absence of a venom gland and the presence of a purely serous and well-differentiated venom gland [1, 18, 19, 32, 33]. Extensive morphological variation is also observed in the posterior maxillary teeth associated with the venom glands [6, 19, 87]. These glands are located in the postocular region and frequently reach the angle of the mouth posteriorly, displacing or reducing the supralabial locally (see Fig. 18.4A; [3]). As described in § 18.2.1, there is no evidence of venom glands in acrochordids, xenodermids or pareids [12, 34]. Moreover, the absence of anterior and posterior ridges on the posterior maxillary teeth in these three families (see [17]) suggests that the major modifications related to the acquisition of the venom system (i.e., venom glands and differentiated maxillary teeth) are restricted to Endoglyptodonta rather than Colubroides, with secondary losses occurring many times in elapoids and colubroids.

Saint-Girons [27] noted that in non-front-fanged snakes, the venom gland may occur or not at subfamilial, tribal, and even genus levels. He also suggested that the absence of the gland in many species probably results from secondary regression. The secondary loss or reduction of venom delivery system components is often associated with a few front-fanged and several non-front-fanged lineages in which constriction has evolved as a strategy for prey capture, or the diet has shifted to eggs or molluscs [33]. In front-fanged snakes, only the lizard egg-eating *Brachyurophis*, and fish-egg-eating *Aipysurus eydouxii* and *Emydocephalus* have substantially reduced venom systems [15, 33, 83, 84]. In non-front-fanged snakes, 30 to 40 per cent of the species have venom glands [20]. Examples of non-front-fanged snakes with reduced or absent components of the venom delivery system include the egg-eating *Dasypeltis scabra*, some constrictor colubrids (*sensu* [17]) of the genera *Pantherophis, Elaphe, Lampropeltis, Rhinocheilus, Spilotes,* and *Pituophis,* and the 'goo-eaters', a Neotropical lineage of dipsadids that feed exclusively on soft-bodied invertebrates [18, 19, 32, 33, 41, 42, 52].

Muscle fibres insert into the venom glands of some non-front-fanged snakes (e.g., *Dispholidus typus, Mehelya poensis,* and *Brachyophis revoili*), and may act as a rudimentary compressor muscle [1, 15, 34]. Muscle fibres also exhibit some association with the venom glands of some elapomorphine and tachymenine dipsadids [16, 38, 61]. A central lumen has been preliminarily reported in the venom gland of some *Apostolepis* [88], in addition to that previously reported in *M. poensis, D. typus,* and *Gonionotophis capensis* (see [15, 82]).

18.8 Summary and Conclusions

Snakes exhibit a wide variety of oral glands whose functions remain poorly known. The taxonomic distribution of various oral glands in snakes, other toxicoferans and other lizards is summarized in Table 18.2 (and Supplementary Table 18.S1). Rictal glands are particularly

Figure 18.5 (*cont.*) opening of its duct (d-il) close to the dentary tooth (dt). Sections b, c, d, and e were stained with H&E. Note that none of these species have supralabial glands. Abbreviations: a-slg, anterior sublingual gland; d, dentary; dt, dentary tooth; hg, harderian gland; mx, maxilla; mxt, maxillary tooth; vno, vomeronasal organ. The panel in the upper left corner denotes the position of the sections. See Supplementary Figure 18.S1 for colour version.

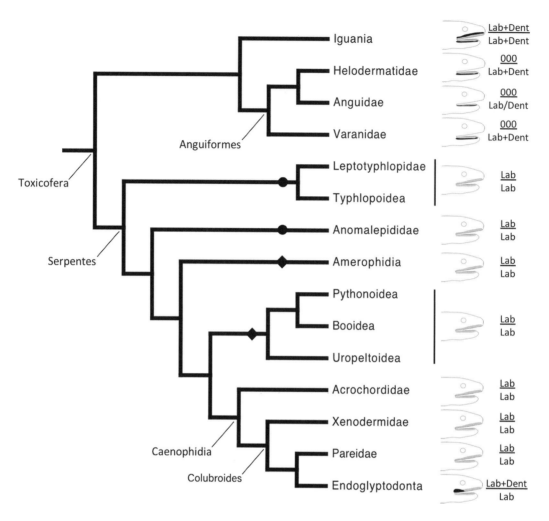

Figure 18.6 Summary phylogenetic tree (based on [23]) showing the condition of the dental glands in Toxicofera. In snakes, the dental glands are represented by the upper jaw glands (venom glands) present in Endoglyptodonta (which includes the major lineages viperids, elapids, atractaspidids, colubrids, dipsadids). In lizards, the dental glands are represented by the upper and lower jaw glands present in Iguania and the specialized lower jaw glands present in Anguiformes (see text for more details). In the head drawings, the dental glands are depicted in black and the typical labial glands in grey. Abbreviations: Dent, dental gland; Lab, labial gland; 000, absence of labial or dental glands. Circles denote clades referenced in the text as 'scolecophidians'; diamonds denote clades referenced in the text as 'henophidians'. Figure modified from [1, 3] and [7, 70, 71].

poorly known in terms of diversity and function. Their presence in leptotyphlopids indicates that they are more widely distributed in snakes than previously expected, and were possibly present in the ancestral snake. The sublingual glands are composed of two distinct (anterior and posterior) parts. The anterior part occurs in all snakes studied so far, but the posterior part is restricted to Caenophidia, where it has evolved into salt glands three times

Table 18.2 Summary of the general condition of oral glands in some lineages of lizards and snakes.

yes = present; no = absent; ? = condition unknown and/or unclear; ant. = anterior part of gland; post. = posterior part of gland. Notes: 1. With some rare exceptions; 2. With a notable exception in *Pseudopus apodus*, which has glands in upper and lower jaws; 3. Preliminary observation indicates infralabial glands are also present in Dibamidae; 4. Bolyeriids appear to have lost rictal structures, including glands and rictal pockets; 5. Rictal glands have been described in Viperinae, but they are virtually unknown in Crotalinae and Azemiopinae; 6. Clearly documented only in Dipsadidae and Colubridae; 7. Known only in Dipsadidae; 8. Clearly documented only in Elapidae; 9. Clearly documented only in Boidae; 10. Some argue these are simply infralabial glands. A single species has been described with upper and lower jaw dental glands; 11. Only in elapids and the atractaspidids *Homoroselaps* spp. An expanded version of this table with finer-scale taxonomic subdivisions is available as Supplementary Table 18.S1.

	Supralabial (SL)	Infralabial (IL)	Temporo-mandibular	Superior rictal	Inferior rictal	Sublingual	Premaxillary	Dental	Accessory	References
Other lizards	no[1]	yes[3]	no	no	no	ant.	no	no	no	[3, 28, 30; present study]
TOXICOFERA										
Iguania	yes	yes	no	no	no	ant.	?	upper + lower jaw	no	[3, 7, 28, 30]
Helodermatidae	no	yes	no	no	no	ant.	no	lower jaw	no	[3, 30, 70, 71]
Anguidae	no[2]	yes	no	no	no	ant.	no	lower jaw[10]	no	[28, 30, 70, 71]
Varanidae	no	yes	no	no	no	ant.	no	lower jaw	no	[3, 28, 30, 70, 71]
SERPENTES										
'scolecophidians'										
Leptotyphlopidae	Yes	yes	no	yes	no	ant.	not distinct from SL	no	no	[28, 47; present study]
Typhlopidae	Yes	yes	yes	no	no	ant.	not distinct from SL	no	no	[3, 8, 57; present study]
Anomalepididae	Yes	yes	no	no	possibly	ant.	not distinct from SL	no	no	[12, 29, 46]

Table 18.2 (cont.)

'henophidians'

Anomochilidae	Yes	yes	no	yes	?	ant.	distinct from SL	no	no	[62]
Tropidophiidae	Yes	yes	no	yes	?	ant.	distinct from SL	no	no	[12, 60; present study]
Aniliidae	Yes	yes	no	yes	yes	ant.	?	no	no	[12, 57, 58]
Pythonoidea	Yes	yes	no	yes[4]	no	ant.	?	no	no	[5, 12, 48]
Booidea	yes	yes	no	yes	?	ant.	distinct from SL[8]	no	no	[12, 31, 48, 57, 58]
Cylindrophiidae	Yes	yes	no	yes	yes	ant.	?	no	no	[12, 15, 24, 65]
Uropeltidae	Yes	yes	no	yes	no	ant.	?	no	no	[12, 24, 48, 57, 58]

CAENOPHIDIA

Acrochordidae	Yes	yes	no	?	no	ant.+post.	?	no	no	[12]
Xenodermidae	Yes	yes	no	yes	no	ant.+post.	?	no	no	[5, 45, 60]
Pareidae	Yes	yes	no	yes	no	ant.+post.	?	no	no	[12, 34, 50, 51]

ENDOGLYPTODONTA

Viperidae	Yes	yes	no	yes[5]	yes[5]	ant.+post.	distinct from SL	upper jaw	yes	[1, 3, 5, 15, 18]
Homalopsidae	Yes	yes	no	possibly	?	ant.+post.	distinct from SL	upper jaw	no	[18, 60, 69]
Elapoidea	Yes	yes	no	yes	yes	ant.+post.	distinct from SL[9]	upper jaw	yes[11]	[15, 61, 63, 66, 81-84]
Colubroidea	Yes	yes	no	yes[6]	yes[7]	ant.+post.	distinct from SL	upper jaw	no	[5, 18, 31, 41, 42, 60, 80]

independently in lineages that underwent an evolutionary transition to marine life. The homology of the premaxillary gland remains poorly resolved. In 'scolecophidians', premaxillary glands are indistinct from the supralabial glands, whereas in many other snakes they are visibly differentiated. The homology of dental glands in lizards and snakes has not been adequately tested. Although the development of dental glands seems similar in snakes and Iguania, almost no comparative studies have studied their development in Anguiformes. Moreover, the lack of dental glands in non-caenophidian snakes does not support the hypothesis that these glands evolved only once in Toxicofera. In snakes, dental glands appear restricted to Endoglyptodonta, because Xenodermidae and Pareidae also lack such glands. Dental glands are highly derived in front-fanged species and present little evidence of secondary reduction. In non-front-fanged snakes, dental glands have been secondarily lost several times and show a wide morphological variation, with species showing glands associated with muscles and lumens for secretion storage. The cellular composition of labial glands (serous, mucous, or mixed) seems to be independent of the presence or absence of dental and/or rictal glands. It is also not a good basis for homology, but it may have some relevance for inferring function. The location of the opening of the ducts (and, the embryonic origin) of the labial glands are a better basis for distinguishing labial from dental and rictal glands. The presence of typical toxins in oral glands other than dental glands (e.g., rictal and infralabial) suggests that venom function may not be restricted to the homologous dental gland tissues, but is instead more broadly dispersed in other glandular tissues.

Acknowledgements

We thank Patrick Campbell (NHM, UK), Francisco Franco (IBSP), and Luis Diaz (MNKNC) for allowing and facilitating the study and preparation of specimens under their care and Alberto B. Carvalho, Vanessa Yamamoto, and Rosely Rodrigues da Silva (Laboratory of Microtomography of the Museum of Zoology, University of São Paulo, Brazil), Mark Wilkinson (NHM, UK), and Vincent Fernandez and Brett Clark (Imaging and Analysis Centre, NHM, UK) for helping to acquire and process CT images. We also thank Marcelo Garrone for providing slides of the heads of some species of snakes and lizards, Roberta Graboski for helping with the tomography of some species, and Henrique Braz, Abigail Tucker, Timothy Jackson and David Gower for constructive reviews. This research was supported by grants from Fundação de Amparo à Pesquisa do Estado de São Paulo (FAPESP) to LO (FAPESP 2014/14364-7 and 2018/09301-7) and HZ (BIOTA/FAPESP 2011/50206-9).

References

1. E. Kochva, The origin of snakes and evolution of the venom apparatus. *Toxicon*, 25 (1987), 65–106.

2. A. Deufel and D. Cundall, Functional plasticity of the venom delivery system in snakes with focus on the poststrike prey release behavior. *Zoologischer Anzeiger*, 245 (2006), 249–267.

3. E. Kochva, Oral glands of the reptilia. In C. K. Gans and A. Gans, eds., *Biology of the*

Reptilia, Vol. 8 (London and New York: Academic Press, 1978), pp. 43–162.

4. D. Cundall, Functional morphology. In R. A. Siegel, J. T. Collins, and S. S. Novak, eds., *Snakes, Ecology and Evolutionary Biology* (New York: MacMillan, 1987), pp. 106–140.

5. G. Underwood, An overview of venomous snake evolution. In R. S. Thorpe, W. Wüster, and A. Malhotra, eds., *Venomous Snakes. Ecology, Evolution and Snakebite*, n. 70 (Oxford: Clarendon Press, 1997), pp. 1–13.

6. K. Jackson, The evolution of venom-delivery systems in snakes. *Zoological Journal of the Linnean Society*, 137 (2003), 227–354.

7. B. G. Fry, N. Vidal, J. A. Norman, et al., Early evolution of the venom system in lizards and snakes. *Nature*, 439 (2006), 584–588.

8. A. M. Taub, Ophidian cephalic glands. *Journal of Morphology*, 118 (1966), 529–542.

9. A. G. Pearse, *Histochemistry: Theoretical and Applied.* Volume 2. 4th ed. (Edinburgh: Churchill Livingstone, 1985).

10. J. A. Kiernan, *Histological and Histochemical Methods: Theory and Practice.* 3rd ed. (London: Oxford University Press, 2001).

11. B. D. Metscher, MicroCT for comparative morphology: simple staining methods allow high-contrast 3D imaging of diverse non-mineralized animal tissues. *BMC Physiology*, 9 (2009), 1–14.

12. G. Underwood, On the rictal structures of some snakes. *Herpetologica*, 58 (2002), 1–17.

13. H. Zaher, Comments on the evolution of the jaw adductor musculature of snakes. *Zoological Journal of the Linnean Society*, 111 (1994), 339–384.

14. S. A. Weinstein, *'Venomous' bites from non-venomous snakes: a critical analysis of risk and management of 'colubrid' snake bites.* (Waltham, MA: Elsevier, 2011).

15. T. N. W. Jackson, B. Young, G. Underwood, et al., Endless form most beautiful: the evolution of ophidian oral glands, including the venom system, and the use of appropriate terminology for homologous structures. *Zoomorphology*, 136 (2017), 107–130.

16. H. Zaher, F. G. Grazziotin, J. E. Cadle, et al., Molecular phylogeny of advanced snakes (Serpentes, Caenophidia) with an emphasis South American Xenodontines: a revised classification and description of new taxa. *Papéis Avulso de Zoologia*, 49 (2009), 115–153.

17. H. Zaher, R. W. Murphy, J. C. Arredondo, et al., Large-scale molecular phylogeny, morphology, divergence-time estimation, and the fossil record of advanced caenophidian snakes (Squamata: Serpentes). *PLoS ONE*, 14 (2019), e0216148.

18. A. M. Taub, Comparative studies on Duvernoy's gland of colubrid snakes. *Bulletin of the American Museum of Natural History*, 138 (1967), 1–50.

19. N. Vidal, Colubroid systematics: evidence for an early appearance of the venom apparatus followed by extensive evolutionary tinkering. *Journal of Toxinology: Toxin Review*, 21 (2002), 21–41.

20. K. V. Kardong, Colubrid snakes and Duvernoy's 'venom' glands. *Journal of Toxicology: Toxin Reviews*, 21 (2002), 1–19.

21. S. A. Weinstein, T. L. Smith, and K. Kardong, Reptile venom glands form, function, and future. In S. P. Mackessy, ed., *Handbook of Venoms and Toxin of Reptiles* (Boca Raton, NY: CRC Taylor & Francis, 2010), pp. 65–91.

22. B. G. Fry, N. R. Casewell, W. Wüster, et al., The structural and functional diversification of the Toxicofera reptile venom system. *Toxicon*, 60 (2012), 434–448.

23. F. T. Burbrink, F. G. Grazziotin, R. A. Pyron, et al., Interrogating genomic-scale data for Squamata (lizards, snakes, and amphisbaenians) shows no support for key traditional morphological relationships. *Systematic Biology*, 69 (2020), 502–520.

24. M. Smith and A. A. Bellairs, The head glands of snakes, with remarks on the evolution of the parotid gland and teeth of the Opisthoglypha. *Zoological Journal of the Linnean Society*, 41 (1947), 353–368.

25. P. Gygax, Entwicklung, Bau und Funktion der Giftdrüse (Duvernoy's gland) von *Natrix*

tessellata. Acta Tropica, Zoology, 28 (1971), 225–274.

26. L. Oliveira, R. A. Guerra-Fuentes, and H. Zaher, Embryological evidence of a new type of seromucous labial gland in neotropical snail-eating snakes of the genus *Sibynomorphus. Zoologischer Anzeiger*, 266 (2017), 89–94.

27. H. Saint-Girons, Évolution de la function venimeuse chez les reptiles. In *Compte-rendu du colloque organisé à la Faculté Catholique des Sciences* (Lyon: Societé Herpétologique de France & Fondation Marcel Merieuse, 1987), pp. 9–22.

28. M. Gabe and H. Saint-Girons, Données histologiques sur les glandes salivaires des lépidosauriens. *Memoires du Museum National d'Histoire Naturelle*, 58 (1969), 3–116.

29. G. Haas, Anatomical observations on the head of *Liotyphlops albirostris* (Typhlopidae, Ophdia). *Acta Zoologica*, 1964 (1964), 1–62.

30. H. Saint-Girons, Les glandes céphaliques exocrines des Reptiles. II. – Considérations fonctionnelles et évolutives. *Annales des Sciences Naturelles, Zoologie*, 10 (1989), 1–17.

31. D. C. Penteado, Estudos histológicos das glândulas da cabeça dos ofídeos brasileiros. *Memórias do Instituto Butantan*, 1 (1918), 27–57.

32. A. M. Taub, Systematic implications from the labial glands of the Colubridae. *Herpetologica*, 23 (1967), 145–148.

33. B. G. Fry, H. Scheib, L. van der Weerd, et al., Evolution of an arsenal: structural and functional diversification of the venom system in the advanced snakes (Caenophidia). *Molecular and Cellular Proteomics*, 7 (2008), 215–46.

34. G. Underwood and E. Kochva, On the affinities of the burrowing asps *Atractaspis* (Serpentes: Atractaspididae). *Zoological Journal of the Linnean Society*, 107 (1993), 3–64.

35. R. A. Lopes, C. de Oliveira, M. N. M. Campos, S. M. Campos, and E. G. Birman, Morphological and histochemical study of cephalic glands of *Bothrops jararaca*

(Ophidia, Viperidae). *Acta Zoologica*, 55 (1974), 17–24.

36. I. Ineich and J. M. Tellier, Une glande supralabiale à débouché externe chez le genre *Echis* (Reptilia, Viperidae), cas unique chez les serpents. *Comptes Rendus de l'Académie des Sciences Paris*, 315 (1992), 49–53.

37. H. Saint-Girons and I. Ineich. Donnés histologiques sur la glande labiale supérieure externe des Viperidae du genre *Echis. Amphibia-Reptilia*, 14 (1993), 315–319.

38. A. H. Savitzky, The origin of the New World proteroglyphous snakes and its bearing on the study of venom delivery systems in snakes. Unpublished PhD dissertation, University of Kansas, Lawrence, United States, 1979.

39. L. Oliveira, M. A. Buononato, and H. Zaher, Chapter 12 - The cephalic glands and venom apparatus of coralsnakes. In N. J. Silva Jr., L. W. Porras, S. T. Aird, and A. L. C. Prudente, eds., Advances in coralsnake biology: with an emphasis on South America. (Eagle Mountain, Utah: Eagle Mountain Publishing, LC, Utah, USA, 2021), pp. 371–390.

40. B. Burns and G. V. Pickwell, Cephalic glands in sea snakes (*Pelamis, Hydrophis* and *Laticauda*). *Copeia*, 1972 (1972), 547–559.

41. L. Oliveira, A. L. C. Prudente, and H. Zaher, H. Unusual labial glands in snakes of the genus *Geophis* Wagler, 1830 (Serpentes: Dipsadinae). *Journal of Morphology*, 275 (2014), 87–99.

42. H. Zaher, L. Oliveira, F. G. Grazziotin, et al., Consuming viscous prey: a novel protein-secreting delivery system in Neotropical snail-eating snakes. *BMC Evolutionary Biology*, 14 (2014), 58.

43. A. H. Savitzky, The relationship of the xenodontine colubrid snakes related to Ninia. Unpublished Masters dissertation, University of Kansas, United States, 1972.

44. M. B. Harvey, G. R. Fuenmayor, J. C. R. Portilla, and J. V. Rueda-Almonacid, Systematics of the enigmatic dipsadinae snake *Tropidophis perijanensis* Alemán

(Serpentes: Colubridae) and review of morphological characters of Dipsadini. *Herpetological Monographs*, 22 (2008), 106–132.

45. G. Underwood, *A Contribution to the Classification of Snakes*. (London: British Museum (Natural History), 1967).

46. G. Haas, Anatomical observations on the head of *Anomalepis aspinosus* (Typhlopidae, Ophidia). *Acta Zoologica*, 49 (1967), 63–139.

47. A. Martins, P. Passos, and R. Pinto, Unveiling diversity under the skin: comparative morphology study of the cephalic glands in threadsnakes (Serpentes: Leptotyphlopidae: Epictinae). *Zoomorphology*, 137 (2018), 433–443.

48. G. Haas, Über die Kaumuskulatur und die Schädelmechanik einiger Wühlschlangen. *Zoologische Jahrbücher (Anatomie)*, 52 (1930), 95–218.

49. L. D. Brongersma, Some features of the Dipsadinae and Pareinae (Serpentes, Colubridae). *Proceedings van de Koninklijke Nederlandse Akademie van Wetenschappen Section C*, 61, (1958), 7–12.

50. G. Haas, A note on the origin of solenoglyph snakes. *Copeia*, 1938 (1938), 73–78.

51. H. Zaher, Hemipenial morphology of the South American xenodontine snakes, with a proposal for a monophyletic Xenodontinae and a reappraisal of colubroid hemipenes. *Bulletin of the American Museum of Natural History*, 240 (1999), 1–168.

52. J. E. Cadle and H. W. Greene, Phylogenetic patterns, biogeography, and the ecological structure of Neotropical snake assemblages. In R. E Ricklefs and D. Schluter, eds., *Species Diversity in Ecological Communities: Historical and Geographical Perspective* (Chicago: University of Chicago Press, 1993), pp. 281–293.

53. I. L. Laporta-Ferreira and M. G. Salomão, Morphology, physiology and toxicology of the oral glands of a tropical cochleophagous snake, *Sibynomorphus neuwiedi* (Colubridae – Dipsadinae). *Zoologischer Anzeiger*, 27 (1991), 198–208.

54. M. G. Salomão and I. L. Laporta-Ferreira, The role of secretions from the supralabial, infralabial, and Duvernoy's glands of the slug-eating snake *Sibynomorphus mikanii* (Colubridae: Dipsadinae) in the immobilization of molluscan prey. *Journal of Herpetology*, 28 (1994), 369–371.

55. L. Oliveira, C. Jared, and A. L. C. Prudente, Oral glands in dipsadinae 'goo-eater' snakes: Morphology and histochemistry of the infralabial glands in *Atractus reticulatus*, *Dipsas indica*, and *Sibynomorphus mikanii*. *Toxicon*, 51 (2008), 898–913.

56. P. F. Campos, L. Oliveira, F. G. Grazziotin, et al., Transcriptomic analysis of snake infralabial glands highlights a plasticity in the site of expression of venom genes. *Toxicon*, 158 (2019), pp. S48.

57. M. Phisalix, *Animaux venimeux et venins*, Vol. 2 (Paris: Masson & Cie, 1922).

58. M. Phisalix and R. Caius. L'extension de la fonction venimeuse dans l'ordre entière des ophidiens et son existence chez des familles ou elle n'avait pas été soupçonnée jusqu'içi. *Journal de Physiologie et de Pathologie Générale*, 17 (1918), 923–964.

59. E. Kochva, The development of the venom gland in the opisthoglyph snake *Telescopus fallax* with remarks on *Thamnophis sirtalis* (Colubridae, Reptilia). *Copeia*, 1965 (1965), 147–154.

60. S. B. McDowell, The architecture of the corner of the mouth of colubroid snakes. *Journal of Herpetology*, 20 (1986), 353–407.

61. M. Wollberg, E. Kochva, and G. Underwood, On the rictal glands of some atractaspid snakes. *Herpetological Journal*, 8 (1998), 137–143.

62. D. Cundall and D. A. Rossman, Cephalic anatomy of the rare Indonesian snake *Anomochilus weberi*. *Zoological Journal of the Linnean Society*, 109 (1993), 235–273.

63. L. Oliveira, M. A. Buononato, and H. Zaher, Glândulas cefálicas e aparato de veneno das cobras-corais. In N. J. Silva Jr., ed., *As cobras-corais do Brasil: Biologia, taxonômica, venenos e envenamentos* (Goiânia: Editora

de Pontifica Universidade Católica de Goiás, Brasil, 2016), pp. 217–241.

64. M. W. Dix, A venom gland in the lower jaw of the coral snake (*Micrurus nigrocinctus mosquitensis* Schmidt). In P. Rosenberg, ed., *Toxins. Animal, Plant and Microbial* (Oxford: Pergamon Press, 1978), pp. 16–28.

65. B. G. Fry, E. A. B. Undheim, S. A. Ali, et al., Squeezers and leaf-cutters: differential diversification and degeneration of the venom system in toxicoferan reptiles. *Molecular and Cellular Proteomics* 12 (2013), 1881–1899.

66. L. S. Babonis and F. Brischoux, Perspectives on the convergent evolution of tetrapod salt glands. *Integrative and Comparative Biology*, 52 (2012), 245–56.

67. W. A. Dunson, R. K. Packer, and M. K. Dunson, Sea snakes: an unusual salt gland under the tongue. *Science*, 173 (1971), 437–441.

68. W. A. Dunson and M. K. Dunson, Convergent evolution of sublingual salt glands in the marine file snake and true sea snakes. *Journal of Comparative Physiology*, 86 (1973), 193–208.

69. W. A. Dunson and M. K. Dunson, Possible new salt gland in a marine homalopsid snake (*Cerberus rhynchops*). *Copeia*, (1979), 661–673.

70. B. G. Fry, *Venomous Reptiles and Their Toxins: Evolution, Pathophysiology and Biodiscovery* (New York: Oxford University Press, 2015).

71. B. G. Fry, K. Winter, J. A. Norman, et al., Functional and structural diversification of the Anguimorpha lizard venom system. *Molecular and Cellular Proteomics*, 9 (2010), 2369–2390.

72. A. S. Tucker, Salivary gland adaptations: modification of the glands for novel uses. In A. S. Tucker and I. Miletich, eds., *Salivary Glands. Development, Adaptations and Disease. Frontiers of Oral Biology,* Vol. 14 (Basel: Karger, 2010), pp. 21–31.

73. H. M. I. Kerkkamp, N. R. Casewell, and F. J. Vonk, Evolution of the snake venom delivery system. In. P. Gopalakrishnakone and A. Malhotra, eds., *Evolution of Venomous Animals and Their Toxins, Toxinology* (Berlin: Springer, 2017), pp. 303–315.

74. E. Kochva, Development of the venom gland and trigeminal muscles in *Vipera palaestinae*, Acta Anatomica, 52 (1963), 49–89.

75. M. Shayer-Wollberg and E. Kochva, Embryonic development of the venom apparatus in *Causus rhombeatus* (Viperidae, Ophidia). *Herpetologica*, 23 (1967), 249–259.

76. F. J. Vonk, J. R. Admiraal, K. Jackson, et al., Evolutionary origin and development of snakes fangs. *Nature*, 454 (2008), 630–633.

77. G. A. Boulenger, Remarks on the dentition of snakes and on the evolution of the poison-fangs. *Proceedings of the Zoological Society of London*, 64 (1896), 614–618.

78. S. P. Mackessy, Morphology and ultrastructure of the venom gland of the Northern Pacific Rattlesnake *Crotalus viridis oreganus. Journal of Morphology*, 208 (1991), 109–128.

79. F. Sakai, S. M. Carneiro, and N. Yamanouye, Morphological study of accessory gland of *Bothrops jararaca* and its secretory cycle. *Toxicon*, 59 (2012), 393–401.

80. E. Kochva and C. Gans, Salivary glands of snakes. *Clinical Toxicology*, 3 (1970), 363–387.

81. H. I. Rosenberg, Histology, histochemistry, and emptying mechanism of the venom glands of some elapid snakes. *Journal of Morphology*, 123 (1967), 133–156.

82. E. Kochva and M. Wollberg, The salivary glands of Aparallactinae (Colubridae) and the venom glands of *Elaps* (Elapidae) in relation to the taxonomic status of this genus. *Zoological Journal of the Linnean Society*, 49 (1970), 217–224.

83. C. J. McCarthy, Morphology of elapid snakes (Serpentes: Elapidae). An assessment of the evidence. *Zoological Journal of the Linnean Society*, 83 (1985), 79–93.

84. P. Gopalakrishnakone and E. Kochva, Venom glands and some associated muscles in sea snakes. *Journal of Morphology*, 205 (1990), 85–96.

85. P. Gopalakrishnakone, Structure of the venom gland of the Malayan Banded Snake *Maticora intestinalis. Snake*, 18 (1986), 19–26.

86. E. Kochva, *Atractaspis* (Serpentes, Atractaspididae) the burrowing asp; a multidisciplinary minireview. *Bulletin of the Natural History Museum of London (Zoology)*, 68 (2002), 91–99.

87. B. A. Young and K. V. Kardong, Dentitional surface features in snakes (Reptilia: Serpentes). *Amphibia–Reptilia*, 17 (1996), 261–276.

88. M. G. Salomão and H. Ferrarezzi, A morphological, histochemical and ultrastructural analysis of the Duvernoy's glands of elapomorphine snakes: the evolution of their venom apparatus and phylogenetic implications. Abstract (Campinas: III Congresso Latino Americano de Herpetologia, 43, Instituto de Biociências da Universidade Estadual de Campinas, 1993).

19

Macrostomy, Macrophagy, and Snake Phylogeny

DAVID CUNDALL AND FRANCES IRISH

19.1 Introduction

Snakes are limbless animals that generally ingest prey whole using ratcheting movements of their jaws achieved by a variety of different mechanisms [1–3]. All are carnivorous and none uses the tongue or inertial acceleration of the whole head to move their prey through the oral cavity. The clade includes many species capable of swallowing prey much larger than their heads, correlated in most to an increase in relative length of the mandibles. Among living snakes, however, the most 'basal' members of the clade have tiny gapes and eat large numbers of tiny prey [4]. Following a review of snake apomorphies, Rieppel [5] resurrected Müller's Macrostomata [6] for a clade of snakes essentially defined by features allowing swallowing of prey large in relative cross-sectional area and that appeared to have evolved from ancestors lacking this ability. Recent analyses of molecular sequence data [7, 8] and Cretaceous fossils [9] do not support the monophyly of Macrostomata and provide equivocal support for an origin of large-gaped snakes from small-gaped ancestors. Given that macrostomy characterizes many extant snakes, understanding its morphological correlates and taxonomic distribution are critical to unravelling the origins of macrostomy. We reconsider the structures and functions of the trunks and heads of extant macrostomous snakes to determine (1) which features correlate most closely with macrostomy, and (2) how these features inform hypotheses about snake origins.

19.2 Defining Macrostomy

What macrostomy means in a broad sense is large hole, or large mouth, i.e., large in relative cross-sectional area, as distinct from macrophagy, eating a large food item – which might be large in relative mass, large in relative diameter, or both [1, 10]. The application of the term macrostomy should require an objective definition, but that does not appear to exist. Diverse interpretations of fossils regarded as snakes have added to the complexity of the problem [9, 11]. It is useful to begin by emphasizing that macrostomy is conferred by a snake's structural and functional attributes. Snakes possessing the attributes are macrostomous, whereas Macrostomata is a taxon name that corresponds to a clade in which all members may or may not exhibit these phenotypic attributes. Taxa included within the clade are macrostomatans. Our focus is on how macrostomy relates to Macrostomata.

One reason that snakes are interesting is because the macrostomy of some critical clades is not a permanent structural condition that could be measured at any time. Despite a common assumption that macrostomy derives primarily from increased lengths of the mandibles and their suspensorial elements (e.g., [12]) in many taxa it depends on movement potentials of soft tissue assemblages. If macrostomy was defined solely by structures of fixed sizes, it could be quantified and analysed relatively simply. Snakes turn out to be excellent examples of the diversity of evolutionary solutions to two problems – catching and subduing prey nearly as heavy as, or heavier than themselves, and then enlarging gape sufficiently to engulf the prey whole. The success of snakes in achieving both increased the diversity of prey that could be consumed and, because heavier prey provide more nutrition, decreased the time spent searching or waiting for it [3, 4, 11]. The evolution of macrostomy was contingent on the prior availability of larger (heavier, bulkier, or both) prey that could be caught and restrained without undue risk of severe injury. Hence, macrostomy presumably co-evolved with the mechanisms to catch and restrain those prey. But most snakes, macrostomous and non-macrostomous, swallow prey whole using jaws with recurved teeth that move like a ratchet, a toothed device allowing movement in one direction only. The problem becomes one of upper and lower jaw ratchets and how prey size, macrostomy, and ratcheting mechanisms are interrelated.

19.3 Methodological Foundations

Our views of macrostomy have developed from observations of feeding behaviour and of head and trunk anatomy in several hundred extant species of snakes (Supplementary Appendix 19.S1). Although we have examined thousands of skeletons and dissected parts of hundreds of specimens representing most major alethinophidian snake clades, the sum of our observations convinces us that the rare, observed event in the field (e.g., Fig. 19.1, showing both macrostomy and macrophagy, and [13, 14] showing macrophagy without macrostomy) can recast the nature of the problem [15, 16]. A brief review of Figure 19.1, showing a python swallowing a duiker, and some attendant details of that event provided by B. Maritz (pers. comm.) may help to clarify how snakes achieved macrostomy. After the python started swallowing, it took it more than three and a half hours to reach the stage shown in the figure, but intraoral transport [17] finished approximately 30 minutes later. The snake swallowed the duiker and crawled away, suggesting that the body wall and gut suffered no irreparable damage from the degree of distension demanded by this prey item.

Figure 19.1 Southern African Python (*Python natalensis*), approximately 3 m long and an estimated mass of approximately 15+ kg swallowing a Common Duiker (*Sylvicapra grimmia*), also with estimated mass of 15+ kg. Photo taken at 15:30 on February 16, 2014 near Mbombela, eastern Mpumalanga Province, South Africa (photo and details courtesy of Bryan Maritz, University of the Western Cape, South Africa). See Supplementary Appendix 19.S2 for colour version.

It is clear in this case that the length of the mandibles has little direct correlation to gape size, which must be defined by the viscoelastic properties and folding patterns of all the intermandibular tissues, only a few of which have been tested [18, 19] or described at any level [20, 21]. During swallowing, the duiker maintained its position relative to the background – the snake's head moved over it, and the snake passed body coils forward, over and under the duiker's head and down its body continuously, expressing gas from the prey, and possibly aiding lower jaw progression under it.

Many examples like the case in Figure 19.1 have been witnessed but they are rarely accompanied by prolonged recording to allow proof that the snake successfully swallowed the prey item and moved away. Even this extraordinary record has one weakness – the snake could not be found the following day (B. Maritz, pers. comm.), so there is no evidence of its long-term survival after this event. Many critical snake clades remain virtually unexplored behaviourally and anatomically. Most of our behavioural data have come from captive individuals, but we have incorporated examples and records of extreme performance from the literature and the web.

Studies of gape size and feeding performance in snakes, based on various behavioural and morphological measures, have flowered over the last 20 years and have been thoroughly reviewed [3]. Most studies have focused on caenophidian species and suggest that features of head size and shape correlate only loosely to some measures of feeding performance.

The process of ingesting prey with relatively large cross-sectional areas stretches critical soft tissues of the lower jaw and body wall cranial to the pylorus to levels beyond experimentally predicted extension limits of epithelia, muscles, and most other soft tissues, with the possible exception of elastin [16]. Unfortunately, detailed histological studies of critical joints and soft tissues in regions of the body of snakes that experience extraordinary distension are rare [20–24]. None has recorded the condition of most soft tissues near known maximum distensions and only one [21] employed both light and electron microscopy to examine limited cellular details.

We gained information on jaw and soft-tissue behaviour from cine and video records of feeding in approximately 190 species of captive snakes (Supplementary Appendix 19.S1) recorded over 40 years with a range of temporal (18–2000 fps) and visual (8 and 16 mm cine, 220 X 292 – 840 X 1084 pixel) resolutions. Apart from a few records made in the field [25], most records were of snakes in a glass-fronted wooden arena striking at live prey and then restraining and eating them. The majority of sampled viperid species were recorded in zoos [26]. The research was done under a number of approved IACUC protocols. In addition, a few species (*Cylindrophis ruffus*, *Xenopeltis unicolor*, *Calabaria reinhardtii*) were tested for potential palatomaxillary mobility by manipulating anesthetized individuals. To gain some idea of the potential movements of individual bones, shortly following the natural deaths of four snakes (*Boa constrictor*, *Python brongersmai*, *Nerodia rhombifer*, *Ahaetulla nasuta*) we protracted one palatomaxillary arch and held it in place while fixing the snake in 10% buffered formalin. MicroCT scans of these heads provided a better understanding of the elements contributing to motion of the upper jaws.

We use as foundations for our discussion recently published phylogenies derived from many characters in both extinct and extant squamates on which macrostomy is explicitly mapped [7, 8]. To begin, however, we explore traits of the anterior trunk and head in snakes that affect their functions. Then we consider previous hypotheses of the sequence of events giving rise to macrostomy in snakes.

19.4 Basic Snake Features: The Trunk

The evidence from extant snakes and available fossils suggests that the earliest animals that could be called snakes had achieved a critical feature of the snake body – namely, trunk elongation through increased numbers of precloacal segments combined with the loss of the pectoral girdle and all pregastric ventral skeletal elements connected in any way to the vertebral column [27]. In all extant snakes, there is complete loss of pectoral remnants, except possible superficial pectoral muscle homologues [28–31]. Associated with the loss of the pectoral girdle is the loss of the sternum and all costal cartilage connections between the sternum and ribs. Reducing skeletal constraints allowed selection to act on all soft tissues surrounding the gut lumen, including the tissues connecting the tips of the mandibles, to permit elastic expansion in circumference. Apart from all the tissues in the gut wall, these include tissues forming the body wall ventral to the axial musculoskeletal system anterior to the intestine – skin, the skeletal muscles associated with the skin, and their associated connective tissues, blood vessels, and nerves [16, 20, 32]. The skin in most snakes varies considerably in histological and functional properties depending on the position along the trunk and the distance from the middorsal line [33]. The skin of the gastric and pregastric trunk appears to have greater folding of interscalar skin. Complex arrangements of intrinsic cutaneous muscles become associated with the ventral scales (scutes) and the more ventral of the dorsal scale rows [34]. Three layers of hypaxial muscles surround the gut ventral to the tips of the costal cartilages, a superficial layer formed by caudally directed superior and cranially directed inferior costocutaneous muscles that attach to the dermis of the ventral scale rows and to the lateral edges of the scutes, and the internal oblique and transverse abdominal muscles that attach to a midventral thickening of the dermis of the scutal and interscutal skin (Fig. 19.2). In addition to the skin, the gut wall, nerves and blood vessels, the hypaxial muscles, and the intersquamal and costocutaneous muscles must all stretch in macrostomous snakes during ingestion of very large prey.

In summary, the critical developmental and structural changes associated with macrostomy in snakes are loss of abaxial skeletal elements of the anterior trunk [39] and acquisition of novel tissue capabilities. Head and Polly [40] noted regional differences in the axial skeletons of snakes that correlate loosely to those in other tetrapods and even more loosely to skeletal muscle patterns [32, 41]. In our view, a critical feature of axial muscles in snakes is the extension of the internal oblique and transverse abdominal muscles to the anteriormost or near-anteriormost rib accompanied by coelomic and lymphatic sinus extensions into the anterior trunk and lower jaw [42, 43]. We agree with the profusion of past literature suggesting that, like modifications of the eye and multiple vision genes (Chapter 15), and the vestibule of the ear (Chapter 13, [44]), the loss of anterior abaxial skeletal derivatives was plausibly driven by mechanical and environmental demands of burrowing (see [9] for citations and robust disagreement). As a result of these changes, snakes appear to be tetrapods without a neck or the region of anterior thorax associated with pectoral limbs.

19.5 Basic Snake Features: The Head

Compared to lizards, crown-clade snakes display numerous derived differences in skull form that converge on features seen in a number of fossorial lizard clades (e.g., [2, 6, 27, 45, 46]). These include loss of some of the dermal posterior circumorbital bones, jugal, and squamosal, and the endochondral epipterygoid. In snakes, unlike all fossorial lizards except amphisbaenians, downgrowths of the dermal parietals and frontals meet the parabasisphenoid and the anterior edge of the prootic, removing potential motion at both meso- and metakinetic cranial joints. Apart from an optic fenestra and foramina for blood vessels and nerves, the orbital and postorbital braincase in crown-clade snakes completely encloses the brain and olfactory bulbs except in some small leptotyphlopids in which the

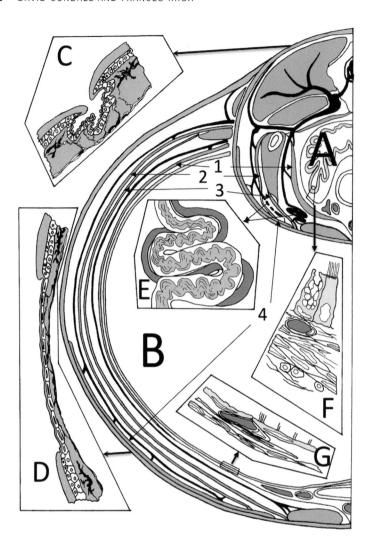

Figure 19.2 Diagrammatic views of the anterior trunk of a macrostomous snake in resting (A) and expanded (B) conditions, showing some structures and tissues that must extend to encompass large prey. Scale and interscale regions of the epidermis and dermis (C, D) differ in keratin deposition, folding, and collagen and elastin arrangements [20] that allow extensive unfolding of interscale epidermis and extension of dermal collagen and elastin in manners that allow passive recovery of folded form after stretching. The distal ends of segmental anterior and posterior nerves of the medial branch of the ventral ramus [35] have large folds ventral to the rib tips with elastin oriented along the long axis of the nerve in the peri- and epineurium and axonal bundles secondarily folded within (E) in a manner similar to that of the ventral grooved blubber nerves of rorqual whales [36]. The nerves are typically accompanied by arteries and veins shown as a single vessel in the diagram. The lining of the oesophagus appears as a typical simple or pseudostratified columnar epithelium of mucus and ciliated cells (F). What happens to this critical boundary layer epithelium during distension beyond levels allowed by unfolding remains unknown but the apical junctional rings of adjacent cells [16, 20] are assumed (G) to behave in a manner similar in cellular details to that of umbrella cells of the mammalian urothelium [37, 38] but on a longer timescale. What happens to the connective tissues,

roof of the dermal braincase does not develop [47–49]. If derived from one of the extant lizard lineages, one would expect the walls of the braincase in snakes (absent in lizards except amphisbaenians) to be developmental additions to the intramembranous frontals and parietals that arise dorsal to the brain in lizards. Instead, the lateral regions develop first in snakes [50–53], one of many changes in developmental timing differentiating snakes from lizards [2, 54, 55]. The resulting braincase, apart from providing solid attachment surfaces for all the muscles moving the upper jaws, also prevents compression of the brain in macrostomous taxa when swallowing very large prey.

Among many other novelties, snakes have exploited antero-posterior movement potentials of the jaw apparatus made possible by the loss of both temporal arches to evolve a number of ratcheting mechanisms linked to the kinds of prey available. Recurved pointed teeth on right and left jaws with independent mobility made ratchet mechanisms possible but limited snakes to ingesting most prey whole. A possible scenario during the Mesozoic could have been the early appearance of a lower jaw ratchet allowed by loss of the mandibular symphysis. Availability of small arthropod prey evidently drove selection for short, transversely oriented upper jaws in typhlopoids and anomalepidids and for transversely oriented tooth-bearing dentaries in the lower jaws of leptotyphlopids. Longer invertebrate or vertebrate prey would have driven selection for longitudinally oriented tooth rows with varyingly effective ratchet mechanisms dependent on the degree of jaw kinesis and undoubtedly correlated with prey type. Such mechanisms may have characterized pachyophiids and occur in various extant snakes, including leptotyphlopids [56] and distantly related *Calabaria*, *Dipsas* (unpublished data), and probably many others. In taxa with lower jaw ratchets, as the relative size of prey increases and the angle of the lower jaw to the braincase increases, the teeth become increasingly ineffective and the jaw adductor muscles stretched to extents permitting little force generation. Hence, as far as we know, there are no living macrostomous taxa utilizing lower jaw ratchets alone.

Exploitation of left-right lower jaw independence in snakes evolved in apparent haphazard fashion. Loss of the mandibular symphysis may have occurred early, but the structures connecting the tips of the dentaries vary considerably among snakes [23, 24, 57], restricting independent antero-posterior movements of the mandibular tips in some [58, 59]. Taxa lacking freedom of movement of the mandibular tips could not be macrostomous, regardless of the length of the mandible, unless another ratcheting mechanism was available.

The mechanical ability of a lower-jaw ratchet decreases with increasing gape angle, thus limiting the cross-sectional diameter of a prey item that can be effectively transported. This

---

Figure 19.2 (*cont.*) vasculature, smooth muscle layers, and enteric nerves remains unknown. Muscle layers surrounding the expanded gut wall are (1) ventral oblique and transverse muscles, (2) superior costocutaneous, and (3) inferior costocutaneous muscles. Label 4 points to intrinsic intersquamal muscles between scale rows. (A black and white version of this figure will appear in some formats. For the colour version, please refer to the plate section).

limitation was overcome through the evolution of upper jaw ratchets of varying mechanisms and capabilities linked to a variety of skeletal modifications of the lower jaws, the latter achieved in most taxa after birth or hatching [12]. These skeletal modifications must have co-evolved with soft-tissue modifications of the prepyloric trunk. Transversely oriented upper and lower jaw ratchets work by powering movement of prey through the oral cavity, whereas longitudinally oriented upper jaw ratchets can work by powering movement of either the prey or the snake's head over the prey. The transverse arrangement works with prey small enough to be moved by muscle-powered rotations of the maxillary (typhlopoids, anomalepidids) or dentary teeth (leptotyphlopids [59]). Prey size for longitudinal tooth arrangements (alethinophidians) is potentially limited only by abaxial adaptations allowing distension.

19.6 Macrostomy and Its Morphological Correlates

We have argued that macrostomy is dependent on an effective upper jaw ratchet, which is dependent on upper jaw mobility. This is because once the lower jaw nears maximum abduction and intermandibular soft tissues are distended, the head advances over the prey primarily or entirely by upper jaw ratcheting. The morphological innovations that underlie palatomaxillary mobility in alethinophidian snakes are essentially similar to those in typhlopoid and anomalepidid snakes – loss of direct bony contact between the dermal upper jaw bones (maxilla, palatine, pterygoid, and ectopterygoid) and the braincase and snout. Potential kinematic effects of these changes are remarkable (Fig. 19.3). Recognizing the patterns and potentials of upper jaw motion makes the search for its structural basis more focused and imposes limits on interpretations drawn from fossils [9].

Release of the caudal ends of the upper jaws and its relation to streptostyly have been dealt with in many publications (e.g., [1, 3, 60] and references therein). Movements of the pterygoid and quadrate appear in many lizards powered by dorsal constrictor muscles innervated by the V_4 branch of the trigeminal nerve. In lizards, these muscles effect dorsoventral movement of the facial skull anterior to the orbits. The key to liberation of the upper jaws in snakes lies in the erosion of dermal bones in the preorbital skull. Frazzetta [61, 62] redefined prokinesis as a movable cranial joint anterior to the orbit, essentially a joint between the snout and the braincase. Cundall and Shardo [63] coined the term 'rhinokinesis' to name the complex movements possible among so-called snout bones (premaxilla, nasals, septomaxilla, and vomers). Lost in this simple two-term dichotomy is the fact that what actually becomes movable is all of the upper toothed bones – maxillae, palatines, and pterygoids. How this was achieved is partially reflected in living taxa, but the details of most remain undescribed. The following is a brief account of our current understanding.

If early snakes were fossorial and exploited burrowing prey in the Jurassic or early Cretaceous, there is little fossil evidence of a diverse soil fauna of a size suitable as prey for fossorial snakes, for which there is also no fossil evidence. However, sophisticated exploration techniques are greatly increasing the rate at which new fossil discoveries are

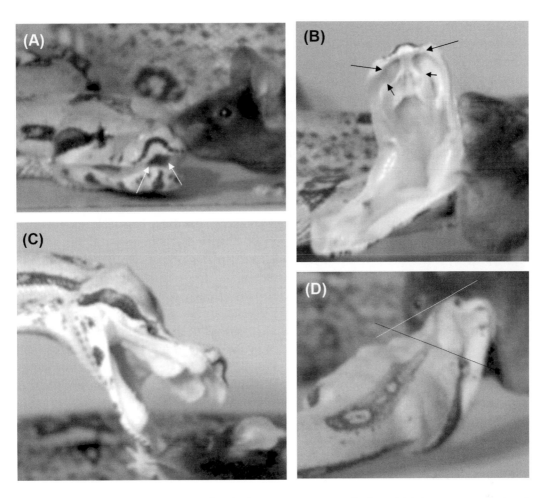

Figure 19.3 Four views of an unusual strike by a *Boa constrictor* during which its prey retreated and the snake continued its strike in a 'searching' mode with full protraction of both upper jaws. Image resolution is compromised by cropping from original video frames and movement of the snake and mouse during the 1/1000 s exposure of each frame. (A) 20 ms after beginning of strike, arrows at tips of maxillae; (B) 70 ms after beginning with upper jaws near maximum protraction, long arrows at tips of maxillae, short arrows at tips of palatines; (C) 158 ms after beginning – left dentary teeth in contact with the mouse's nose; (D) 304 ms after the beginning, showing the extraordinary flexibility of the snout, black line showing the plane of the interorbital braincase and white line showing the plane of the premaxilla as the snake protracts and elevates the anterior end of its right maxilla to contact a mouse partially snared by the left dentary. The anterior snout has rotated approximately 45° around its long axis.

made (e.g., [64]). Caldwell's [9] prediction that snakes (his snake-lizards) appeared in the Permian or Triassic may yet be supported by fossils. Available Mesozoic fossils (Chapters 2 and 3) provide tantalizing but sometimes confusing evidence of snakes or snake-like organisms from aquatic and terrestrial sites. One of the largest (*Dinilysia*) has otic vestibuli like those of extant fossorial snakes (Chapter 13; [44]). On the basis of living scolecophidians and many non-caenophidian alethinophidians (uropeltids, *Anomochilus, Anilius, Xenopeltis, Calabaria*), some fossorial snakes retained or regained maxillary–premaxillary contact, robust nasal bones firmly connected to the frontals, and palatines intimately associated with the vomers but, unlike *Dinilysia* [65] and *Calabaria* [2], lost the postorbital bar supporting the rear end of the maxilla.

Of the extant snakes listed above, we know little of the feeding behaviour and movements of the jaw bones of all except *Xenopeltis* and *Calabaria* (unpublished data) and three scolecophidian species [59]. Collectively, studies of these taxa suggest that effective upper jaw ratcheting mechanisms in extant non-caenophidian snakes with longitudinal maxillary tooth rows correlate loosely with the presence of teeth on the palatine. On these grounds, all phylogenetic analyses suggest that *Melanophidium*, which has small palatine teeth (in at least some species) and separated premaxillae and maxillae [66], is sister to all other extant uropeltids and hence that most uropeltids, although still far outside Caenophidia, are likely secondarily specialized for fossoriality. If booid, pythonid, madtsoiid, and pachyophiid snakes appeared in the Jurassic or Cretaceous, available evidence [9] fails to confirm that either madtsoiids or pachyophiids were macrostomous although their recurved tooth form suggests that all used a ratchet mechanism of some kind to transport prey. In booids and pythonids, release of the anterior ends of both the maxillae and the palatines from the snout co-evolved with rhinokinesis, prokinesis, and kinesis of both the prefrontal and postorbital to allow independent anteroposterior movement of right and left palatomaxillary arches. Simple lateromedial movements of the rear ends of each palatomaxillary arch do allow ratcheting, but become progressively more limited as the relative cross-sectional area of prey increases [1]. The pterygoid walk of Boltt and Ewer [67], used by all macrostomous snakes for intraoral prey transport, was built on rhino-, pro-, and prefrontal kinesis.

Independence of right and left upper jaws in *Boa constrictor* is clear (Fig. 19.3B–D). Unfortunately, we have the technical capabilities to visualize these skeletal movements only in large living snakes. However, some approximation can be gained from microCT scans of heads of recently dead snakes whose jaws have been manipulated using thread to protract one upper jaw and hold it in place during fixation (Fig. 19.4). A palatomaxillary arch cannot be both protracted and elevated by this method (pulling on palatine or pterygoid teeth usually depresses the arch) but it allows some estimation of the range of motion permitted by soft tissues at the various movable joints in the head. Histological studies of other species have revealed that many of the ligaments connecting most movable bones are complex arrays of collagen and elastin [22]. Hence, extensive sliding occurs at 'joints' among the maxilla, palatine, and prefrontal and between the prefrontal and frontal bones. Ligamentous connections between the maxilla and premaxilla and between the palatine and vomer are complexed with other tissues (skin, buccal lining, vomeronasal cartilages) to make their viscoelastic mechanical properties difficult to predict [19, 20]. What is clear, however, is that in macrostomous snakes, a range of suspensorial arrangements evolved affecting relations of the upper jaw to the braincase, lower jaw and its suspensorium, circumorbital bones, snout, and dorsal constrictor muscles.

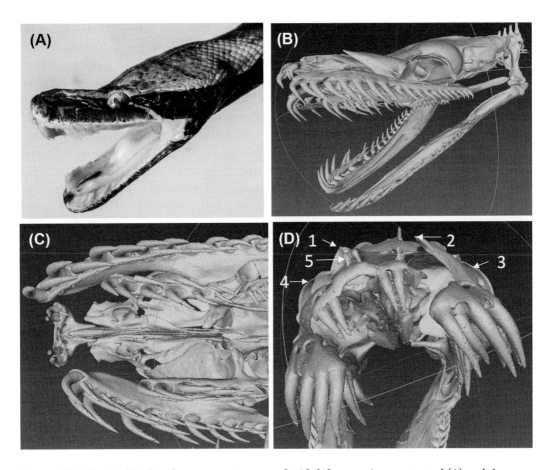

Figure 19.4 Head of *Python brongersmai* preserved with left upper jaw protracted (A) and three microCT images of the same specimen showing (B) extent of left prefrontal rotation, (C) protraction of the left maxilla, palatine, and vomer, and (D) rotation of the snout, elevation of the postorbital, and depression of the quadrate: the posterior roof of the braincase is horizontal and oriented perpendicular to the plane of the image – hence, the anterior snout is rotated to the right of the midline and 36° around the long axis. 1 = dorsal edge of right quadrate, 2 = anterior tip of left prefrontal (dorsal edge of left quadrate is just visible below the number 2), 3 = edge of left postorbital, 4 = edge of right postorbital, 5 = anterior tip of right nasal. See Supplementary Appendix 19.S2 for colour version.

19.7 Macrostomy and Macrostomata

It has been clear for some time that skeletal morphology alone often does not allow accurate inferences of behaviour. This is particularly true for snakes. The evolution of pterygoid-walk capability allowed the exploitation of an extraordinary range of prey [10], but macrostomy became evolutionarily relevant only after the appearance of restraint mechanisms for immobilizing or killing large prey before ingestion. Current evidence [8, 68] suggests that constriction preceded envenomation in snakes. We are aware that genes associated with venom production occur in clades currently thought to have evolved much earlier than snakes (Suranse et al., this volume), but no fossil evidence of venomous snakes has been found prior to the Neogene [69]. Clade names such as Macrostomata and Toxicofera [70] are useful hypotheses of the appearance of novel synapomorphies. However, the actual terms intimate far more than a node or stem in a tree. Because Toxicofera implies the use of toxins in prey restraint or defensive behaviour, its application has been questioned because many toxicoferan taxa show none of the morphological, functional, ecological, or behavioural properties associated with the supposed synapomorphy [7, 70, 71]. In the case of Macrostomata, the collection of traits required to engulf animals with cross-sectional areas much larger than the resting cross-sectional areas of a snake's head and anterior trunk appear to relate to ventral soft tissues in the lower jaw and trunk, and an upper jaw ratchet. However, exploiting large, live prey also required rapid restraint mechanisms.

For snakes attempting to engulf vertebrate prey, constriction can restrain many ectotherms [68], including elongate, snake-like ectotherms such as eels and elongate amphibians, but may not kill them. Although macrostomy is not necessarily required to ingest many elongate prey, some apparently non-macrostomous dipsadid snakes (e.g., *Farancia*) use constriction to restrain large *Amphiuma* and ingest them using the pterygoid walk alone (pers. obs.). On the other hand, *Cylindrophis*, a non-macrostomous non-macrostomatan, ingests elongate prey faster using snout shifting and trunk compression [13, 58]. Records of uropeltids swallowing worms and of laticaudines transporting eels in seconds [72] suggest that there remain very poorly known behavioural capabilities in many snakes.

There is good reason to think that macrostomy co-evolved with the appearance of bulky (large cross-sectional area relative to mass) endothermic prey. Whereas bulky ectothermic prey can be difficult to kill by constriction, bulky endotherms die quickly when constricted. For animals incapable of ripping apart or chewing vertebrate prey prior to ingestion, however, exploiting bulky prey depended on the evolution of a mouth capable of ingesting the prey whole and of a body capable of transporting it to the stomach, a suite of characteristics that may have evolved once (tentatively supported by some molecular analyses [73]) or independently a number of times, as suggested by most recent analyses [7–9, 74]. Macrostomy solved only part of the problem. But constricting snakes presumably coevolved with diverse endothermic clades in the late Mesozoic and Palaeogene. Those with an effective upper jaw ratchet could potentially exploit a wide diversity of prey if ventral soft tissues and/or the mandibles and their suspensoria adapted to allow macrostomy.

The evolution of dental glands capable of releasing venoms (Chapter 18) and the co-evolution of a variety of dental and palatomaxillary innovations to increase the efficacy of venom effects on internal tissues of prey made possible the killing of large prey by small snakes, but 'large prey' is a relative term and really refers only to prey mass. A number of snake species are known to ingest prey with a mass equal to or even slightly larger than their own but for some, the prey are elongate vertebrates that do not demand macrostomy. However, most require some type of restraint prior to ingestion and hence both macrostomy and macrophagy entailed evolutionary adaptations of the trunk for striking or constricting or both. Macrostomy required adaptations for trunk distension as well, and these involved abaxial extensions of nervous and vascular tissues, whereas striking and constriction primarily entailed adaptations of primaxial tissues [40].

Current fossil evidence supports phylogenetic analyses of behavioural variation [66] suggesting that the evolution of macrostomy was initially driven by constriction. Cretaceous booid and pythonid fossils, unfortunately based primarily on vertebrae [75, 76] presumably lacked the pachyostotic vertebral and rib regions of Cretaceous pachyophiids [77]. Our lack of useful functional-morphological data on the feeding mechanics of most rear-fanged caenophidians and the absence of Mesozoic fossils of rear-fanged caenophidians or front-fanged viperids or elapids limits conjectures on the primacy of constriction or envenomation for prey restraint. However, the striking behaviour of some extant booids and pythonids matches the neurophysiological capabilities of living viperids specialized for releasing prey after envenomation [26], suggesting that the ability to capture large, elusive prey could have evolved in the Mesozoic. To demonstrate that an extinct snake-like organism was capable of macrostomy, however, requires evidence of palatomaxillary mobility. For this, palatine and maxillary associations with the snout and prefrontal, and the latter's relation to the frontal, are useful indicators. Complete postorbital bars and prefrontal-frontal joints that are oriented longitudinally suggest very limited upper jaw mobility and unlikely potential for macrostomy. Available data on the cranial anatomy of *Najash* suggests that mandibular suspensorial adaptations associated with aquatic or fossorial living may have evolved earlier than liberation of the palatomaxillary arch [78, 79]. Whether the aquatic lineages of snake-like squamates survived the K–Pg extinction remains unclear. Inasmuch as extant snakes display collections of features most parsimoniously explained by a fossorial ancestry [80], fitting the Mesozoic fossils into the story really requires more fossil data, not rewriting everything we know about the snakes from which we can directly acquire behavioural and soft-tissue data [9].

We contributed to one of the recent molecular studies showing Macrostomata is not monophyletic [8], and agree with that interpretation, but we note that branch lengths separating critical 'basal' alethinophidian clades are very short, and so may be particularly sensitive to input data and analytical parameters. Although little effort has been expended on defining the anatomical and physiological foundations of macrostomy, we know that some of the morphological and functional features defining 'basal' macrostomous taxa (e.g., nature of intermandibular soft tissues, mobility of the snout and palatomaxillary arches) are not expressed in some caenophidian lineages. Hence, the problem is distinguishing reversals from plesiomorphies – a common problem in studies of possibly rapidly radiating clades. Like lizard feeding and phylogeny [81], we hope the challenges of reconciling genomic and more-broadly based data sets will ultimately allow us to understand how macrostomy evolved in snakes.

Acknowledgements

We thank the editors for inviting us to contribute to this volume. Harry Greene and Tom LaDuke provided critical perspectives underlying the ultimate phrasing of relationships between macrostomy and prey size. We thank David Gower, Kurt Schwenk and Alan Savitzky for providing additional critical reviews of a penultimate draft that helped us to recast some important elements of our argument. We have been recording feeding behaviour in snakes since 1978 and have been helped by many people, most acknowledged in prior publications. For help with the figures in this chapter we thank Gillian Cundall and Alan Savitzky for help with Figure 19.2 and Morgan Hill Chase and Andrew Smith at the Microscopy and Imaging Facility, AMNH, for the microCT scan images in Figure 19.4.

References

1. D. Cundall and H. W. Greene, Feeding in snakes. In K. Schwenk, ed., *Feeding: Form, Function, and Evolution in Tetrapod Vertebrates* (San Diego, CA: Academic Press, 2000), pp. 293–333.

2. D. Cundall and F. Irish, The snake skull. In C. Gans, A. S. Gaunt and K. Adler, eds., *Biology of the Reptilia*, Vol. 20, *Morphology H.* (Ithaca, NY: Society for the Study of Amphibians and Reptiles, 2008), pp. 349–692.

3. B. R. Moon, D. A. Penning, M. Segall, and A. Herrel, Feeding in snakes: Form, function, and evolution of the feeding system. In V. Bels and I. Q. Whishaw, eds., *Feeding in Vertebrates: Evolution, Morphology, Behaviour, Biomechanics* (Switzerland: Springer Nature, 2019), pp. 527–574.

4. H. W. Greene, *Snakes: The Evolution of Mystery in Nature* (Berkeley, CA: University of California Press, 1997).

5. O. Rieppel, A review of the origin of snakes. *Evolutionary Biology*, 22 (1988), 37–130.

6. J. Müller, Beiträge zur Anatomie und Naturgeschichtte der Amphibien. *Zeitschrift für Physiologie*, 4 (1832), 190–275.

7. S. Harrington and T. Reeder, Phylogenetic inference and divergence dating of snakes using molecules, morphology and fossils: new insights into convergent evolution of feeding morphology and limb reduction. *Biological Journal of the Linnean Society*, 121 (2017), 379–394.

8. F. T. Burbrink, F. G. Grazziotin, R. A. Pyron, et al., Interrogating genomic-scale data for Squamata (lizards, snakes, and amphisbaenians) shows no support for key traditional morphological relationships. *Systematic Biology*, 69 (2020), 502–520.

9. M. W. Caldwell, *The Origin of Snakes: Morphology and the Fossil Record* (Boca Raton, FL: CRC Press, 2020).

10. H. W. Greene, Dietary correlates of the origin and radiation of snakes. *American Zoologist*, 23 (1983), 431–441.

11. A. Y. Hsiang, D. J. Field, T. H. Webster, et al., The origin of snakes: revealing the ecology, behaviour, and evolutionary history of early snakes using genomics, phenomics, and the fossil record. *BMC Evolutionary Biology*, 15 (2015), 87.

12. A. Scanferla, Post-natal ontogeny and the evolution of macrostomy in snakes. *Royal Society Open Science*, 3 (2016), 160612.

13. M. R. Bezuijen, Field observation of a large prey item consumed by a small *Cylindrophis ruffus* (Laurenti, 1768) (Serpentes: Cylindrophiidae). *Hamadryad*, 34 (2009), 185–187.

14. C. Kusamba, A. Resetar, V. Wallach, K. Lulengo, and Z. T. Nagy, Mouthful of snake: An African snake-eater's (*Polemon fulvicollis graueri*) large typhlopid prey. *Herpetology Notes*, 6 (2013), 235–237.

15. K. Jackson, N. J. Kley, and E. L. Brainerd, How snakes eat snakes: the biomechanical challenges of ophiophagy for the California kingsnake, *Lampropeltis getula californiae* (Serpentes: Colubridae). *Zoology*, 107 (2004), 191–200.

16. D. Cundall, A few puzzles in the evolution of feeding mechanisms in snakes. *Herpetologica*, 75 (2019), 99–107.

17. K. Schwenk and M. Rubega, Diversity of vertebrate feeding systems. In J. M. Stark

and T. Wang, eds., *Physiological and ecological adaptations to feeding in vertebrates* (Enfield, NH: Science Publishers, 2005), pp. 1–41.

18. B. C. Jayne, H. K. Voris, and P. K. L. Ng, How big is too big? Using crustacean-eating snakes (Homalopsidae) to test how anatomy and behavior affect prey size and feeding performance. *Biological Journal of the Linnean Society*, 123 (2018), 636–650.

19. N. H. Gripshover and B. C. Jayne, Crayfish eating in snakes: testing how anatomy and behavior affect prey size and feeding performance. *Integrative Organismal Biology*, 3 (2021), obab001.

20. M. Close and D. Cundall, Snake lower jaw skin: Extension and recovery of a hyperextensible keratinized integument. *Journal of Experimental Zoology*, 321A (2014), 78–97.

21. M. Close, S. Perni, C. Franzini-Armstrong,, and D. Cundall, Highly extensible skeletal muscle in snakes. *Journal of Experimental Biology*, 217 (2014), 2445–2448.

22. B. A. Young, The arthrology of the head of the Red-sided Garter snake, *Thamnophis sirtalis parietalis. Netherlands Journal of Zoology*, 38 (1988), 166–205.

23. B. A. Young, The comparative morphology of the intermandibular connective tissue in snakes (Reptilia: Squamata). *Zoologischer Anzeiger*, 237 (1998), 59–84.

24. B. A. Young The comparative morphology of the mandibular midline raphe in snakes (Reptilia: Squamata). *Zoologischer Anzeiger*, 237 (1998–1999), 217–241.

25. D. Cundall and S. J. Beaupre, Field records of predatory strike kinematics in rattlesnakes, *Crotalus horridus. Amphibia-Reptilia*, 22 (2001), 492–498.

26. D. Cundall, Viper fangs: Functional limitations of extreme teeth, *Physiological and Biochemical Zoology: Ecological and Evolutionary Approaches*, 82 (2009), 63–79.

27. J. A. Gauthier, M. Kearney, J. A. Maisano, O. Rieppel, and D. B. Behlke, Assembling the squamate tree of life: Perspectives from the phenotype and the fossil record. *Bulletin of the Peabody Museum of Natural History*, 53 (2012), 3–308.

28. A. E. Greer, Limb reduction in squamates: Identification of the lineages and discussion of the trends. *Journal of Herpetology*, 25 (1991), 166–173.

29. T. Tsuihiji, M. Kearney, and O. Rieppel, First report of a pectoral girdle muscle in snakes, with comments on the snake cervico-dorsal boundary. *Copeia*, 2006 (2006), 206–215.

30. T. Tsuihiji, M. Kearney, and O. Rieppel, Finding the neck-trunk boundary in snakes: Anteroposterior dissociation of myological characteristics in snakes and its implications for their neck and trunk body regionalization. *Journal of Morphology*, 273 (2012), 992–1009.

31. F. Leal and M. J. Cohn, Developmental, genetic, and genomic insights into the evolutionary loss of limbs in snakes. *Genesis: The Journal of Genetics and Development*, 56 (2018), e23077.

32. J.-P. Gasc, Axial musculature. In C. Gans and T. S. Parsons, eds., *Biology of the Reptilia*, Vol. 11, *Morphology F* (London: Academic Press, 1981), pp. 355–435.

33. G. Rivera, A. H. Savitzky, and J. A. Hinkley, Mechanical properties of the integument of the common gartersnake, *Thamnophis sirtalis* (Serpentes: Colubridae). *Journal of Experimental Biology*, 208 (2005), 2913–2922.

34. P. Buffa, Ricerche sulla muscolatura cutanea dei serpenti e considerazioni sulla locomozione di questi animali. *Atti della Accademia scientifica veneto-trentino-istriana*, 1 (1904), 145–237.

35. J. R. Fetcho, The organization of motor neurons innervating the axial musculature of vertebrates. II. Florida water snakes (*Nerodia fasciata pictiventris*). *Journal of Comparative Neurology*, 249 (1986), 551–563.

36. A. W. Vogl, M. A. Lillie, M. A. Piscitelli, et al., Stretchy nerves are an essential component of the extreme feeding mechanism of rorqual whales. *Current Biology*, 25 (2015), R345–R361.

37. G. Apodaca, The uroepithelium: Not just a passive barrier. *Traffic*, 5 (2004), 117–128.

38. A. F. Eaton, D. R. Clayton, W. G. Ruiz, et al., Expansion and contraction of the umbrella cell apical junctional ring in response to bladder filling and voiding. *Molecular Biology of the Cell*, 30 (2019), 2037–2052.

39. A. C. Burke and J. L. Nowicki, A new view of patterning domains in the vertebrate mesoderm. *Developmental Cell*, 4 (2003), 159–165.

40. J. J. Head and P. D. Polly, Evolution of the snake body form reveals homoplasy in amniote *Hox* gene function. *Nature*, 520 (2015), 86–89.

41. W. Mosauer, The myology of the trunk region of snakes and its significance for ophidian taxonomy and phylogeny. *Publications of the University of California at Los Angeles in Biological Sciences*, 1 (1935), 81–120.

42. G. Ottaviani and A. Tazzi, The lymphatic system. In C. Gans and T. S. Parsons, eds., *Biology of the Reptilia*, Vol. 6, *Morphology E* (London: Academic Press, 1977), pp. 315–464.

43. D. Cundall, C. Tuttman, and M. Close, A model of the anterior esophagus in snakes, with functional and developmental implications. *Anatomical Record*, 297 (2014), 586–598.

44. H. Yi and M. A. Norell, The burrowing origin of modern snakes. *Science Advances*, 1 (2015), e1500743.

45. M. S. Y. Lee, Convergent evolution and character correlation in burrowing reptiles: towards a resolution of squamate relationships. *Biological Journal of the Linnean Society*, 65 (1998), 369–453.

46. S. B. McDowell, Jr., The skull of Serpentes. In C. Gans, A. S. Gaunt, and K. Adler, eds., *Biology of the Reptilia*, Vol. 21, *Morphology I.* (Ithaca, NY: Society for the Study of Amphibians and Reptiles, 2008), pp. 467–620.

47. A. M. Abdeen, A. M. Abo-Taira, and M. M Zaher, Further studies on the ophidian cranial osteology: the skull of the Egyptian blind snake *Leptotyphlops cairi* (Leptotyphlopidae). I. The cranium. A – The median dorsal bones, bones of the upper jaw, circumorbital bone and occipital ring. *Journal of the Egyptian-German Society of Zoology*, 5 (1991), 417–437.

48. A. M. Abdeen, A. M. Abo-Taira, and M. M Zaher, Further studies on the ophidian cranial osteology: The skull of the Egyptian blind snake *Leptotyphlops cairi* (Leptotyphlopidae). I. The cranium. B – The otic capsule, palate and temporal bones. *Journal of the Egyptian-German Society of Zoology*, 5 (1991), 439–455.

49. D. G. Broadley and S. Broadley, A review of the African worm snakes from south of latitude 12°S (Serpentes: Leptotyphlopidae). *Syntarsus*, 5 (1999), 1–36.

50. J. C. Boughner, M. Buchtova, K. Fu, et al., Embryonic development of *Python sebae*. I. Staging criteria and macroscopic skeletal morphogenesis of the head and limbs. *Zoology*, 110 (2007), 212–230.

51. S. M. Boback, E. K. Dichter, and H. L. Mistry, A developmental staging series for the African house snake, *Boaedon (Lamprophis) fuliginosus*. *Zoology*, 115 (2012), 38–46.

52. K. M. Polachowski and I. Werneburg, Late embryos and bony skull development in *Bothropoides jararaca* (Serpentes, Viperidae). *Zoology*, 116 (2013), 36–63.

53. E. R. Khannoon and S. E. Evans, The development of the skull of the Egyptian cobra *Naja h. haje* (Squamata: Serpentes: Elapidae). *PLoS ONE*, 10 (2015), e0122185.

54. F. J. Irish, The role of heterochrony in the origin of a novel bauplan: evolution of the ophidian skull. *Geobios, Mémoires Special*, 12 (1989), 227–233.

55. I. Werneburg, K. M. Polachowski, and M. N. Hutchinson, Bony skull development in the Argus Monitor (Squamata, Varanidae, *Varanus panoptes*) with comments on developmental timing and adult anatomy. *Zoology*, 118 (2015), 255–280.

56. N. J. Kley and E. L. Brainerd, Post-cranial prey transport mechanisms in the black pinesnake, *Pituophis melanoleucus lodingi*:

an x-ray videographic study. *Zoology*, 105 (2002), 153–164.

57. U. Kiran, A new structure in the lower jaw of colubrid snakes. *Snake*, 13 (1981), 131–133.

58. D. Cundall, Feeding behaviour in *Cylindrophis* and its bearing on the evolution of alethinophidian snakes. *Journal of Zoology, London*, 237 (1995), 353–376.

59. N. J. Kley, Prey transport mechanisms in blindsnakes and the evolution of unilateral feeding systems in snakes. *American Zoologist*, 41 (2001), 1321–1337.

60. K. Schwenk, Feeding in lepidosaurs. In K. Schwenk, ed., *Feeding: Form, Function, and Evolution in Tetrapod Vertebrates* (San Diego, CA: Academic Press, 2000), pp. 175–291.

61. T. H. Frazzetta, A functional consideration of cranial kinesis in lizards. *Journal of Morphology*, 111 (1962), 287–320.

62. T. H. Frazzetta, Morphology and function of the jaw apparatus in *Python sebae* and *Python molurus*. *Journal of Morphology*, 118 (1966), 217–296.

63. D. Cundall and J. Shardo, Rhinokinetic snout of thamnophiine snakes. *Journal of Morphology*, 225 (1995), 31–50.

64. D. E. Shcherbakov, T. Tim, A. B. Tzetlin, O. Vinn, and A. Y. Zhuravlev, A probable oligochaete from an Early Triassic Lagerstätte of the southern Cis-Urals and its evolutionary implications. *Acta Palaeontologica Polonica*, 65 (2020), 219–233.

65. H. Zaher and C. A. Scanferla, The skull of the Upper Cretaceous snake *Dinilysia patagonica* Smith-Woodward, 1901, and its phylogenetic position revisited. *Zoological Journal of the Linnean Society*, 164 (2012), 194–238.

66. D. J. Gower, V. Giri, A. Captain, , and M. Wilkinson, A reassessment of Melanophium Günther, 1864 (Squamata: Serpentes: Uropeltidae) from the Western Ghats of peninsular India, with description of a new species. *Zootaxa*, 4085 (2016), 481–503.

67. R. E. Boltt and R. F. Ewer, The functional anatomy of the head of the puff adder, *Bitis arietans* (Merr.). *Journal of Morphology*, 114 (1964), 83–106.

68. H. W. Greene and G. M. Burghardt, Behaviour and phylogeny: Constriction in ancient and modern snakes. *Science*, 200 (1978), 74–77.

69. J. J. Head, K. Mahlow, and J. Müller, Fossil calibration dates for molecular phylogenetic analysis of snakes 2: Caneophidia, Colubroidea, Elapoidea, Colubridae. *Palaeontologia Electronica* 19.2.2FC (2016), 1–21.

70. A. D. Hargreaves, M. T. Swain, D. W. Logan, and J. F. Mulley, Testing the Toxicofera: Comparative transcriptomics casts doubt on the single, early evolution of the reptile venom system. *Toxicon*, 92 (2015), 140–156.

71. S. S. Sweet, Chasing flamingos: Toxicofera and the misinterpretation of venom in varanid lizards. In M. Cota, ed., *Proceedings of the 2015 Interdisciplinary World Conference on Monitor Lizards* (Bangkok, Thailand: Institute for Research and Development, Suan Sunandha Rajhabat University, 2016), pp. 123–149.

72. C. W. Radcliffe and D. A. Chiszar, A descriptive analysis of predatory behavior in the yellow-lipped sea krait (*Laticauda colubrina*). *Journal of Herpetology*, 14 (1980), 422–424.

73. A. Miralles, J. Marin, D. Markus, et al., Molecular evidence for the paraphyly of Scolecophidia and its evolutionary implications. *Journal of Evolutionary Biology*, 31 (2018), 1782–1793.

74. H. Zaher, and K. T. Smith, Pythons in the Eocene of Europe reveal a much older divergence of the group in sympatry with boas. *Biology Letters*, 16 (2020), 20200735.

75. J.-C. Rage, Serpentes, In P. Wellnhofer, ed., *Handbuch der Paläoherpetologie/Encyclopedia of Paleontology*, Part 11. (Stuttgart: Gustav Fischer, 1984), pp. 1–80.

76. J.-C. Rage, Fossil snakes. In R. A. Seigel, J. T. Collins, and S. S. Novak, eds., *Snakes: Ecology and Evolutionary Biology* (New York: Macmillan, 1987), pp. 51–76.

77. O. Rieppel, H., Zaher, E. Tchernov,, and M. J. Polcyn, The anatomy and relationships of *Haasiophis terrasanctus*, a fossil snake with well-developed hind limbs from the mid-Cretaceous of the Middle East. *Journal of Paleontology*, 77 (2003), 536–558.

78. H., Zaher, S. Apesteguía, and C. A. Scanferla, The anatomy of the upper cretaceous snake *Najash rionegrina* Apesteguía & Zaher, 2006, and the evolution of limblessness in snakes. *Zoological Journal of the Linnean Society*, 156 (2009), 801–826.

79. F. F. Garberoglio, S. Apesteguía, T. R. Simōes, et al., New skulls and skeletons of the Cretaceous legged snake *Najash*, and the evolution of the modern snake body plan. *Science Advances*, 5 (2019), eaax5833.

80. D. Cundall, Review of M. W. Caldwell, the origin of snakes: morphology and the fossil record. *Herpetological Review*, 51 (2020), 364–368.

81. N. M. Koch and J. A. Gauthier, Noise and biases in genomic data may underlie radically different hypotheses for the position of Iguania within Squamata. *PLoS ONE*, 13 (2018), e0202729.

82. P. Uetz, P. Freed, and J. Hosek, eds., *The Reptile Database*. www.reptile-database.org (accessed 1 February 2021).

Index

Systematics Association Special Volumes

1. The New Systematics (1940)[a]
   Edited by J. S. Huxley (reprinted 1971)

2. Chemotaxonomy and Serotaxonomy (1968)*
   Edited by J. C. Hawkes

3. Data Processing in Biology and Geology (1971)*
   Edited by J. L. Cutbill

4. Scanning Electron Microscopy (1971)*
   Edited by V. H. Heywood

5. Taxonomy and Ecology (1973)*
   Edited by V. H. Heywood

6. The Changing Flora and Fauna of Britain (1974)*
   Edited by D. L. Hawksworth

7. Biological Identification with Computers (1975)*
   Edited by R. J. Pankhurst

8. Lichenology: Progress and Problems (1976)*
   Edited by D. H. Brown, D. L. Hawksworth and *R. H. Bailey*

9. KeyWorks to the Fauna and Flora of the British Isles and Northwestern Europe, fourth edition (1978)*
   Edited by G. J. Kerrich, D. L. Hawksworth and *R. W. Sims*

10. Modern Approaches to the Taxonomy of Red and Brown Algae (1978)*
    Edited by D. E. G. Irvine and *J. H. Price*

11. Biology and Systematics of Colonial Organisms (1979)*
    Edited by C. Larwood and *B. R. Rosen*

12. The Origin of Major Invertebrate Groups (1979)*
    Edited by M. R. House

13. Advances in Bryozoology (1979)*
    Edited by G. P. Larwood and *M. B. Abbott*

14. Bryophyte Systematics (1979)*
    Edited by G. C. S. Clarke and *J. G. Duckett*

15. The Terrestrial Environment and the Origin of Land Vertebrates (1980)*
    Edited by A. L. Panchen

16. Chemosystematics: Principles and Practice (1980)*
    Edited by F. A. Bisby, J. G. Vaughan and *C. A. Wright*

17. The Shore Environment: Methods and Ecosystems (two volumes) (1980)*
    Edited by J. H. Price, D. E. C. Irvine and *W. F. Farnham*

18. The Ammonoidea (1981)*
    Edited by M. R. House and *J. R. Senior*

19. Biosystematics of Social Insects (1981)*
    Edited by P. E. House and J.-L. Clement

20. Genome Evolution (1982)*
    Edited by G. A. Dover and R. B. Flavell

21. Problems of Phylogenetic Reconstruction (1982)*
    Edited by K. A. Joysey and A. E. Friday

22. Concepts in Nematode Systematics (1983)*
    Edited by A. R. Stone, H. M. Platt and L. F. Khalil

23. Evolution, Time and Space: The Emergence of the Biosphere (1983)*
    Edited by R. W. Sims, J. H. Price and P. E. S. Whalley

24. Protein Polymorphism: Adaptive and Taxonomic Significance (1983)*
    Edited by G. S. Oxford and D. Rollinson

25. Current Concepts in Plant Taxonomy (1983)*
    Edited by V. H. Heywood and D. M. Moore

26. Databases in Systematics (1984)*
    Edited by R. Allkin and *F. A. Bisby*

27. Systematics of the Green Algae (1984)*
    Edited by D. E. G. Irvine and D. M. John

28. The Origins and Relationships of Lower Invertebrates (1985)‡
    Edited by S. Conway Morris, J. D. George, R. Gibson and H. M. Platt

29. Infraspecific Classification of Wild and Cultivated Plants (1986)‡
    Edited by B. T. Styles

30. Biomineralization in Lower Plants and Animals (1986)‡
    Edited by B. S. C. Leadbeater and R. Riding

31. Systematic and Taxonomic Approaches in Palaeobotany (1986)‡
    Edited by R. A. Spicer and B. A. Thomas

32. Coevolution and Systematics (1986)‡
    Edited by A. R. Stone and D. L. Hawksworth

33. Key Works to the Fauna and Flora of the British Isles and Northwestern Europe, fifth edition (1988)‡
    Edited by R. W. Sims, P. Freeman and D. L. Hawksworth

34. Extinction and Survival in the Fossil Record (1988)‡
    Edited by G. P. Larwood

35. The Phylogeny and Classification of the Tetrapods (two volumes) (1988)‡
    Edited by M. J. Benton

36. Prospects in Systematics (1988)‡
    Edited by J. L. Hawksworth

37. Biosystematics of Haematophagous Insects (1988)‡
    Edited by M. W. Service

38. The Chromophyte Algae: Problems and Perspective (1989)‡
    Edited by J. C. Green, B. S.C. Leadbeater and W. L. Diver

39. Electrophoretic Studies on Agricultural Pests (1989)‡
    Edited by H. D. Loxdale and J. den Hollander

---

[a]Published by Clarendon Press for the Systematics Association

[*]Published by Academic Press for the Systematics Association

[‡]Published by Oxford University Press for the Systematics Association

[**]Published by Chapman & Hall for the Systematics Association

[‡‡]Published by CRC Press for the Systematics Association